International
Association
of Fire Chiefs

National
Fire Protection
Association

# Water Rescue

## Principles and Practice to NFPA 1006 and 1670: Surface, Swiftwater, Dive, Ice, Surf, and Flood

### SECOND EDITION

## Steve Treinish

Lieutenant
Columbus, Ohio Division of Fire
Columbus, Ohio

JONES & BARTLETT
LEARNING

**Jones & Bartlett Learning**
World Headquarters
5 Wall Street
Burlington, MA 01803
978-443-5000
info@jblearning.com
www.jblearning.com

**National Fire Protection Association**
1 Batterymarch Park
Quincy, MA 02169-7471
www.NFPA.org

**International Association of Fire Chiefs**
4025 Fair Ridge Drive
Fairfax, VA 22033
www.IAFC.org

Jones & Bartlett Learning books and products are available through most bookstores and online booksellers. To contact Jones & Bartlett Learning directly, call 800-832-0034, fax 978-443-8000, or visit our website, www.jblearning.com.

Substantial discounts on bulk quantities of Jones & Bartlett Learning publications are available to corporations, professional associations, and other qualified organizations. For details and specific discount information, contact the special sales department at Jones & Bartlett Learning via the above contact information or send an email to specialsales@jblearning.com.

18604-8

**Production Credits**
General Manager, Safety and Trades: Kimberly Brophy
VP, Product Development: Christine Emerton
Executive Editor: Bill Larkin
Senior Editor: Amanda Brandt
VP, Sales, Public Safety Group: Matthew Maniscalco
Director of Production: Jenny L. Corriveau
Associate Production Editor: Robert Furrier
Director of Marketing Operations: Brian Rooney
Marketing Manager: Jessica Carmichael
Production Services Manager: Colleen Lamy

VP, Manufacturing and Inventory Control: Therese Connell
Composition: S4Carlise Publishing Services
Cover Design: Kristin E. Parker
Text Design: Scott Moden
Director of Rights & Media: Joanna Gallant
Rights & Media Specialist: Thais Miller
Media Development Editor: Shannon Sheehan
Cover Image: Courtesy of Steve Treinish
Printing and Binding: LSC Communications
Cover Printing: LSC Communications

**Library of Congress Cataloging-in-Publication Data**
Names: Treinish, Steve, author.
Title: Water rescue principles and practice to NFPA 1006 and 1670 : surface, swiftwater, dive, surf, and flood / Steve Treinish.
Description: Second edition. | Burlington, MA : Jones & Bartlett Learning, 2017.
Identifiers: LCCN 2017043584 | ISBN 9781284042054
Subjects: LCSH: Rescue work. | Lifesaving--Study and teaching. | Training. | Search and rescue operations.
Classification: LCC TH9402 .T75 2017 | DDC 363.12/381--dc23
LC record available at https://lccn.loc.gov/2017043584

6048
Printed in the United States of America
21 20 19 18 17   10 9 8 7 6 5 4 3 2 1

# Brief Contents

# Contents

CHAPTER **3**

# Watercraft Rescue Technician

# 81

CHAPTER **4**

# Surface Water Rescue Operations

# 101

CHAPTER **5**
# Surface Water Rescue Technician    134

# Ice Rescue Operations    235

# Ice Rescue Technician    256

CHAPTER **11**
# Surf Rescue Operations

CHAPTER **12**
# Surf Rescue Technician

# CHAPTER **13**
# Dive Equipment and Use    **311**

# CHAPTER **14**
# Dive Rescue Operations    **329**

CHAPTER **15**

# Dive Rescue Technician 372

CHAPTER **16**

# Helicopter Rescue Support 408

# Skill Drills

# Acknowledgments

Jones & Bartlett Learning, the National Fire Protection Association, and the International Association of Fire Chiefs would like to thank all of the authors, contributors, and reviewers of *Water Rescue Principles and Practice to NFPA 1006 and 1670: Surface, Swiftwater, Dive, Ice, Surf, and Flood, Second Edition.*

## Author

### Steve Treinish

Steve Treinish started his fire service career in 1987 with the Millersport Fire Department, and he is currently a Lieutenant with the Columbus Division of Fire, having served there for over 28 years. He is the manager of the division's Dive and Rescue Team and spearheaded its formation in 2004. Treinish also serves as an instructor at the Training Academy when needed, and he serves as a rescue technician and dive supervisor, as well as an advanced peer with the division's critical incident stress management team. He is also a member of the Central Ohio Strike Team, which offers search and rescue services statewide.

Treinish leads Fairfield County's Special Operations Dive Unit and Water Rescue Team, which offers dive, ice, and swiftwater rescue to central and southeast Ohio. He is a swiftwater instructor trainer and a fire training officer for the Ohio Fire Academy, and an instructor trainer with the Public Safety Diving Association. He serves as a member of NFPA's 1006 rescuer qualification committee, as well as the State of Ohio's Water Delivery Tactical Advisory Committee. He is currently pursuing a degree in emergency management from Franklin University.

Treinish enjoys being married to his wife, Monica, and has two sons, Tate and Cash. The fire service has been influential in the family; his father, Jack, served for 29 years with the CFD, and his brother, Jack Treinish, Jr, is an Assistant Chief with the West Licking Joint Fire District. Both his wife and his sister Jill are nurses. During his free time, Treinish enjoys numerous water sports and spending time with family and friends.

Treinish would like to thank his family—his wife and sons as well as his parents—who kept their kids playing in the water by boating, waterskiing, and fishing from a very, very young age, often at the expense of other things. This led to his niche in the fire service, and, ultimately, satisfied one of his career goals—to get to a position in which he could help protect and mentor other rescuers.

Treinish would also like to thank his colleagues (especially the water people) on the 1006 committee; it has been an absolute pleasure and privilege to work and learn from them all. Pete Schecter, Jeff Mathews, and Francis Brennan have the admirable ability to field questions and opinions without ego or judgment. The folks at the Columbus Division of Fire, especially Station 2, have a sincere professionalism and dedication to the job, no matter what. They have been family to the author for over 28 years, and he appreciates the gift of working with and learning from some of the best. To the OFA instructors and staff, the State of Ohio Water Delivery Tactical Advisory Committee, the CFD DART team, and the Fairfield County special ops, Treinish is humbled and inspired to serve you doing your work.

Finally, the author would like to thank Amanda Brandt and Bill Larkin for guiding him through this edition with tremendous skill, and the staff at Jones & Bartlett Learning for making a tough task enjoyable. Your teamwork and guidance were second to none. A special thank you as well to the reviewers from around the country for the feedback and constructive criticism that helped to shape this book.

In the course of doing this work, the author has met many people from around the world and developed friendships with many of those who helped along the way. This text is dedicated to them and to the many rescuers operating around the world. Some give up a night's sleep, some have given their lives, and the world is a better place because of it.

## Contributors and Reviewers

### John Burley
Lieutenant
Jackson Township Fire Department
Grove City, Ohio

### Jason W. Burrow
Lieutenant
Hanover Fire EMS
Hanover, Virginia

### Kevin B. Clendenin, AS
Fire Chief, Clendenin Volunteer Fire Department
Clendenin, West Virginia
Lead Water Instructor, Spec Rescue International
Virginia Beach, Virginia

**Kent Courtney, Paramedic**
Fire/Rescue/EMS Educator
Emergency Specialist, Peabody Western Coal Company
Kayenta, Arizona
Owner, Essential Safety Training and Consulting
Rimrock, Arizona

**Stephen DeFranco, Jr.**
Chief
Blue Valley Rescue Squad
Bangor, Pennsylvania

**Michael Dick**
Deputy Chief
Bryan County Emergency Services
Georgia Search and Rescue Task Force 5
Bryan County, Georgia

**Gerry Dworkin**
Consultant, Aquatics Safety & Water Rescue
Lifesaving Resources
Kennebunkport, Maine

**Michael Foley**
Deputy Chief
East Windsor Rescue Squad District II Inc.
East Windsor, New Jersey

**Rob Gibbons**
Engineer/Paramedic
Scottsdale Fire Department
Scottsdale, Arizona

**Christopher Hall**
Brevard Fire Department
Brevard, North Carolina
West Buncombe Fire Department
Asheville, North Carolina

**Tony Hargett**
Developer/Master Instructor
Aqua 7 Rescue
Roseville, California

**Ryan L. Harrell**
Captain
Mooresville Fire Rescue
Mooresville, North Carolina

**Shawn Haynes**
Fire Rescue Training Specialist
North Carolina Department of Insurance
Office of State Fire Marshal
Raleigh, North Carolina

**Robert Hevia, CFO**
District Chief
City of Miami Fire-Rescue
Miami, Florida

**Ken Holland**
Senior Emergency Services Specialist, NFPA

**Mike Hudson, Paramedic**
Captain
Sea Bright Ocean Rescue
Surf Rescue Team 43-88
Borough of Sea Bright, New Jersey

**Shawn Kelley**
International Association of Fire Chiefs
Fairfax, Virginia

**Jared Kimball, Paramedic**
Rescue Projects Coordinator
Gulf States Dive and Rescue
New Orleans, Louisiana

**Brad Kueneman**
Program Coordinator
Conestoga College, Pre-Service Firefighter Program
Kitchener, Ontario, Canada

**May Lauzon**
Police Officer/Firefighter, Lifeguard Coordinator
North Myrtle Beach Department of Public Safety
North Myrtle Beach, South Carolina

**Daniel Manning, PhD**
Assistant Fire Chief of Special Operations, Laughlin AFB Fire
    Emergency Services
Professor, Adler University
Professor, Colorado State University—Global Campus
Professor, Pima Community College
Laughlin, Texas

**Jack McGovern**
Lieutenant
Dive and Water Rescue Team Leader
Fredericksburg Fire Department
Fredericksburg, Virginia

**Timothy Mentrasti**
Assistant Chief, Dive/Water Rescue Unit
Yorktown Heights Fire District
Yorktown Heights, New York

**Tony Myers**
Pennsylvania State Fire Instructor
Fire Chief
Shrewsbury Volunteer Fire Company
Shrewsbury, Pennsylvania

**Wayne Papalski, BS, NR-P, FP-C, TP-C**
Search and Rescue Flight Paramedic
U.S. Navy
Oak Harbor, Washington

**Jonathan Morgan Parr**
Water Rescue Coordinator
Sterling Volunteer Rescue Squad
Sterling, Virginia

**Christopher N. Reynolds**
Middle River Volunteer Fire and Rescue Company
Baltimore County, Maryland

**Todd Rishling**
Lieutenant/Paramedic, Water Rescue Team Leader
MABAS Division 1
Elk Grove Village Fire Department
Elk Grove Village, Illinois

**Peter M. Schecter**
Independent Training & Operations Consultant
Hollywood, Florida

**Melvin Shepard, BS, Paramedic**
Division Chief
Orange Beach Surf Rescue
Orange Beach, Alabama

**Robert Shields**
Captain
Cumberland Rescue Service
Cumberland, Rhode Island

**C. W. Sigman**
Director
Kanawha County Emergency Management
Charleston, West Virginia

**Adam Snyder**
Atlantic Beach Fire & Rescue
Atlantic Beach, North Carolina

**Bo Tibbetts**
President, Technical Rescue International
Instructor Trainer, Ice Rescue Systems
Grand Junction, Colorado

**Scott VanPatten**
Fire Fighter, Paramedic, Rescue Technician
City of El Dorado Fire Department
Butler County Rescue Squad
Technical Rescue Instructor, Butler Community College
El Dorado, Kansas

**Connor Wells**
Technical Rescue Team
Toms River Fire
Toms River, New Jersey

**Charles "Andrew" Wilson**
Dawson County Emergency Services
Dawsonville, Georgia

**James R. Woods, MS, Paramedic**
Union County College
Cranford, New Jersey

# New for the Second Edition

## Updated Content and Organization

The second edition of *Water Rescue Principles and Practice to NFPA 1006 and 1670: Surface, Swiftwater, Dive, Ice, Surf, and Flood, Second Edition*, was significantly updated to better serve water rescue instructors and students. Mapped to the water rescue chapters of NFPA 1006, *Standard for Technical Rescue Personnel Professional Qualifications*, this edition also meets all water rescue competencies in NFPA 1670, *Standard on Operations and Training for Technical Search and Rescue Incidents*.

There are many ways to perform water rescue work, and water rescue can vary widely based on area, environment, and need. This edition keeps that in mind and acknowledges that it would be impossible to cover all possible methods currently in use. Your department or team will have to research, adjust, and evaluate the equipment, evolutions, and techniques that work best for your particular hazard and the agency charged with responding to it.

The book reflects the unique makeup of NFPA 1006 and is divided into seven sub-disciplines: watercraft, surface, swiftwater, flood, ice, surf, and dive. Each sub-discipline is clearly divided into Operations and Technician levels of training, with all Awareness level training covered in Chapter 1. The dive section comprises three chapters, with one chapter examining and explaining the equipment used above and below water to effect a safe search.

The book merely touches on patient care around the water, and only basic care is discussed. Each agency should have its own proper medical protocol and certification requirements.

## Updated Design and Online Learning

The new interior design was developed to embrace online learning. Throughout the textbook, there are additional online learning opportunities that can be accessed by redeeming the corresponding Navigate 2 access code in the front of the printed textbook. These activities are earmarked with the Navigate 2 logo NAVIGATE2 and include the following:

**Technical Rescue Incident**—Reinforce comprehension of course materials and hone critical thinking skills with additional practice activities.

**Knowledge Check**—Access the interactive eBook to evaluate understanding of course materials with additional knowledge check questions and answers. To make the most of the eBook, create personal and collaborative notes, highlight key concepts, and bookmark selected content for easy reference, even when you're offline!

**On Scene**—Review answers to the On Scene case studies and check out other valuable resources, such as flashcards, progress and performance analytics, and more.

Beyond enhancing the student learning experience, Navigate offers a full suite of instructor tools that will help ensure a successful program. In addition to comprehensive lecture materials, instructors can assign and customize homework and assessments, access a robust gradebook, and tailor course plans using the real-time, actionable data found in the analytics dashboard.

Improve learning outcomes by taking full advantage of these engaging, accessible, and effective solutions and tools!

# Foreword

## Creating NFPA's Technical Rescue Standards

I became involved in the effort to create technical rescue standards while working as a fire and rescue training coordinator for a regional public safety academy in Bucks County, Pennsylvania. We had just completed applying for and obtaining accreditation through both IFSAC and the Pro Board® and, in attending meetings with other training professionals from around the country, there was a great deal of discussion about why there was not something similar to NFPA 1001 (that gave us Fire Fighter I and Fire Fighter II and established a training and certification base line) for the world of technical rescue.

Many certification entities wanted to be able to certify rescue disciplines, but in the absence of a recognized standard, it was not going to happen. One of the most vocal advocates was Chief Hugh Pike, then head of training and certification for all of the U.S. Department of Defense fire services. I met Pike during my time working to obtain accreditation, as he was involved in both IFSAC and the Pro Board®. It was largely through this interaction that I became aware of activity that would lead to the creation of not one, but two standards addressing technical rescue.

During a meeting of the International Association of Fire Chiefs in Dallas, in the early 1990s, one session on the schedule was a town meeting to discuss the creation of an NFPA standard addressing technical rescue. I went to the meeting and saw a lot of heavy hitters in the rescue world there, from large cities and departments all over the country.

A brief presentation was made on the need for a technical rescue professional qualifications standard, after which those attending were invited to speak and offer their opinions as stakeholders. The proposal drew support from all of the larger departments (who had the budgets to pursue this).

Being a freshly minted Battalion Chief and Training Coordinator (of a much smaller department and without the budget to do this), I spoke in favor of the concept of a standard, but offered a word of caution not to exclude those agencies that were not as large as those present. The standard needed to work for all of us, not just the elite few. My comments drew some harsh criticism from the rescue gods: "If you can't afford it, then get out of the business" kind of stuff.

After the meeting wound down, I was approached by a gray-haired guy who spoke with a distinctive New York City accent. He told me that I made some good points, that the standard *did* need to be inclusive, and that if I felt that strongly I should get involved in writing it. When I confessed my ignorance of how to do that, he took the time to explain it to me (in words even I could understand), and also told me to get with Hugh Pike if I really wanted to get involved. We finished our conversation and headed on to the next program of the day, going our separate ways.

At the end of the day, during rehab at the hotel bar, I was speaking with my boss (director of the training center) and other staff attending the conference, relating the day's events, including the criticism by the elite of the rescue world at the time. As I was finishing up, I saw the gentleman who spoke with me after the meeting and pointed him out to my boss who replied, "Do you have any idea who that is? That's Ray Downey from FDNY who was talking to you. You better listen to what he told you." I never got to thank Chief Downey; he was lost on September 11, 2001.

A lot of people pushed hard for a rescue standard, and the first to get started became what we now know as NFPA 1670. It was a fight to get on the committee, as all of the movers and shakers of the rescue world were applying. I quickly learned what "held for balance" means when NFPA rejects your committee application.

Speaking further with Hugh Pike and lobbying him to get on the committee, he told me to be patient, that the "real" committee was still being formed, and that it would be the Professional Qualifications document we know today as NFPA 1006.

I showed up for my first ever NFPA technical committee meeting in 1995 and met a group of people that would become some of my closest friends, mentors, tormentors, and devil's advocates in the unique process that created this document. Some even had to leave that meeting to respond to the Oklahoma City bombing terrorism event.

The process that created the first 1006 document involved a lot of arguing, compromise, education, and hard work (and a lot of fun), but to an individual, our focus was on what was best for the *rescuer*, and this remains true to this day. We continually challenged each other to come up with skills and job performance requirements (JPRs) that met the challenges of each specialty area/discipline and were measurable and relevant.

From the start, there was hard and sharp delineation between 1670 and 1006; while some of the committee wanted the documents to be better aligned from the start, we were prevented from collaborating or cooperating. This was challenging, as some members of 1670 also served on 1006 (spies among us). We are working hard to better align the documents and make them complementary—as opposed to contradictory—wherever possible. This effort will continue.

Members of the committee have come and gone, but the group remains fully committed to continually looking at the dynamic art and science that is technical rescue, and ensuring that the standard (NFPA 1006) is a tool that can be used to measure skills and knowledge across the broad spectrum of providers, both in the United States and around the world.

I encourage you to be the next agent of change—step up, get involved, and apply. These standards are only as good as the people who are tasked with writing them. Look at me; anyone can do this if they want to.

Peter M. Schecter
Chief (Ret.), Warrington Township Fire Department
Warrington Township, Pennsylvania
Independent Consultant, Public Safety Training & Operations
Oakland Park, Florida

# CHAPTER  1

# Awareness Level

## KNOWLEDGE OBJECTIVES

After studying this chapter, you should be able to:

- Identify and describe the levels of water rescue personnel. (**NFPA 1006: 16.1.4, 17.1.4, 18.1.4, 19.1.4, 20.1.4, 21.1.3, 22.1.4**, pp. 4–5)

- Explain how water rescue personnel at the awareness level can identify hazards at an incident. (**NFPA 1006: 16.1.1, 16.1.2, 16.1.3, 16.1.4, 17.1.1, 17.1.2, 17.1.3, 17.1.4, 18.1.1, 18.1.2, 18.1.3, 18.1.4, 19.1.1, 19.1.2, 19.1.3, 19.1.4, 20.1.1, 20.1.2, 20.1.3, 20.1.4, 21.1.1, 21.1.2, 21.1.3, 21.1.4, 21.1.6, 22.1.1, 22.1.2, 22.1.3, 22.1.4**, pp. 5–6, 8–10)

- Explain how to determine whether an incident has all the resources it needs and how to secure additional resources if necessary. (**NFPA 1006: 16.1.1, 16.1.3, 16.1.4, 17.1.1, 17.1.3, 17.1.4, 18.1.1, 18.1.3, 18.1.4, 19.1.1, 19.1.3, 19.1.4, 20.1.1, 20.1.3, 20.1.4, 21.1.1, 21.1.2, 21.1.3, 22.1.1, 22.1.3, 22.1.4**, pp. 10–11)

- Describe the importance of an incident command system in water rescue. (**NFPA 1006: 21.1.3**, p. 11)

- Describe the purpose and components of an incident action plan (IAP). (**NFPA 1006: 16.1.1, 16.1.3, 16.1.4, 17.1.1, 17.1.3, 17.1.4, 18.1.1, 18.1.3, 18.1.4, 19.1.1, 19.1.3, 19.1.4, 20.1.1, 20.1.3, 20.1.4, 21.1.3, 22.1.1, 22.1.3, 22.1.4**, pp. 11–13)

- Describe how to ensure scene security. (**NFPA 1006: 16.1.1, 16.1.2, 16.1.4, 17.1.1, 17.1.2, 17.1.4, 18.1.1, 18.1.2, 18.1.4, 19.1.1, 19.1.2, 19.1.4, 20.1.1, 20.1.2, 20.1.4, 21.1.2, 22.1.1, 22.1.2, 22.1.4**, pp. 13–14)

- Explain when and how to implement active and passive search measures. (**NFPA 1006: 16.1.1, 16.1.4, 17.1.1, 17.1.4, 18.1.1, 18.1.4, 19.1.1, 19.1.4, 20.1.1, 20.1.4, 21.1.2, 21.1.4, 22.1.1, 22.1.4**, pp. 14–16)

- Identify and describe basic personal protective equipment (PPE) used at water rescue incidents. (**NFPA 1006: 16.1.2, 16.1.3, 17.1.2, 17.1.3, 18.1.2, 19.1.2, 20.1.2, 21.1.2, 21.1.4, 21.1.7, 22.1.2**, pp. 16–22)

# Understanding and Managing Water Rescue Incidents

- Identify and describe the operational support positions used at water rescue incidents. (**NFPA 1006: 16.1.1, 16.1.3, 17.1.1, 17.1.3, 18.1.1, 18.1.3, 19.1.1, 19.1.3, 20.1.1, 20.1.3, 21.1.1, 21.1.3, 21.1.5, 21.1.6, 21.1.7, 21.1.8, 22.1.1, 22.1.3, pp. 22–26**)
- Discuss the role of incident safety officers at water rescue incidents. (**NFPA 1006: 16.1.1, 17.1.1, 18.1.1, 19.1.1, 20.1.1, 21.1.1, 22.1.1, pp. 26–27**)
- Discuss the issue of normalization of deviance and describe how to combat it at water rescue incidents. (p. 27)
- Describe how to assist in terminating a water rescue incident. (p. 27)

## SKILLS OBJECTIVES

After studying this chapter, you should be able to:

- Support the incident, specifically operations and technician rescuers, as needed. (**NFPA 1006: 16.1.1, 16.1.3, 17.1.1, 17.1.3, 18.1.1, 18.1.3, 19.1.1, 19.1.3, 20.1.1, 20.1.3, 21.1.1, 21.1.5, 21.1.6, 22.1.1, 22.1.3, pp. 4–27**)
- Size up a water rescue scene: Determine the number of victims, establish last seen points, and secure and identify witnesses. (**NFPA 1006: 16.1.4, 17.1.4, 18.1.4, 19.1.4, 20.1.4, 21.1.2, 21.1.4, 21.1.6, 22.1.4, pp. 5–6, 8–10**)
- Transfer pertinent information to incident commanders so an incident action plan can be formalized. (**NFPA 1006: 16.1.3, 17.1.3, pp. 11–13**)

# Technical Rescue Incident

You are a fire fighter in a fire department that does not deal with floods or even water rescue on a large scale. Even with the heavy rains and storms you have seen in your area in the past, it just seems as if nothing has ever really happened to raise eyebrows. In the last few days, however, rain has pounded your jurisdiction, hardly stopping for more than a few hours. You hear a "500-year flood" mentioned on the news, and then a really vague call comes in: Someone is stranded in a house due to rising water and cannot get to safety.

**1.** What information can you obtain to start planning this response?

**2.** Can specific resources be called in, and how do you start asking them to assist in the response?

**3.** What training and equipment do the teams around your area have, and how can they help meet your current needs?

Access Navigate for more practice activities.

## Introduction

Water rescue differs from the other technical rescue disciplines in that it can often overlap with and fill in gaps between the subdisciplines defined by the National Fire Protection Association (NFPA). Parameters such as weather, victim ability and movement, and scene dynamics can often turn a surface mission into a dive mission, or an ice mission into a surface mission (**FIGURE 1-1**). Storms on larger waters can cause an incident to start as a surf rescue, be suspended until conditions allow a surface search, and end with a dive recovery. Rescuers often choose to specialize in a subdiscipline, but they should have a good understanding of all types of waters and rescues that may be encountered in their area of response.

**FIGURE 1-1** Weather, victim ability and movement, and scene dynamics can change the rescue environment and the expertise needed to effect a rescue.
Courtesy of Steve Treinish.

## Rescue Levels

The NFPA has long been active in the fire service world, creating standards that cover an enormous number of fire-related activities. Issues ranging from fire alarm panels, to fire station planning, to the basic fire fighter professional qualification standards are addressed in the various documents that the NFPA publishes. These standards and codes are updated every 3–5 years by committees whose members are appointed based on their expertise in the specific area.

Two standards specifically address technical rescue: NFPA 1670, *Standard on Operations and Training for Technical Search and Rescue Incidents*, and NFPA 1006, *Standard for Technical Rescue Personnel Professional Qualifications*. These documents help guide an **authority having jurisdiction (AHJ)** in forming rescue teams and in determining what individual rescuers should know and be able to do. AHJs can be fire departments, rescue teams, or any team that may respond to a rescue. Written **job performance requirements (JPRs)** detail rescue evolution jobs and tasks, list the needed equipment to accomplish them, and outline a way to evaluate a person performing them.

Within both NFPA 1670 and NFPA 1006, rescuer training, skills, and qualifications are divided into three levels:

- **Awareness level:** Responders have no formal training in water rescue, but may still be dispatched, especially at the beginning of an incident.
- **Operations level:** Rescuers generally have training that prepares them to operate around the incident, but not physically enter the hazard.
- **Technician level:** Rescuers have completed operations-level training, and will often be exposed to the same hazard in which the victim is trapped.

For example, in an incident in which a person is trapped on a cell phone tower, awareness-level personnel should recognize the need for rescuers who have specific gear and training to perform a high-angle rescue, and know where and how to get those rescuers to their location. The operations-level rescuers on the team that is called in may operate from the ground or from a position of safety by operating rope systems. By comparison, the rescuers who physically put on harnesses and go up the tower must be trained to the technician level.

Water rescue is set up the same way. Operations-level rescuers working from the bank or even a boat are inherently protected by not being physically in the water. The personnel who don water personal protective equipment (PPE) and perform entry into the water must all be trained at the technician level, since they will be exposed to the same hazard in which the victim is trapped.

## NFPA Documents and Water Rescue

Both NFPA 1670 and NFPA 1006 break water rescue into several subdisciplines:

- **Surface:** Water areas that are not moving faster than 1.15 miles per hour, are not frozen, do not present wave action, and present with victims on the surface of the water (not submerged)
- **Ice:** Any frozen water surface or area, even if surrounded by open water
- **Swiftwater:** Any water flowing faster than 1.15 miles per hour
- **Surf:** Water areas that contain waves or wave action
- **Dive:** Water that contains submerged victims
- **Flood:** Any areas that do not commonly hold or contain water runoff or flow
- **Watercraft:** Any vessel—either hand or motor powered—that is used to effect rescue, both directly and in support roles

NFPA 1670 addresses team operations and guides rescue *teams* by establishing functional capabilities for teams at the three levels of response. This standard is essentially a guide as to which services a team should be able to provide at the awareness, operations, and technician levels. Personnel should also have an understanding of the general requirements for technical rescue.

NFPA 1006 addresses the qualifications and requirements for *individual* rescuers, and offers more guidance on specific skills, requisite knowledge and skills, and JPRs that rescuers must demonstrate during an evaluation in a technical rescue class. This standard defines the skills and knowledge needed to operate as a team member at each level, and it can be used to ensure that individual rescuers operate to the level the team requires to provide the service outlined in NFPA 1670.

It is no accident that NFPA 1006 and NFPA 1670 are so closely aligned. AHJs that choose to offer water rescue services need a very detailed and thorough guide regarding what both the individual rescuer and the rescue team must bring to the table if those services are to operate at a certain level of expertise and capability.

## Incident Size-Up

**Incident size-up** is the process of observing, gathering information, and determining a course of action and needs to be met to mitigate an incident (**FIGURE 1-2**). An easy way to grasp the process is to imagine yourself being in charge of a first-due engine company and dispatched to a house fire. En route, you check the mobile data terminal for remarks and listen intently for extra information from dispatch. You go over the potential needs and tactics mentally and advise your crew of their roles based on what you may find. Even as you

**FIGURE 1-2** Incident size-up involves observing the scene, gathering information, and determining a course of action.
Courtesy of Steve Treinish.

think over your options, you remain free to adjust your plans based on what you see when you pull into the driveway at the scene. If you spot a very small campfire that is located far from any structure and do not see anyone with injuries, you can almost instantaneously transmit on the radio that the incident is handled: You know that you have a capable crew, water, and hose, and you recognize that the fire is certainly not big. This incident can be handled with the incident management system that was dispatched—namely, *you*.

At the other end of the spectrum, what if dispatch radios you en route that this call may be the result of a meth lab explosion? When you reach the driveway, you see a well-involved house, two severely burned victims in the front yard, and numerous chemical drums in the open garage on fire. In this scenario, you must quickly adjust your tactics to keep your crew safe. In addition to calling for a hazardous materials team, you should probably call for law enforcement assistance and an arson or investigation squad as well. This incident management will be complex and will require specific training and the ability to mitigate the problem. Upon sizing up this scene, you decide to call in specific additional help to combat the criminal and hazardous materials aspects of the incident. The hazardous materials response team may need its own radio channel and other resources, emergency medical services (EMS) responders may be taxed with two severely burned patients, and the fire is large. Because of the expanded control and command needs associated with this complex response, you also call for an additional chief and an incident safety officer to help handle the incident management aspects of this response. The better the training and the more extensive the experience the initial response team has, the more quickly these decisions can be made.

## Recognizing the Hazards

Awareness-level personnel should be able to recognize both the immediate hazard in which the victim is trapped and any potential hazards that may be found in the area, such as utilities like electrical services and natural gas or propane supplies, hazardous materials releases such as gasoline or sewage, and risks of injury in reaching the water such as cliffs or slippery slopes or ramps. They are also tasked with limiting further harm to both civilians and other rescuers, and with providing operational assistance to the additional rescuers whom they call in. At least to some extent, they should be able to recognize the need for more specialized crews or equipment, and start gathering information for an **incident action plan (IAP)**. An IAP is the

blueprint formulated to mitigate the incident at hand, taking into consideration safety, methods, personnel, and equipment. IAPs are discussed in more detail later in this chapter.

Awareness-level personnel do not operate rescue equipment, but rather function in a support role. They may contribute to the rescue operation by securing safety zones, organizing rehabilitation for rescuers and crews, securing scene lighting, providing ventilation, documenting data, and providing other assistance as needed.

## Scene Safety

Awareness-level personnel are expected to secure a scene so as to prevent harm to others, and these steps to protect everyone involved start with the response. Sometimes getting to the actual scene of the water emergency may require waiting for additional help or staging at a safe distance (**FIGURE 1-3**).

For example, a report may state that a car has run off a cliff and fallen into the quarry below. Rescuers may need to establish a lowering system with ropes just to approach the point at which the car went into the water, before they can even think about getting help to the car and its passengers. This setup would require a means of securing the cliff edge long before the water aspect comes into play.

In another example, floodwaters might still be rising when the initial dispatch is made. If you attempt to access a car in the water only to soon require rescue yourself, then you have made things worse: You have created a danger not only to yourself but also to the crews now tasked with rescuing *you*. Rescuers must be cognizant enough to at least consider these kinds of peripheral safety hazards, which may necessitate changing the response location.

**FIGURE 1-3** Getting to the scene of the water emergency may require waiting for additional help or staging at a safe distance.
Courtesy of Steve Treinish.

100% 

# Voice of Experience

It was a blistering hot summer day in mid-July when our water rescue team was dispatched to an incident in the river. Dispatch reported that a man was in the river pinned against a tree. The tree had fallen into the river during high water runoff season, which is typically in the month of June. Immediately my thoughts were, "Wow, this guy's in trouble. This is serious." Anytime I hear of a person trapped against something in a water current, I know what the victim is feeling—like they're fighting an 800-pound gorilla.

Upon arrival at the scene, I noted that incident command had been established. A plan was being developed as to how best to extricate this individual, who was trapped in a strainer. As I sized up the scene I knew this would most likely require additional resources, including qualified manpower and additional equipment.

I found myself going back to the basics, identifying multiple hazards in the river and on the banks while constantly analyzing the risks versus benefits of the rescue scenario. My goal is to always keep personnel safe while executing the rescue, so I was checking to make sure everyone had the proper PPE donned, buckles fastened, and helmets properly secured.

As the plan was being established, the victim had already been in the water pressed against a fallen tree for over 45 minutes and he was losing strength by the second. The water temperature was about 65 degrees, and the victim's head was slowly descending into the water below the branches of the tree. As I stood there evaluating the resources we had available to us, and how to utilize them, I must say I felt helpless. At that moment, a rafter was floating up on the scene and he had the idea to cut the hose off his air pump and give it to the victim to use as a snorkel. As the rafter was handing it off, he nearly got caught in the strainer himself. Thankfully, the makeshift snorkel worked and the victim had air as his head went under water.

Plan A, was to rig a rope extrication system. While this was being staged and prepped, another plan was taking shape. It required the use of a johnboat and a 35-horsepower tiller unit with a skilled boat driver and a swiftwater technician. This plan required the rescuers to make a sweep upon the individual while trying to move him laterally away from the debris. After several attempts, this plan was called off due to multiple safety issues that almost capsized the watercraft.

I personally have always been a proponent of employing the KISS methodology—keep it simple and safe. We try to take the "stupid" out of the equation, replacing it with safety. Our team was proficient in simple, low-angle rigging systems that were tried and true. We were able to get a line down into the water with a sash weight, low enough to begin the process of pulling the victim backwards away from the strainer, which gave him some measure of relief. He then was able to pull himself, aided by us and the current, up onto the tree. At this point we were able to get a PFD on him and start the process of getting him safely to shore personnel.

This incident was beneficial with respect to gaining knowledge and understanding the force of water, in-water and shore hazards, proper implementation of the ICS, risk-versus-benefit analysis, and ultimately thinking outside the box when evaluating available resources. The biggest lesson I learned is not to overanalyze the situation, but to continue to go back to the basics of general awareness, understanding, and managing a water rescue incident.

**Bo Tibbetts**
Technical Rescue International
Public Safety Dive Services
Grand Junction, Colorado

## Team Safety

Before any operation begins, a **basic safety plan** must be in place, and all personnel involved must understand the hazards on scene (both presently and potentially), how the plan will be implemented, and who will play which roles. Other safety parameters to be covered include who backs up the backup rescuers, how communications will take place, and how and if the original rescue attempt will be suspended, continued, or attempted by another method or from another direction.

### LISTEN UP!

As with many fireground rapid intervention or removal procedures, placing the backup crews on a separate radio channel when activated or needed may help the entire incident response proceed more smoothly. The rescuers and those being rescued can all benefit from reduced radio traffic, and the potential confusion caused by two or more sets of rescuers simultaneously describing the conditions, actions, and needs of two separate rescue operations will be reduced. Most firegrounds keep the trapped rescuer(s) on the original radio channel and place subsequent radio traffic on another predetermined channel.

A good question to ask during any phase of a rescue incident is "If something goes catastrophically wrong, how will we deal with it?" In water rescue, this question is normally posed as "What will happen if the rescuers end up in the water?" and "If this system fails, what will happen?" In a very basic incident such as a person near the shore of a retention pond, a rescuer falling in wearing a life vest will likely get wet and perhaps cold, but may need one person to assist him or her back out of the water. With fire apparatus nearby, the rescuer can be quickly warmed up, and with such a rapid solution, the rescue can likely go on as planned.

Now imagine a different scenario—a crew of three rescuers in a boat on a fast-moving river, with this boat being capsized and stranded against a tree. In this case, it may take a specialized team of rope and water rescuers to rescue the rescuers. The force of the river may injure them all and require other rescuers to participate in the operation. And, of course, you cannot forget about the original victim who started this event. If the "safety net" to mitigate this catastrophic failure is not in place, the evolution may need to wait until proper backup is ready.

Awareness-level personnel must ensure that the entire cadre of response personnel know the general safety hazards present before a more specialized plan is formed, so as to protect each individual. Specific hazards related to the water, such as swiftwater features, must also be recognized and dealt with by the rescuers trained in those subdisciplines.

## Individual Safety

When called to an incident, awareness-level personnel should ask, "How bad is it?" These responders must be able to recognize that the problem presented to them is beyond their training and capabilities, and they must know where to find rescuers and specialists who are sufficiently trained and capable of handling the incident (**FIGURE 1-4**). A very honest evaluation of the personal skills that a rescuer possesses is required based on the conditions in which the victim is trapped. A non-swimmer should not jump into moving water to try to save someone because of the high risk of harm the aspiring rescuer faces from this action. Likewise, someone who considers himself or herself a strong swimmer yet has no flotation device at hand should not allow emotion to outweigh cognitive thinking and try to effect a rescue.

Water rescue is often very emotional, dramatic, and fast paced, especially in close quarters. Responders at the awareness level must have enough training and knowledge to stay safe by not placing themselves in the hazard, but also have a basic understanding of the behavior and parameters of both the water and the potential victims. Simply put, know the hazards' behavior and figure out who is missing, where they likely are headed, and what is needed to rescue them.

**FIGURE 1-4** At the awareness level, responders must be able to recognize the limitations in their training and capability, and they must know where to find rescuers and specialists who are sufficiently trained and capable.
Courtesy of Steve Treinish.

## Spatial Awareness

**Spatial awareness** is the rescuer's ability to see the proverbial forest, rather than just the trees. The water itself may be just one part of the overall hazard matrix. Other potential hazards—such as air and water temperature, utilities, and automobiles—may be present in or around the water. Thermal issues may be present, such that responders may end up treating a hypothermic patient in warm sunshine. The same rescue could involve gasoline leaking downstream in a creek: Awareness-level personnel should know that gas floats on water, that the incident now includes an environmental concern, and that hazardous material teams may be needed.

Recreational waters such as populated lakes, swimming pools, and retention ponds may all have electric components present at the shore or underwater, and these components' safety systems may potentially be damaged. As an example, suppose a homeowner working on a dock falls in a lake and drops an electric drill he was using into the water with him. You believe the victim is unconscious, perhaps from striking his head during the fall. However, if no one checks the breaker or notices the electric cord, the first rescuer in may receive an electrical shock.

The ability to not just look, but "see" a scene is paramount to rescuer safety. Often individual rescuers develop tunnel vision when working around rescue scenes. Sometimes, with a highly technical rescue, the risk of tunnel vision is understood, and sometimes it is even needed. **FIGURE 1-5A** shows a typical river, just under a bridge that necessitates several rescues each year. What do you see? **FIGURE 1-5B**

is a close-up of a piece of rebar that can injure swimmers or damage boats. Awareness-level personnel can help the incident progress safely by having a good knowledge of hazards and concerns around the water environment and making these risks known to everyone operating on scene.

## Rescue Versus Recovery

Rescuers must start thinking about safety during the dispatch message, and must ensure the scene is safe to approach. Once they arrive, the first decision to be made—even if the call ends up with a victim found safe and sound—is whether the call is a rescue or a recovery. A **survivability profile** must be done upon arrival. Survivability profiling entails taking a realistic look at the conditions surrounding the victim—including the time he or she has been underwater, the water conditions, the temperature, and outside factors such as a fall from a high place or even a plane crash—and determining the victim's chance of survival under those conditions.

## Timelines

A time window of 60 minutes from the time of submersion is generally viewed as the consensus standard for determining survivability in a water rescue, although some teams use 90 minutes under ice or cold water as the cutoff. Whatever standard your AHJ uses, do not be drawn into a potentially fatal incident by making decisions solely based on the time of the call. A call in which a person reports witnessing a drowning right off a dock only a few minutes ago would

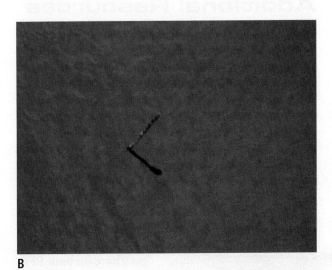

A
B

**FIGURE 1-5 A**. Typical river view. **B**. Look for hazards that might be missed at first glance. This rebar is just barely visible in the lower right edge of the photo to the left.

warrant operating in rescue mode. In contrast, in an incident in which a car entered the water, ice formed over the hole, and the event was not reported until sunrise, the team will likely operate in recovery mode. The amount of time it takes to form ice over previously open water, along with the elapsed time before the event was reported, will negate any option of rescue.

Rescuers should always give the benefit of the doubt to rescue operations, but once the timeline is known and the point at which a rescue operation might be feasible has passed, rescuers can operate in a much safer manner because the time constraint has been lifted. Beware the adrenaline rush!

Consider the example of arriving at a large lake where the caller observed a boat simply floating offshore. Pertinent questions related to this incident timeline might include the following:

- When did the boat found floating actually launch?
- Is there a trailer in the parking lot?
- If there is, is the car engine warm or cold, or can law enforcement trace the license plate and contact the owner?

Many boats break their tie-up lines in storms, or get washed downstream in flood conditions. This incident would certainly not warrant placing divers in harm's way, because the timeline does not call for diving at this time. A rescuer dying in an attempt to save a victim who is later found safe is a tragedy. In some cases there may not be a victim at all, just some incorrect or assumed information that makes it seem as if there is one.

## Additional Resources

Awareness-level training gives responders the ability to answer two additional key questions concerning the response they have been called to: "Are there resources in place to handle this response?" and "If not, where do we get them?" Note that trained rescuers must be included in the term *resources*, and that although these questions seem easy enough to answer, a few issues must be considered. When dealing with water rescue, often departments and agencies will provide "rescue teams" as part of their operations. It can be problematic, however, when the "water rescue team" shows up with a couple of canoes and three rescuers who do not even have the basic PPE. Resource needs will vary based on

the incident, victim(s), environment, and many other factors. Rescuers must have a solid foundation in the resource needs for different situations.

## Typing Teams

One way to facilitate answering the resource question is to **type** local rescue teams. The National Incident Management System (NIMS) has developed a process for defining exactly which equipment, training, and capability a team must have, and an inspection process to verify these capabilities. The capabilities and equipment for a team are organized on a continuum ranging from basic (Type 4) to major resources and capabilities (Type 1). The requirements for each level are very specific. Once a team believes it can meet the requirements to become a Type *X* team, the team equipment, personnel, and training are verified by inspection personnel impartial to the AHJ.

Typing of rescue teams ensures that the problem presented (e.g., a large but generally docile flood) can be paired most effectively with the resources—both personnel and equipment—needed to mitigate it. Typing is also invaluable in preplanning water responses.

## Getting the Needed Help

Awareness-level training can play an important role in preplanning. Understanding the basics of water behavior, hydrology, and the geography of the response area can be especially helpful for awareness-level personnel, especially those who may wind up being incident or operations commanders. Using the typing system described earlier, resources can be more easily selected and deployed, or plugged into standard operating procedures or automatic dispatching.

During preplanning, questions or concerns can be answered and worked out before an actual run occurs. Problematic areas that can be addressed include the following issues:

- Which methods of communication will be used? Radio types, frequencies, and cell coverage can all be examined. Dispatchers may have to cross-patch or link radio frequencies to ensure effective communications.
- How long will it take to activate the team and for the team to arrive? Many departments need approval from a ranking officer to deploy resources outside their own district. If the rescue team cannot respond until their own trucks are covered by called-in fire fighters, there may be a delay.

- To which location should the rescuers respond, or where should they stage? Rescuers need to know if they are going to work immediately, or if they will stage somewhere and await orders.
- Are the team members self-sufficient? In extreme conditions, Type 1 teams are required to carry their own medical supplies, drinking water, and food, and be self-sufficient for 24 hours.

More importantly, preplanning should consider whether the resources being dispatched are the best tool to fix the problem. A good starting point in making this determination is the AHJ's **standard operating procedures (SOPs)**. SOPs are written guidelines to help incident responders function in a uniform and consistent way during specific evolutions. For example, when a rescuer trained to the awareness level arrives and ascertains that a victim has been underwater for less than 60 minutes, the department SOP may state that calling for a certain dive team is warranted—so the rescuer now knows which action to take. For a certain type of water rescue, a certain team or teams may be dispatched immediately. However, the fix must always fit the problem: A dive rescue team may not be able to effectively handle an ice rescue mission; a swiftwater team with only one boat may need to be supplemented in case of a large flood.

Good directions to staging areas and accurate travel directions to the incident can play a huge part in the mission's success. The optimal way to request additional help is to provide this information as part of the dispatch information. Similar to asking for another fire engine or medic vehicle when needed, asking for a named team or specific resources can be done quickly over the radio. Keep in mind, though, that dispatchers will only be as effective in summoning these resources as the information you give them before the request.

Department SOPs may also be used to call assisting teams to an incident. Some SOPs can be site specific, shaving crucial minutes off responses and reducing confusion. Preplanning, either by written plan or verbal agreement, can also accomplish this task. Having a resource manual loaded into a binder and providing it on the mobile data terminal can also offer ways to directly contact agencies or teams if needed. Often, a good conversation on a cell phone between responders or AHJs can help clarify dispatching notification and communication issues. If a cell phone is used for this type of communication, ensure everyone understands the result of the phone call and is aware of incoming agencies' resources.

## LISTEN UP!

In extreme cases in very rural areas, people who know the district can even be placed at each intersection to direct incoming crews. It might sound funny, but directions such as "turn right past the Johnson farm" are not uncommon. At night or if road signs are missing, critical minutes can be spent trying to find the location. Do not depend on cell service or rely on moving maps on smartphones.

# Incident Command System

Awareness-level personnel must be familiar with the incident management system (IMS) used by the AHJ. As a result of the NIMS training required for all technical rescue personnel, every responder should be familiar with the concept of the **incident command system (ICS)**. The ICS is a method of making control of an incident more manageable. It focuses on delegating each command position as needed as an incident grows. Some very minor water rescue runs are dispatched and completed by a single incident commander; others require just a few crews, each commanded by a squad officer or crew leader; and still others require a complex IAP managed by a joint command staff of local agencies and responders.

Responders should be aware that water rescue will often have its own operational branch, with an operations section chief, logistics, rehabilitation, decontamination, and staging groups all added if the incident grows in size. ICS is a very effective management tool when used properly.

# Developing the Incident Action Plan

Incident action plans can be developed for mental use, verbal use, or command use. Although typically considered a command-level task, IAPs are created for almost every task in the fire service, whether explicitly or not. This section touches on IAPs that may be required for larger incidents, and delves further into IAPs that the individual rescuer or smaller company-type rescue crews may use when operating up close and personal at rescue scenes.

## Large-Scale IAPs (Written)

Large-scale incidents often have written action plans for expected or potential risks, such as severe weather moving through an area, or perhaps large-scale flooding requiring preplanned actions or assignments. IAPs at this level are frequently broken into operational periods of hours, with details on who is working, where personnel are working, what will be done, and other parameters. These documents can be developed at a command center and disseminated through the appropriate channels to street-level supervisors.

## Tactical Sheets

Larger, more complex incidents can have a written worksheet or checklist—often referred to as a *tactical sheet*—to support operations. Needs such as PPE lists, radio frequencies, operational area boundaries, and individual rescue companies or team rosters may all be included in worksheets used by the incident command staff. The Federal Emergency Management Agency (FEMA) offers many helpful tactical worksheets in the Incident Command Section, including the ICS 215, a resource sheet that is commonly used with many different incident types.

## Medium-Scale IAPs (Verbal)

Much more familiar to most rescuers are the "bread and butter" jobs we see every day on tours or during calls. Incident commanders who oversee these operations often verbalize their IAPs over the radio, and most of these initial IAPs start with company officers reporting what they are actually seeing and what they are going to do about it. For example, at a typical house fire, an engine company might give the following radio report:

> Engine 9 to dispatch, Engine 9 on scene and has a two-story wood-frame residence with fire showing from the front door. Residents are reporting an all clear. Lt. Mack is in charge. Engine 9 is attacking with a 1¾-inch preconnect. Ladder 1, search the second floor when you arrive. Engine 1, lay a supply line. Battalion 1, take command on the Alpha side upon arrival.

In this example, Lt. Mack quickly developed an IAP upon arrival and verbally relayed it over the radio so that all parties responding knew what their initial actions should be, and what was expected from the next companies arriving. When Battalion 1 arrives, conditions may require the chief to change the initial IAP to add attack lines, add search crews, or even withdraw companies. Changes to or continuation of the IAP in place should also be verbalized over the radio to dispatch so that everyone hears this information and makes the necessary changes.

## Smaller-Scale IAPs (Mental)

At the smallest scale, IAPs are developed by individual rescuers who physically perform rescue work, or by their immediate supervisors. These mental IAPs are often relayed or transmitted to the incident commander, who then uses them to help develop a broader IAP that may include additional resources, equipment, command needs, or other peripheral aspects of an incident.

Even when developing these mental IAPs, rescuers use several sources to develop the IAP that best fits the incident at hand:

- Standard operating procedures (SOPs) (or standard operating guidelines [SOGs]). SOPs are requirements and directions that the AHJ develops for handling various tasks. They take into account safety, department equipment and capabilities, personnel training and capabilities, and needed results. Although SOPs provide uniformity in terms of how an incident is handled, they may also allow some variation, provided that straying away from these guidelines can be justified.
- Experience. It is knowledge sometimes gained right after it was needed, but experience may be the "glue" that can hold together a plan once formulated, the "calculator" used to refigure things for the glue to work, or the "broom" that moves the plan to the brain's trashcan. Experience can also drive recognition-primed decision making, which enables seasoned rescuers to determine which action to take within just a few seconds when faced with a situation.
- Training. Consider which training you, your crew, your department, and any specialty responders have. Are the skills to safely perform rescues mastered?
- Size-up. This survey of the scene should include the following questions:
  - What will this problem take to solve (e.g., more trucks, people, higher level of expertise)?
  - How do I keep others and myself safe while we are solving it (e.g., PPE, moving farther away

from the scene, heavy supervision or backups, additional resources)?

- Will things get worse or better from my (our) action or lack of action (e.g., can I rescue someone now, or make a scene safe)?
- Which booby traps might I expect from this mission once I go to work (e.g., if a catastrophic failure occurs, what is in place to mitigate that outcome)?

Ideally, the ideas formed from the SOPs, experience, training, and size-up will come together to create an IAP. IAPs can be formulated within seconds, or they can require much discussion and much more time to develop. Water rescue—or any fire service mission, for that matter—can be very dynamic, and rescuers and command personnel should work together to develop or change the IAP as needed. Combining all the facets of size-up, SOPs, experience, and training leads to a "system" that rescuers can use to develop, execute, and evaluate a plan for any water rescue.

## Scene Security

Another key pillar in incident awareness is the prevention of further harm, be it to civilians or to fellow rescuers. Many scenes can be emotionally charged, especially when victims are visibly in distress. Knowing when you are "outgunned" is a crucial risk-versus-benefit decision, and that decision may not be unanimously accepted by all those persons present at the scene. Civilians do not usually have the same training as rescuers do, so they often have a very different view of the risks involved in rescue. These persons must be kept out of the hazard, to prevent them from becoming victims themselves. Failing to keep them away from the scene simply makes the entire scene worse.

Another aspect of scene and rescuer security is the growing need for personal protection while operating, especially in disasters such as massive floods or hurricanes. It is a sad but true fact that people often try to steal equipment, supplies, and goods from stores, other victims, and even rescue crews. Just as when they are operating in or around a criminal scene, rescuers must be safe from these potential threats while they operate at a rescue scene (**FIGURE 1-6**).

## Scene Tape

At the basic level, scene tape is recognizable to almost every person in the world. Yellow scene tape labeled

**FIGURE 1-6** Some rescue situations have an increased likelihood of violence or other criminal activity. Be sure to call for the appropriate law enforcement resources.
Courtesy of Eric Reed.

**FIGURE 1-7** Using scene tape can prevent unapproved entry to an area, and help rescuers concentrate on the task at hand with no outside distractions.
Courtesy of Steve Treinish.

"CAUTION" is often used, but from a legal standpoint, tape marked with more severe wording, such as "DO NOT ENTER," may be desired. Either type of tape can assist with cordoning off the scene (**FIGURE 1-7**). Unfortunately, scene tape can easily be torn down or bypassed by people who want to get closer to the action.

## Vehicle Blockades

Consider the use of fire apparatus to block streets, lanes, driveways, or other avenues of travel. Many people will drive right around danger signs or traffic cones, but will yield to flashing emergency lights and large vehicles. Be aware of your surroundings: Just as on highways and streets, some civilians may try to squeeze by the scene in their vehicles, possibly injuring or killing rescuers. Be firm in blocking traffic. Many civilians do not understand the dangers of water responses, and they may be distraught or very angry at the idea of not being able to access or try to save their property.

## Law Enforcement

Very often, police officers will be needed to prevent civilians from entering a hazardous area, using both police vehicles and physical officer presence. Operations on bridges or along roads or highways present dangers to rescuers from traffic, and law enforcement presence is much more effective in thwarting travel than orange cones alone. If law enforcement cannot stage at the road closure location, consider positioning a piece of fire apparatus to physically block the road to ensure safety.

Closer to a scene, distraught family members can sometimes be so problematic that, for their own safety, they need to be placed in a police cruiser. Police officers have the tools, training, and authority to safely restrain people, and it is not unusual for them to have to do so. Remember, these may not be bad folks, but simply folks overwhelmed by a bad situation. Regardless, if not controlled, their actions could affect both the rescuers' safety and the outcome of a potential rescue. The presence of family members within eyesight or earshot of rescuers on scene can add considerable amounts of stress as the rescuers attempt to resolve the incident.

## Public Address Use

Awareness-level personnel may be tasked with helping in an evacuation. Floods, especially flash floods, may lead to groups of people who are willing to leave but do not know where to go. Consider the use of bullhorns or siren-based public address (PA) systems to verbally guide evacuees to safety. Using the PA system on the fire apparatus can be especially helpful, as most people will be drawn to the flashing lights. They see

fire department personnel as representing safety, so use that to your advantage.

## Active and Passive Search Measures

When witnesses help rescuers determine a point last seen (PLS), which is the approximate location of victims when they were last seen, the operation will normally require an **active search measure**. With an active search measure, rescuers immediately work to rescue a victim, provided the victim is in sight (or the PLS is established), the timeline supports action, and the proper resources are in place (**FIGURE 1-8**). Recall the house fire example discussed earlier: If fire is showing from a structure, we will promptly go to work. In contrast, if nothing is showing from the house, we will not necessarily pull lines or place the pump in gear; instead, we will investigate.

---

**LISTEN UP!**

An easy way to differentiate between active and passive search measures is to think of a typical fire response. Hearing "working fire" means crews are going to work. Hearing "nothing showing" means there will be some investigation by first-arriving crews before attack lines are pulled or supply lines put in service.

---

If the incident is reported by a concerned party, as in the case of a missing or overdue boater or paddler, there may be no witnesses. In this kind of scenario, **passive search measures** would be a better tactic

**FIGURE 1-8** Witnesses state that a fisherman fell in the water. With evidence such as a pizza box, fishing gear, and mud and silt showing at the water's edge, rescuers would be warranted in performing an underwater search.
Courtesy of Steve Treinish.

**Restricted**
**Turbulent Water**
**Sudden Water Discharge**
**No Swimming or Wading**
**No Boats**

**FIGURE 1-9** The same bank as in Figure 1-8, but without the supporting evidence of the missing fisherman. Crews would be much better advised to check parking lots, restrooms, and similar locations, and make phone calls to family.
Courtesy of Steve Treinish.

**FIGURE 1-10** A proper witness interview can help provide the point last seen, narrowing down search areas.
Courtesy of Steve Treinish.

(**FIGURE 1-9**). Passive search measures can be more frustrating to rescuers because they involve staging resources, including rescuers who are ready to go while investigations are being undertaken. Passive search measures usually do not involve an intense, quick search, as there are generally no witnesses to these events. People not reporting in from a boat ride at a certain time or a lone boat floating in a lake would be enough to warrant investigation, but most likely those initial scenarios would not provide enough information to be able to decide exactly where to begin an active search.

Passive search measures may entail calling a missing person's phone, sending law enforcement to any known address, or even calling friends and family of a missing person. Good descriptions of victims, which kind of vehicles they are driving, places they typically visit, and any possible alternative locations where they may have traveled should all be considered. For example, a missing boater, when no obvious distress is noted in the area, might be located by searching gas stations, bars or restaurants, marinas, or bait shops.

Active and passive search measures can also change with the dynamics of a response. The overdue boater incident may start in a passive mode, but change to active mode if another person reports an event or provides additional information that fits the rescue timeline.

## Witnesses on Scene

Generally speaking, witnesses should be interviewed from the location where they watched the event occur, so better lines of sight can be established (**FIGURE 1-10**). Let the witness tell the story. It is very easy for a rescuer to accidentally lead a witness into details because the rescuer wants to move quickly. Instead of prompting the witness with a question like "Over there by the big tree?", a better method might be to ask the witness to pick out a close landmark or an object across the water.

Interviewing witnesses separately from each other will also help the team establish a better PLS. When a large group is questioned collectively, the largest or loudest personality often dominates the others, even if the other group members might put a better location in play.

Children can make good witnesses, but be aware of the effect of suddenly putting children into a spotlight with many loud and authoritative adults demanding information. Get down on children's level and speak their language. Chapter 14, *Dive Rescue Operations*, provides further details on interviewing witnesses.

**LISTEN UP!**

Decisions made early in a rescue are made based on the information available at that time. Rescues of any type requiring any kind of search frequently obtain additional or more detailed information later that may change where operations occur.

**LISTEN UP!**

Be wary of witnesses who are trying to sway others or acting suspiciously. If a witness tries to remove himself or herself and get away from the scene, it could be a sign of criminal activity. Law enforcement can be an important asset here, as police officers have much greater expertise in speaking to suspicious persons. Some witnesses may be interviewed multiple times to ensure the story is the same, or if something glaringly wrong stands out.

## No Witnesses

Consider a call that reports a missing teenager at a house with a farm pond. A frantic mother who has not seen her child in more than an hour would certainly elicit a hurried response, and perhaps even start some types of searches. A better choice, however, may be to operate in passive rescue mode and start a detailed shore investigation. Shore investigations in smaller waters could entail trying to ascertain the victim's last location, looking for exit marks on the shorelines, searching for evidence of water activity (e.g., disturbed silt, algae beds broken up), and examining other areas that kids might find attractive. Finding an area with disturbed silt, fresh climb marks up the bank, and wet soil would point to the possibility that the victim found his or her way out and is somewhere other than the water. These passive search measures may result in better target locations or even discovery of the victim, but can also evolve into active search measures if better information comes forth.

## Information Documentation

All witnesses' information should be documented in writing, and their stories jotted down. Many prolonged water responses will require the witnesses to be interviewed again, possibly multiple times. Knowing where to find these people over the coming days can help investigators even after the rescue or recovery is complete.

A **scene sketch** can go a long way toward processing witness statements and help establish the lines of sight and events they witnessed (**FIGURE 1-11**). This sketch does not need to be perfect, but landmarks and features drawn into it will help the rescuers sort through the information coming in. Rescuers should be wary of letting witnesses see other sketches, as they may start to doubt themselves. Ideally, each witness will have a corresponding sketch that is completed while talking with him or her.

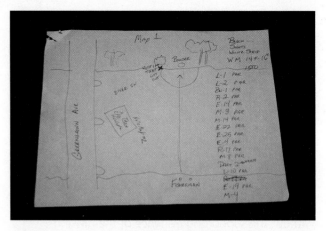

**FIGURE 1-11** A scene sketch can go a long way toward processing and comparing witness statements.
Courtesy of Steve Treinish.

# Personal Protective Equipment

At any incident—rescue, fire, or EMS—responders are equipped with **personal protective equipment (PPE)**, which is meant to protect them from discomfort, injury, or even death. Various pieces of water PPE can protect the responder from submersion, thermal, and contamination risks. The world of water rescue PPE is rather large, and the needs are different for agencies doing different tasks. A lifeguard in San Diego may not wear a drysuit and life vest when working on the beach, but in a flood response a rescuer will require it. The basics of water PPE are covered in this chapter, with subdiscipline PPE examined in each discipline-specific chapter as needed.

---

**KNOWLEDGE CHECK**

Which general statement about passive search measures is correct?

**a.** They are based on the assumption that the victim cannot move or respond.

**b.** They are used when the location of the victim is known.

**c.** They typically involve a rapid search of the immediate area.

**d.** They are usually implemented when there are no witnesses to the actual event.

*Access your Navigate eBook for more Knowledge Check questions and answers.*

---

**LISTEN UP!**

Many water rescuers come from the fire service, where the more common uniform is fire gear. Fire fighters essentially live in this type of gear, and many will wear it around the water by habit. Nevertheless, it is not recommended to wear structural firefighting gear around the water. If a rescuer falls in the water while wearing this type of gear, the clothing will quickly become so heavy that it can impede swimming action. Trying to swim or tread water can also expel any trapped air, and the weight of the gear, coupled with the tools and equipment that many fire fighters carry in their pockets, can quickly tire a rescuer to the point of trouble.

The best way of preventing this outcome is to avoid wearing bunker gear around water incidents. If it is worn and an accidental immersion occurs, using the vapor barrier to float can buy a rescuer time on the surface. Air will be trapped in the barrier; by bending the knees and tightening the coat collar, the rescuer can take advantage of this air to help flotation (**FIGURE 1-12**). Care should be taken to lean back and not expel this air.

**FIGURE 1-12** While it is not recommended to wear structural firefighting gear around water, the vapor barrier in this gear will trap quite a bit of air if someone falls into the water. That air can then be used for flotation.
Courtesy of Steve Treinish.

## Thermal Protection

As many rescuers know, water wicks body heat away 25 times faster than air does. For this reason, rescuers should be protected from hypothermia (abnormally low body temperature) as well as hyperthermia (abnormally high body temperature). Many rescuers greatly underestimate how cold they will get when operating in the water, and how fast it happens. A variety of drysuits and thermal protection layers are available for ice, swift, dive, and watercraft operations, and teams should ensure rescuers are protected from the cold air or water.

In hot weather, drysuits do not let evaporative sweating occur, and direct sunlight often makes wearing them unbearable for long amounts of work time. Rescuers should be monitored for hydration problems and heat or cold emergency symptoms as work time in the water progresses.

## Contaminated Water Protection

Water contamination is a major issue in water rescues, especially when rescuers are in direct contact with it (**FIGURE 1-13**). The two main types of contamination that water rescuers must deal with are biological and chemical.

## Biological Contamination

**Biological contaminants** include human and animal remains and wastes, sewage, and parasites and algae. Look at any ice-covered pond that supports waterfowl,

**FIGURE 1-13** Water contamination can be a significant issue for rescuers.
Courtesy of Steve Treinish.

and you will see duck and geese droppings; in warmer weather, these droppings can be seen along the shore. It is common for water rescuers to be called on to search for and recover human remains. Bodies in various states of decomposition will be problematic for rescuers not properly equipped to guard against contamination from this source.

Because rivers are giant funnels, they often collect contaminated runoff from sources such as farms, industrial operations, and sewer treatment plants. Even properly functioning storage and treatment facilities can malfunction or overflow. Static ponds or lakes can be worse yet, as the slow inflow and outflow of water may limit the cleaning action that flowing water can provide. For example, a clear quarry lake may look refreshing and attract recreational users, yet still host a variety of parasites. *Giardia* is one such parasite that is found in many areas of the world; exposure to it may result in diarrhea, upset stomach, and possibly dehydration. Although no studies have been published that definitively tie this parasite to water rescuers, many rescuers experience adverse symptoms from exposure and simply do not recognize them as deriving from an exposure to the water environment.

## Chemical Contamination

**Chemical contamination** in water is somewhat harder to pin down than biological contamination. Chemical contamination can change with current flow, rainfall, construction, or illegal dumping (**FIGURE 1-14**). Many chemicals are heavier than the water itself, so they tend to settle into silt or bottom material. This contamination can be present on PPE that rescuers

**FIGURE 1-14** Many discharges carry legal wastes into local waterways. Storm drains carry any pollution found in surface areas directly into the water with no treatment.
Courtesy of Steve Treinish.

**FIGURE 1-15** Any rescuer working near water must wear an appropriate personal flotation device (PFD).
Courtesy of Steve Treinish.

use during the course of a rescue, and it can be transferred by touching rescuers who have not undergone sufficient decontamination. Rescuers assisting or handling wet gear may accidentally expose themselves to these materials by wiping their eyes or noses, or by handling food before they wash their hands. Summertime means sweat in many areas, and a wipe of the eyes can introduce water contaminants into the body. Good hand-washing technique with soap and water or using alcohol gel disinfectants help to limit this risk.

Floodwaters can be the worst brew of contaminated water that rescuers experience. It is difficult to know exact levels of contamination in this type of moving water, as huge amounts of chemicals can be released into floodwaters from gas stations, sewers, factories, farms, and residences. For larger flood responses in urban areas, imagine walking through a local home improvement store and seeing everything in it released into the water, plus gasoline and diesel fuel—that may be the reality faced by rescuers in their actual responses.

More information on water contamination and the PPE specific to specific types of rescues can be found in the subdiscipline-specific chapters.

## Personal Flotation Devices

Once a scene requires rescuers to move closer to the water, they must have the ability to stay afloat. No one arrives at a water response planning to be submerged, but slips and slides do occur all too frequently. Remember, too, that during the various tasks around the water scene everyone exposed to the water hazard—regardless of task or ability—needs protection from falling in. AHJs can specify how close people involved can be to the water without having PPE donned, but a typical range is 15–25 feet. Keep in mind that operating on bridges, steep banks, or slippery surfaces may require PPE even when farther away from the water.

Undoubtedly, the most important part of rescuer personal protection is the **personal flotation device (PFD)**. A PFD is the core PPE for rescuers doing all types of water rescue work. All awareness-level personnel should have a good understanding of the types and uses of PFDs, and there is no excuse for not wearing a PFD if you are around the water environment (**FIGURE 1-15**). For professional rescuers, there is no justification in the world for not being properly protected around any environment that can cause harm.

The United States Coast Guard (USCG) divides PFDs into five types based on the protection they give to the wearer. Most devices are actually vests that are worn by the user, and zipped or buckled to ensure proper fit. Some devices are throwable, such as seat cushions and ring buoys. The use of a vest PFD is paramount to rescuer protection, as it is worn, not carried. Carrying a throwable device to provide personal flotation does not guarantee that it will save your life, as it could be forgotten, set aside, or lost, or it could slip from your hands.

When vests are properly used and worn, they enable the wearer to float in the water with no effort. Some will even turn an unconscious wearer face-up to protect the airway. Nevertheless, the vest must be properly worn to be effective—and that requires practice in donning this equipment. Without such training, some fire fighters may struggle to close the zipper of the PFD over their heavy turnout gear, or try to put the PFD on upside down, inside out, or backward.

Following are the USCG PFD types, along with a brief explanation of their benefits and uses.

## Type I PFDs

Type I PFDs are the big, bulky PFDs commonly found in ships' lockers. These horse-collar life jackets hang around the wearer's neck and are secured with a waist-band strap (**FIGURE 1-16**). Type I PFDs are used primarily on commercial and offshore craft (for example, cruise ships) and in heavy seas when rescue possibilities may not come quickly. They should turn an unconscious victim face-up while floating, and must deliver a minimum of 22 pounds of buoyancy. They are not very comfortable or user-friendly, and not considered rescue crew PFDs.

## Type II PFDs

Type II PFDs have the same basic design as Type I PFDs but provide only 15 pounds of buoyancy for an adult (**FIGURE 1-17**). These smaller PFDs are commonly used on smaller, noncommercial pleasure craft.

## Type III PFDs

Type III PFDs are the typical "ski-vest"-style PFD (**FIGURE 1-18**). Also included in this class are auto-inflated suspender-type PFDs, which are often

**FIGURE 1-17** Type II PFDs are smaller than Type I PFDs and are used on smaller, noncommercial craft.
Courtesy of Steve Treinish.

**FIGURE 1-16** Type I PFDs are common on cruise ships and commercial watercraft.
© Pavel L Photo and Video/Shutterstock.

**FIGURE 1-18** Type III PFDs include the typical "ski-vest"–style PFDs as well as auto-inflated suspender-type PFDs.
Courtesy of Steve Treinish.

favored by sailors and hobby fishermen, although newer technology can place these devices into other classes as well. Check the equipment tag to confirm whether a particular PFD is classified as Type III. Type III PFDs are not required to turn unconscious wearers face-up, but they are comfortable to wear and provide the same 15-pound buoyancy as Type II PFDs.

## Type IV PFDs

Type IV PFDs include throwable ring buoys, seat cushions, and the like (**FIGURE 1-19**). Ring buoys

**FIGURE 1-19** Type IV PFDs include devices that can be thrown to a victim, such as ring buoys and seat cushions.
Courtesy of Steve Treinish.

are circular foam flotation devices with rope around the outside for gripping. These PFDs are not actually worn, but rather are thrown to a victim in distress to provide auxiliary flotation until a very quick rescue is performed. Most watercraft more than 16 feet of length are required to have a Type IV PFD on board.

## Type V PFDs

Type V PFDs are job-functional PFDs, such as float coats, work vests, paddling PFDs, and swiftwater rescue PFDs (**FIGURE 1-20**). These specialized PFDs are suited for particular environments. They provide the buoyancy of a Type I, II, or III PFD, and the buoyancy will be marked on the device. Type V PFDs are considered USCG approved only if they are worn for the purpose identified on their tag.

## Choosing a Rescue PFD

The PFD most commonly carried by frontline rescuers in fire or EMS companies or in police cruisers is a Type III vest-style PFD. Such a device provides

**SAFETY TIP**

Type V PFDs used in swiftwater rescue may be equipped with a short tether and carabiner to be used in tethered rescue. This very specific capability is used only by swiftwater technicians, and the tether–carabiner combination should not be used or attached to other styles of PFDs, or used by untrained rescuers. Doing so can result in a rescuer being trapped in the water with no release, creating a very dangerous and potentially fatal situation.

**A**

**B**

**FIGURE 1-20** Type V PFDs are specialized, job-functional PFDs. **A**. Type V PFD. **B**. Float coat.
Courtesy of Steve Treinish.

buoyancy to a rescuer, provides ample room for the wearer to work, and is fairly inexpensive. Some companies and most water rescue teams will purchase and use a Type V swiftwater rescue vest, but the higher cost of these devices makes them somewhat less common in the field. Regardless of which type PFD a team buys, the best PFD is the one that will be worn!

The use of automatic, inflatable suspender-style PFDs is not preferred for water rescue work when the rescuer is faced with a high probability of entering the water, simply because the inflation system, while having a redundant backup manual inflator, is not failsafe. Nevertheless, members assigned to boats who do not expect to enter the water may find they are much more comfortable to wear.

If you purchase your own PFD, consider one with a quick-release buckle system, and a cutting tool or knife at the minimum (**FIGURE 1-21**). This equipment may not be worn at all times or used in every situation, but does give some "room to grow" into more training. For example, all PFDs can be equipped with a knife, but the quick-release buckle becomes an important tool in swiftwater technician evolutions. Becoming entangled in a line puts a rescuer in a very bad position, and rescuers must be able to cut themselves or other rescuers loose in such a predicament.

Some rescuers also provide themselves with a short tag leash and carabiner to use as a quick, short tether. There is nothing wrong with this setup, but be aware of the entanglement hazard if the slack is not kept controlled. Reflective striping, whistles, and even dye packs are other possible add-ons to PFDs to make them better suited for the user or the mission at hand.

## Water Rescue Helmets

Another piece of PPE available to rescue crews is the water rescue helmet (**FIGURE 1-22**). There is always the possibility that a rescuer may move quickly into the water, either by choice or by accident, and face hazards such as rocks, docks, or debris, especially in moving water. Water rescue helmets are light weight equipment that will drain water.

Structural firefighting helmets may be used to protect rescuers operating away from the water itself, such as setting up lighting or clearing egress paths, but their use is generally frowned upon in or on the water. The large bills and heat shields of these helmets can cause injury if they are subjected to moving water, which can snap the rescuer's head around.

Some water rescue helmets have built-in short visors. While ball caps are commonly worn under the helmet, they can occlude vision and put a great deal of leverage on the neck in the water. Shorter, glued-on visors are available that give some glare protection and reduce the snapping back of the neck if a rescuer goes into the water face first. Mounted lights on the helmet are great, but carry a caveat: Anywhere that your head is pointed, the light will also be pointed, which can blind other searchers or rescuers.

## Gloves

Good work gloves may not necessarily protect rescuers from cold water, but they protect the user from rope burns from the various ropes and lines that may

**FIGURE 1-21** Wear a knife or other cutting tool with your PFD to ensure you can disentangle yourself from lines or ropes if needed.
Courtesy of Steve Treinish.

**FIGURE 1-22** A water rescue helmet is an important piece of protective equipment.
Courtesy of Steve Treinish.

**FIGURE 1-23** Gloves should be worn whenever ropes may be handled; they also provide protection from scrapes and abrasion.
Courtesy of Steve Treinish.

**FIGURE 1-24** Rescuers often creep toward the water's edge in an effort to help others, but proximity to the water requires the use of appropriate personal protective equipment.
Courtesy of Steve Treinish.

be used during a rescue, and also help the user maintain a better grip on these ropes (**FIGURE 1-23**). Awareness-level personnel may not have specific training on rope systems, but may be asked to hold a rope or pull a rope in certain situations. Any rope around water may become water soaked; thus, if your grip slips, a good set of gloves may protect your hands from friction burns as the rope slides through them. Also, many water responses require fighting through brush or trees on the bank of the water, or using hand tools or chainsaws, and gloves offer hand protection during this work.

## PPE Use

Most agencies develop and enforce SOPs dealing with when to use water rescue PPE. Anyone within a certain distance of the water must have at least the proper PFD, helmet, and line-handling gloves donned. A common distance specified in SOPs is 15 feet from the water, but rescuers must remember they can be exposed to the hazard from farther distances, such as steep banks or bridges over water. Simply put, if there is any chance—no matter how slight—you may end up in the water, you need protection. One all too common safety violation is rescuers who "creep" up closer and closer to a water area, either to try to help or just to see the incident details better (**FIGURE 1-24**).

## Operational Support Positions

Numerous positions need to be filled and many tasks must be accomplished during a water rescue, all of which are crucial to a successful outcome and help

provide rescuer and scene safety. Some operational support positions will require rescuer PPE, but others may not, depending on where the rescuer is located on scene.

## Boarding and Exiting Watercraft

Many of the support positions that awareness-level personnel may be asked to assume will require operations from watercraft. While not actually considered a crew member or operator, awareness-level personnel should be able to get on board and get out of watercraft and should follow all laws pertaining to use of that watercraft. The safest means to board is to be guided by the crew or operator of the boat. Generally, though, the smaller the boat, the less stable it may be when loading or unloading. Placing body or equipment weight on the edges of the boat may cause it to tip uncomfortably, or even capsize it. Center of gravity also plays into this balancing act, as loads kept lower in the boat result in more stability.

Another aspect of safe boarding and exit of watercraft concerns stepping to the dock or boat too quickly, while lines are not secure. Ensure the boat will not be blown by wind or current, or pushed by your own legs away from the dock or pier, when

**LISTEN UP!**

When in doubt, ask the crew specifically how and when to board and exit the boat, and at which physical location on the watercraft. Some boats offer more stability in certain areas. Watercraft crews can also offer guidance on how best to load and stow any needed gear such as lights or emergency medical kits.

stepping onto or off the watercraft. It is entirely possible for a responder to do a "split" and fall in the water when the boat moves away as the person is stepping onto a boat with loose lines.

## Spotters

**Spotters** are personnel placed in strategic positions to watch for victims, watercraft that might be missing, hazards to rescuers, or anything else deemed appropriate (**FIGURE 1-25**). A detailed description of the missing person or watercraft, the approximate time it went missing, and where it was last headed can all affect the placing of spotters. Victims who are reported to have been traveling with current may require spotters in different locations, such as bridges, outlooks, or areas downstream.

Spotters should also have direct communication with the team, binoculars (if available), and any safety gear needed for the environment. If spotters will be working near the water or in watercraft, they should have sufficient training for that environment and wear an appropriate PFD.

## Lighting

Nighttime lighting is one of the biggest concerns of water responses, both for ensuring rescuer safety and actually looking for victims. Scene lighting can range from aerial ladders providing stadium lighting to crews using small hand-held lights at the water's edge of a steep riverbank.

## Rescuer EMS and Monitoring

During extreme heat or cold, or extended operations, rescuers may need or desire rehabilitation operations.

**FIGURE 1-25** Spotters may be placed to look for victims or objects or for conditions that may negatively affect rescuer safety.

Courtesy of Steve Treinish.

Rehab crews must have a good medical understanding of what the PPE may do to the rescuers' physical condition. Drysuits are used in many evolutions, but do not let the users' bodies cool by sweating, so hyperthermia can occur. Plenty of fluids should be given during breaks, and vital signs monitored. Conversely, in extreme cold weather, rescuers may show signs of hypothermia and must be warmed before they are released to further work.

Longer-term deployment may mean rescuers need not only hydration, but also body fuel. Something as simple as a granola bar is a welcome addition to hot coffee after hours in the field, and teams should not discount the caloric needs of their personnel. Offering lunches or dinners during long-duration incidents can help keep rescuers working effectively and give rescuers a chance to "disconnect" for a mental break. Ensuring crews are well rested, fed, and hydrated is essential to their well-being.

Many other injuries are possible in the field, ranging from simple scratches to sprains to fractures, and even worse. Water rescue in tough terrain can physically tax responders, and cardiac events in the fire service are—unfortunately—not uncommon. Qualified EMS crews should be able to treat rescuers quickly, and incident commanders should ensure that EMS capabilities are in place for both victims and responders. EMS crews should remain in place until rescuers are clear of any unusual risk of injury.

## Victim EMS and Triage

Water rescue incidents may sometimes involve mass casualties, so awareness-level personnel should be familiar with emergency medical triage. Triage is simply determining how many victims are on scene and prioritizing care for them. Larger scenes require additional resources, and these resources should be sought very early in the incident. At smaller incidents, a scene might be sorted out very quickly; larger incidents are more complex, however, and rescuers should ensure that the help available on scene gets to where it is needed most.

Most agencies operating EMS and rescue units are familiar with **triage tags**. Triage tags help sort casualties by identifying wounded and deceased victims by the tag color:

- Green tags signify victims who are noncritical and need very minor treatment. These "walking wounded" can move under their own power. Care for green-tagged victims may be delayed to allow treatment of others.

- Yellow tags signify victims who have serious, but not life-threatening, injuries. Broken bones and lacerations are common. Care for yellow victims is urgent, but not immediately needed.
- Red tags signify victims with life-threatening, severe injury. These victims are the highest priority for care and transport, and areas specific for treatment of these victims must be designated.
- Black tags identify deceased victims—those who are pulseless and not breathing. It is often hard to give up on victims, but these individuals are beyond hope, and efforts and supplies should be directed toward those who might live.

---

**KNOWLEDGE CHECK**

Which type of PFD is a ring buoy?

a. I
b. II
c. III
d. IV

*Access your Navigate eBook for more Knowledge Check questions and answers.*

---

## Helicopter Support

Helicopters can be used to provide transport to trauma centers, especially in rural areas. In addition, they may be utilized by rescue teams to help search for victims, light a scene, or offer incident command a bird's-eye view of the entire scene. Awareness-level personnel are often tasked with finding suitable landing areas and assisting with patient transfer to the helicopter.

### Landing Zones

Most AHJs have SOPs related to helicopter operations that were based on what the helicopter crews require. The **landing zone (LZ)** for most touch-down areas should be at least 100 feet by 100 feet; clear of any ground debris; nearly level; clear of any air obstructions, such as power lines or trees; and close to the incident. Marking the corners of the LZ is recommended, but only with objects that will not be blown by rotor wash and become a projectile or hazard to the aircraft.

When lighting LZs at night, ensure the lights are not directed at the incoming helicopter, so you do not blind the pilot. Most EMS helicopters can communicate with fire, EMS, and rescue crews, and most desire a rundown of the patient information

and LZ setup before they land. Fire suppression equipment such as handlines or extinguishers may be needed as a precaution in case of helicopter landing, and one person should be placed near the tail rotor to prevent inadvertent approaches to that danger zone.

## Working Around Helicopters

Obviously, rescuers face major hazards when working around a running helicopter, with the rotors being the greatest threat. Spinning rotors are hard to see and unforgiving in their effects. Making visual contact with the helicopter pilot or crew before approaching is paramount, and an approach should never be made from the rear of the craft. Helmets, gloves, eye protection, and snug-fitting clothing should be worn.

Most helicopters rely on a clock-style zone of approach. If the helicopter is facing 12 o'clock, most approaches should be done between 9 and 3 o'clock, but under visual approval of the pilot or crew.

## Public Information Officer

Appointing a **public information officer (PIO)** ensures that there is a channel through which information can be shared with the public during an incident (**FIGURE 1-26**). Identifying this single point of contact for the media and the public allows rescuers to focus on the mission at hand, and creates a gateway for protecting sensitive information from being leaked or released improperly. No family should ever hear about tragedy over the airwaves. All personnel operating in or around the incident should be aware of who the PIO is, and should direct any public inquiry to this person.

**FIGURE 1-26** The public information officer disperses incident information as needed, including to the media.
Courtesy of Steve Treinish.

In smaller, dramatic rescues, information reported by the media will often be what they can witness and film, what the AHJ provides, or both. Often a proactive stance by the PIO can clarify responses, risks, and procedures; then, with a good understanding of the challenges faced by rescue teams, the public view of the operation may become very positive. Larger-scale incidents may need evacuation directions or public danger warnings broadcasted, and working with the media can enable that to be done quickly and effectively.

While many rescuers do not welcome the media around a scene, they should recognize that members of the media have a job to do—just as the rescuers do. Sometimes over-eager media personnel may try to use different routes in or around blockades to "get the shot." No matter how they gain access to the scene, always remember that these individuals are civilians with cameras, so they need the same water PPE if they are allowed near the hazard zone (15–25 feet from the water is a general recommendation), no matter how benign it seems.

Many media outlets will feature some kind of follow-up or safety-based story after a dramatic incident. These reports can be extraordinary opportunities to showcase the capabilities of rescuers, and to provide public education that can save lives in the future.

> **LISTEN UP!**
>
> Rescuers need to remember that cameras are often rolling even when nothing dramatic is happening. Video filler for news stories may include footage of crewmembers staging, resting, or waiting to be deployed. Even if it does not occur in sight of a search or a rescue, laughing or joking can be taken out of context by the viewing public.

## Social Media Concerns

The explosion of smartphones, tablets, and social media accounts has created something of a problem for public safety workers. Patient privacy and care can be violated with photos and videos posted on social media sites and live feeds. While photos taken during incidents can be of great value later for education or analysis of the mission, they can also lead to a huge liability for the rescuer personally, and the department he or she works for. Social media networks may be used very effectively by departments and agencies, but a specific social media policy that each rescuer acknowledges and understands should be in place.

> **LISTEN UP!**
>
> Pictures and an explanation of procedures can help alleviate public misconceptions and incorrect information about incidents. Very often, the public does not know why a rescue or recovery operated in the fashion it did, or why it may take longer than most people think it should. Privacy should be respected, and the administration of the agency posting any messages for the public should know and approve the media reporting or the social media post.

## Accountability Officer

Crew **accountability** may not be the uppermost thought in most rescuers' minds in regard to water rescues, but water rescue can turn into a hugely complicated endeavor in times of floods or large area searches. Proper accountability is crucial for ensuring all rescuers' locations are known. Since water-covered areas can be quite large, specific terms may need to be used to identify a particular region or water area. Regardless of which accountability system your agency uses, being aware of who is where is crucial to the safety of all involved. Operations covering larger areas can benefit from having a printed satellite picture with the areas to which crews are assigned labeled on the map. Simply put, the larger the area and the more complex the mission, the more accountability should be expanded to handle it.

## Personnel Accountability Reports

Even the best accountability system does no good if it is not used properly. A personnel accountability board merely tracks who is assigned to what and where, but does not track the state of the crews (**FIGURE 1-27**).

A **personnel accountability report (PAR)** is used not only to identify where the rescue crews are, but also to confirm that they are safe and accounted for. This information must be communicated to the command staff—typically to the incident commander, the operations commander, or the accountability officer. Face-to-face communication, hand or body signals, or radios can all be used to relay information, but most PARs are done through a roll-call–type radio question-and-answer sequence. That is, an announcement is made of an upcoming PAR, with each crew listed announced. As crews are announced, the crew leader states "OK" or "Have PAR," or something similar. If a group has a member or members not accounted for, search efforts can then begin to find them.

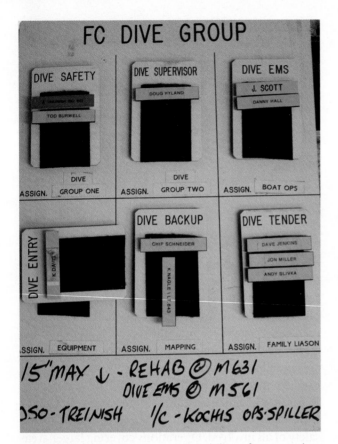

**FIGURE 1-27** Proper accountability—for example, through a personnel accountability board—is crucial to ensuring all rescuers' locations are known.
Courtesy of Steve Treinish.

PARs are generally done after a certain period of time elapses, such as every 30 minutes, during a shift in operations, or after a critical event. Any benchmark desired can be used for PARs, and scene safety is greatly increased by the use of a functional system that is used by everyone involved in the incident.

## LISTEN UP!

Local landmarks can be used to help identify areas where crews may be. For example, suppose Team A has four members who are searching the lakeshore by a boat for a missing person. If they are overdue to return to the command post, and you were tasked with finding them, how would the last known location sound if it were just "On the lake"? Perhaps logging their accountability as "Searching the north shore between Fisher's Marina and Seller's Point" would help. This more precise description of the area greatly narrows down where the team should be found.

## Other Agencies

Many agencies can help rescuers, especially in large-scale events. The American Red Cross, military units,
civilian specialty groups such as the Civil Air Patrol, private search and rescue groups, and, in really pressing times, even non-affiliated civilians can be placed in service in various positions. While these support roles do not generally place individuals in the hazard zone, working in this area is sometimes possible. To ensure their safety, these support personnel must have donned PPE and have a basic understanding of the hazard and how to stay out of it.

## Incident Safety Officers at Water Rescues

The past few years have seen an explosion in scene safety measures. **Rapid intervention crews (RICs)** (placed near where an actual rescue is taking place and tasked only with an immediate rescue of the rescuers, if needed) should be trained to the technician level and have their own rescue equipment.

Also helping ensure scene safety are accountability boards and full-time **incident safety officers (ISOs)**, whose presence is now common at most fire service responses that utilize multiple crews or place anyone in a hazard zone. Any safety officer supervising at a water rescue incident should be trained to at least the awareness level for water rescue (**FIGURE 1-28**). This level of preparedness lets the ISO focus on the big picture, so to speak, and to back up crews who may be focusing on a smaller portion of the call.

Another great position to fill during a mission would be an additional safety officer who is trained to the technician level and supervises the specific rescue evolution. This safety officer should know exactly what

**FIGURE 1-28** Any safety officer supervising at a water rescue incident should be trained to at least the awareness level for water rescue.
Courtesy of Steve Treinish.

is happening in the rescue, as well as the specific hazards associated with these actions.

# Normalization of Deviance

A significant problem in the fire service is **normalization of deviance**—that is, accepting a risk, even if it is well known, because a previous risk has not resulted in a near miss, injury, or death. Not wearing seat belts is good example. Many fire fighters complain about seat belt use, but know full well that not wearing the belt can easily cost their own or a fellow crew member's life. Yet many fire fighters still do not wear seat belts, claiming, "I've never been in a wreck," "The ladder will win against any car," or "I trust my driver." After a few thousand responses with no wreck or dire results, it becomes normal practice not to wear seat belts. Then, when a crash happens, the seat belt is not in place as a protection system—and the rescuer's own injuries, or injuries to others in the vehicle, could be severe.

## Lack of PPE by Rescuers

Crews have a tendency to creep closer and closer to the water during a rescue, even if they are not wearing PFDs. The rule of wearing PPE within 15 feet of the water often gets forgotten in the heat of the moment (**FIGURE 1-29**). Most members of the fire service are "wanna do"-type people. Unfortunately, when they violate the rules in this way time after time and live to tell about it, this practice becomes normal, then accepted.

Likewise, such violations of SOPs often occur when an additional set of hands is needed at the water's edge. An unprotected rescuer may hurry to help when something falls off another rescuer and is dragging and causing a trip hazard, or when the victim turns out to be heavier than expected and the rescuers carrying a stretcher begin to struggle. Often, unprotected rescuers not wearing PPE will dash in and help at the water's edge. Their exposure to the water hazard may

**FIGURE 1-29** Crews have a tendency to creep closer and closer to the water during a rescue, even if they are not wearing PFDs.
Courtesy of Steve Treinish.

be limited by the short duration of their time there, but the risk is still present. Everyone on a water response needs to be cognizant of this risk, and look out for one another.

# Incident Termination

Although **incident termination** is an operations-level skill, awareness-level personnel can certainly contribute to incident termination activities. An incident is not over until all personnel and gear are accounted for, all the necessary documentation has been completed, reports have been made to the appropriate parties, and incident debriefing has occurred.

The primary concern in incident termination is safety—both for responders and for bystanders. Once a rescue or recovery is complete, responders may be drained from the physical and emotional stress, and bystanders may consider the area safe before it truly is. Water scenes have inherent dangers, and anyone on scene must be cognizant of these dangers and must stay focused on keeping themselves and others safe, even during cleanup.

# *After-Action* REVIEW

## IN SUMMARY

- Rescuers often choose to specialize in a subdiscipline, but should have a good understanding of all types of water and rescue that may be encountered in their area of response.
- Two standards specifically address technical rescue: NFPA 1670, *Standard on Operations and Training for Technical Search and Rescue Incidents*, and NFPA 1006, *Standard for Technical Rescuer Professional Qualifications*.
- Awareness-level personnel should be able to recognize both the immediate hazard in which the victim is trapped and any potential hazards found in the area, such as utilities, hazardous materials releases, or even risks of injury associated with getting to the water, such as cliffs or slippery slopes or ramps.
- Awareness-level personnel should answer two key questions concerning the response they have been called to: "Are there resources in place to handle this response?" and "If not, where do I get them?"
- After completing the National Incident Management System training required for all technical rescue personnel, every responder should be familiar with the concept of the incident command system.
- Incident action plans can be developed for mental use, verbal use, or command use, and are done for almost every task in the fire service, whether explicitly or not.
- Civilians do not normally have the same training as rescuers do, so they often have a very different view of the risks involved in rescue; consequently, they must be kept out of the hazard.
- Bullhorns or siren-based public address systems can be used to help guide civilians during an evacuation.
- In an active search, rescuers immediately begin working to rescue a victim, provided a victim is in sight (or the point last seen is established), the timeline supports action, and the proper resources are in place.
- Various pieces of water personal protective equipment can be donned to protect the responder from submersion, thermal, and contamination risks.
- Numerous positions need to be filled and many tasks must be accomplished during a water rescue, all of which are crucial to a successful outcome.
- The past few years have seen an explosion in scene safety measures. Rapid intervention crews, accountability boards, and full-time incident safety officers are now included in most fire service responses that utilize multiple crews or place anyone in a hazard zone.
- A significant problem in the fire service is normalization of deviance—that is, accepting a risk, even if it is well known, because a previous risk has not resulted in a near miss, injury, or death.
- An incident is not over until all personnel and gear are accounted for, all the necessary documentation has been completed, reports have been made to the appropriate parties, and incident debriefing has occurred.

## KEY TERMS

*Access Navigate for flashcards to test your key term knowledge.*

**Accountability** A system or process to track resources at an incident scene. (NFPA 1561)
**Active search measure** An operation with a witnessed submersion or immersion, with rescuers having direct evidence or witnesses placing the victim in the water.

**Authority having jurisdiction (AHJ)** An organization, office, or individual responsible for enforcing the requirements of a code or standard, or for approving equipment, materials, an installation, or a procedure. (NFPA 1006)

**Awareness level** This level represents the minimum capability of individuals who provide response to technical search and rescue incidents. (NFPA 1006)

**Basic safety plan** The layout of the present and potential hazards found around an incident, the strategy to mitigate them, and assignment of duties to rescuers involved.

**Biological contaminants** Water pollution from living or dead organic tissue or matter, including microorganisms and toxins derived from natural substances.

**Chemical contamination** Water pollution from human-made substances found in the water column or the bottom materials such as mud or silt.

**Dive** Type of rescue where victims are submerged in a liquid, normally water.

**Flood** Type of rescue where victims are in areas not normally inundated with water.

**Ice** Type of rescue where victims are in frozen surfaces, even if surrounded by surface water.

**Incident action plan (IAP)** A verbal or written plan containing incident objectives reflecting the overall strategy and specific control actions where appropriate for managing an incident or planned event. (NFPA 1026)

**Incident command system (ICS)** A standardized on-scene emergency management construct specifically designed to provide for the adoption of an integrated organizational structure that reflects the complexity and demands of single or multiple incidents, without being hindered by jurisdictional boundaries. ICS is a combination of facilities, equipment, personnel, procedures, and communications operating within a common organizational structure, designed to aid in the management of resources during incidents. It is used for all kinds of emergencies and is applicable to small as well as large and complex incidents. ICS is used by various jurisdictions and functional agencies, both public and private, to organize field-level incident management operations. (NFPA 1006)

**Incident safety officer (ISO)** A member of the command staff responsible for monitoring and assessing safety hazards or unsafe situations and for developing measures for ensuring personnel safety. (NFPA 1521)

**Incident size-up** The ongoing observation and evaluation of factors that are used to develop strategic goals and tactical objectives. (NFPA 1006)

**Incident termination** The act of safely returning crews and equipment to a safe area, documenting any damage or injury to crew or property, analyzing the mitigation effort, and rendering a scene safe before vacating it.

**Job performance requirements (JPRs)** A written statement that describes a specific job task, lists the items necessary to complete the task, and defines measurable or observable outcomes and evaluation areas for the specific task. (NFPA 1000)

**Landing zone (LZ)** An area utilized for landing helicopters.

**Normalization of deviance** The acceptance of risk based on the premise that previous similar circumstances did not result in any problems, normally in the form of injury. Rescuers often know risk is present, but choose to continue an evolution based on the fact that nothing happened in the past.

**Operations level** This level represents the capability of individuals to respond to technical search and rescue incidents and to identify hazards, use equipment, and apply limited techniques specified in this standard to support and participate in technical search and rescue incidents. (NFPA 1006)

**Passive search measures** Operations in which considerably more research and investigation are needed before an active search begins; usually unwitnessed or overdue types of calls, with victims not directly seen entering the water.

**Personnel accountability report (PAR)** Periodic reports verifying the status of responders assigned to an incident or planned event. (NFPA 1026)

**Personal flotation device (PFD)** A device manufactured in accordance with U.S. Coast Guard specifications that provides supplemental flotation for persons in the water. (NFPA 1006)

**Personal protective equipment (PPE)** The equipment provided to shield or isolate a person from the chemical, physical, or thermal hazards that can be encountered at a specific rescue incident. (NFPA 1006)

**Public information officer (PIO)** A member of the Command Staff responsible for interfacing with the public and media or with other agencies with incident-related information requirements. (NFPA 1026)

**Rapid intervention crew (RIC)** A minimum of two fully equipped personnel on site, in a ready state, for immediate rescue of disoriented, injured, lost, or trapped rescue personnel. (NFPA 1006)

**Scene sketch** A rough drawing of an area involved in a rescue, detailing landmarks, area probability, rescuer progress, and any safety hazards.

**Spatial awareness** A deliberate and focused mindfulness of the entire scope and picture of the environment where a rescuer is working in or around during an incident.

**Spotters** Persons looking for an object or person that is missing, or potential hazards that may affect rescue safety during an incident.

**Standard operating procedures (SOPs)** A written organizational directive that establishes or prescribes specific operational or administrative methods to be followed routinely for the performance of designated operations or actions. (NFPA 1670)

**Surf** Type of rescue where victims are in water areas with waves or wave action.

**Surface** Type of rescue where victims are in areas of open water with no current, ice, or waves, and water moving slower than 1 knot.

**Survivability profile** Determination, based on a thorough risk-benefit analysis and other incident factors, that addresses the potential for a victim to survive or die with or without rescue intervention.

**Swiftwater** Type of rescue where victims are in water moving faster than 1 knot.

**Technician level** This level represents the capability of individuals to respond to technical search and rescue incidents and to identify hazards, use equipment, and apply advanced techniques specified in this standard necessary to coordinate, perform, and supervise technical search and rescue incidents. (NFPA 1006)

**Triage tags** Tags used in the classification of casualties according to the nature and severity of their injuries. (NFPA 1006)

**Type** The level of a team's capability before an incident callout, which takes into account the team's training, equipment, and response times. The authority having jurisdiction can benefit by determining which team will best suit the response needs based on these credentials.

# On Scene

You are in charge of the ladder company today, and your crew is driving through your district to look at some target buildings. As you are passing a bike path that runs alongside a river in your downtown district, a pedestrian flags you down. The river is almost 200 yards wide and is running with a visible current in the middle. There are bridges 400 yards apart, and it is known that a sizable homeless population frequents the area along the banks. The pedestrian states that she saw a person trying to swim across the river, but her cell phone died and she had to run into a nearby office building to use a phone there to call for help. She hung up before the call connected when she saw your truck approaching and ran out to get your attention. There is no obvious evidence of anyone in the water in distress, but this woman is upset, and your gut tells you there could be a real emergency. A few minutes later, another witness rides up on a bicycle and tells you he also saw someone in the water. Your dispatcher radios you and say someone called in a person in the water from the sixth floor of a building that overlooks the river.

The emergency lights on the ladder truck are attracting more people, and since the fire fighters are scanning the water, the crowd quickly realizes someone is missing. Darkness is coming quickly, and the water temperature is approximately 45°F. Your ladder company has a driver and two fire fighters in the back and is equipped with two very old life jackets and some ½-inch kernmantle search rope.

**1.** Which of the following pieces of information would best help you determine whether this is a rescue or a recovery?

    **A.** The time of submersion

    **B.** The water temperature

    **C.** The victim's age

    **D.** The time the call was received by the dispatcher

**2.** You decide that you will need to limit the number of people on the shore. What is the best way to handle this task?

    **A.** Call for additional law enforcement officers and ask for a perimeter to be set up

    **B.** Yell for everyone to back up a few steps

    **C.** Send one fire fighter with scene tape to create a visual barrier from the water

    **D.** Watch the crowd and ensure no one else enters the water

# On Scene Continued

3. You have two witnesses who do not know each other. What is the best way to obtain information from them?

   **A.** Have the police question them together in the cruiser to save time

   **B.** Have a fire fighter interview them separately at the location(s) where they saw the swimmer, trying to triangulate the approximate point last seen

   **C.** Bring them down to your location and have them point to where they think the victim is

   **D.** Get their identification, ask them to wait by the ladder, and have the next unit in question them

4. You decide this will be a working rescue, and you will need an underwater rescue team. You ask dispatch to call in a neighboring department that offers this capability. What is your next best move?

   **A.** Set up a command structure, secure the scene, interview the witnesses, and ensure the incoming team knows your status

   **B.** Stage the witnesses close to the parking lot so the dive team can question them as they arrive

   **C.** Set up the aerial ladder to obtain an elevated view of the water

   **D.** Pick the best swimmer on your crew and start searching the shallow water near shore.

*Access Navigate to find answers to this On Scene, along with other resources such as an audiobook and TestPrep.*

CHAPTER

**2**

## Operations Level

# Watercraft Rescue Operations

## KNOWLEDGE OBJECTIVES

After studying this chapter, you should be able to:

- Identify and describe personal protective equipment (PPE) available to rescuers operating or acting as crewmembers of watercraft-based operations. (**NFPA 1006: 21.2.7,** pp. 35–37)
- Define common terms associated with watercraft. (**NFPA 1006: 21.2.2,** pp. 37–38)
- Identify and describe types of watercraft common in water rescue. (**NFPA 1006: 21.2.1, 21.2.2,** pp. 38–46)
- Identify and describe the types of propulsion systems and motors used with watercraft employed in the water rescue process and support operations. (**NFPA 1006: 21.2.1, 21.2.2,** pp. 46–53)
- Discuss the risks and benefits of including various types of equipment on the rescue boat. (**NFPA 1006: 21.2.2, 21.2.3, 21.2.4, 21.2.14, 21.2.16, 21.2.17,** pp. 53–61)
- Identify and describe the components of towing operations. (**NFPA 1006: 21.2.9, 21.2.10, 21.2.15,** pp. 61, 63–67)
- Explain the procedures and parameters of launching and recovering different types of watercraft. (**NFPA 1006: 21.2.4, 21.2.10,** pp. 65, 68)
- Discuss the general guidelines for piloting watercraft. (**NFPA 1006: 21.2.5, 21.2.8, 21.2.11,** pp. 68–73)
- Discuss the effects of wind and current on watercraft operations. (**NFPA 1006: 21.2.1, 21.2.8, 21.2.9, 21.2.10,** p. 73)

- Identify and describe key considerations when deploying and retrieving rescuers. (**NFPA 1006: 21.2.3, 21.2.6, 21.2.12, 21.2.13,** pp. 73–75)
- Describe how to care for watercraft and related equipment after its use. (**NFPA 1006: 21.2.18,** pp. 75–76)

## SKILLS OBJECTIVES

After studying this chapter, you should be able to:

- Flip a capsized inflatable boat. (pp. 40–41)
- Extricate a victim using parbuckling. (**NFPA 1006: 21.2.14,** p. 59)
- Tie a cleat hitch. (**NFPA 1006: 21.2.9, 21.2.10,** pp. 66–67)

# Technical Rescue Incident

Just after you got back to bed after your third run after midnight, your company is called to a reported person in the water at a local reservoir. The reservoir is a very large section of river that has been dammed to provide drinking water to the city, and it has a very high recreational use. It is almost 7 miles long and approximately 0.5 mile wide. Tonight, someone fishing on the bank noticed a boat floating in the middle of the reservoir and called 911.

When you arrive, you find the fisherman at one of the boat ramps, and he points you to the area where he saw the boat floating. You make out a shape in the dark against the lights on the other shore. No verbal response is gained from shouting to it. You decide to launch your department's rescue boat to investigate.

**1.** How will you mark and communicate the locations of your boat and the floating boat?

**2.** Which equipment will you need to take with your boat and crew?

**3.** Which other means of searching can be done during this response?

 **NAVIGATE 2** *Access Navigate for more practice activities.*

## Introduction

NFPA 1006, *Standard for Technical Rescue Personnel Professional Qualifications*, addresses **watercraft** in many of the chapters pertaining to water rescue. This is a very large area in which to provide guidance (**FIGURE 2-1**). While some agencies or departments may not own or operate boats of any kind, many situations call for some type of watercraft to help rescuers properly and safely operate at a water rescue incident, either as a primary means of rescue or as peripheral support to other shore-based techniques. A water rescuer trained to the operations level is expected to know basic boat layout and operation, including such activities as anchoring, preparing the boat for launch,

departing and arriving at docks, and rescuing both civilians and rescuers from the water. At the technician level, rescuers are charged with the actual operation of the boat and all parameters within it.

Obviously, with a very small watercraft like a personal watercraft (PWC), a single operator is very common. By comparison, on the larger boats found in larger water areas such as the Great Lakes or the oceans, a multiple-person crew is not only common, but also needed for safe operation of the vessel. NFPA 1006 addresses this issue, with the operations level explaining the position of "helmsman." Many crews on larger watercraft have a designated helmsman who only steers the vessel as ordered; in other words, this rescuer drives the boat, given a place to go, under the

A

B

**FIGURE 2-1** NFPA watercraft requirements address the types, operation, uses, and safety of boats large enough to handle large waters and small enough to handle creeks.

Courtesy of Steve Treinish.

technician's guidance. Technician-level training would be necessary to plan the route, plan for boat and crew safety, and plan for mission-specific equipment needs and contingency plans. As an analogy, think of a large ship: The captain (technician) is ultimately responsible for the operation of the craft, but directs the crew (operations) in their duties.

According to NFPA 1006, for all water rescue subdisciplines (surface, swift, ice, dive, and surf), a rescuer must be able to use watercraft in a manner consistent with the procedures established by the authority having jurisdiction (AHJ), which may include launching and recovering the AHJ's specific watercraft; operating the boat; and deploying rescue swimmers, ice rescue technicians, divers, and other rescue personnel. In addition, the NFPA 1006 Annex indicates that training should occur in waters that are expected within the jurisdiction, so varied experience in those environments is a must.

Rescuers must be familiar with boat jargon, construction, and use on the water, and must be well versed in emergency procedures on the water. An operations-level rescuer might be assigned to a different team or task during a water rescue mission, and/or might be assigned to function as part of a boat crew even if his or her usual team or agency does not operate one (**FIGURE 2-2**). Technician-level rescuers need enough knowledge that if they find themselves operating a boat with which they are not familiar, they have the training and skills to act as a leader of the boat, the crew, and other rescuers.

Boats can be used for obvious reasons, such as providing a rapid response to a rescue scene or working in offshore areas not accessible from land or too deep to wade or walk, but they can also be used as part of a rope system, completely controlled from shore, and lowered in a current to a stranded victim. In addition, they can be used during ice missions as a safe area that actually travels to the victim with the rescuers (**FIGURE 2-3**). Hazardous ice that breaks when rescuers are trying to operate on it can be navigated by pulling ropes or using poles to slide a boat over the ice; in such a case, the boat, with a tagline attached, becomes a quick transportation mode back to shore.

### SAFETY TIP

Some agencies use old boats, canoes, and even donated fishing boats as frontline equipment and operate them in water conditions where they were not meant to be used. While this equipment is inexpensive, and these agencies may think the vessels will work for water incidents, their use can very easily lead to major problems. Trying to make a boat work in situations for which it is not intended is neither appropriate nor effective. In some drastic cases, boats have been used in rescue work that would not be safe for fishing, let alone for rescuing someone. "Better than nothing" does not apply with these watercraft.

This chapter examines the different kinds of watercraft available, the ways in which they are used, and some benefits and problems with each type. Also covered are the basics of various waterways, the rules of the road, and some minor maintenance.

**FIGURE 2-2** Water rescuers must be able to perform or even command watercraft, even if their organization does not possess one.

Courtesy of Steve Treinish.

**FIGURE 2-3** Boats can offer protection on frozen surfaces, such as ice.

Courtesy of Steve Treinish.

# Personal Protective Equipment Related to Watercraft Use

You will notice a common theme related to water rescue personal protective equipment (PPE) in this text: the use of a properly fitting and mission-specific personal flotation device (PFD). Using this equipment correctly is the most effective way to stack the deck in rescuers' favor as it relates to rescuer and victim safety (**FIGURE 2-4**). PFDs are discussed further in Chapter 1, *Understanding and Managing Water Rescue Incidents*.

All PPE should be used in accordance with the manufacturers' instructions, including donning and doffing and acceptable situations for use. Some agencies use a checklist to ensure the gear is properly donned and ready to protect the user. This approach is especially helpful for those personnel who may not wear certain PPE as part of an operational cache. For example, some rescuers on larger ships may not actively use or practice donning cold-water immersion suits, but might very well need to use them in an emergency. Many cold-water immersion suits are equipped with personal locator beacons and/or strobe or light-emitting diode (LED) lighting, and some older immersion suits are equipped with inflatable supplemental flotation around the head.

The quick use of a checklist helps ensure each item of PPE is operational and actually being used to protect the user. Checklists can be small enough that the manufacturer places one on the PPE item, or large enough to be printed and used by rescuers as part of a buddy system and read aloud to ensure no crucial steps are missed when placing the PPE in service.

## Rescue Helmets

In addition to the standard PFDs, specialized helmets are required PPE for watercraft crews. Riding in a watercraft might not seem to carry much of a risk for a head injury, but there is always the possibility that a rescuer may move quickly into the water, either by choice or by accident, and then face hazards such as rocks, docks, or debris, especially in moving water. While ball caps are commonly worn under the helmet, they can occlude vision and put a great deal of leverage on the neck in the water. As an alternative, some helmets have built-in short visors. Shorter visors are also available that attach to the helmet with Velcro; they can provide some glare protection and reduce the snapping back of the neck if a rescuer goes in the water face first. Mounted lights on the helmet are handy, but can blind other searchers or rescuers if pointed directly at them.

## Wetsuits

A **wetsuit** is a form-fitting suit made of neoprene, ranging in thickness from 3 mm to 7 mm (**FIGURE 2-5**). The thicker the suit, the colder the water in which the rescuer can operate, although thicker suits may restrict some movements a bit.

In a wetsuit, a thin layer of water becomes trapped between the wearer's skin and the material. The rescuer's body heat then warms this water up to bearable temperatures. One problem with wetsuits is that the warm water trapped against the skin is forced out and replaced with the colder, surrounding water during hard or fast motion. This can cause the rescuer to lose body heat, and can be quite uncomfortable if enough water is flushed in this fashion.

Another problem with wetsuits arises with chemical exposure. Contaminated water will be trapped

**FIGURE 2-4** A properly fitting and mission-specific PFD is the most effective way to improve rescuer and victim safety.
Courtesy of Steve Treinish.

**FIGURE 2-5** Wetsuit.
Courtesy of Steve Treinish.

against the skin, and could result in irritation or chemical burns. The presence of such chemicals also makes it difficult to clean the suit properly, and can damage the material used to make the suit.

## Wet Gloves and Shoes

**Wet gloves** and **wet shoes** give some thermal protection to the wearer, usually through their neoprene construction, as in wetsuits (**FIGURE 2-6**). Wet shoes may also be called **booties** or **river shoes**. Look for heavy rubber soles, a zipper or laces to make putting the shoes on and removing them easier, and a rubber area over the foot and toe. If the top area is rubberized, the boot or shoe will last much longer, as toenails or dive fins will not wear through the neoprene nearly as quickly as they would through an unprotected fabric. An old pair of high-top tennis shoes may also work very well in water that is not considered cold. Some rescuers use heavy work boots, but these shoes absorb a lot of water and will result in leg fatigue when they are water-laden.

## Drysuits

A **drysuit** is a type of water PPE that resembles a coverall; it is very similar to the Tyvek suits that many hazardous material teams and construction workers use (**FIGURE 2-7**). Drysuits are normally fitted with latex rubber neck and wrist seals to keep water out, thereby keeping the rescuer dry. Thermal insulation may be required depending on the water temperature and rescuer comfort, and some flotation is inherent with these suits, since air can become trapped in them while they are being worn. This flotation should not be taken for granted, however, as the air may be forced from the suit; thus, PFDs should be worn to bolster this protection.

**FIGURE 2-7** This drysuit can be used for dive, swift, ice, flood, and watercraft rescue.
Courtesy of Steve Treinish.

Drysuits are very effective on cold and windy days. Even when the surface air feels warm, the spray on the water might be much colder.

### SAFETY TIP

Most drysuits are factory equipped with latex neck and wrist seals, and often include latex booties meant to be worn under boots. Diving drysuits may have latex hoods to ensure complete encapsulation protection as well. Given the materials used in these parts of the drysuit, rescuers should address any latex allergies before touching standard suits.

## Eye Protection

Eye protection can be a very important part of boat-based water rescues (**FIGURE 2-8**). Eye protection may be needed in water rescue work for two reasons: to protect against foreign objects and to ward off glare. Water spray or debris hitting the eye at fast speeds can damage the eye very easily. Glare is a concern because facing westward on a sunny late afternoon can be blinding on the water, as can operating on a lake surrounded by snow. Many good water goggles are now available that are specially made for the PWC market, and they work well for water rescuers. A set of tinted safety glasses can work as well.

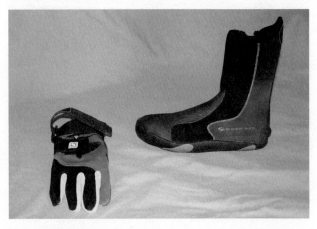

**FIGURE 2-6** Wet gloves and wet shoes give some thermal protection to the wearer.
Courtesy of Steve Treinish.

**FIGURE 2-8** Eye protection may be needed to protect the eyes against foreign objects and glare.
Courtesy of Steve Treinish.

# Watercraft Terminology

Boats and rescue watercraft encompass a broad scope of vessels, but, unfortunately, there is no perfect craft yet. Rescue boats range from small **kayaks** to small ships powered by diesel-driven screws. No matter what their size, all boats share some common terminology:

- **Bow**—The forward part of a watercraft.
- **Stern**—The rearward part of a watercraft.
- **Transom**—The rear, vertical edge of a watercraft, if constructed in such a manner.
- **Port**—The left side of a watercraft, looking forward to the bow.
- **Starboard**—The right side of a watercraft, looking forward to the bow.
- **Abeam**—A position directly perpendicular to either the port or starboard sides of a watercraft.
- **Keel**—The bottom "edge" in the centerline in the bottom, running from the bow to the stern. It is generally the main structural part of the hull.
- **Drain plug**—A small rubber or threaded plug that is installed into a port to drain a watercraft of unwanted water (**FIGURE 2-9**).
- **Beam**—The width of the vessel at its widest part.
- **Length**—The total length of the vessel, from the bow to the transom.
- **Hull**—The basic "form" of a watercraft.
- **Topsides**—The sides of the watercraft that fall between the water and the deck.
- **Gunwale**—The edges of the watercraft, on top of the topside.
- **Waterline**—The area where the surface of the water meets the gunwale.
- **Sail area**—The surface of a watercraft above the waterline, which can be affected by wind.

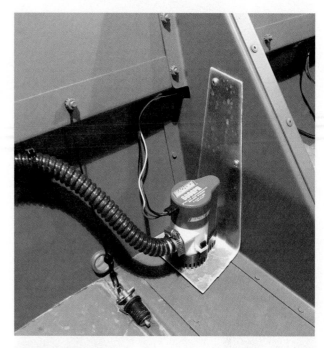

**FIGURE 2-9** Drain plug.
Courtesy of Steve Treinish.

- **Draft**—The amount of the boat that sits in the water while stationary. It is typically measured in feet from the waterline to the lowest part of the boat or propulsion system.
- **Outboard motor**—A motor in which the engine, gears, and prop are all mounted on the transom (**FIGURE 2-10**). It can be as small as 2 horsepower (HP) or as large as 325 HP.
- **Lower unit**—The section of an outboard motor that encloses the gears, drive shaft, and propeller. It includes basically everything from the engine down (**FIGURE 2-11**).
- **Prop**—The propeller of a motor.
- **Inboard motor**—A motor system in which the motor and propeller are both mounted and plumbed through the hull.
- **Inboard/outboard motor**—A hybrid unit, in which the engine is mounted in the boat in front of the transom, and the propeller behind the transom.
- **Tilt and trim unit**—A small hydraulic unit that changes the angle of the lower unit for best ride, speed, and shallow-water operation. It is also used to tilt the motor up for transport or maintenance, and is normally found on motors larger than 50–75 HP.
- **Jet drive**—A system in which, instead of a propeller, propulsion comes from a fan-type pump (**FIGURE 2-12**).
- **Tiller steer**—A smaller outboard motor (less than 50 HP) in which the power and steering controls appear on a short handle attached to the motor itself (**FIGURE 2-13**).

**FIGURE 2-10** Outboard motor.
Courtesy of Steve Treinish.

**FIGURE 2-11** Lower unit and propeller.
Courtesy of Steve Treinish.

**FIGURE 2-12** Jet drive.
Courtesy of Steve Treinish.

**FIGURE 2-13** Tiller arm
Courtesy of Steve Treinish.

# Watercraft Types

The most common boats seen in rescue work in smaller inland waterways are **inflatable boats**, metal-hull johnboats, and tailored watercraft powered by an outboard or jet-drive motor. On larger bodies of water, such as harbors, larger lakes, or the open ocean, rescuers commonly use rigid-hull runabouts, rigid-hull inflatable boats, or even larger oceangoing ships. All are very capable of water rescue, and each type of craft has advantages and disadvantages.

## Capacity Plates

Most boats less than 20 feet in length are required to display a capacity plate, which shows the number of passengers allowable, a total weight rating including the motor and all gear, and a maximum recommended horsepower limit on the watercraft (**FIGURE 2-14**).

This information is merely a guide, and mariners must be aware of the need to stack the weight on the boat, whether passengers or gear, properly. Too much weight toward the bow, and the boat may plow through or even dive into the water; too much weight on either side, and the boat may tilt to the point of capsizing. Water conditions also play into this planning, as larger chop or waves can cause a boat to capsize or swamp.

## Inflatables

Inflatable boats are exactly what their name implies—boats inflated with air (**FIGURE 2-15**). They may have either solid or air-inflated flooring, and inflated "speed tubes" along the underside of the boat may help stabilize the boat at high speed. Inflatable boats have shorter gunwales, which can make it easier both for a victim or rescuer to be parbuckled or retrieved into the boat, and to deploy rescuers into the water. Some

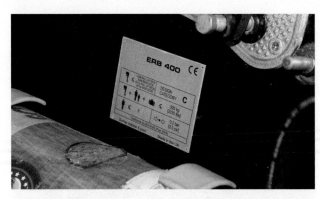

**FIGURE 2-14** Capacity plates show the number of passengers allowable; a total weight rating, including the motor and all gear; and a maximum recommended horsepower limit on the watercraft.
Courtesy of Steve Treinish.

**FIGURE 2-15** Inflatable boat.
Courtesy of Steve Treinish.

inflatables are equipped with an onboard inflation system plumbed into them as a unit with built-in pressure relief valves. These systems are generally supplied with a self-contained breathing apparatus (SCBA) or a self-contained underwater breathing apparatus (SCUBA) cylinder mounted in the boat somewhere, and an inflator hookup hose to be used with that cylinder.

### LISTEN UP!

Some models of inflatable boats have pressure relief valves that can be covered by a plastic cap. Rescuers should be keenly aware of this feature, as occluding these valves during inflation may rupture the boat.

Inflatable craft are generally very light, so they can be carried to remote locations by rescuers. Moreover, even though they are smaller, they have a lower center of gravity compared to most rigid-hull boats. This low-riding operation, coupled with the boat's main flotation being around the craft's outside, makes an inflatable boat very stable in many types of water while doing many types of rescue work. Even if full of water, inflatables will continue to float as long as they are structurally intact. Inflatable boats offer supreme flotation, including the ability to rescue multiple victims at once if rescuers are faced with dire circumstances. In short, if the water is deep enough to prevent tears and punctures, they are the boat type of choice for many teams.

Some inflatable boats are equipped with **scuppers**, which are one-way valves that allow water to drain out of the boat, but prevent water from coming in (**FIGURE 2-16**). Scuppers give the inflatable an advantage when approaching a boil line or taking on water from a current.

Inflatable boats can carry anywhere from two to six people, depending on the length and width of the boat, and many are outfitted with tiller-controlled outboard motors. Most have a smaller outboard motor horsepower rating, from 15 to 40 or 50 HP, but some **rigid-hull inflatables (RHIs)** are larger boats and have twin 150-HP motors. RHIs are inflatable boats that have solid hulls with inflatable sides (**FIGURE 2-17**).

All inflatable boats have one major drawback: They can be ruptured or torn by sharp objects, either during transport to the water's edge or while in the water. The risk to safety posed by this scenario has been somewhat alleviated today, as most inflatable construction now relies on multiple air cells instead of one giant bladder. Thus, if one air cell is punctured on the water, the boat will not immediately lose its flotation, allowing the boat to make its way to safety.

Another aspect of air inflation is that pressure can change when the boat is exposed to heat or cold. The air cells in an inflatable that is allowed to sit in the warm sun may expand and build pressure, then contract and lose pressure if the boat is placed in service in cold waters.

The longevity of inflatable boats can be increased by checking the manufacturer's recommendations, as keeping these boats at full inflation may stress glue-closed seams. With onboard filling, they can be topped off or fully inflated in a matter of seconds.

### SAFETY TIP

In urban flooding situations, many streets will be used as impromptu waterways. Street signs and mailboxes may be struck by boats passing overhead, and such collisions will destroy inflatable boats.

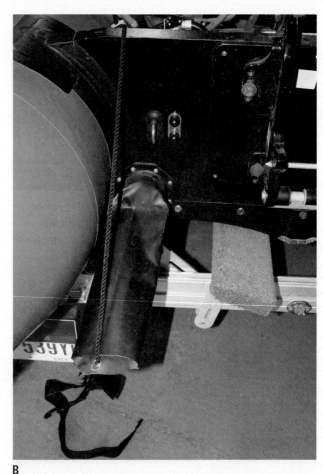

A

B

**FIGURE 2-16 A.** In the up position, water will neither enter nor drain. **B.** In the down position, water will drain out. Water pressure will seal the flexible tube when it is submerged, preventing water flowing into the boat.
Courtesy of Steve Treinish.

**FIGURE 2-17** Rigid-hull inflatable boat.
Courtesy of Steve Treinish.

## Johnboats

Many rescue teams use **johnboats**. These craft are most commonly made of aluminum, but may be constructed of fiberglass, or even steel in older models (**FIGURE 2-18**). Aluminum presents the best durability: Short of dropping the boat on a Halligan point or running something else sharp through it, it is difficult to beat up an aluminum johnboat to the point that it becomes unusable. Johns are commonly constructed by either welding or riveting the seams of the metal. Welded seams usually hold up better in the long run, since the rivets can become leaky over time and with use.

Johnboats can range from 10 feet to 16 or 18 feet in length, and they can be fairly easily carried by rescuers over tough terrain. Given the naturally light weight of aluminum, johnboats made of this material are also easy to slide, both on land and over ice. Most johnboats are powered with small outboard motors in the 5–20 HP range, but can be effectively oared or paddled by a practiced rescuer.

One major drawback of johnboats is that they are relatively easy to flood or capsize. While this may not create a life-threatening situation for rescuers, it can cause several problems. The victim may not have time on his or her side, and the response may be delayed. If the boat flips in the current, rescuers must consider where it will travel, and if it capsizes, what will happen to the gear. Some of the team's gear will naturally float, but other equipment will sink. In addition, the possibility of rescuer injury must be considered.

**FIGURE 2-18** Johnboat.
Courtesy of Steve Treinish.

**FIGURE 2-19** Foam seats in johnboat.
Courtesy of Steve Treinish.

Most johns include built-in foam blocks under the seats, but if this foam has been removed, the boat will sink like a stone (**FIGURE 2-19**). A capsized boat equipped with foam flotation under the seats will float while full of water, and can be easily righted from a capsized position with practice. To right a capsized johnboat, two rescuers must work as a team.

## Rigid-Hull Boats

**Rigid-hull boats** (also known as runabouts) can be specially equipped or manufactured for rescue operations, and can handle bigger water more easily than inflatable boats or johnboats (**FIGURE 2-20**). These watercraft are constructed of solid materials, usually fiberglass or aluminum. They are not nearly as portable as the smaller craft, so launch areas must be pre-planned and used or, in extreme cases, made from scratch during the rescue. These larger craft are usually faster over the water, but a tradeoff might be their inability to handle shallow water. They can also be set up to favor more permanent modifications, including the installation of fire pumps, backboards, equipment, radios, and motor controls. They make a much more workable dive platform due to their roomier outlay, and can accommodate more rescuers or multiple victims if needed. These bigger boats may need more horsepower to propel them, ranging from single motors with approximately 35 HP to twin outboards with 200 HP or more for each motor.

**FIGURE 2-20** Rigid-hull boat.
Courtesy of Steve Treinish.

## Flip Lines

**Flip lines** should be attached to small boats before putting them in the water. A flip line is a short, leash-style rope or strap that is attached to a brace on the gunwale, at the side of the boat. In case of a capsized johnboat or inflatable boat, rescuers try to get to the side from which the flip lines are hanging and throw them back over the exposed bottom. They then move to the other side, and place their feet on the side of the boat while pulling the lines. The dual actions of pulling the flip line and pushing with the feet should flip the boat over. The rescuers can then re-enter the boat and bail it out. Follow the steps in **SKILL DRILL 2-1** to flip a capsized inflatable boat.

# SKILL DRILL 2-1
## Flipping an Inflatable Boat NFPA 1006: 21.2.16

**1** Rescuers reach underneath the boat, pulling the flip straps and throwing them over the exposed keel. They then proceed to the other side of the boat.

**2** Rescuers pull on the straps and push with their feet, which starts to roll the boat.

**3** One rescuer braces the boat as it rolls over. This controls the flip so the boat does not injure other rescuers in the water.

# SKILL DRILL 2-1 Continued
## Flipping an Inflatable Boat NFPA 1006: 21.2.16

**4** The boat will be able to float enough for a rescuer to enter it.

**5** Entering the boat from the stern allows the weight of the bow to be used as leverage when pushing up and into the boat. Now the rescuer can start bailing it out, or can stay with the boat until assistance responds.

Courtesy of Steve Treinish.

## Additional Watercraft

Some other types of watercraft, though less commonly encountered in rescue missions, have specific uses in water operations. These boats work very well in the specific conditions for which they were designed.

## Pontoon Boats

**Pontoon boats** are basically a giant deck supported by two or three large barrel-like floats running from bow to stern under the deck (**FIGURE 2-21**). They have a lot of deck space, which translates into a really nice platform for dive operations. They generally offer good stability, except in moving water or wind, where the size and surface area of the boat that is presented to the current or wind make them very susceptible to being pushed around. Many can handle larger payloads. Most have some sort of top or canopy, which can be a welcome addition in a cold rain or under hot sun, and which supports overhead lighting during night

**FIGURE 2-21** Pontoon boat.
Courtesy of Steve Treinish.

operations. Due to their size, teams have used pontoon boats offshore for diver staging and rehabilitation. On some occasions, civilian pontoon boats have been used to transport multiple patients from watercraft accidents, or fire equipment to the scene of island house fires. Pontoons may have outboard motors as small as 10 HP, or be powered by inboard/outboard motors as large as 300 HP.

## Sit-In Kayaks

**Sit-in kayaks** are very small, paddle-powered boats in which the operator actually sits down in the boat itself, rather than on a seat (**FIGURE 2-22**). They are paddled by one person and can be used in very fast water. Most have a spray curtain that fits around the hip or lap area when the paddler is sitting in the boat, forming a fairly watertight seal around the operator. They are easily paddled, can maneuver around and even over obstacles as no other boat would, and are great tools for searching for floating victims in waterways, especially smaller creeks and streams. Because a rescuer can paddle and use the current to propel the kayak, a responder using this mode of travel can cover more ground than a walking searcher. On the down side, special training is required to teach the operator what to do in the event that the kayak capsizes, as he or she may be trapped underwater.

## Sit-on-Top Kayaks

**Sit-on-top kayaks** are plastic molded watercraft that paddlers sit on, instead of in (**FIGURE 2-23**). Sit-on-top kayaks, which are relatively new to the market, can lead to a wetter ride than sit-in kayaks, but they can hold up to three rescuers at the same time. The larger kayaks can even be outfitted with fresh water, first aid kits, and other small items needed for a longer search.

## Canoes

Although not as stable as other watercraft, the common **canoe** can be used by a two-person crew to search calmer waters. These boats can be made of aluminum or fiberglass, and are fairly common sights on rivers and creeks. Canoes rely on paddles for propulsion, but a canoe equipped for one person can be outfitted with a very small outboard motor. They do have a tendency to capsize easily, but in experienced hands can go many places.

## Banana Boats

A **banana boat**—or a "rapid deployment craft," as one manufacturer labels it—is actually an inflatable, but is specific to water rescue (**FIGURE 2-24**). They are called banana boats, or snout rigs, because they resemble a large, inflated banana on the water, or an inflatable boat with a large snout on each end. These boats are somewhat limited in their carrying capability, but are extremely effective on fast water or around low-head dams. A **low-head dam** is a short, vertical wall in the water that creates a hazardous water condition known as a hydraulic that can trap people in the water. (Hydraulics are covered in more detail in Chapter 6, *Swiftwater Rescue Operations*.) Banana boats can be equipped with a small motor, paddled by the rescuers, or used in conjunction with rope systems to move them around.

## Inflatable Rafts

Much more common in the western U.S. states on the larger and faster rivers, **inflatable rafts** are very large

**FIGURE 2-22** Sit-in kayak.
Courtesy of U.S. Air Force/Maj. Marnee A.C. Losurdo.

**FIGURE 2-23** Sit-on-top kayak.
Courtesy of Steve Treinish.

**FIGURE 2-24** A banana boat is a fast, deployable inflatable.
Courtesy of Steve Treinish.

inflatable watercraft often used in whitewater rafting trips (**FIGURE 2-25**). They can be as much as 30 feet in length and are normally powered by hand. These rafts are very stable, but not as portable as small inflatable boats.

## Personal Watercraft

The popularity of **personal watercraft (PWC)** in rescue calls has exploded in recent years. They are especially attractive options in surf and swiftwater rescue, due to the PWC being lightweight, quick, and capable of operating in very rough or shallow water (**FIGURE 2-26**). PWC rely on impeller-driven jet drives, so there are no spinning prop blades to injure rescuers or victims. This setup is safer for the rescuer, but the impeller or nozzle can become plugged in heavy vegetation or a rope tangle, and the impeller may not deliver peak performance in **aerated water**.

Personal watercraft cannot carry a very large load, but may transport up to three rescuers. In addition, trailer-type sleds are available and commonly used, especially in surf rescue. It requires a

good deal of practice to operate PWC, but in the hands of a capable operator, they are invaluable tools in surf or surface rescue. These craft are powered by motors up to 1300 cc in volume, and can achieve speeds as high as 60 miles per hour. While smaller and somewhat challenging to use when transporting victims, they can handle extremely rough waves or surf with ease.

Some PWC manufacturers have established loaner programs to assist public safety agencies on a yearly basis. These loaner programs can be a win-win for rescue teams and the manufacturers.

## Airboats

**Airboats** are very loud, costly, and usually large, but offer one clear advantage over other types of watercraft: They can travel over almost any surface under their own power, including open water, solid and broken ice, vegetation, and moving water. Airboats are powered by an automotive-type engine directly attached to a very large fan blade on the stern (**FIGURE 2-27**). They may also be equipped with hard tops, heaters, and other comforts.

## Hovercraft

**Hovercraft** resemble airboats, but, instead of having a hull, they rely on a fan system that blows a stream of air downward through a skirt to provide a small cushion of clearance for travel (**FIGURE 2-28**). Like airboats, they can travel over almost any surface, and can be used for water, peat, mud, or any thickness of ice. These craft can be large enough to carry several passengers and crew, or small enough for two or three rescuers to rapidly travel across challenging terrain to reach a victim. They are very effective on long-distance, rotten ice missions where the ice is not thick enough to slide across without breaking it.

**FIGURE 2-25** Inflatable raft.
Courtesy of Saturn Rafts.

A

B

**FIGURE 2-26** Personal watercraft.
**A:** Courtesy of Shawn Alladio, K38 Rescue. **B:** Courtesy of Steve Treinish.

**FIGURE 2-27** Airboat.
Courtesy of Gary Robertshaw.

**FIGURE 2-28** Hovercraft.
Courtesy of Terry Messer, Mansfield Fire-Rescue.

## Ships

Coast Guard cutter types are probably the ships that most readily spring to mind, though they are not common in inland waters (**FIGURE 2-29**). Larger cruisers and ships are primarily used on the Great Lakes and oceans of the world. Although they are not part of most agencies' equipment cache, many departments do operate in conjunction with them. Larger fire departments in coastal or large inland water areas commonly have large fireboats in service, and these vessels can play a part in water rescue. Rescuers in these departments or members of teams that operate in collaborative responses with

**FIGURE 2-29** Coast Guard cutter types are probably the ships that most readily spring to mind, though they are not common in inland waters.
Courtesy of Steve Treinish.

these boats must know how they are best integrated into the rescue process.

# Propulsion Methods

Other than when using poles to push a boat out onto an ice surface, or when using a watercraft in a rope-controlled system, most every watercraft will rely on one of two types of propulsion: muscle or machine. Keep in mind that at any time during these boats' operation, muscle power may be required to back up motor power.

## Human-Powered Propulsion

Using human-powered watercraft for rescue is much more effective than the untrained person might expect, especially when the craft is paddled or oared by someone who knows how to use moving water to his or her advantage. **Paddles** are held by the person operating them and can be used only one at a time, whereas **oars** are held on the boat by **oarlocks** and used by the operator together at the same time (**FIGURE 2-30**). Kayakers use a double-bladed paddle that can be operated with the arms, via a crawling motion (**FIGURE 2-31**).

Oars or paddles have one large advantage over motors: They operate without gasoline, spark plugs, or propellers, so they can serve as a backup propulsion system to any motor. Both oars and paddles can be used as tools to effect a reach rescue, as sounding poles and search poles, and even as impromptu depth finders. Efficient paddling or oaring takes training and practice, and water rescue personnel would be wise to keep these skills up, rather than relying too heavily on a motor. Motors can fail, and in some environments in which they operate, that failure is a matter of when, not if.

A

B

**FIGURE 2-30 A**. Paddle. **B**. Oars.

Courtesy of Steve Treinish.

**FIGURE 2-31** Kayakers use double-bladed paddles.

Courtesy of Steve Treinish.

**LISTEN UP!**

One very common problem for rescuers is the lack of immediate deference to paddles or oars when a motor fails in current. A boat that has lost propulsion in current can be placed in extreme danger within just a few seconds. Rescuers should have a paddle at the ready, and designate who will paddle on which side. Controlling the boat and reaching a safe location in the current must take place before any other task is attempted. Once a safe place is reached, then motor issues can be addressed.

## Motor Power

Motors take all the chores of propulsion from a rescuer's back and arms, and put them literally in the palm of the hand. A twist grip throttle, called the tiller handle, usually controls smaller outboard motors (refer back to Figure 2-15). This handle can control both engine speed and the direction of a turn. On some models, the tiller handle also controls forward and reverse, with the boat being moved in these directions by twisting the throttle from the neutral position. On other models, forward and reverse movement are controlled by a separate lever on the motor, usually on the starboard side of the engine.

Using a tiller handle takes a bit of practice, especially if the operator has never piloted a boat with this type of control. It is somewhat unnatural to steer a tiller handle at first, because the control must be moved in the opposite direction of the desired turn. With experience, though, this operation becomes very natural.

## Smaller Motors

Smaller outboard motors, with less than 25–40 HP, will likely have a pull starter. Electric start motors are available, but add weight to the motor owing to the starter itself, the battery required to spin it, and the electrical system. Smaller, pull-start motors are much more portable, and most can be carried to the place of launch by just one or two rescuers.

Smaller boats that use tiller control motors must be leaned into turns if they are to stay upright. Bigger boats lean, too, but the boat is generally too large to be affected dramatically by the operator's body weight. Leaning with body weight will greatly affect smaller boats, especially johnboats or fishing boats. The tighter the turn, the greater the lean, as the boat needs to lean to overcome the centrifugal force of the turn (**FIGURE 2-32**). As an analogy, picture a motorcycle going around a curve. The harder the turn, the more the rider must lean the bike over to keep it from "high siding" and flipping over. Boats, especially smaller boats, behave similarly. The harder and faster the turn, the more pronounced this lean to the inside will be. By comparison, inflatables essentially "slide"

**FIGURE 2-32** Turning forces.
Courtesy of Steve Treinish.

**FIGURE 2-33** Larger motors have a control system that is relatively easy to master quickly.
Courtesy of Steve Treinish.

over the water, as their flotation and center of gravity are very low on the water.

**Small Motor Submersion.** Although rescue teams try to avoid it, a motor may be submerged if a boat capsizes or the motor is dropped. Generally, if the submersion lasts for only a few minutes, and if the proper procedures are followed, the motor can usually be made to run again. Upon getting the motor back to shore, the spark plugs should immediately be removed. The starter rope should then be hand-pulled, expelling as much water from the cylinders as possible. A good dose of WD-40 or fogging oil should be sprayed directly in the cylinders, and the starter rope pulled again. This action coats the cylinder walls and piston with oil, which reduces rust and corrosion. The rest of the motor should be blown dry as possible, and a spray of WD-40 on the engine will do no harm.

Fuel tanks should be drained and refilled with fresh gasoline, and the entire motor inventoried for other damage. Shifters, throttles, kill switches, and other components must all be checked for damage and presence. It is not recommended to try to fire up the motor without an engine technician checking it first and returning it to service.

After submersion, a four-stroke motor will require a couple of oil changes and a professional diagnosis. This effort is recommended because trying to save an engine is something that can be done to show responsibility for rescue gear, and two oil changes are certainly less expensive than replacing the motor altogether.

## Larger Motors

Larger motors have a control system that is relatively easy to master quickly (**FIGURE 2-33**). A steering

wheel turns the motor, just as in an automobile, and a control stick is used to select forward, neutral, and reverse. The steering control may be a cable-and-pulley system on older units, or an enclosed sheath-and-cable unit on more modern and smaller motors, or even have a stand-alone hydraulic unit connected to a hydraulic cylinder that turns the motor.

**SAFETY TIP**

Boats do not have brakes—just reverse thrust. Reverse thrust is not always an option, however, and trying to reverse too early while the boat is in high-speed operation may damage the motor. Thrust is examined in detail later in this chapter.

## Inboard/Outboard Motors

Inboard/outboard motors (I/Os) are seen on some rescue watercraft, but mainly larger runabouts. This kind of motor is quite often an automotive-type engine that drives a lower unit that is very similar to an outboard. The difference is that the I/O engine sits in a covered position within the boat rather than outside it. I/Os are quieter than outboard motors, but harder to maintain.

## Inboard Motors

Inboard motors are found on larger, commercial-type boats and many top-end watersport boats. With this type of motor, the engine is cased above the boat floor and covered with a box-type cover, but the propeller and shaft are entirely under the keel. Inboards supply very good torque and power, but can be damaged in shallower water because they do not offer tilt and trim adjustments.

## Surface Drive Motors

Surface drive motors are a unique type of motor that is used primarily in very shallow or debris-laden water (**FIGURE 2-34**). These units usually offer less than 35 HP, and feature a long, trailing lower unit with a guarded propeller. They are a favorite of waterfowl hunters and have been successfully adapted to flood rescue work by some agencies.

# Motor Characteristics and Minor Maintenance

Both large and small outboard motors can have a few other differences between them. Each rescuer should be well versed in the operation, maintenance, and checklist that the manufacturer recommends for the motor to be used by the rescue team. The majority of boat motors now on the market are four-cycle, or four-stroke, units, so they run on straight automotive gasoline. The other type of combustion motor used in watercraft is a two-cycle motor.

Special attention should be paid to the gasoline used in any marine motor, as ethanol has a very destructive effect on fuel filters and tubing in some engines. Straight unleaded gasoline is desirable if it can be found. If not, consider use of a stabilizing treatment to make starting and storage easier on the motors and the users.

## Four-Cycle Motors

Emissions regulations have pushed the marine industry toward greater use of four-cycle engines, and great strides have been made in their power-to-weight ratios.

Four-cycle motors are quieter than two-cycle motors, and use unmixed gasoline. Their internal crankcase is filled with oil, just like in a car engine. Many now come with fuel injection, making chokes obsolete.

One major problem currently faced by the marine industry is the inclusion of ethanol additives in automotive fuel. Ethanol can destroy certain filter media, and cause fuel to become unusable. When a watercraft has a four-cycle motor, consider procuring gasoline that is 100% gasoline (i.e., without any ethanol). Commercial fuel additives that address ethanol problems are readily available and should be used if necessary.

## Two-Cycle Motors

Two-cycle motors require mixing the gasoline with the outboard oil used for internal lubrication (**FIGURE 2-35**). If you run a two-cycle motor without supplying two-cycle oil, either by premixing it in the gasoline or by using the oil injection tank, you will do major damage to the motor, and very likely destroy it. Newer models of larger two-cycle motors have an oil injection system in which a computer controls the amount of oil delivered to the engine, but their oil tank still needs to be checked and kept full. Many newer motors also have sensors that will kill the engine if oil is not present in the system, or if the level in the oil tank falls below a level sufficient to supply the motor.

## Cooling Water Intake

Another potential cause of outboard damage is a plugged intake for cooling water. Outboard motors

**FIGURE 2-34** Surface drive motor.
Courtesy of Go-Devil.

**FIGURE 2-35** Two-cycle motors require mixing the gasoline with the outboard oil used for internal lubrication.
Courtesy of Steve Treinish.

**FIGURE 2-36** Outboard motors circulate water throughout the engine to maintain the proper operating temperature.
Courtesy of Steve Treinish.

use an impeller to pull ambient water surrounding the lower unit up through the water jacket on the motor; this heated water is then discharged somewhere on the back or rear corner of the motor (**FIGURE 2-36**). If a motor is started but does not pump water within a few seconds, shut it down and check the intake grates and the flow port.

///// **SAFETY TIP**

A spinning propeller can take off a finger or hand in the blink of an eye. Even if the motor is placed in neutral, shut the motor off before checking the grates!

**FIGURE 2-37** Water intake grates.
Courtesy of Steve Treinish.

The intake grates are usually located in front of the propeller and are fairly small (**FIGURE 2-37**). They often become plugged, although a piece of debris, vegetation, or plastic sheeting in the water could also block intake flow. Another place to check for problems is the small tube on the rear of the engine that the cooling water flows from—spiders and insects like to build nests in this tube. Using a small piece of wire to clean out this tube may very well get cooling water flowing through the motor again. If the intakes and cooling flow tube are clear and water still does not flow, shut down the motor and have it serviced. Running an outboard motor without any flowing water to cool it is a sure death sentence for this equipment.

**Water Supply Muffs.** Running a motor on dry land for maintenance or repair is entirely possible, and can be done right on the trailer if the boat is kept on this type of equipment. Many departments start and run the motor until operating temperature is reached as part of a weekly maintenance schedule. A garden hose and a set of water supply muffs can be used to supply the motor with sufficient water for cooling during maintenance (**FIGURE 2-38**). Be sure to check the manufacturer's recommendation on how to attach the muffs properly.

## Larger Cooling Systems

Larger motors are frequently equipped with either a larger raw-water cooling system, which is similar to a smaller motor's cooling system, or a closed system as found on most automobiles. Larger raw-water systems pick up ambient water from the surrounding area, pump it through the motor, and discharge it out a port above the water line. This visible discharge helps verify the system is pumping, and also keeps water from flooding the engine compartment. Hose failures around these motors could potentially flood the boat, so operators should ensure they know how to check the system for proper operation.

## Fuel Delivery

Another source of problems that may prevent start-up of outboard motors is the fuel supply hose. While newer gas tanks and hoses have fittings that can prevent a wrong-way hookup, the gasoline supply hose on some older outboards must be hooked up in a certain direction. A small, squeezable **fuel bulb** is used to prime the fuel system; this bulb contains a small, one-way check valve (**FIGURE 2-39**). If the gas hose is hooked up backward in such a system, the motor will become starved for fuel and stall. A hasty deployment by a rescuer who overlooks the proper setup will encounter problems, especially given that most motors will have enough fuel in the system to get a crew out on the water, and possibly into a danger zone. The same goes for the tank vent on portable gasoline tanks: A closed vent will not allow air to flow in and displace the fuel used, and the motor will simply starve for fuel.

Most new motors have a decent track record of reliability, and if they suddenly fail for no obvious reason, check the fuel system. Larger gasoline tanks that are permanently mounted in a boat have a preplumbed vent as well as a permanent fuel line.

## Kill Switches

Most modern motors have a **kill switch** or **tether switch**. This kind of switch is connected to the operator with

**FIGURE 2-38** Water supply muffs are used to supply cooling water when running motors on shore.
Courtesy of Steve Treinish.

**FIGURE 2-39** A small, squeezable fuel bulb is used to prime the fuel system.
Courtesy of Steve Treinish.

a leash: One end clips to the operator, and the other connects into a switch wired into the motor ignition (**FIGURE 2-40**). If the driver falls overboard or gets too far away from the motor, the leash pulls a clip from the switch, killing the motor and preventing runaway boats. Of course, the switch system will work as intended only if the user remembers to attach the leash to himself or herself.

**SAFETY TIP**

Failure to use the kill switch is a frequently encountered problem. It may be comical to imagine a boat zipping across the lake after the operator falls out, but that outcome can very well happen. Choosing to use this device and using it correctly also give the operator the instant ability to kill the engine if needed.

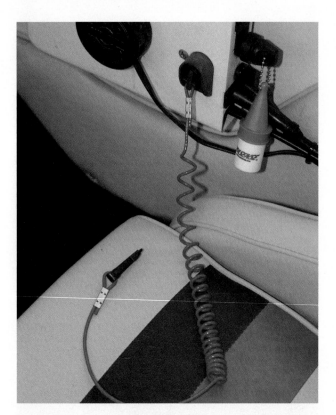

**FIGURE 2-40** With a kill switch, one end is clipped to the operator via a leash, and the other end is connected to a switch wired into the motor ignition.

Courtesy of Steve Treinish.

**LISTEN UP!**

Every season, many mariners haul their boats to a shop because of the same complaint: The engine will not start. Once the kill switch is properly inserted, these motors start right up. Some motors can be operated without a tether, just by placing the switch into the "run" position, but this is a safety issue. Ensure the switch is there and properly inserted before attempting to start the motor.

## Propeller Dangers and Problems

The edge of a propeller is sharp enough to lacerate a rescuer even if it is not moving. Damaged propellers also present a cut hazard when rocks or debris chip a blade, and sharp burrs are frequently found on damaged propellers. To protect against these hazards, a round guard in the shape of a ring that surrounds a propeller, or a cage-like device, similar to a fan guard (commonly called a **prop guard**), can be installed on some motors. The spinning of the propeller also presents a potential problem in operation, in that striking

**FIGURE 2-41** Aerated water.

Courtesy of Steve Treinish.

a hard object can result in a bent or broken blade. At best, this deformation of the propeller will slightly affect performance on the water. At worst, it will internally damage the motor, putting it out of service immediately.

## Jet Drives

One alternative to propellers is a jet drive. Jet drives use an encased impeller to force water through a cone-shaped housing where the propeller would normally reside, thereby creating thrust (see Figure 2-14). Jet drives and prop guards can both be affected when rubbish or vegetation plugs up the intake on the unit, but jet drives do not have any exposed blades spinning underwater.

## Aerated Water

Aerated water, which is commonly encountered in swiftwater rescues, can greatly affect the operation of any motor, be it propeller, jet, or PWC. Aerated water is water that is churning and contains air bubbles (**FIGURE 2-41**). This phenomenon is found in fast-moving water, low-head dams, or anywhere water is violently agitated.

Air does not transmit thrust in the same way that liquids do. The propeller or pump will not "hook up" and deliver thrust in air and, in extreme cases, all forward thrust can be lost in aerated water. Boaters call this "running away from water," and the condition is obvious—evidenced by an immediate lack of thrust and a motor suddenly increasing its revolutions per minute (RPM). This problem is something to be aware of in any water, but especially in low-head dams, where the backflow may pull a watercraft back into the face of the dam when propulsion is lost. It can be stressful when you are trying to get out of a difficult situation quickly, grab a handful of throttle, and hear the motor wind up with no thrust with it.

# Rescue Boat Load-Out

Many different tools might be needed on any given water rescue mission involving boats. However, the equipment load-out can be broken down into sub-categories, and some equipment is considered crucial across the board. Not every mission will require every tool, and missions may require the same tools for different reasons. Consider PFDs: Every rescuer may be wearing one, but did you remember to bring one of these devices for each victim you are rescuing? And did you bring PFDs for the occupants of that stranded boat you found during your training on the river?

## Safety Gear

Safety gear is watercraft equipment considered to be the minimum cache to allow the watercraft to operate in a safe manner, with all aboard properly protected from emergencies. Some state laws pertaining to safety gear apply to every watercraft, even rescue boats. Be sure to review the laws that apply in your state.

## A PFD for Everyone on Board

This safety gear requirement can be changed slightly in the water rescue environment to include PPE for all rescuers onboard. Such a cache could include helmets, PFDs, cutting tools, and a small light. At a bare minimum, go with the PFDs, both for the rescuers starting out on the mission and for all victims expected to be rescued or placed in the boat, and then add a few extra PFDs. The ones sitting in the floor can be quickly thrown to someone who falls overboard. Too much flotation on the water is like having too much water in a line in a fire—it is not

really a problem at all, and it may even be a stroke of luck.

## Means of Communication

Portable radios are mostly reliable, but they do fail. In addition, the dreaded gravity robs many boaters of many things, including portable radios. Whistles, flares, and distress flags can all be used as backup communication if the portable radios are lost or fail. Hand or body signals can also be used for boat-to-boat or boat-to-shore communication, although these signals must be worked out in advance.

Most people today carry cell phones, and these devices can be used for communication during a response if the rescuers have the pertinent numbers stored or written down. Often, however, water basins are naturally lower than the surrounding terrain, so a cell phone signal could be lost or degraded—just another thing to check out when preplanning for responses.

**LISTEN UP!**

Cell phones mixed into water rescue are only as good as the cases that protect them. Do not rely on an unprotected cell phone as a means of communication. Drybags or dryboxes can offer protection from the water surface, splash, or precipitation **(FIGURE 2-42)**.

**Marine Radios.** The maritime industry has established a complete radio channel system with designated, universally used working, hailing, and emergency channels **(FIGURE 2-43)**. Specifically, Marine Channel 16 is set aside by the International Maritime Organization as an emergency hailing channel, and is monitored by

**FIGURE 2-42** Drybags offer protection from the water surface, splash, or precipitation.

Courtesy of Steve Treinish.

**FIGURE 2-43** The maritime industry has established a complete radio channel system with designated, universally used working, hailing, and emergency channels.

Courtesy of Steve Treinish.

many agencies and stations worldwide as a distress channel.

Most larger rescue watercraft will have a marine radio onboard in addition to whatever fire service radio is needed. This system allows for good communication with other boaters, other rescue boats, marine law enforcement, and even helicopters that may be involved in searches or rescues. Small, hand-held marine radios can be purchased for less than $100 and are a great asset, even in smaller craft. Many of these small radios can transmit for several miles, making them a good option if marine channels prove easier for crews to use.

## Emergency Signaling Equipment

Even in the age of modern radios, cell phones, and global positioning system (GPS) locators, these devices can still fail. Batteries can discharge, units can be accidentally dropped in the water, and the utilities that support these devices can fail. Boaters should be familiar with the typical emergency signaling equipment available.

In case of a maritime emergency, if the boat is drifting with the wind or moving water, the motion should be stopped if possible. Only after safety is secured should communication be considered. Pilots are taught to follow an adage in aviation emergencies: Aviate, navigate, and communicate. That is good advice for mariners who face an emergency, too: Control the boat, find the route needed to stay safe, and then focus on getting help.

**Flares.** Visible in day time or nighttime conditions, flares can be hand-held, launched like a rocket from the flare tube, or fired from a pistol (**FIGURE 2-44**). With any flare, care should be taken to keep it dry, and caution should be used when lighting or firing it. Molten material can easily burn human skin—or burn through the rubber of an inflatable boat. Consider launching two flares if you believe help to be close so that the rescuers can confirm and triangulate your location.

**Distress Flag.** A small, bright orange flag flown by stranded boaters, a distress flag is an international signal that is safe, clean, and reusable (see Figure 2-44). The downsides of using such a flag are twofold: It cannot reach the same height as a launched flare, and it tends to blend into the landscape.

**Smoke Canisters.** Similar to the "landing zone marking smoke" used by the military, smoke canisters are commercially available for emergency signaling. Windy conditions will blow smoke away faster than lighter winds, but canisters are still a viable option in the emergency toolbox.

**Personal Locator Beacons.** Today, many mariners, aviators, hikers, climbers, and other outdoor types carry personal locator beacons (PLBs). A very small unit that can be pinned to a PFD or boat and that even floats, the PLB can send out a signal on a radio frequency that can be tracked by rescue apparatus equipped to home in on it.

**Man Overboard Units.** Man overboard units are small devices that can signal the operator of a larger vessel that a crewmember has fallen overboard, thereby alerting everyone to the event. Some of these devices interact with the GPS mapping display and actually place a pin when the unit is triggered. The boat can then navigate back to this location and rescue the person who fell overboard.

**FIGURE 2-44** Flares and distress flag.
Courtesy of Steve Treinish.

## Fire Extinguisher

The United States Coast Guard (USCG) issues specific standards for the type and size of fire extinguishers to be included on watercraft. For most watercraft, a type ABC portable extinguisher is considered sufficient. Check your local or state boating laws for more information.

## Essential Gear

Essential gear is equipment necessary for the proper operation of a boat. These items are required to provide a relatively smooth watercraft deployment in most circumstances, but are not considered absolutely crucial to life safety.

## Gasoline Tank and Boat Plug

In the heat of the moment, both gasoline tanks and boat plugs have been left on shore or not installed upon launching for a water rescue. Overeager fire fighters might push a boat out onto the water before the motor is started to give the boat crew a head start; while that action is meant well, if a current grabs the boat and starts it downstream, "no gas" will quickly change to "no go." Make sure the tank is on board, the fuel line hooked up (remember to check the direction of the bulb), and the motor starts before the boat leaves the shore. It is also not unusual to witness drain plugs left out as a boat leaves. Eventually, the crew will notice the large amount of water on the floor, but that may be later rather than sooner if their adrenaline is pumping hard.

Gasoline tanks and boat plugs are considered to be the minimum load on any boat operation, at its simplest. Consider what else might be loaded to make a mission easier on the rescuers, and better for the victim, provided the boat is not overloaded and you have workable room.

## Anchor

In case of an engine failure, it is helpful to be able to stop the boat and stay put. Mariners do not want their boat to move unless they make or let it move, especially in current. An **anchor** can be deployed to stop movement, and it is normally used to hold a boat in one certain spot. Most single-anchor systems deploy from the bow, putting the bow "V" into the current, or into the wind and waves.

## Oars or Paddles

One paddle in a boat with three or four rescuers leads to a somewhat comical situation if they must try to share this single piece of equipment. If room allows, take two paddles, or even four. Paddles with collapsing poles can be placed in the boat and do not take up much room. Motors fail on water missions more often than might be expected, simply due to the amount of abuse they have to withstand, so rescuers must be ready to provide human propulsion.

One facet of training is getting rescuers to instantly grab a paddle or oar if a motor fails, especially if they are in moving water. It is normal to have multiple rescuers fixate on a repair or troubleshooting, when the first priority should be to steer the boat to a position of safety before work on the problem can begin. In swiftwater, such a delay can lead to capsizing a boat or having a crewmember knocked overboard if the boat gets pushed into strainers or objects.

## Lights

Smaller, watertight flashlights provide very good light. Dive lights are smaller, and also waterproof, provided they are sealed properly. Larger spotlights can provide millions of candlepower and, with LED bulbs becoming common, can offer a very low current draw and great durability. Remote controls can be installed on some lights to operate them from interior cabins.

## Dewatering Equipment

Obviously, watercraft that fill with unwanted water can present an issue. Even in normal operating conditions, however, water can fill the bilge area through spray, waves, or even rescuers bringing water in the boat when victims are retrieved. Bilges are simply spaces that allow water to drain to a low space to be removed, either with pumps or manually. The most commonly used bilge pumps are operated by 12-volt battery power, and are controlled with automatic float or manual switches (**FIGURE 2-45**). Think of a sump pump in a house—that is the essence of the bilge

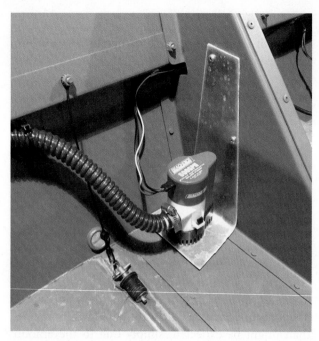

**FIGURE 2-45** Bilge pump.
Courtesy of Steve Treinish.

system found in most watercraft. Dewatering pumps operated by hand may be used as well. Although they are somewhat labor intensive, these pumps do operate regardless of whether the battery or power supply fails. Water can also be removed by a dipper, or bailer, which can be made very easily using a plastic coffee can found in groceries. This approach works well with inflatables or johnboats.

**Bilge Discharges.** With the process of pumping water from the bilge comes the possibility that other substances may be introduced to the water. Fuel, oils, and other hydrocarbons often find their way into the bilge, since it is the low point and located close to the engine. Larger boats may also have power steering, tilt and trim, and other hydraulic systems in use. Fuel spills of even a few ounces can wreak havoc on marine environments.

The best cure for this ill is prevention. Keeping a clean bilge, using the proper chemicals to do so, and disposing of waste properly is the best way to protect waterways. Bilges should be cleaned properly, in an environmentally responsible way, with wastes being sent to the proper runoff piping or drains. Using absorbent cloths or pads to soak up waste is also recommended. In case of waterway spills, contact the local authorities; they may include the local Environmental Protection Agency (EPA), the USCG, or drinking water authorities.

## Depth Sounders

While it may not be required for operating smaller boats, the use of an electronic depth sounder can be advantageous in preventing damage to motors and out-drive equipment. Striking a propeller or lower unit on a hard surface can easily do enough damage to cause a repair bill in the thousands of dollars or, even worse, damage the hull to the point that the boat sinks. An electronic depth sounder uses radio frequencies to measure the depth of the water, and most units also feature alarms that can be set for various depths of water. For example, a watercraft that drafts 3 feet of water may be saved from damage by setting the alarm to 6 or 7 feet, allowing the operator time to take evasive measures.

## Rescue Gear

Equipment needed or taken on board the boat to directly effect rescue is considered rescue equipment. In addition, you must consider the possibility that you end up rescuing yourself or one of your own. Crew overboard (COB) events are a watercraft emergency that must be factored into the operation, and are discussed later in this chapter.

## Throw Bag

A **throw bag** is a small pouch containing rope that is thrown to a victim, while the rescuer holds one end (**FIGURE 2-46**). One throw bag is a true minimum; many times on the water, you will want several in the boat. Instead of one bag being reloaded or coiled for another shot if the first attempt fails, the other throw bags can be deployed.

## Reach-Type Tool

Pike poles, shepherd's crooks, or special-use water jaws can all be used to extend assistance to a victim (**FIGURE 2-47**). They also can be used to probe the bottom of a body of water while searching for a submerged victim. Water jaws use hydraulic force to open and close a set of large jaws, enable a rescuer to pull up a victim, and have an extendable pole. With some practice and experimentation, crews can locate and remove bodies from almost 20 feet of water, using the jaws alone.

## First Aid or EMS Equipment

If the victim needs immediate medical assistance, even just a pocket mask, an Ambu-Bag, or an airway kit is better than nothing, and will let rescuers at least start rescue breathing for a recovered victim. Many larger water areas permit rescuers to start EMS treatment while en route to the shore. Alcohol-based sanitizers are very handy to rescuers, as wiping water on the hands often transmits bacteria and germs to the eyes, nose, or ears in the course of operation. Mosquito

**FIGURE 2-46** Throw bag.
Courtesy of Steve Treinish.

**FIGURE 2-47** Reaching tools can be extended to victims.
Courtesy of Steve Treinish.

**FIGURE 2-48** Long water rescue lines can be stored in bags or on spools.
Courtesy of Steve Treinish.

repellent will also be needed in insect-ridden areas, especially at dusk or nighttime.

## Longer Water Rescue Lines

Longer water rescue lines are stored in bags or on spools (**FIGURE 2-48**). A mission may require longer lines and rope gear to be transported or deployed via watercraft. These lines can also function as longer tag lines for "go" rescues. In addition, longer lines are sometimes stretched across a river or stream to create a **strainer** when trying to snag floating bodies as they drift downstream. Make sure all personnel on the river are aware of their presence if such rescue lines are deployed.

**Tow Equipment.** During many incidents, a rescue watercraft will be tasked with towing in another boat. A small, premade bridle can be attached to tow points found on the transom of most watercraft, and a line run from the bow of the stricken vessel to this bridle. Placing padding under the towline may help prevent damage to the boat and the line itself, as chafing can

quickly cause a failure in any rope. Tarps, blankets, or even a cheap PFD can be used for this purpose.

Vessels being towed should always be towed from the bow, as the shape of the bow breaks the waves and lets the boat cut through or glide over the water. Towing another watercraft stern to stern will almost certainly swamp or capsize one of the vessels. Lengthening the towline can also limit the amount of play that may be experienced by the vessel being towed.

It is important to ensure all personnel on the stricken boat have proper flotation prior to getting under way. Speed should be limited to a no-wake speed to maintain control of both vessels. Impact protection should be considered as well, in case of the failure of the boat that is towing. Losing propulsion, and thus maneuvering ability, may cause the towed boat to strike the stern of the towing boat. Buoys or fenders hung off both watercraft may fend off damage during such an event. Oars or paddles may also be extended to push off a boat coming close, but this action should be attempted only with smaller and lighter watercraft.

## Retrieval Gear

Retrieval gear includes any equipment that is needed to bring people in the water onto the watercraft. This operation could include loading rescue swimmers or divers, or loading a deceased victim for transport to shore. Either way, retrieval must be conducted properly to minimize the risk of injury to crewmembers and biological-contaminant exposure from the victims. Swim ladders, webbing, Stokes baskets, or body bags may all be needed.

## In-Water Extrications

Whether the incident involves a planned rescue in still water or an unplanned rescue of a crewmember overboard in fast-moving water or raging surf, rescuers still need a way to safely and efficiently get people back onboard their watercraft. NFPA 1006 states that operations-level rescuers should be familiar with the various extrication methods used to recover rescuers who may be part of a COB event.

## Simple Methods

**Double Bounce.** A quick way to extricate people who are capable of helping themselves is to simply grab the person by the shoulders of the PFD and pull him or her into the boat. This is very easy in inflatable watercraft, as the huge amount of flotation does not let the boat tip much. In contrast, in smaller craft such as johnboats, weight must be shifted to the side opposite the extrication to prevent capsizing. A double bounce using the flotation of the PFD can help manage this feat: Do a 1–2 count, push the person down slightly in the water, and use the buoyancy of the PFD to get an assist up.

**Stirrups and Loops.** Another easy extrication method is to create a loop in a rope and hang it down in the water to create a stirrup, just like the stirrup on a saddle for a horse (**FIGURE 2-49**). Throw bags work very well for this purpose, as do the bow and stern lines found in most boats. Body belaying or using a friction belay over the gunwale can help secure the rope. With some practice, persons who fall overboard can learn to self-extricate themselves with very little outside assistance.

## Extrication of Injured, Unconscious, or Deceased Victims

### Parbuckling

**Parbuckling** is a method of bringing out a victim using a natural 2:1 haul system. It works very well to get a victim out of the water quickly, and helps manage any

victim who is naturally heavy. A person weighing 300 pounds who is unconscious in the water will present rescuers with a lot of dead weight, and two rescuers in a small boat will have trouble loading him or her if they do not use some sort of method to make this task more manageable. Unconscious or floating deceased victims can be handled much more easily when extricated from the water via parbuckling. To extricate a victim using parbuckling, follow the steps in **SKILL DRILL 2-2**.

## Stokes-Type Litters

As discussed in Chapter 5, *Surface Rescue Technician*, if the victim has a known or suspected cervical spine injury, you must take cervical spine precautions. To do so, you can equip a plastic, floating backboard with the normal strapping and footplates. Consider using a longer hand strap attached down by the legs of the victim to make hauling the packaged victim into the boat easier. The rescuer can use that strap to pull the board vertically up the side of the boat. Stokes baskets or litters also work well when using the same technique. Either way, victim flotation is absolutely critical: It would not look good to let a perfectly packaged victim sink to the bottom of a lake after dropping a basket litter with no flotation.

On a related note, you must protect the victim's airway from splashing water or submersion. Handle hypothermic patients gently, so as to reduce the risk of abnormal heart rhythms. With nonbreathing patients, ensuring a functional airway is the first concern, so it becomes a case of "doing what has to be done, quickly."

**FIGURE 2-49** Self-extrication may be facilitated with a loop in a rope that hangs down in the water to create a stirrup.
Courtesy of Steve Treinish.

# SKILL DRILL 2-2
## Extricating a Victim Using Parbuckling NFPA 1006: 21.2.14

**1** Either with flotation or with assistance, bring the victim into a position parallel with the side of the boat. Rescuers kneel on the inboard edge of the straps.

**2** One section of the fence droops down into the water from the boat, around the victim, and back up to the boat crew.

**3** Form a 2:1 advantage system, much like a traveling pulley in a rope system.

**4** With minimal effort and, more importantly, minimal force on the side of the boat, bring the victim up and into the boat. This is a very quick and efficient extrication technique, but does not lend itself to taking cervical spine precautions.

Courtesy of Steve Treinish.

In victims with these kinds of conditions, tradeoffs must often be made to preserve life.

## Extrication of Swimmers or Divers

Most larger watercraft have some kind of swim platform, which can range from a small, permanent wooden platform attached to the stern beside the motor, to a molded fiberglass deck at the stern of the boat (**FIGURE 2-50A**). Portable folding ladders can also be acquired from any boating supply house or marina (**FIGURE 2-50B**). At least one watercraft manufacturer has developed a fold-down front rack that can be used to deploy and retrieve rescuers (**FIGURE 2-50C**).

Some boats are even equipped with small cranes to perform a lifting operation with a victim packaged in a Stokes litter. Commonly seen in high-performance racing safety boats, these devices have also been fitted to public safety boats to facilitate extrication of living and deceased victims.

A

B

### SAFETY TIP

The motor should always be off before allowing anyone around the area of the motor. Securing the key or kill switch on larger vessels can also prevent injury. Swimmers should *not* use the lower unit of the motor as a step: It is all too easy to slip and suffer quite the laceration on a foot or leg from a propeller.

## Miscellaneous Tools

Many of the tools that work for one particular type of boat in one particular evolution may or may not be needed in another evolution, so common sense should be used when selecting equipment. For example, a non-motored johnboat being towed by hand by flood rescuers will probably not have a tool kit on board, with good reason. Similarly, a watercraft launched for a confirmed floating body recovery will not have life support equipment ready to go at the gunwale. In many cases, however, specific items may make life easier on the water for rescue crews.

## Body Bags

Floating deceased or confirmed victims can be bagged offshore to eliminate the "gawk factor" at the exit point. This practice also helps to treat the victim with a measure of respect by reducing the number

C

**FIGURE 2-50 A**. Swim platform. **B**. Swim ladder. **C**. Front rack.
Courtesy of Steve Treinish.

of people who will see the remains. However, local law enforcement agencies may need to be consulted as to the needs of the investigation, and their advice

**FIGURE 2-51** Marker buoy.
Courtesy of Steve Treinish.

**FIGURE 2-52** Boating equipment is not a cut-and-dried list. Equipment needs should be modified and examined depending on what works for the team and what the response needs.
Courtesy of Steve Treinish.

should guide rescuers. Grossly decomposed bodies can be parbuckled straight into body bags to limit the amount of biological contaminant splashing around the boat.

## Marker Buoys

Marker buoys are smaller buoys about the size of a soda can that can be used to mark evidence, locations, or bodies (**FIGURE 2-51**). They are commonly called pelican buoys, in reference to one of the manufacturers of these types of buoys. In case of a victim submersion, they can also be used to mark the point last seen (PLS) of the victim.

## Rescuer Supplies

Some simple things can make life easier for rescuers during a water deployment, and can be tailored for the environment faced. Drinking water, snacks, towels or dry gloves or shirts, and even insect repellant can be welcome additions that enhance rescuer comfort. Many areas of the United States have incredible bug problems. Moreover, given the various viruses transmitted by mosquitos, protection against these pests is a sensible health precaution.

## Tool Kit

A basic tool kit should be taken on larger boats or boats operating far enough offshore that the tools cannot be readily brought in if trouble arises. Even a small, canvas roll-up pouch containing the basics is better than nothing, and can be stowed in a small space. Larger boats with bigger motors may carry spare props, fuel filters, lines, and other boat supplies, along with the tools to install them.

**Tool Storage and Placement.** It is fairly common to have all of the aforementioned equipment and more stored in the boat, or on the trailer. For example, many boats have a toolbox on the trailer that stores drag hooks, water muffs (which are used to safely run the motor on dry land by hooking a garden hose to the water intake), extra flashlights, strobe lights, body bags, and tools. Many boat companies also keep a few ice rescue suits, a line gun, ice awls, swiftwater helmets, and several more PFDs in bags in the boat itself while on the trailer; when a water rescue call comes in, they simply leave equipment not needed onshore at the launch point. Some even go so far as to load the throw bags on the top of the gear pile, at the front. A water run may sometimes be mitigated with a quick toss of a throw bag; if the throw bags are lying on top, easy to access, and ready to go, they can be thrown in seconds. When equipment is stored on the boat and its trailer, the boat and the boatload can be tailored to the specific mission in a matter of seconds, and all the gear needed for almost any type of mission is sitting there, ready to go.

Boating equipment is not a cut-and-dried list. Teams may require different equipment or setups, and these setups should be modified and examined periodically to make sure they are doing what they should, and to determine which are working for the team. Different water levels or geographic locations can also change the level of watercraft response, or the equipment that comes with it (**FIGURE 2-52**).

## Towing Operations

Many watercraft used in inland areas are used on multiple waterways, and are most often stored and

# Voice of Experience

We knew the day would keep us busy; it was the 25th running of Eppies Great Race along the American River, a triathlon consisting of a 6-mile run, 12.5-mile bike ride, and a 6-mile kayak paddle. With the paddling being the last leg, many competitors were tired and making small mistakes at that stage. There were hundreds of paddlers, but as the day went on, rafters started putting in and mixing with the competitive paddlers. I had set up my stand-by position 25 yards downstream of a well known hydraulic with a very fast chute of standing waves that many paddlers miscalculate and roll their craft to river right. I was on a current, three up PWC, but didn't have a rescue board because in those days we used Stokes baskets, which didn't provide a rescue platform. Our job was to make sure overboard paddlers either got back into their kayaks or made it to shore.

We were kept just as busy as in previous years. Many different types of rafts were coming down, and most riders were drinking. Now, in addition to dealing with the competing kayakers, we were dealing with intoxicated people dropping over the hydraulic who weren't ready for the drop and the action of the reversal and would fall into the water. The rafts would wash out, but the people could get stuck because the action pulls you back to the rock, drops you under the water, and pushes you forward and then back up, and then back to the rock. Fortunately, the hole is only about 4 feet deep, so most people can stand up and just push against the reversal and back into the current. The width of the rock and the flow over it totals 5–6 feet wide, so even going to one side or the other, you can meet a current and get out of the hole.

I kept my eyes upstream, and put my craft into a slight current where most rafters never get to due to the other river features in the rapids. I held my position using idle and a little reverse, when I noticed a smaller raft with two larger women sitting low in the raft and a small girl sitting on the rear tube coming directly toward the hole and reversal. I knew that if they came over that rock and hit that hole, the little girl would be thrown forward; I was glad to see that the little girl was wearing a PFD.

As the flow set them up, they came directly over the rock and dropped into the hole. The raft shot right off the reversal, almost as if it had picked up speed. This acceleration and the abrupt movement of the raft caused the little girl to drop over the back of the raft and into the reversal. Seeing this, and that the women did not realize what had happened, I quickly accelerated to the hole. I couldn't see the girl and quickly determined that she was shorter than 4 feet tall. The PFD had come off and was floating just behind the raft.

I knew I would have to reach in and feel for her, but first I needed full control of the PWC. Using reverse, I quickly flipped the PWC and ran the boarding platform right up against the rock so the reversal would hold the craft in place. I shifted most of my weight to the port side, reached down, and grabbed the girl by her hair and pulled her up.

I brought the PWC up alongside the raft and helped the little girl get back into it. She was still crying and spitting out water. I retrieved the PFD for them and strongly encouraged them to take the girl to the hospital to get her checked out.

The ladies thanked me and I left to get back into position. The PWC was probably the only craft that could have maneuvered that quickly, be small enough to get into the hole, and be low enough to the water to allow a person to reach in far enough and find a small child. Even with the craft right over her, there were no moving parts that presented any danger to her. I believe that if any other type of craft had been on standby in that position, they would have been diving in for the girl or pulling her little body from the river.

**Tony Hargett**
Developer/Master Instructor
Aqua 7 Rescue
Roseville, California

transported on trailers specifically mated and used with a specific boat. Many rescuers have never towed or even hooked up a trailer of any kind, let alone responded to a scene with a rescue craft. Some information on trailers and trailering follows.

## Trailer Makeup

Most trailers are very basic, having one to three axles. The watercraft rests on wooden or plastic bunks or a system of rollers, and is secured to the trailer with a winch at the bow and possibly some straps at the transom (**FIGURE 2-53**). These rollers and bunks can be tailored to help a boat center itself on the trailer when retrieving it from the water. Lighting for the entire trailer is supplied from the tow vehicle's system through wiring that plugs into the tow vehicle near the coupler.

## Hitches, Couplers, and Safety Chains

Most trailers are attached to the tow vehicle with a coupling that fits over a ball on the bumper or hitch. These balls range from 1⅞ to 2⁵⁄₁₆ inches in diameter. Care must be taken to match the hitch size to the ball size, or the trailer may physically disconnect from the tow vehicle. A latched coupler might sit atop the ball and look very much attached, but bounce off when under way. Ensure the trailer coupler is unlocked when dropping it over the ball, properly latches, and has a safety pin installed in it before moving. Safety chains or cables hook into the hitch on each side of the ball to provide further insurance that the trailer stays attached to the vehicle in case of hitch malfunction or breakage.

## Trailering Skills

One skill that complements watercraft skills is trailer backing. Set up a course with cones, find a driver who is experienced and can provide tips on this skill, and simply practice. Most trailers, especially the smaller ones, will react very quickly to steering input, and many drivers will end up doing a tight "S" course, or even worse, jackknifing the trailer. Back slowly, and start with very small inputs to the steering wheel.

**FIGURE 2-53** Boat trailer.

Courtesy of Steve Treinish.

> ## LISTEN UP!
>
> If you are not used to backing trailers, you may find some relief in this hand positioning on the tow vehicle's steering wheel. Place your palms face up, at the five and seven o'clock positions on the steering wheel. Point your thumbs outward. Whichever way the trailer needs to go, your thumbs will now point in the direction the steering wheel should be turned. Fire fighters who are adept at backing tiller ladders will remember this trick very well, as it works for that apparatus as well.

Positioning a spotter behind the trailer is highly recommended. Each trailer needs to be backed into the water a certain distance to let the boat float or slide off of it, but backing it to where the axles are just above the water is usually a good starting point. The trailer lighting plug can be disconnected if time permits. This step keeps bulbs and fuses from blowing if a light should flood, but remember to hook them up once out of the water.

> ## SAFETY TIP
>
> Marine-specific trailer lights are available and should be considered for purchase. Many times, unplugged lights are forgotten when retrieving a boat, and there may be some liability if trailers with inoperative lights are struck by another car. Forgetting to plug in the lights when retrieving a boat renders the lights useless.

When backing an empty trailer, be aware that the driver may not be able to see the trailer at all unless it is moving slightly to one side or the other of the tow vehicle. Some trailers enable the boat to be driven onto the trailer, but others do not. If the trailer is too far down the ramp, the stern will have a tendency to float to either side because it is floating, rather than sitting on the bunks. If the trailer is too far up the ramp, the winch may pull at the wrong angle, making the boat much tougher to pull up and onto the bunks of the trailer.

## Tie-Downs

Most boats are strapped to the trailer with a bow winch, and possibly a set of stern tie-downs, or one larger strap running from side to side of the trailer frame, toward the stern of the boat (**FIGURE 2-54**). A helpful practice is to undo the transom strapping, if the boat is so equipped, before launching it. This sounds very elementary, but if the tie-downs are not removed when the trailer is backed down the ramp, the boat may fill up with water as its stern is forced underwater, or damage can be done to the hull from the buoyancy of the boat pulling on the strapping. The bow winch can be used to safely lower the boat into the water while maintaining control.

## SAFETY TIP

Carpeted bunks create a lot of friction when dry, and many boat owners launch their boats in one movement. All straps and the winch are undone, and the boat launches directly from the trailer once it hits the water. The plastic or Delrin blocks that some boat trailers use are *extremely* slippery, even when dry. Bow winches must be kept attached to keep the boat from sliding onto concrete as soon as it angles down while backing down the ramp.

## Hand Launching

Rescuers may simply carry smaller boats and their motors, if so equipped, to the launch area. Of course, this kind of hand launch is truly a team effort, and coordination of the crew members' activities is of the highest importance, to protect both the rescuers and the expensive watercraft they carry. Many missions present with a victim who is a long way from a boat ramp. In these cases, it may be advantageous to have a crew clear brush, trees, or debris from the bank, and launch closer to the victim. Rescuers can always

choose to retrieve the boat, crew, or equipment at a different location once the urgency has been eliminated from the incident. In fact, this choice can often be much safer.

Normally, inflatables and johnboats will be hand launched. Rescuers should place themselves around the boat, while keeping in mind that the motor adds weight at the stern (**FIGURE 2-55**). Doubling the number of rescuers present toward the stern can make life easier on everyone. Place the taller, stronger rescuers on the stern to increase clearance over ground. One rescuer should be at the bow with the bow rope, and all rescuers should be in proper PPE. The team lifts the boat from the trailer or ground, with the stern traveling in front. Rescuers at the stern give the orders to pick up and step, and also call out hazards such as rocks or drop-offs. Anyone can stop movement, but that command should be verbalized as "Stop!" "Whoa" sounds like "go," and things like "hold on" or "hang on" may not be understood.

Any boat should be carried above ground—not dragged or slid—to prevent damage. When reaching the water's edge, usually two or four rescuers will enter the water slightly. Awareness of potential rocks or debris just out of sight in the shallows must also play into hand launching. When the boat's stern floats, the person holding the bow can lift (usually the bow is very light, compared to the stern), and the crew can pass the boat hand by hand down the line into the water. This keeps rescuers' actual entry into the water at a minimum.

When technicians are equipped with proper PPE and the water is shallow enough, a boat can be carried in, and the motor attached after the launch. Depth is key here, as injury may occur and motors can be damaged if the bottom drops off or rescuers stumble and drop the boat.

A    B

**FIGURE 2-54 A.** Tie-downs. **B.** Winch.
Courtesy of Steve Treinish.

**FIGURE 2-55** Hand launching.

Courtesy of Steve Treinish.

## Docking

Getting under way and returning from a mission may require rescuers to assist with lines and bumpers so as to protect the dock and the boat, as well as secure the boat from being pushed or pulled by wind or wave action. Approaches or departures from docking or mooring areas may require the operator to adjust the course slightly to travel in a straight line. For example, when faced with a stiff wind from the starboard side, the boat will need to be turned to the right enough to counter the wind pushing it from right to left.

Bumpers are small, inflated buoys that absorb the smashing that can occur between the dock and boat. Securing the boat to the dock may require multiple lines or bumpers. The boat may be secured and tied from bow to stern or from stern to bow depending on currents, tides, wind, and water conditions. Generally, watercraft are tied up at docks or slips using cleats—that is, the anvil-looking hardware attached to the floor of the dock. Some docks offer four-point tie-offs so that the boat is kept away from the dock on both sides and bumpers are not needed. Mariners should not underestimate the power of wind or water when tying up, especially when the boat may not be tended or used for several days.

### Cleat Hitches

Many different knots are used in maritime operations, but the most basic is the cleat hitch. By wrapping the rope around the cleat before tying it off, a good friction belay can be gained while the operator adjusts the position of the boat. A stiff breeze broadsiding a big boat will create a sail effect that can easily pull people in, or even let the boat be blown away from the dock. Follow the steps in **SKILL DRILL 2-3** to tie a cleat hitch.

### Bow Lines

When launching a boat, someone must hold a rope attached to the boat—normally a line attached to the bow—to prevent the boat from floating away as it comes off the trailer. When boats are launched in haste, it is not unusual to see them just keep going once they float off the trailer. Granted, the new fire fighter could get wet and bring the boat back, but these common-sense mistakes are readily avoided through practice and with greater confidence. In windy or current conditions, another line at the stern may be required to keep the boat under control at the dock.

## Operator Support

NFPA 1006 distinguishes between the boat's crew and its skipper based on their tasks. Crews should know how certain activities will commence and proceed once they are needed or encountered, and the operator should be able to operate the watercraft safely while undertaking these tasks. NFPA 1006 mentions a **helmsman** who works at the operations level; this person steers the boat under the guidance of another individual, usually the person in charge of the vessel. The operator of the vessel may give directions to the helmsman such as the course to steer and the location to target.

A typical fire company offers a good analogy for the various roles on the watercraft: All persons on the truck should know how to handle typical and emergency situations, and at least one will be driving. But the officer acts as the operator of the truck, giving guidance and orders as needed.

### Hazardous Conditions

Even if a rescuer is acting as a crewmember, and is not actually the operator of the boat, everyone can support safety and efficient operations by simply looking around when the watercraft is under way. During both operations and training, many crews encounter other boaters in various forms of distress. The most common maladies found on the water are breakdowns, boaters out of fuel, and sinking watercraft. Each agency should have SOPs that pertain to boats needing mechanical work or running out of fuel. Many navigable waterways will have the support of a local, state, or federal agency, such as a Harbor Patrol, Natural Resources department, or the USCG that can assist boaters, and larger waters often have private boat towing and salvage services available. All of these organizations usually monitor marine radios for alerts, although a phone call from

# SKILL DRILL 2-3
## Tying a Cleat Hitch NFPA 1006: 21.2.9, 21.2.10

**1** From the watercraft side, wrap the rope around the far end of the cleat, loop it back over the top, and wrap it around the close end, pointing away from the watercraft.

**2** The rope should be moving away from the watercraft, by laying back onto the cleat, forming a figure 8. This provides excellent friction.

**3** Form another complete figure 8 by repeating steps 1 and 2. At this point there should be enough friction to hold the watercraft on the cleat if wind or currents pull against the boat.

# SKILL DRILL 2-3 Continued
## Tying a Cleat Hitch NFPA 1006: 21.2.9, 21.2.10

**4** Use a half hitch to complete the tie-up, ensuring the rope does not slip or pull free.

**5** Loop the half hitch over one side of the cleat. The knot gets its holding power from the friction of the wraps, not the half hitch.

**6** Ensure the remaining line does not constitute a trip hazard.

Courtesy of Steve Treinish.

rescuers or the dispatcher may still be needed. Of course, any person or persons in danger should be removed from the boat, or the boat secured by anchoring or even towing to a safe place.

## Sinking Boats

In the event of a sinking boat on open water, the first priority is life safety. Any victims in the water should be provided flotation and warmth if needed, and the location of the boat quickly determined. Fuel is likely to leak through the sinking boat's fill hoses and vents, with this fuel then floating on the water. Absorbent pads may be quickly overwhelmed, so notifying the other agencies that can help clean up the spill is paramount. Watercraft or Natural Resources departments often handle or assist with these incidents, as can the USCG. Private marine salvage or dive operators can also be of assistance. If rescue crews must leave the scene, perhaps for patient evacuation, marking the location with marker buoys may help other boats avoid the area. Other boaters may also be enlisted to voluntarily wait for help and try to warn off other watercraft.

## Hazards to Navigation

Sunken boats, trees, and debris can all cause trouble for mariners, especially for those with large or less maneuverable watercraft. Just as with sinking boats, other agencies are often equipped to mitigate these hazards. Removal of the hazard may involve steps ranging from a barge and crane actually removing it for disposal, to blasting underwater to destroy it and letting watercraft pass over it. Most marine agencies have access to larger buoys, and can add marker buoys as needed to close the area or create a safe passage around it.

## Start the Motor First!

Before the watercraft pulls away from the dock, all systems must be operational. Batteries on larger boats are often controlled by a master switch, which lets the user pick from individual batteries or the entire group for starting the engine. Watercraft with helm-based controls (steering wheel and throttle controls) normally have a neutral safety switch in place, which prevents the motor from starting or cranking if the engine is in forward or reverse, similar to an automotive shifter. If the motor will not crank, check these two items.

If the motor cranks but will not start, the issue may be fuel flow. Ensure that the fuel valve is turned on, if applicable, and that the tank vent is open. Also, check the kill switch leash. These leashes are notorious for slipping loose or being inadvertently pulled out where the clip fastens to the switch. It is also a good idea to keep spare leashes on hand. A few civilian boaters have pulled their boats on a trailer into a marina complaining of a nonstarting motor, only to have the marine technician simply put the leash in place and send the skipper back out thoroughly embarrassed. These kinds of mistakes are committed by crews that do not practice enough, and each of them costs valuable time during an emergency response.

GPS units, depth finders, and any other ancillary equipment should also be turned on and operating prior to launch if this equipment is crucial to the mission.

# Piloting the Boat

Once the watercraft is under way, it is important to realize the major difference between boats and automobiles—namely, boats have no braking systems. Boaters who drastically underestimate how far a boat will coast once they let off the throttle seriously damage many docks every year. Some may think shifting to reverse and applying throttle will stop the boat. In reality, once a boat is on **plane**, or operating "on top" of the water, the gears should not be shifted at all, except to neutral, until the boat comes off plane or settles back down into the water (**FIGURE 2-56**). Immediately shifting the motor into reverse while on plane will most likely severely damage the gearing in the motor.

## No-Wake Speed

**No-wake speed** is a boat speed that is slow enough not to cause large waves at the rear of the transom (**FIGURE 2-57**). **Wake** is the larger waves caused by a boat moving through the water. No-wake speed is the appropriate speed at which to shift gears, if needed, or to idle into shore. Many waterways have a 300-foot buffer zone around the shore, or protected areas that are **no-wake zones** (i.e., areas where no-wake speed is required). In nonrescue operations, these limits should be respected.

**FIGURE 2-56** Once a boat is on plane, the gears should not be shifted at all, except to neutral, until the boat settles back down into the water.
Courtesy of Steve Treinish.

**FIGURE 2-57** No-wake speed is slow enough not to cause large waves at the rear of the transom.
Courtesy of Steve Treinish.

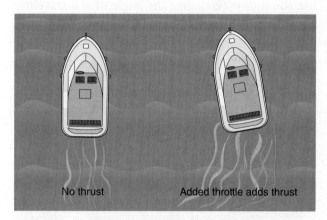

No thrust          Added throttle adds thrust

**FIGURE 2-58** Boats turn very little with no applied power. Thrust from the motor will swing the transom in the desired direction when the motor is in gear.
© Jones & Bartlett Learning.

## Steering

Most boats will not steer well in neutral. Thus, to steer effectively, especially at low speeds, thrust must be used, which means the motor must be in gear. A boat is mostly turned by directing the thrust from either the prop or the jet unit, or by a rudder, in the case of an inboard motor. The turning force a motor provides to turn a boat is negligible unless the prop is turning (**FIGURE 2-58**).

Many newer boaters who find themselves on a collision course will cut the throttle and turn the wheel or motor full lock in one direction or the other, but still go straight ahead into the collision. This is especially true with operators of PWCs, which have absolutely no rudder or lower unit gears to provide any sort of steering when no power is applied. On many PWCs, an operator can chop the throttle and turn the handlebars all the

way to one side, but the craft will continue straight as an arrow. Driving or turning a larger, rigid-hull boat, by comparison, is much easier once under way.

## Tilt and Trim

Most outboards larger than approximately 50 HP and all inboard/outboard motors have tilt and trim capabilities. Tilt simply picks up the lower unit of the motor from vertical; it is used for shallow water operation, trailer operations, and raising the motor for repair or routine maintenance. By comparison, trim lets the operator adjust the angle of the boat on the water as it is running at speed.

Most tilt and trim controls are located on the throttle control, which the operator's thumb controls. Some are stand-alone units mounted on a dash or sidewall around the operator's seat. The control is usually a single switch, with the up position raising the motor up, and down lowering it. The middle is a stop position, and the switch will rocker into that position if the switch is released. A typical unit will slowly raise the motor through the trim ranges, and then suddenly speed up as the tilt range is reached. Dropping the motor has just the opposite effect: It is lowered fairly rapidly, but then slows down as it cycles down into the trim range.

After a boat accelerates and gets up on plane, the angle of the motor helps control how much of the boat is actually riding on the water surface. If the bow is angled too far down, much of the boat is needlessly in the water, resulting in drag, slowing the boat, and making for a rougher ride. At the other end of the scale, if the bow is trying to rise too far up, the boat cannot beat gravity, so the bow will drop again. This gives the boat a very porpoise-like ride, which also affects speed and smoothness. Using the trim unit to use the motor to change the angle of the traveling boat can greatly improve the ride, speed, and handling of any boat (**FIGURE 2-59**). A trim gauge is usually available to help the operator trim the boat properly. When operating in shallow water, the trim and tilt controls can be used to raise the propeller or lower unit up to provide a shallow water drive, but you should expect a mushier feel steering, since the thrust angle is changed.

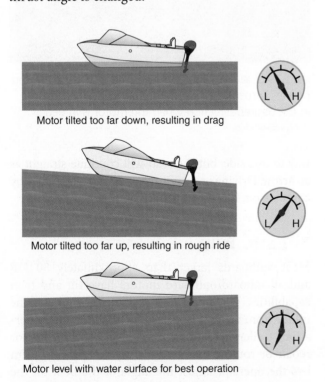

Motor tilted too far down, resulting in drag

Motor tilted too far up, resulting in rough ride

Motor level with water surface for best operation

**FIGURE 2-59** Motor angle plays a large part in the ride and performance of a boat. Most boats include an indicator gauge to assist the operator in judging trim.
© Jones & Bartlett Learning.

## Navigation

While local waterways are commonly known well by community members, some missions will require navigation to go to, return from, or search locations away from the launch point. Modern electronics, including GPS units, have made marine (as well as aerial and overland) navigation as easy as keeping an arrow on a line shown on a display screen. Weather radar, forward-looking radar, and map cartography and landmarks can also be downloaded to a multiple-use display, putting all the information onto one screen to show the "big picture."

### Electronic Navigation

GPS units are the mostly widely used navigation systems. In fact, most people probably have one on their smartphone. With GPS, satellites in Earth's orbit constantly triangulate the user's position and show it on a screen (**FIGURE 2-60**). Navigation information— such as speed, distance to or from a point, headings, and estimated time of arrival—can also be shown on the GPS unit, depending on its size and features.

For rural searches using small watercraft, a small, hand-held unit may suffice, especially if the user can put in waypoints to assist navigation. Waypoints are simply markers that the GPS unit tracks as points on a map. Waypoints can also be marked, and the longitude/latitude coordinates navigated to, by other GPS users. For example, this kind of navigation might be employed by rescuers using a boat to search for and locate a victim, but who then need to call for other resources such as a helicopter or land search unit to meet them for extrication. The GPS coordinates can be broadcast or sent to the responding units, and their GPS units will show them the way to the specified location.

GPS units can also mark waypoints of obstructions or features that might harm other watercraft or

**FIGURE 2-60** GPS unit.
Photo by U.S. Coast Guard Petty Officer 3rd Class Stephen Lehmann.

responders. Seeing the hazard on a moving map will allow the user to stay clear of the hazard.

Marine GPS units are frequently used in conjunction with depth and contour sonar units, thereby identifying the depth below the boat. Side-scan sonar units can actually sketch out features under or to the sides of a boat, rendering an incredibly detailed picture of the bottom. Dive teams may use GPS units to ensure that they travel in a grid layout, taking bottom pictures of the water, while searching for a target. Once an object is noted that needs to be checked by other means, such as diving or underwater vehicles, the searchers can navigate to its location within a couple feet of distance.

When map cartography is uploaded to the GPS unit, a moving map (also called a chart plotter) can be displayed with navigation features, depths, hazards, and even other boats shown on the display. Go-to navigation just requires inputting the coordinates desired, and the GPS unit then points the way. Putting this track on a moving map lets the operator literally follow the arrow to the destination.

## Compasses

A compass requires no electricity, downloads, or updates to operate. It simply does its job at all times. In this ancient, yet still practical, device, a magnet system is used to point a free-turning dial at magnetic north at all times, giving the user his or her direction of travel. Around the compass are 360 marks, or degrees: North is 0 degrees, east is 90 degrees, south is 180 degrees, and west is 270 degrees. A lubber line is marked on the compass, usually pointing straight ahead, and cannot be adjusted or moved.

**LISTEN UP!**

Well-versed navigators will immediately recognize that there is also a "magnetic north" and magnetic variation factors that play into precise compass navigation. While this text acknowledges the existence of these, we prefer a simpler approach to describe basic compass work.

**Course.** The watercraft's course is the direction it must travel, relative to the compass reading. For example, a skipper wanting to travel directly north will turn the boat until the "N" or "0" aligns on the lubber line. If there is no current, wind, or wave action to move the watercraft laterally, the boat will then travel directly north.

**Heading.** Of course, there are very few times when the direction of travel on watercraft is *not* affected by the current, wind, or waves. The heading is the direction in which the boat must point to counter these forces and continue on the desired course. The stronger the force pushing across the watercraft, the greater the difference between the heading and course numbers.

## Natural Navigation

Of course, electronics can always fail, and not all boats are equipped with GPS. Natural navigation may be used in these cases—that is, simply using landmarks, eyeballs, and notation to keep a rough track of travel. Cell towers, dead trees, buildings, and other land features can all be used as markers. Travel can be timed to give a very rough estimate of distance covered, but remember that waves and wind conditions can affect this timing. Landmarks can be jotted down on a pad and used as waypoints.

**LISTEN UP!**

Any electronic device can fail, including GPS and chart plotters. Mariners should ensure paper charts are also available for use, particularly when they are operating on larger or unfamiliar waters. Many lakes have been mapped in terms of their depths and landmarks, and these very inexpensive maps are readily available at outdoor supply stores.

## Night Operations

All watercraft operating at night *must* show some lighting, even at anchor. Specific regulations address watercraft lighting, especially in ocean or offshore areas, but in this text, we will consider only the requirements in inland waters. The definition of *night* varies from state to state, but is generally thought of as 30 minutes before sunset until 30 minutes after sunrise. Check your local boating laws for the specific definition in your area, but in this text, we will stick with the more general terms "day" and "night."

At night, all watercraft not anchored must show **navigation lights**. Navigation lights comprise three colors and are specifically located on the boat. As a general rule, with watercraft less than 39 meters long, one white light must be visible for 360 degrees around the boat. On the port side of the watercraft, a red light must be visible from the bow to an angle of 22½ degrees behind a point 90 degrees off the port side. The starboard side gets a green light in the same fashion. This system allows other mariners to figure out which side of a boat they are looking at, and which way the boat is headed. Less-experienced mariners may find it easier to think of the bow of the boat being the 12 on a clock. **FIGURE 2-61** shows this lighting range.

## Buoys and Day Boards

A generally consistent system has been established for channel navigation. Quite often, channel or passage markers comprise either unlit or lighted red and green buoys or day boards. Day boards are simply numbered signs on posts. Watercraft should pass between them, keeping the red on the starboard side when traveling inland—think "Red on the right when returning." Inland waters often use smaller, white-and-orange unlit buoys to signal danger, no-wake zones, special zones, and other messages (**FIGURE 2-62**). Public

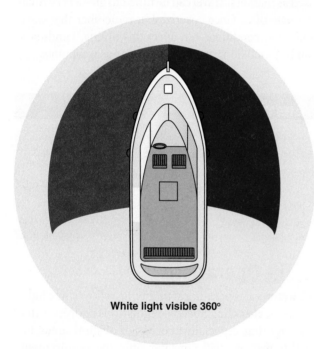

**White light visible 360°**

**FIGURE 2-61** A common lighting range found on vessels less than 39 meters long.

© Jones & Bartlett Learning.

**FIGURE 2-62** Navigational buoy.

Courtesy of Steve Treinish.

safety personnel using waterways with these types of navigation methods should be well versed in their use in the waters in which they will be operating.

## Anchoring

Most rescuers would understand that an anchor would be deployed if they want the boat to remain stationary, but there is a bit more to use of anchors than that simple directive. The type of anchor plays a large part in its effects, but the amount of anchor line played out is the biggest factor in successfully stopping a boat from drifting.

## Types of Anchors

Two types of anchors are commonly used: **mushroom anchors** and **fluke anchors** (**FIGURE 2-63**). Mushroom anchors are primarily found on small, lighter-weight boats where a simple weight will stop the boat from moving. They look like a traffic cone and can weigh up to 20 pounds. By comparison, fluke anchors look like the anchors that hang off the bow of most ships, but they are available in much smaller sizes to hold much smaller boats. They dig into the bottom surface when pulled, and can actually hold incredible amounts of weight relative to their size and weight when properly deployed. A **grappling hook** can be used as an emergency anchor for an out-of-control boat, or when all else fails, but it relies on grabbing something to hold, and will pull along a smooth bottom uselessly.

**Anchor Scope.** **Anchor scope** is the term used to describe the amount of anchor line deployed into the water relative to the water depth (**FIGURE 2-64**).

**FIGURE 2-63** Mushroom anchors look like traffic cones, and fluke anchors have the stereotypical "anchor" shape. Fluke anchors are meant to dig into the bottom to supply holding power, while mushroom anchors are actual dead weight.

Courtesy of Steve Treinish.

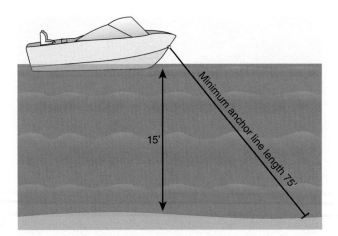

**FIGURE 2-64** Anchor scope is the amount of anchor line played out relative to the depth of the water. A minimum of 5:1 is preferred, but in rough or windy conditions, this may increase to 7:1 or more.
© Jones & Bartlett Learning.

A minimum of 5:1 is normally used, but the scope can be increased to increase holding power in rough or windy conditions, or when the boat will not hold position. An anchor scope of 7:1 or even 10:1 is recommended for heavier surf or wind conditions, and even more anchor line can be added if necessary.

**Three-Point Anchoring.** Another anchoring option used in watercraft-based dive operations is **three-point anchoring**, also known as **hurricane anchoring** (**FIGURE 2-65**). It entails using three anchors, each played out with the proper amount of scope, to keep the boat stationary. Wind or current tends to move a boat around a single anchor in sweeping arcs, and divers' swimming patterns will pull the boat around the same way, thereby jeopardizing the search pattern integrity. Three-point anchoring provides a very stable, nondrifting platform that preserves the search pattern.

# Wind and Current Effects

Wind, currents, and tides can all affect boat operations. This is especially true at lower speeds, such as when docking or in a no-wake zone. Many boats present a fairly large surface for water and winds to push on. Backing away from a dock, pulling onto a trailer, or trying to maneuver at a fixed object may all require a team effort, and require multiple dock lines, fenders, or bumpers, or even calling for assistance before reaching the dock. Coordination with the operator, dock crews, and boat crews is essential. Have a plan before arriving at the dock or shore.

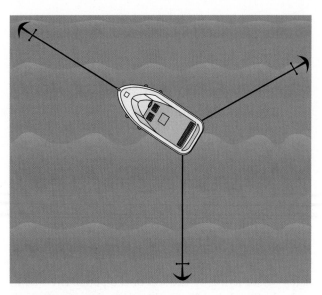

**FIGURE 2-65** Three-point anchoring results in a very stable platform that wind, waves, or divers cannot move easily. Three anchors must be used if this system is to be effective.
© Jones & Bartlett Learning.

# Deploying and Retrieving Rescuers

Boat operations with rescuers deploying from them usually take the form of tending and deploying divers or rescue swimmers. A few key considerations must be taken into account when operating around rescuers. First and foremost, you must be concerned about a propeller striking an in-water rescuer, even if you are not operating the boat. The use of a spotter in the bow area is highly recommended, and this spotter *must* have verbal or signed communication with the operator.

## SAFETY TIP

Newer operators often underestimate how far a boat can travel when slowing down. When retrieving rescuers, operators are cautioned against overshooting the swimmer, then shifting to reverse in an effort to stop the watercraft's progress. After overshooting the rescuer in the water, the spinning propeller might get pointed toward the rescuer, and will then travel to the rescuer as the boat backs up. Circling around under control for a slower, more controlled attempt may be a better method. Sight and verbal contact should be maintained with the rescuer by a crewmember not piloting the boat.

A team might be leaning toward not allowing any motorized operations with rescuers in the water, but if the proper procedures are followed, such an operation can be done safely (**FIGURE 2-66**). For example, a dive operation offshore may not bring divers up every time a boat approaches, because the boats and the dive crew should know before the dive begins what, where, and how things will happen. In a well-oiled and -rehearsed team, the divers will know the boats are approaching, and from what direction; the tenders will know where the divers are at all times; and the safety officer or dive supervisor will have direct contact with the boat crew. The team will have previously decided where the arriving boat will dock with the boat being used as the dive platform, and will know which direction and procedures will be used when the boat leaves. Once again, planning and executing a preplanned procedure are far more effective—and safer—than flying by the seat of the pants. Even hand or body signals can be used in the event of radio failure, but all teams that may be involved must agree on them beforehand.

## Crew Overboard

Crew overboard is the generic term that NFPA uses for a situation in which one or more persons fall into the water from a watercraft. Although every skipper plans to keep all mariners safe and sound on his or her vessel, accidents do happen. The best way to ensure safety is simply to watch out for everyone else while on board. Work can be done in pairs on larger boats, and no person should ever be without flotation. **Ring buoys** or other throwable flotation can be kept along rails or decks.

> **SAFETY TIP**
>
> On larger boats, ring buoys are often placed for emergency use without lines attached. This arrangement is intended to provide flotation to the person who has fallen in, but removes the problem of an attached rope either pulling the flotation away or becoming fouled in a propeller. Larger vessels cannot simply stop, and some are tall enough that a line on a ring buoy would either pull the flotation from the crewmember or pull the crewmember through the water.

## Crewmember in the Water

If you find yourself in the water unexpectedly, you can take a few actions to improve your situation. First, make yourself known. Shouting or waving your arms will help alert and orient crewmates still on board to your location. Having a set of hand signals known to the entire crew can greatly reduce stress on both sides of this issue, as the COB can signal that he or she is

**FIGURE 2-66** With proper procedures in place, motorized operations may be conducted with rescuers in the water.
Courtesy of Steve Treinish.

"OK" once sight contact is made. This signal allows the crew left on the boat to relax somewhat, as they now know the rescuer in the water is safe for the time being. The COB in the water can, in turn, breathe much easier if he or she sees the OK signal given back, as the COB now knows that the crew has spotted him or her and that help will arrive soon.

Ideally, the COB will be wearing flotation. Marker strobes that automatically flash when water is encountered can be added to PFDs and will greatly help in locating a COB at night. Attaching and activating a chemical light stick to helmets or PFDs for any nighttime operation may also create a good visual target at night. Keep in mind that the crew needs to have a line of sight to this beacon, so, if you are the COB, make an effort to help them see it. In waves or surf, visibility may suffer as the COB may drop out of sight in the trough of the wave. Any movement or color variation against the water is helpful to others watching for a signal, even at night. Waving arms or even a brightly colored helmet, especially at the top of a wave, can help spotters locate a COB. PFD and helmet colors of orange and yellow seem to stand out best against water surfaces.

> **SAFETY TIP**
>
> While it might seem a good idea to pull off a light or strobe and wave it at the boat, beware of the perils of doing so. If the light or strobe is dropped into the current, tide, or waves, it may lead potential rescuers to the light itself, rather than to the person to whom it should be attached. And no matter what, do not remove the PFD to wave it.

Floating in cold water can cause hypothermia quickly. As the COB, relaxing and not thrashing in the water will let your clothing trap layers and pockets of ambient water, and body heat can warm this water. Excessive movement will flush out this trapped water and allow it to be replaced with colder water, thereby robbing the COB of more body heat. Pulling the knees up toward the chest and hugging them with the arms can help conserve body heat. In the event more than one rescuer falls in, huddling as close as possible will help to retain body heat for all persons in the water.

### KNOWLEDGE CHECK

Which action is acceptable when a boat is operating on plane?
a. Shifting to any gear
b. Shifting to neutral
c. Shifting to reverse
d. None of the above

*Access your Navigate eBook for more Knowledge Check questions and answers.*

## Crewmembers Left on Board

If a COB event is witnessed, the witness's first action should be to alert the crew remaining on the boat. Ideally, one set of eyes should be kept on the COB at all times. The person who is actually seeing the COB can verbally guide the watercraft back to the area, even if these directions must be relayed to the helmsman. In choppy water, once the eyes move to another area, it is often difficult to relocate the original object. If the shore or other landmarks are seen, triangulating the COB with that landmark can help provide an approximate location if needed. Placing an immediate mark or waypoint on a GPS unit is also advisable if one is in use. Be on the lookout for movements such as waving, and color variations such as bright orange or yellow against the water.

### SAFETY TIP

An immediate accountability report or roll call must be done to confirm the number of COBs. This number must be relayed to the crew and to command as quickly as possible, but must be accurate.

Ensure all areas are searched, especially for an unwitnessed COB event. Binoculars can be used if needed, and a good, slow examination of the water areas will ensure a complete and thorough scan of the water takes place. Tides, currents, and wave action will also move a person floating in the water, and these

factors should be considered. Air resources such as search planes or helicopters should be considered and deployed earlier, rather than later. Once a potential object is found, it must be identified completely before deemed something else. Also, signaling to the COB that he or she has been sighted is important, whether through hand signals, flares, or verbally. It is important to control and know which watercraft are in an area searching, so that if one identifies the COB, all will get and understand the message and the location. Once again, common terminology, proper training, and preplanning can go a long way toward making a COB event just an event, and not a tragedy.

### SAFETY TIP

Be aware of other vessels coming into an area at high speed if the COB event is broadcast over a radio. Once the COB is spotted and confirmed, watercraft can be used to effect a rescue, but until then, slow speeds and visual scanning are essential.

## Maintaining Watercraft and Equipment After Their Use

Proper treatment of watercraft and the related accessories will go a long way toward making the next use go smoothly. Any issues or breakages should be noted and reported through the proper channels. Just as when placing fire or EMS equipment back in service, care should be used, as the next life it saves might be yours.

## Cleaning

Watercraft should be cleaned and allowed to dry, with special attention paid to areas subject to salt-water spray, as it is corrosive to metals, and to flushing the lower unit with fresh (normally domestic) water to prevent infestation of other waterways with plants and animals. Various mussels and other organisms, including invasive species that upset the balance of native fish and plants, may be present on the trailer bunks, PFDs, PPE, and anything else that may absorb water. They can also be sucked into the cooling systems and lower unit, and then transferred to other waters when the boat is launched. Washing all equipment exposed to the water with hot water and soap—and ensuring the entire vessel, trailer, and contents are thoroughly dry—can help prevent this problem. Ensure all water is removed from the bilge.

Proper drying will reduce mold and mildew, which can thrive under canvas if items are stowed wet. Most of the time air drying is fine, provided it is of long enough duration to allow full drying. Avoid drying anything in the sun if possible, to prevent damage from the sun's ultraviolet light.

**LISTEN UP!**

During cool seasons, placing small boats under bay heaters often dries them very quickly and effectively. The use of fans can also speed up the drying process.

## Trailer

One item frequently neglected is the trailer (if the boat sits on such an apparatus). The wheel bearings of a boat trailer take a lot of abuse from being dunked underwater time and time again. A spring-loaded bearing packer holds bearing grease in a reservoir right on the bearing itself, which can help keep grease throughout the bearing and can displace any water that seeps in. A rusted bearing is a failure waiting to happen.

The trailer's electrical connections should be checked for debris or mud, and tail and marker lights checked for proper operation. A light spray-coat of corrosion block will help prevent rust and corrosion on any metal parts such as the winch, rollers, or hitch locks. The application of this agent is also a good time to check safety chains or cables for the next call.

## Ancillary Equipment

A few systems may be enabled on watercraft for storage and preparation for the next use. Bilge pumps are often kept in an automatic state, to protect against leakage or rainwater. Batteries can be kept at full charge with the use of maintainers or chargers, but must be supplied with AC voltage, usually in the form of a shoreline (**FIGURE 2-67**). Some solar chargers are now on the market that will keep batteries charged and are environmentally friendly. Covers for the watercraft can range from a small cover for the helm to an entire mooring cover. Any cover used should be fitted tight enough to stay put during wind or rain, but allow good airflow underneath.

**FIGURE 2-67** Small battery tender, powered by AC voltage.
Courtesy of Steve Treinish.

## In-Water Storage

When AHJs leave watercraft in a dock space or tied up in the water, lines, knots, and cleats should be double-checked for security, and any theft-deterrent chains or locks placed in service. Bumpers may be needed to protect the boat from movement against docks or walls from waves or wind—remember, this movement may come from any direction. Multiple lines and fenders may be required to prevent damage in the event of a wind or water directional shift.

## On-Trailer Storage

AHJs that store watercraft on a trailer also have to ensure covers are in place and secure if used. A boat on an unlocked trailer can be fairly easy to steal, if the boat is not secured indoors or in a secure lot. A trailer coupler lock might prevent such problems. The trailer tongue safety latch and pin should be ready to rehook to the tow vehicle if needed, and lights should be checked for proper operation. Just as with fire apparatus, any breakage or issues should be noted and repaired, and the vessel should be made ready for another call.

# *After-Action* REVIEW

## IN SUMMARY

- While agencies or departments may not own or operate boats of any kind, in many situations some type of watercraft will be needed to help rescuers properly and safely operate at a water rescue incident.
- The use of a properly fitting and mission-specific personal flotation device (PFD) is the most effective way to ensure rescuer and victim safety.
- Rescue boats range from small kayaks to small ships powered by diesel-driven screws, but common terminology is typically used for all watercraft.
- The boats most often used in rescue work in smaller inland waterways are inflatable boats, metal-hull johnboats, and tailored watercraft powered by an outboard or jet-drive motor.
- On larger bodies of water, such as harbors, larger lakes, or the open ocean, rescuers commonly use rigid-hull runabouts, rigid-hull inflatable boats, or even larger oceangoing ships.
- Most watercraft have one of two types of propulsion: muscle or machine.
- Not every mission will require every tool, and missions may require the same tools for different reasons.
- Certain problems commonly happen on or around the water with newer boaters, but these risks can usually be minimized with training and familiarization.
- One major difference between boats and automobiles is that boats have no braking systems.
- Wind, currents, and tides can affect boat operations, especially at lower speeds, such as when docking or in a no-wake zone.
- When operating watercraft around rescuers, the first and foremost concern is a propeller striking an in-water rescuer.
- Proper treatment of watercraft and the related accessories will go a long way toward making the next use go smoothly.

## KEY TERMS

*Access Navigate for flashcards to test your key term knowledge.*

**Abeam** A point directly perpendicular to the port or starboard side of a boat.

**Aerated water** Water mixed with a great deal of air, resulting in bubbles or foam. This water does not provide very good thrust from the propeller.

**Airboats** Rigid-hull watercraft driven by a large fan blade powered by a car engine. There is no propeller system to foul, and these boats can cross water, ice, swampland, and ground.

**Anchor** A device designed to engage the bottom of a waterway and, through its resistance to drag, maintain a vessel within a given radius. (NFPA 1925)

**Anchor scope** The amount of line deployed relative to the depth of the water while using an anchor. The stronger the pull of the boat, the more scope is used.

**Banana boat** An inflatable rescue craft specific to rescue work. It is very stable and resembles a large banana, with rescuers sitting inside the craft. Also called a *snout rig*.

**Beam** The breadth (i.e., width) of a ship at its widest point. (NFPA 1405)

**Booties** See *wet shoes*.

**Bow** The front end of a boat or vessel. (NFPA 1005)

**Canoe** A small watercraft made of metal or fiberglass; it has a very small beam compared to its length, resulting in poor stability. Canoes are usually human powered, but can accept very small motors if equipped properly.

**Draft** (1) The vertical distance between the water surface and the lowest point of a vessel. (2) The depth of water a vessel needs in order to float. (NFPA 1005)

**Drain plug** A small rubber or threaded plug that installs into a port or hole to drain a boat of unwanted water.

**Drysuit** Clothing designed to operate on or around wet conditions, keeping the wearer dry. It may include rubber seals in the neck and wrist area and may require thermal insulation.

**Flip lines** Shorter ropes or straps used to right a capsized boat from in the water.

**Fluke anchors** Traditional-looking anchors consisting of two spears on either side of a metal bar. The spears will dig into the bottom mud or snag an object, anchoring a boat.

**Fuel bulb** A small, squeezable bulb in the fuel line used to prime a motor. The bulb is squeezed to provide priming fuel for start-up of the motor; it contains a one-way valve to prevent backflow of the line, making restarting easier.

**Grappling hook** A metal object with curved ends that is used to snag objects; it may be used as an anchor or victim retrieval tool in water rescue.

**Gunwale** The upper edge of a side of a vessel or boat designed to prevent items from being washed overboard. (NFPA 1405)

**Helmsman** A rescuer who works at the operations level and steers the boat under the guidance of another individual, usually the person in charge of the vessel.

**Hovercraft** Watercraft carried on a cushion of air blown from a motor-driven fan and propelled by a fan in the stern; similar to an airboat, but without friction on the hull.

**Hull** The main structural frame or body of a vessel below the weather deck. (NFPA 1005)

**Hurricane anchoring** See *three-point anchoring*.

**Inboard motor** A motor in which both the motor and the propeller are mounted and plumbed through the hull.

**Inboard/outboard motor** A hybrid motor in which the engine is mounted in the boat in front of the transom and the propeller is located behind the transom, similar to an outboard motor.

**Inflatable boats** Any boat that achieves and maintains its intended shape and buoyancy through the medium of inflation. (NFPA 1925)

**Inflatable rafts** Larger watercraft commonly used for whitewater rafting, which rely on air for flotation.

**Jet drive** A propulsion unit that generates thrust in reaction to a water stream. (NFPA 1925)

**Johnboats** Field name for smaller (usually less than 16 feet) metal boats with a square bow shape.

**Kayaks** Smaller canoe-type watercraft, which usually hold one person. The operator can sit in or on the kayak; most are human powered.

**Keel** The principal structural member of a ship, running fore and aft on the centerline, extending from bow to stern, forming the backbone of the vessel to which the frames are attached. (NFPA 1405)

**Kill switch** A safety measure in which a small clip is attached to the motor operator, with a leash connecting the clip to a switch on the motor that will stop the motor from running if the operator is thrown from the operating position.

**Length** The total length of the boat from the bow to the transom.

**Low-head dam** A low-height dam, which has no effective means of regulating water flow. Water frequently flows over a low-head dam and creates a hydraulic, which will recirculate objects or people in them for long periods of time.

**Lower unit** The section of an outboard motor enclosing the gears, drive shaft, and propeller.

**Marker buoys** Small, can-shaped floating objects used to mark locations of objects underwater or to provide visual reference to an area needing to be marked; also called *pelican buoys*.

**Mushroom anchors** Boat anchors resembling a mushroom shape, which are mostly used in smaller watercraft.

**Navigation lights** Lighting that must be displayed while a watercraft is on the water during nighttime operations or anchoring.

**No-wake speed** A boat speed slow enough as to not make large waves behind the craft; generally, an idle speed.

**No-wake zones** Areas where speeds faster than an idle, or causing waves behind a boat, are prohibited.

**Oarlocks** A pin-and-hole system used to hold oars at the side of a boat; it gives the user leverage to operate the oars.

**Oars** Long, paddle-like planks, which can also be used as a lever to provide propulsion; typically used two at a time, with oarlocks.

**Outboard motor** A motor for which the engine, gears, and prop are all mounted on the transom.

**Paddles** Stick-like instruments used to propel watercraft by hand; they are held by the user, with a wide blade used in the water.

**Parbuckling** A technique for moving a load utilizing a simple 2:1 mechanical advantage system in which the load is placed inside a bight formed in a length of rope, webbing, tarpaulin, blanket, netting, and so forth that creates the mechanical advantage, rather than being attached to the outside of the bight with ancillary rope rescue hardware. (NFPA 1006)

**Personal watercraft (PWC)** A vessel less than 13 ft (4 m) in length that uses an internal combustion engine powering a water jet pump as its primary source of propulsion and is designed to be operated by a person or persons sitting, standing, or kneeling on rather than within the confines of the hull. (NFPA 302)

**Plane** A boat moving at a faster speed, such that it operates "on top" of the water; marked by making waves behind the boat during operation.

**Pontoon boats** Larger watercraft with a deck area floated on two or more sealed cylinders underneath. They do not handle current well, but provide great dive or staging platforms.

**Port** The left side of the boat, looking forward to the bow.

**Prop** The propeller of a motor.

**Prop guard** An attachment to the lower unit that encircles the propeller, encasing it somewhat to reduce prop strikes.

**Rigid-hull boats** Watercraft made mostly of a solid, rigid material, usually fiberglass or metal, that achieve buoyancy from water displacement.

**Rigid-hull inflatables (RHIs)** Solid-shaped hull mated with a flexible multicompartment buoyancy tube(s) at the gunwale. (NFPA 1925)

**Ring buoys** Circle-shaped flotation devices that may be thrown to, and provide buoyancy for, victims in the water.

**River shoes** See *wet shoes*.

**Sail area** The area of the ship that is above the waterline and that is subject to the effects of wind, particularly a crosswind on the broad side of a ship. (NFPA 1405)

**Scuppers** An opening in the side of a vessel through which rain, sea, or fire-fighting water is discharged. (NFPA 1405)

**Sit-in kayaks** Small, one-person, paddle-powered watercraft that the user sits in, with the shell of the craft covering his or her legs and lap.

**Sit-on-top kayaks** Paddle-powered watercraft that the user sits on, instead of in. They can be large enough to hold three boaters, all on top of the boat.

**Starboard** The right-hand side of a ship as one faces forward. (NFPA 1405)

**Stern** The rear section of a boat or vessel.

**Strainer** An object that lets water through it, but not other objects.

**Tether switch** See *kill switch*.

**Three-point anchoring** Using three anchors off a watercraft to provide a solid, nondrifting platform in the water for operations.

**Throw bag** A water rescue system that includes 50 ft to 75 ft (15.24 m to 22.86 m) of water rescue rope, an appropriately sized bag, and a closed-cell foam float. (NFPA 1006)

**Tiller steer** Smaller outboard motors (less than 50 HP) that have throttle and steering controls on a short handle attached to the motor itself.

**Tilt and trim unit** A small hydraulic unit used to change the angle of the boat motor to provide better performance, shallow water clearance, and maintenance angles.

**Topsides** The sides of a boat between the water surface and the deck.

**Transom** The rear vertical surface of a boat.

**Wake** Waves formed at the stern of a watercraft, caused by the boat moving through the water at a speed higher than idle.

**Water muffs** A small suction cup tool that supplies water to the lower unit of an outboard motor.

**Watercraft** Manned vessels that are propelled across the surface of a body of water by means of oars, paddles, water jets, propellers, towlines, or air cushions and are used to transport personnel and equipment while keeping their occupants out of the water. (NFPA 1006)

**Waterline** The area where the surface of the water meets the gunwale.

**Wet gloves** A hand covering offering thermal protection; usually made of neoprene, with varying thicknesses.

**Wet shoes** Neoprene shoes that provide traction and thermal protection to the wearer; also called *booties* or *river shoes*.

**Wetsuit** A body-hugging suit that provides thermal protection to the wearer; it ranges in thickness depending on thermal protection desired.

# On Scene

Your water rescue team has been called to assist with water rescues owing to a major flood in a small city. This city is a little more than 75 miles away, and while you know the general location, you are not very familiar with the area. Rainfall has caused the failure of a levee, and water is rising through the town. You arrive at the command post, are given a basic map of the area, and are asked to search for victims in a 6-square-block area. Residents were given instructions yesterday to evacuate, but officials believe that many of them ignored this order.

You are given a radio that will patch into the frequency the operations commander is using, but your team must supply all other equipment. Your team has traveled in with a 16-foot johnboat equipped with a 20-HP outboard motor, a 12-foot inflatable equipped with a 40-HP motor, and a 14-foot johnboat that is rated for 20 HP, but has only oars. Other equipment includes several hand-held GPS units, basic life support (BLS) equipment, numerous throw bags, 12 extra PFDs, and other typical peripheral rescue equipment. The weather forecast is for more rain.

**1.** Which boat would you consider the best type for this mission?

   **A.** A runabout trailered in with a four-wheel drive truck

   **B.** An inflatable boat with a 40-HP outboard motor and on-board inflation system

   **C.** A 16-foot johnboat with a 20-HP outboard motor

   **D.** A twin PWC, which is capable of carrying two rescuers

**2.** What is the best load-out for the reconnaissance mission?

   **A.** Flotation devices for any expected victims, drinking water, BLS gear, an anchor, and oars

   **B.** A backboard, cervical spine gear, and an ALS kit

   **C.** A hand-held GPS, the basic map, structural marking equipment, drinking water, extra flotation devices for victims, a cell phone, a BLS kit, and an extra radio if available

   **D.** A marine radio, extra PFDs, extra fire extinguishers, and extra throw bags

**3.** What is the minimum safety equipment to be taken?

   **A.** Oars, two radios, an anchor and rope, extra throw bags, and signaling equipment

   **B.** Extra gas, a distress flag, search rope, and a cell phone

   **C.** A BLS kit, drinking water, the map, and a GPS unit

   **D.** Oars or paddles, two radios, a cell phone, signal equipment, searchlights, anchor and rope, and drinking water

**4.** What would be a good way to navigate in this unknown area?

   **A.** GPS tracking and mapping, backed up by logging known landmarks and travel times

   **B.** Sending signal flares at pertinent places along the route

   **C.** Radioing landmarks to the incident commander or person in charge

   **D.** Preplanning a route through the city based on street maps available at the command post

*Access Navigate to find answers to this On Scene, along with other resources such as an audiobook and TestPrep.*

# Watercraft Rescue Technician

## KNOWLEDGE OBJECTIVES

After studying this chapter, you should be able to:

- Describe how to prepare a watercraft for launch and operation. (**NFPA 1006: 21.3.1**, p. 82)
- Describe how to launch and dock a watercraft. (**NFPA 1006: 21.3.5**, pp. 82–84)
- Define typical tasks performed to safely operate watercraft. (**NFPA 1006: 21.3.2, 21.3.4, 21.3.6**, pp. 84–87)
- Describe the basic operation of various types of navigation aids found on watercraft. (**NFPA 1006: 21.3.3, 21.3.11**, pp. 85–87)
- Describe the operation of watercraft while anchoring and recovering a watercraft. (**NFPA 1006: 21.3.4, 21.3.5, 21.3.6**, pp. 86–87, 90–91)
- Discuss the operation of watercraft while towing other watercraft. (**NFPA 1006: 21.3.4, 21.3.10**, pp. 90–91)
- Explain the effects of wind and current on rescue operations. (**NFPA 1006: 21.3.2, 21.3.4**, pp. 91–96)

- Explain how to safely deploy technician-level rescuers into the water environment from a watercraft. (**NFPA 1006: 21.3.8**, p. 96)
- Explain how to safely operate watercraft to recover technician-level rescuers, crew overboard, or incapacitated victims. (**NFPA 1006: 21.3.7, 21.3.8, 21.3.9**, pp. 96–97)
- Explain how to shut down and secure watercraft after operations. (**NFPA 1006: 21.3.12**, pp. 97–98)

## SKILLS OBJECTIVES

After studying this chapter, you should be able to:

- Safely deploy a watercraft from dockage, moorage, or a trailer. (**NFPA 1006: 21.3.2, 21.3.5, 21.3.6**, pp. 82–84)
- Pilot watercraft. (**NFPA 1006: 21.3.2, 21.3.4, 21.3.6**, pp. 84–87)
- Use navigation aids while operating a watercraft. (**NFPA 1006: 21.3.3, 21.3.11**, pp. 85–87)
- Deploy and retrieve an anchor used to hold watercraft stationary. (**NFPA 1006: 21.3.6**, pp. 86–87, 89)
- Operate watercraft while towing other watercraft. (**NFPA 1006: 21.3.4, 21.3.10**, pp. 90–91)
- Approach victims, shores, or other areas or watercraft when faced with wind or water action. (**NFPA 1006: 21.3.4**, p. 95)
- Deploy and retrieve rescuers or victims in the water. (**NFPA 1006: 21.3.7, 21.3.8, 21.3.9**, pp. 96–97)

# Technical Rescue Incident

The response to the boat versus personal watercraft (PWC) crash on the busy lake came just as dusk was fading to full-blown dark. It was really dark—a heavy layer of clouds blocked any moonlight or starlight from illuminating any part of the lake. To make matters worse, a cold front had worked its way through the area, causing fog to thicken in the area. You had to respond with your department's rescue boat approximately 4 miles from the launch area, around several points in the water, and back toward the end of a smaller bay. You are proud of the response, and proud of your crew. The crew has packaged the victim very well, and, though she is not suffering from life-threatening injuries, your crew has supplied top-notch care. You are now using the vessel's dash-mounted global positioning system (GPS) unit to track back to the dock to meet emergency medical services (EMS) personnel, properly displaying running lights, and posting your progress and estimated time of arrival (ETA) information to command, which is set up at the dock.

Then the boat motor simply stops. All lights, radios, and GPS go dark. You glance toward the shore on both sides, and realize the fog and darkness have blocked any vision of the shore. You pick up the microphone to call command, immediately realizing how much you rely on the radio—but it is no longer working.

**1.** What are your priorities at this point?

**2.** Which equipment could help make this scenario end with a positive outcome?

**3.** Which means of signaling, navigating, and communicating can be done with the boat not working?

## Introduction

Water rescues on larger bodies of water are reliant on watercraft. The speed and ease of traveling, treatment space, and protection from the elements often require watercraft to be used as a front line tool. Operating boats is often a confusing and trying challenge for new or inexperienced operators. Many states have passed laws requiring any watercraft pilot older than the age of 12 to take and pass a boat operations class, but it is a very good idea for any new boater to take this class, regardless of age. Some problems commonly happen on or around the water with newer boaters, but these issues can often be avoided with training and greater familiarization with boating procedures.

## Making the Watercraft Operational

As a watercraft technician, the operator of the watercraft is ultimately responsible for ensuring each system—both crucial and ancillary—is turned on and operating as desired before departing. Fuel, electrical, navigation, and communication systems should all be fully func-

tional. It does no good to travel to a search site miles away, only to realize the boat's sonar will not boot up. Another common problem in watercraft using portable tanks are fuel lines that are not properly attached to the motor before use. The system usually holds enough fuel to start the motor and run it for a few minutes before fuel starvation occurs. This event can put any boat crew in jeopardy while trying to develop a solution to the problem.

## Deploying the Boat

Generally, watercraft are kept ready for service in one of two ways: (1) in the water at a dock or (2) on a trailer. Boats may also be fastened to mooring balls, which involves a type of permanent anchor marked with a buoy, to which the boat attaches via a bowline. A mooring lets the boat move freely with the water and wind, and may help prevent damage from collisions with docks and walls. However, this type of moorage does not generally lend itself to a speedy response, in that other transportation to the boat is required. In busy harbors, moorage is used to allow more boat "parking" when there is limited dockage, but generally rescue watercraft are best kept at a dock, allowing for speedy response to and from shore (**FIGURE 3-1**).

**FIGURE 3-1** Most rescue watercraft are best kept in the water at the ready. Here, a fireboat shares dockage with law enforcement.
Courtesy of Steve Treinish.

As an operator, you should recognize that any launch or deployment of a watercraft must allow for safe operations while also accounting for current, wave, or wind action. Backing away from the dock, moorage, or trailer may be required, but turning the boat and progressing forward should be done as soon as possible to maintain better control. To overcome wind or water forces acting on the boat, slightly more throttle may be needed when maneuvering.

## Crew Communication

Launching a boat can be difficult enough in calm conditions, but in heavy wind or waves it can be dangerous. The forces acting on the boat can easily pull a crewmember overboard or into the water, and the addition of throttle by an operator places a spinning propeller into this mix. Instructions should be clear and concise. Phrases such as "go" and "whoa" sound very similar, and can confuse anyone. The word "stop," spoken in a loud, clear voice, is preferable to "whoa." "Hold" works well when asking a crewmember to hold a line fast, and "release" works well when the operator is ready for the line to be detached or released.

### LISTEN UP!

Crew resource management (CRM) can be used effectively in boat launching. The driver of the tow vehicle can be questioned before anyone steps in between the vehicle and the trailer as to whether the truck is in "park" or "neutral," and the emergency brake set. Also recognize that, during launch, team members will be operating in a hazard zone that the driver is often blind to. Conversely, when pulling the boat or trailer from the ramp, the driver can ask whether the area is clear before removing the parking brakes.

## Launching from a Trailer

During the approach to the ramp or dock, a great deal of information can be ascertained. Any wind or water action will usually be easy to see, and the boat can be launched to take advantage of these forces. As an example, suppose you arrive to find wind blowing from right to left as you face the ramp, with no wave or current action. Many boat ramps have docks on either side, or even center piers in case of multiple ramps. Launching and securing the boat on the right-hand side of the dock or pier in these conditions will let the boat blow off the dock slightly. Current or tide action can be handled the same way. The crew should ensure they have sufficient ropes ready to tie off when launching, as the boat can be blown away from the dock in such a case (**FIGURE 3-2**).

Another advantage of this launch technique is that the boat will play (drift) off the dock for however long the lines are tied, preventing damage from rubbing or abrading the dock or pier. If this kind of drift is not possible, bumpers or fenders should be used to cushion the boat from solid objects, and additional crewmembers may be required to help keep the boat from bumping and scraping against the dock.

If conditions are extreme, it may be better to launch the boat and hold it offshore while equipment and personnel are readied. Once the load is ready, an approach directly into the wind or water to pick up the cache can be performed with much more control.

## Launching from a Dock

Launching from a dock or pier offers an advantage in that shore crews can assist with the deployment. Experienced operators will let the wind or water push the boat away from the dock, using a person on shore to belay a rope tied to one of the front cleats. This strategy lets the boat

**FIGURE 3-2** Ensure the boat is secure when launching, preferably by tying to a secure anchor, and anyone involved should wear appropriate PPE.
Courtesy of Steve Treinish.

swivel into position with minimal effort. The bow can be allowed to swivel out as well. Be careful, though: The motor or outdrive can easily strike the portion of the dock or pier underwater. This launch technique also places the stern on the upwind side, creating more area that will be pushed by wind or water (**FIGURE 3-3**).

When on the upwind or up-current side, the same maneuver must happen to get the boat out, but you must use the throttle to overcome the forces working against you. In this situation, the crewmember belaying a line from the bow is very helpful, if not required.

## Returning to Dockage or a Trailer

Approaching dockage, moorage, or a launching site from downwind and/or down current allows the skipper to maintain much more control over the boat and the approach speed. Boats can become quite hard to control when approaching with the wind or water direction.

**FIGURE 3-3** Launching from a dock.
Courtesy of Steve Treinish.

Placing the side of the boat near the bow nearest the dock first is usually preferable. This approach allows a crewmember to belay a bow rope to a cleat. In calmer conditions, a line can be tossed from the stern to the dock, and then each line slackened or tightened as needed to place the boat beside the dock. (Do not forget to provide abrasion protection if needed.)

Larger boats may require some maneuvering against the bow line to help align them with the dock. If there is room, an operator has a few options available for accomplishing this task. If there is room forward, applying gentle throttle against the rope and turning away from the dock will let the propulsion push the stern in. Alternatively, the operator can put the motor in reverse and then turn toward the dock. This lets the propulsion pull the stern in.

# Piloting the Boat

For this chapter, we will assume the watercraft's crew does not include a helmsman. Helmsman is a position normally used on large vessels, and NFPA 1006, *Standard for Technical Rescue Personnel Professional Qualifications*, is more easily understood by assuming the operator is the person physically controlling the boat and controls. (For more information on the basics of watercraft operation, see Chapter 2, *Watercraft Rescue Operations*.)

## No-Wake Speed

No-wake speed is used for any approach to an object, person, or hazardous condition such as fog or poor visibility. In addition to the 300-foot buffer imposed in many harbors, no-wake zones protect many areas such as dams, natural resources, bridge abutments, and developments on the water. Operators should use common sense when boating in less than optimal visibility, and always remember the lag time that occurs between hazard recognition, the operator's reaction, and the boat's reaction: It can often be several seconds.

## Tilt and Trim

**Trimming** a boat can adjust the angle of the keel relative to the water, which can be used to advantage by the boat's operator. Trimming the bow down in rough water can let the bow of the boat slice through the waves, increasing the comfort of the ride for passengers and reducing the bouncing of the hull on the waves. Trimming the motor up allows the motor and propeller to work at optimal efficiency, maximizing fuel burn and speed.

Tilt and trim can also be used to navigate shallower waters that might otherwise cause a propeller to strike bottom or debris. Note, however, that control of the vessel may be affected by the change in the thrust angle and the watercraft's reduced responsiveness.

## Electronic Navigation

Modern electronics, including global positioning system (GPS) units, have made marine (as well as aerial and overland) navigation as easy as keeping an arrow on a line shown on a display screen. Navigation over any terrain is quickly becoming as easy as connecting the dots on the screen, and then following the direction indicated by the arrow. GPS units can also calculate speed, wind effect, estimated time of arrival, and current position, and many units will broadcast this vessel information to other watercraft equipped with receivers that, in turn, plot the information on their own screens. Weather radar, forward-looking radar, and map cartography and landmarks can also be downloaded to a multiple-use display, putting all the information onto one screen to show the "big picture." Smartphones and tablets are commonly used in many vessels, and can likewise download map cartography, weather radar, and forecasts to be used in watercraft operation (**FIGURE 3-4**). Although many styles and manufacturers are available, all of the navigation products generally offer the same capabilities and information.

---

**LISTEN UP!**

One easy way to provide a great redundant capability is to place a small, battery-powered GPS in service while under way, even when console-mounted and vehicle-powered units are in operation. If an electrical system failure subsequently occurs, the hand-held GPS may enable the crew to make a safe return. Even smaller units offer tracking, in which the unit simply marks its location every few minutes. Back-tracking is possible by following this trail back to the starting position.

---

## Paper Navigation Aids

Any electronic device can fail, including GPS and chart plotters. Operators should ensure paper charts are also available for use, particularly when they are navigating larger or unfamiliar waters (**FIGURE 3-5**). Even oceangoing ships keep accurate positions, headings, and ETAs as a backup to

total system failure, even if the ship is equipped with multiple electronic navigation units. The basic premise is that the travel route is calculated on paper, with the documentation including a compass heading, distance, and speed. This paper navigation aid can be

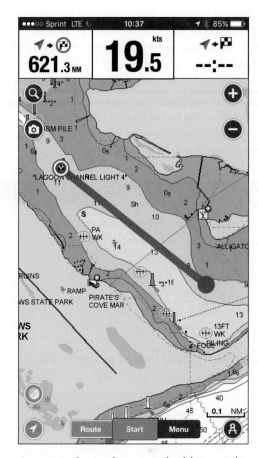

**FIGURE 3-4** Smartphones and tablets can be used to access map cartography, weather radar, and forecasts.
Courtesy of Steve Treinish.

**FIGURE 3-5** Paper charts should be available for use, particularly when navigating larger or unfamiliar waters.
Courtesy of Steve Treinish.

updated for emergency use even as the electronic navigation units do the hard, moment-to-moment work.

**Compass Roses.** Marine charts include a compass rose, detailing the degrees of the compass relative to the land masses, marine features, and landmarks. At the most basic level, a starting point and an ending point are drawn on the chart. The ending point may indicate either the end of a route or a point at which to change direction. A course heading can be determined from the compass rose. This can be done with a chartplotter—a small, ruler-like tool that helps transfer this information from the rose to the marked line. Once the course heading is determined from the chart, the compass is used to ensure the boat travels in that direction.

### LISTEN UP!

Paper charts cannot account for wind and water effects, such as waves and tides. Even though a boat may point directly at its desired heading, it may still be blown off-course laterally by these forces. Landmarks can be used to estimate where the boat is actually traveling, but if no landmarks are available, operators may be moved a considerable distance off-course.

**Distances.** Paper charts also document distances between points. Just as on a street map, distances are scaled so that much longer actual distances are represented on the chart by much shorter mapping distances. For example, 1 inch on the paper chart may be equivalent to 10 or 15 miles of actual distance. A divider is a tool used to plot the total distance of the route. It is adjusted to the proper range on the chart, then moved along the course line to find a total distance. Once the distance is known, calculations for speed can be performed to forecast an estimated time of arrival.

## Anchoring

The crew aspect of anchoring is covered in Chapter 2. Operators must have a good grasp of the types of anchors and anchor scope, which is the amount of line or chain deployed to keep a vessel stationary.

- Claw or plow anchors use the pull of the boat at anchor to dig deeper into the material, providing great holding power. They work in mud, sand, or rocky bottoms.
- Fluke anchors grab at bottom features such as rocks or coral and can supply limited holding power in sand or mud. These anchors are susceptible to wedging into crevices and cracks so tightly they become hard to recover.

- Mushroom anchors are meant to sink into the mud or soft bottom materials and allow suction to help hold the boat. They are usually found on smaller watercraft, such as johnboats or inflatables. Larger boats with a large sail area can easily pull these anchors across the bottom.

In a typical anchoring operation, the watercraft is motored to the location the anchor needs to set so as to place the vessel where desired. The anchor is then lowered into the water to the bottom, and the boat is allowed to drift back with the current or wave action, or reverse gear. Almost every primary anchor will be deployed from the bow, which lets the boat ride the waves head on. Anchoring from the stern may swamp a boat.

The anchor scope is the amount of line played out from the anchor to the boat's location. Scope greatly influences the holding power. Increasing the scope lets the anchor pull more parallel with the bottom, rather than being pulled vertically as well. A section of chain, called a rode, is commonly fastened between the anchor and the anchor rope and helps to keep the pull parallel with the bottom. This technique increases anchor effectiveness. Experienced mariners often have a good idea of how much anchor scope is needed, and this information can be relayed to the crew before anchoring.

### SAFETY TIP

Boat operators should ensure their crews understand the dangers of deploying an anchor. While it is a fairly straightforward process to drop the anchor in the water, a crewmember may inadvertently become tangled in a loop or bight of rope when it is being pulled over with the anchor.

The watercraft operator must be aware of boat movement while anchored. Ensuring the anchor has set, or is holding the boat stationary, can be done electronically with anchor alarms tied into the GPS unit; these alarms will sound if movement across the water is detected. Visual aids can be used as well. Lining up two points onshore provides a range—that is, two objects that can be used to gauge motion. For example, lining up two trees, one in front of the other, will quickly show movement if the trees no longer align, but spread out visually. Large vessels can use range buoys to ensure they are staying in the confines of a channel.

Another aspect of anchoring is accounting for the effects of shifting winds or currents on an anchored vessel. A single bow anchor will let a boat swivel around the anchor point, which reduces strain and lets the bow continually do its job, which is to cut the wave instead of being pushed by it. Anchoring

DANGEROUS CURRENTS

close to shore and then being pushed around in a circular motion, however, may cause the boat to strike a shallower bottom, possibly doing damage to the hull.

## Anchor Recovery

When recovering an anchor, it is generally easier to motor forward to a point directly over the anchor. This step lets the recovering crew get the anchor up from the bottom, reducing the power needed to raise it. Occasionally, an anchor will dig in or stick so hard that it may need to be pulled from the opposite direction to free it. Some fluke anchors can be rigged so that this pull can be done very easily, keeping the expensive anchor ready for another day. Crew members working to pull an anchor should have verbal, signed, or eye contact with the operator, so that they can advise the operator about which direction the boat needs to travel to best position for the retrieval.

## Crew Resource Management

**Crew resource management (CRM)** is a term used to describe the ability of any team member to verbalize concerns, especially when dealing with safety. A concept originating from the airline industry, CRM encourages crews to communicate more effectively by reducing the amount of pressure to *not* say something if someone recognizes a problem or concern, while maintaining the authority given to the person in charge (**FIGURE 3-6**). Humans, especially when under duress, may have trouble with multitasking—for example, watching the water ahead, a navigation system, and weather updates, while also listening to a radio. Allowing the crew to help monitor for potential problems and maintain spatial awareness can make the mission not only safer, but also much more effective.

**FIGURE 3-6** Crew resource management encourages crews to communicate more effectively by reducing the amount of pressure to *not* say something if someone has a concern.
Courtesy of Steve Treinish.

## Sterile Cockpits

Another concept from the aviation industry that the fire service can utilize is the **sterile cockpit** theory. During any operation or condition that may result in a catastrophic event, all members of the crew should eliminate all needless conversation and noise. Evolutions such as a nighttime approach to a person in the water certainly need clear and concise orders and queries, without distraction. No unrelated questions, comments, joking, or other chitchat should be allowed. Maintaining a sterile cockpit during crucial parts of missions allows communication effectiveness to be maximized.

# Advanced Electronics

Any helmsman should be able to hold a given course by using a compass or by watching the arrow on a GPS-equipped smartphone. (The basic compass and GPS units are covered in Chapter 2.) More advanced electronics are often found on larger boats, making their operations safer and easier for the operator and crew.

## Radar

Although they are not found on most watercraft, radar systems are becoming more popular as their prices come down. Many smaller cruising or fishing boats are now equipped with radar, for example. During night or foggy operations, land masses, other boats, or hazards can be tracked on the radar display. While many inland lakes have some nighttime landmarks, fog or atmospheric conditions may occlude vision, and once you are on a dark ocean on an overcast night, you will certainly appreciate being able to "see" the environment with the radar.

### Forward-Looking Infrared Radar

Forward-looking infrared radar (FLIR) shows the heat signature of objects. FLIR, which is often used on police helicopters, provides reasonably good night vision. Fire fighters are often familiar with thermal vision cameras, which are carried by most trucks today. Picture this thermal camera, but with zoom capability and longer range, put the picture on a computer-type screen, and you have the idea of how FLIR works. Crewmember overboard (COB) events can be mitigated much more quickly with FLIR systems when crews are working on dark water.

## Weather Information

Current weather conditions can also be shown with weather reporting technology. Of course, today's smartphones can also be used to check the local precipitation in the user's area. Satellite systems are

# Voice of Experience

There are many things we have learned growing in the fire service. Unfortunately, there are many things we don't pay attention to or merely forget. Situational awareness is one of those biggies that you should never forget about.

Water rescue is truly one of the most dangerous things we can do. If we sit back and think about it, at a run-of-the-mill house fire, we have a rapid intervention crew somewhere close by to come help if we have a collapse, lose water, get overcome with heat, or get lost. At hazardous materials incidents we are protected by a level A suit, clothing, and an SCBA. In water, you either drown or you don't; it's that simple. We shouldn't take our basic and advanced training lightly. Situational awareness is something that must be constantly observed in the water, especially moving water. We must know what is coming from upstream and where we are headed downstream. The elements are constantly changing while we are traveling up or down a river, especially a flooded river with obstructions and other traveling loads that may or may not be beneath the surface of the flood water. One wrong move or second of complacency can lead to injury or death of you or your crew.

While on a deployment to eastern North Carolina during Hurricane Matthew in 2016, three other rescuers and I were tasked with removing two adults from a home being overcome by flood waters. Our crew was competent, confident, and trustworthy. We had all worked together before and we had been to the same classes and trained in the same elements year after year. Our mission was a 4-mile round trip. Two of the miles were against current. As the mission began, the current wasn't much of an issue, but as we got closer to our objective it began to increase in velocity. Initially we were fine, until we were about halfway to where we were going. We were heading due east, and the current was coming from due north. We were ferrying like we were supposed to, the boat was at about ¾ throttle (just enough to keep us ahead of the current while keeping our heading straight), the boat was on plane, and the crew was spaced out in the boat as they should have been. Everything was going as planned until the bowman kicked the fuel line off the fuel tank. The initial problem took us all by surprise. As the engine cut off, quick action by the crew to grab paddles placed us in a natural eddy of trees, allowing us to maintain our position until the fuel line could be replaced. Situational awareness could have kept this problem from happening. If the bowman had realized where his foot was, we wouldn't have had an issue. On the other hand, situational awareness is what kept us from being swept into the thick wood line, potentially flipping our boat and adding four more victims to the equation.

To take something from the experience, no matter how basic or advanced the training is, try to take in every single detail. There is a reason you are being taught a certain way. Someone, somewhere had a problem, and the way you are being shown is the way they learned to solve it. We all are here because this is the best profession in the world. Enjoy it, learn something, then teach it. Always remember to stay safe, and do good work.

**Ryan Harrell**
Captain
Mooresville Fire and Rescue
Mooresville, North Carolina

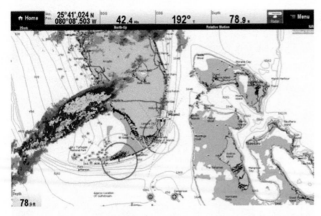

**FIGURE 3-7** Satellite systems are used to broadcast weather information to larger receivers and can update every few minutes.
Courtesy of FLIR Maritime US, Inc.

used to broadcast this information to larger receivers and can update every few minutes (**FIGURE 3-7**). This information can prove very valuable if severe weather is in the forecast in a search area. Local conditions such as wave heights, wind direction and velocity, and sky conditions can also be obtained with these units.

## Night Operations

Requirements for navigation and anchoring lighting are described in Chapter 2. Operating a vessel at night can be challenging, but using common sense goes a long way toward improving outcomes, even in rescue situations. Spotlights can be used, but forward vision in the darkness is often reduced more than most rescuers would think. Objects in the water can often be spotted in smaller lakes by watching the water surface and lights on an opposite shore. As an example, consider a disabled boat with electrical issues at night. It cannot show itself due to the lights not working. Watching those lights may spot this boat, as the silhouette will often be noticed in between your watercraft and the shore.

In some cases, shutting down the motor, asking for quiet, and shouting or using a loudspeaker will let rescuers hear shouts long before they see the target. Sound travels very effectively over a flat water surface, and rescuers can easily hear shouts for help over much longer distances than they might expect, especially on calm nights.

### Emergency Lighting

Rescue watercraft can be equipped with lights and sirens if needed, but beware of passing boaters who will follow a response out of curiosity. Typical emergency response lighting can be used on water, but backup scene and navigation lights should also be considered. Emergency lighting in the marine sense comprises whatever lights can be used in case of emergency. At night, many water areas will be very dark when running, flood, or spotlights fail. For this reason, waterproof flashlights should be provided on any vessel that may be called upon for night operations. Chemical light sticks can go a very long way toward improving lighting conditions if lighting systems fail or if rescuers stumble on other boats with malfunctions. Small, battery-operated stern and bow lights are available and would be a good addition to any larger watercraft that does a lot of night evolutions.

### Infrared Optics

What was science fiction not too long ago is now reality; that is, the ability to "see" in the dark is now affordable for most mariners. Small, hand-held units or binoculars equipped with infrared vision are commonly available. Even a department's thermal imaging camera can be helpful if brought onto a smaller boat and used during a nighttime search.

## Rules of the Waterways

One might think rescue boat operations on the water would be as simple as turning on the red lights and taking off. To a certain extent they are—but just as with fire apparatus on the street, a few rules must be followed, even in response mode. In addition, returning from a mission, driving practice, and patrol missions all put rescuers back into "civilian" mode on the water. This section takes a brief look at some regulations and concerns commonly referred to as the "rules of the road."

**LISTEN UP!**

The International Maritime Organization regulations of 1973 address what all watercraft should do in certain situations on the water. In 1980, the United States adopted rules very similar to the 1973 international regulations, which it amended to address primarily inland waters. Most recreational boaters are subject to these regulations.

## Right of Way

Generally, the less maneuverable boat has the right of way. For example, most sailboats have the right of way over a motorized watercraft, since the motorized craft can operate regardless of wind conditions or direction much more easily than the sailboat. In certain places, such as rivers, the boat traveling downstream has the right of way over an upstream boat, as control can be maintained more easily by the boat running upstream.

When operating in proximity to other watercraft, boaters might find themselves in one of three situations where a right of way is addressed: passing another boat, overtaking another boat, and avoiding collisions. The right of way is not generally an issue until there is a chance of collision, and mariners should strive to stay clear of each other by making course changes long before this happens. When operating on smaller recreational lakes or rivers, however, this can prove very hard to do due to crowds or lack of space on the water.

When motoring by another oncoming boat, both boats should be steered to starboard, so that each boat passes the other on the port side. Keep in mind, not all boat operators are familiar with this rule, and some just ignore it. Be ready to swing a boat to port in such a case. On a practical level, the best thing to do might be simply to slow down when approaching if the other vessel's actions are not clear.

Overtaking a boat while on the same line of travel includes a bit more to consider. The boat coming from behind must yield by directing itself preferably to the port side of the overtaken boat. This situation is akin to driving a car and using a passing lane. The regulations say a short horn blast is to be used during this scenario, but this point is commonly disregarded on the water, especially inland waters. The horn blast is intended to notify the boat being overtaken so that it does not change course direction when the other boat is near it. The passing boat may also elect to swing to starboard instead of to port, and two short horn blasts are to be used in this case. Upon hearing the horn from the passing boat, the boat being overtaken should repeat the horn signals to show understanding. If the overtaken boat sees danger, five blasts of the horn should be given—this is the universal horn signal for "danger." Course crossings, such as two boats approaching each other from 12 o'clock and 3 o'clock, are more dangerous, especially when the bulk of the traffic is moving one direction or the other. The rule says specifically that the watercraft with the other boat on its port side has the right of way (**FIGURE 3-8**).

The red light required on the port side can also work well as a reminder of the proper procedure. If a mariner sees a boat approaching and the red light is or would be seen, he or she must alter course. If a mariner

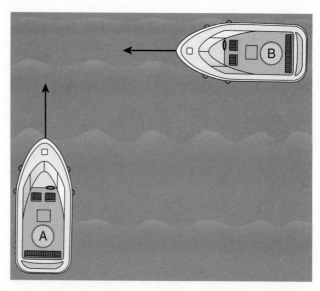

**FIGURE 3-8** The boat with the other boat on its port side has the right of way. In this case, Boat B has the right of way, and Boat A must give way.
© Jones & Bartlett Learning.

sees a green light, he or she has the right of way and can proceed. Caution must be used, however, since all boaters do not universally adhere to this rule. Just as operating with red lights and a siren on the street does not guarantee automotive traffic will yield to a fire apparatus, emergency lights or even running lights on watercraft do not ensure the right of way will be given. As a matter of fact, emergency lights may even attract more people to the scene.

One other challenging situation can occur when navigating channels and narrow rivers. Going around a blind corner can lead to a collision, so watercraft are asked to sound the horn for one long blast when approaching this point. An answering blast would mean another boat is close enough to hear, and both boaters should reduce their speed and try to stay to the starboard side to pass.

## Towing Operations

Any watercraft towing another vessel has the right of way over other watercraft, for a very practical reason: The only thing that may be worse than a boat not having brakes is *two* boats tied together with no brakes.

It is not uncommon for rescue watercraft to be flagged down on the water by other boaters who have experienced a mechanical failure or run out of fuel. After all, it probably says "RESCUE" on the side of your boat. If the authority having jurisdiction (AHJ) allows boat towing, after ensuring each occupant is wearing a personal flotation device (PFD), the most important thing to remember is to *slow down*. Boats in tow may sway laterally back and forth on a towline.

Lengthening the towline may improve handling. Towlines should not be hand-held, no matter how light or small the boat being towed. The towline should be secured and capable of handling the load before it ever gets tightened.

Tilting or lifting the motor of the towed boat out of the water may or may not improve handling, depending on hull design. When approaching land or dockage, ensure that both vessels' operators know what will happen before loosening lines, and that the stricken boat is within range of a thrown line from the dock, or is able to paddle to safety. By using a very slow speed and taking their time, two operators may be able to put the stricken boat very close to a dock.

### KNOWLEDGE CHECK

Motorized operations while a rescue team is in the water should not be permitted.

a. True
b. False

*Access your Navigate eBook for more Knowledge Check questions and answers.*

# Wind and Current Effects

Wind, currents, and tides can affect boat operations. This is especially true at lower speeds, such as when docking or in a no-wake zone. Many boats present a fairly large surface for water and winds to push on. Thus, backing away from a dock, pulling onto a trailer, and trying to maneuver at a fixed object may all require navigating with more applied power than normal. Running upstream in currents can take much more throttle to make forward progress, and an operator may still have to be on the throttle to hold a boat stationary in current.

## Windy Conditions

Running a boat in windy conditions can be easily managed, but the operator must think ahead of the boat and be proactive, not reactive. A strong gust of wind will easily move a boat around at low speeds, and the bigger the surface area presented to the wind and the higher the wind speed, the worse this effect will be. Operating with a perpendicular crosswind or current is the worst-case scenario, as the wind or water pushes directly on the side of the boat. Watercraft can be blown or pushed over long distances if operators do not think ahead and take steps to avoid this outcome. Be aware of the wind direction as it plays against your planned direction of travel. Skippers might have to operate at low speeds with a slightly higher power setting, or even crab, which is steering the boat into the wind while traveling straight ahead (**FIGURE 3-9**).

# Watercraft Operation in Current

Some watercraft river operations use paddles, oars, or motors, or even just a static line to maneuver the watercraft. Before any operation on the river is begun, all rescuers must be on the same page. It can get confusing for everyone if directions are not standardized during planning and operations. Imagine what it would be like if four different agencies were working a scene and they all used terms like "left" or "the right bank." Is that the right (as opposed to left) bank, or the

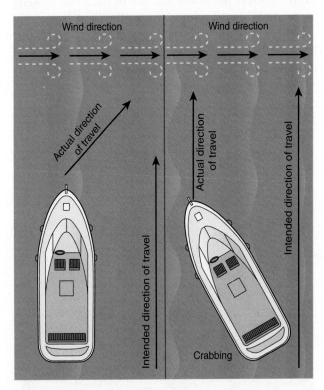

**FIGURE 3-9** Crabbing a boat involves steering a boat into the wind to counteract the wind blowing it off course. The boat may be pointed slightly to port or starboard, but travel in a straight course.

right (proper) bank to use? Moreover, in remote areas the terms "north," "south," "east," and "west" may not mean much, especially to rescuers who are not familiar with the area. It quickly becomes clear that terminology can create major problems, unless all rescuers are on the same page. Thus, it is standard procedure on all moving water rescues or operations to use some basic universal terminology; once the basics are understood, it becomes clear which direction anyone is talking about on any moving water.

## Current Terms

**Downstream (down current)** is fairly simple to understand: It is the direction the water is flowing. **Upstream (up current)** is the direction from which the water is coming. **River left** is the left side of the river, looking downstream, and **river right** is the right bank, still looking downstream (**FIGURE 3-10**). This terminology works on moving waters that range in size from the smallest creek to the mighty Mississippi.

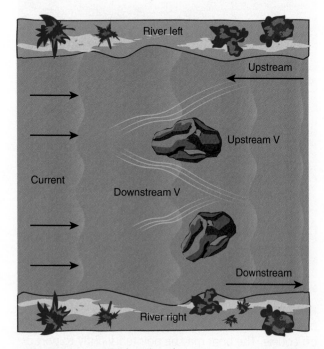

**FIGURE 3-10** Directional terms on moving water, with upstream and downstream *V*s also shown.

© Jones & Bartlett Learning.

## Ferrying

**Ferrying** a boat is using the river current itself to cross the river, simply by controlling lines attached to the bow and stern. Imagine a boat pointed upstream, with the bow attached to a pulley hanging on a static line strung perfectly perpendicular to the current. The stern can be pulled upstream toward the static line, and as the boat becomes angled to the current, it will be pushed diagonally in the direction the bow is pointing. This is called the ferry angle. As the boat moves more toward perpendicular to the current, the greater the force the current puts on the side of the boat, the quicker the lateral movement will be. Caution must be used when ferrying a boat in faster water, as too severe of a ferry angle will put too much pressure on the side of the boat and possibly capsize it. Using a ferry angle is not a new technique; indeed, people figured it out many years ago and used this concept to move supplies and each other across rivers.

In **FIGURE 3-11**, boat A is being operated parallel to the flow, with the current forces equal on both

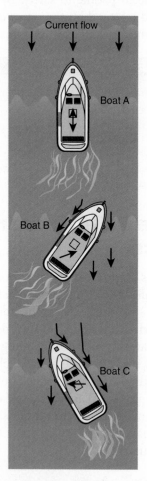

**FIGURE 3-11** Ferrying a boat by changing the angle of the boat relative to the current. As the angle becomes more perpendicular to the current, the boat will move laterally more quickly.

© Jones & Bartlett Learning.

sides. Throttle control can be used to maneuver upstream or to let the boat slide downstream, without moving the bow. If the boat is turned more perpendicular, the resulting forces create ferry angle. Boat B is moving river left, even with the motor not turned. The current forces are pushing down and away from the port side, ferrying the boat to the right. If the throttle is controlled and holds the boat from going upstream or downstream, the current will push the boat directly to the river left bank. This is a quick way to deploy a line to the opposite bank to set up a two-line tether rescue. Boat C is moving river right, under the same setup as boat B. Using ferry angle, boat operators can travel both upstream and downstream, and river left and river right, at will. Ferrying is often used to maneuver watercraft around obstacles or shallow areas, or to deliver equipment and gear to the other side of a river.

Boats can also be ferried across under power, by adjusting the ferry angle by way of the motor. With a nonmotored boat, the oars or paddles can be used to maintain the ferry angle.

## Current Hazards

Smaller boats, such as canoes, johnboats, inflatables, or kayaks, either motored or not, are the usual go-to watercraft in smaller rivers or swiftwater operations. Larger boats can be used on larger waters, but any boat can easily be pushed into or against objects encountered in current. Generally speaking, boats face far less danger when operating parallel to the current, as compared to operating perpendicular to the current. Keeping a boat facing up current is the preferred action, as the crew can see debris or objects coming their way, the operator can ferry river left or right easily, and the boat is not subjected to forces that may cause problems.

Operators should strive to make any turns needed in current very quickly and aggressively. The longer the boat is presented perpendicular to the current, the greater the risk of problems. In the following discussion of current hazards, we will use smaller craft as an example.

**Rotational Turning Force.** Some river hazards can be forecasted and used to rescuers' advantage. **Rotational turning force** is the force of the current acting on the side of the watercraft that is angled to the current. Imagine you are operating a small inflatable and want to turn and proceed downstream. The river is running from your right to your left, and you are entering from a small sheltered area on river left. When the bow of the boat enters the flowing water, it will be pushed to the left (downstream). Knowing this will happen, the crew can anticipate the turn and lean into it, and the operator can give some throttle to complete the turn.

**Rotational Capsizing Force.** **Rotational capsizing force** is found when the force of moving water pushes on the side of a boat that is turned perpendicular to the current and strikes or is pushed against an object. When this happens, most inexperienced boaters will immediately lean into the upstream side of the boat, away from the obstacle, which will just make the boat capsize faster. This is instinct—but this instinct must be overcome by training. The proper move in this situation is to lean toward the strainer, thereby enabling the upstream side of the boat to rise, and allowing water to pass underneath the boat, instead of pushing on it so hard (**FIGURE 3-12**). Once that happens, the crew must stay in that position to keep the water flowing under the boat, while a plan is devised to get off the obstruction. Many times, just changing the angle of the boat will enable it to slide or be pushed off the object by the current that is pinning it there. Even mechanical advantage haul systems are sometimes used with bigger boats to move one side or the other around to achieve this effect.

**Strainers.** A **strainer** is an object in moving water that impedes the water's flow around it, while still allowing water to pass through (**FIGURE 3-13**). Strainers in flooded waters often include fallen trees, and most people are familiar with the pile of wood that gets hung up in front of bridge pilings or islands, and then collects smaller limbs, trash, rope, and other debris. A boat

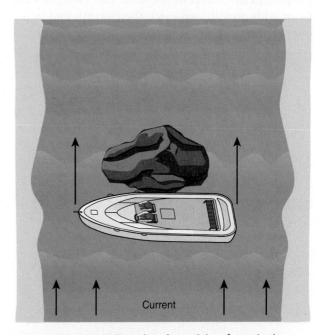

**FIGURE 3-12** Rotational capsizing force is the current acting on a boat that is trapped against an object, which stops the boat's motion. Rescuers must lean downstream to let the water pass under the boat, or it may capsize.

© Jones & Bartlett Learning.

**FIGURE 3-13** Strainers are objects that let water through but stop other objects, including humans.
Courtesy of Steve Treinish.

can easily be sucked into a strainer. When this happens, the boat is often turned perpendicular, submitting it to the rotational capsizing force.

## Other Current Features

**Eddies.** Another critical driving skill on swiftwater is the ability to find a safe spot or a rest area in the current. **Eddies** form when water curls around a large object in the current. While the water behind the object may be moving upstream, an eddy is a very good place in swiftwater to seek shelter from the current (**FIGURE 3-14**). Even river swimmers and people out of control in the water can swim or steer to an eddy and at least catch their breath and figure out their next move.

Getting a boat into an eddy area involves making an **eddy turn**. The boat is piloted downstream, with the bow of the boat then being turned and driven into the upstream flowing water. At this point, the boat is throttled around to complete the turn. This maneuver takes some good instruction and good practice because the stronger the current, the quicker the turn has to be, and the more throttle needs to be applied. Many new boat operators in current are not aggressive enough with the throttle. If the boat drifts down to the water that flows downstream after the eddy because of an ineffective turn, it may be taken out of the protected area. The bow of the boat needs to be as close as possible to the object creating the eddy for maximum effectiveness.

Getting out of an eddy when you are ready to leave requires a **peel-out**. As this name implies, the boat is peeled out of the eddy using the current and throttle. Start by keeping the boat pointed upstream while slowly nosing the bow into the downstream current. When the current grabs the bow, open the throttle to the desired level and lean the boat into the turn. The goal is to get the boat quickly turned downstream, eliminating the amount of time in which the boat is perpendicular to the current (**FIGURE 3-15**). Using the flow to help turn the boat quickly relies on the rotational turning force of the river, rather than the rotational capsizing force. Rotational turning force affects the turning axis, whereas the rotational capsizing force affects the rolling axis.

Both eddy turns and peel-outs can be done in smaller inflatables or johnboats with oars, paddles, or

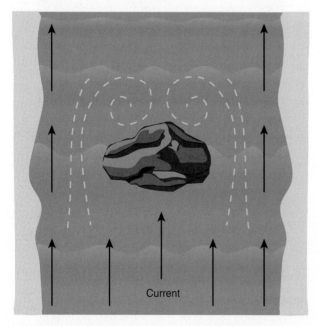

**FIGURE 3-14** Eddies form when moving water "wraps" around an object, resulting in an upstream flow and creating a relatively calmer area.
© Jones & Bartlett Learning.

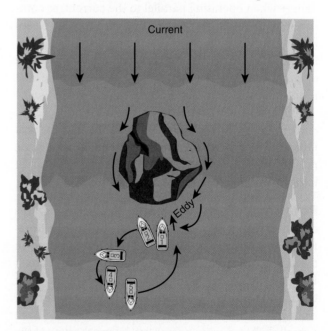

**FIGURE 3-15** The goal of a peel-out is to get the boat quickly turned downstream, eliminating the amount of time in which the boat is perpendicular to the current.
© Jones & Bartlett Learning.

motors. Manual propulsion in current can be done, but the operator needs to look ahead and anticipate the factors associated with the river, the boat, and the currents. This also plays into the idea of a rescuer constantly planning ahead: "What if such-and-such fails or breaks, at this instance?" Running a small boat with the oars in the locks, ready to go, or even checking your anchor rope before leaving can make a huge difference in times of trouble. If your motor suddenly fails, will you be fighting five throw bags, an anchor, seven PFDs, and a ring buoy to find your oars? Will your anchor deploy smoothly since it is stowed neatly up front, or will a huge "spaghetti" ball of rope go with the anchor itself, letting the anchor hang from the boat without touching the bottom (a scenario that is all too common with unpracticed rescuers)?

**Upstream and Downstream V's.** A **downstream V** is seen from upstream as the point of the "V" pointing downstream; it represents the clear path between two side-by-side obstacles. An **upstream V** has the point of the "V" pointing upstream, and is the result of water flowing around the obstruction to either side. Steer clear of upstream V's; something there is making the water move, and it usually will do some kind of damage to your boat.

## Approaches in Moving Water

Any approach to anything in moving water is best done from the downstream area, working against the current. Regardless of the type of boat, any motor-operated vessel should make the approach from downstream, moving upstream. The operator can adjust the throttle in very small increments. Once the thrust moving the boat upstream matches the flow moving downstream, the boat will hold steady in the current, and can then be ferried in either direction laterally. Increasing or decreasing the throttle will move the boat up current or let it slowly drop down current, respectively. Working in this manner, the operator can ensure the boat makes a slow, deliberate approach to a stranded victim.

## Recoveries in Moving Water

Approaching the victim and simply floating with the current with the motor shut down until the victim is recovered is the best way of recovering a victim in moving water. Paddles or oars can be used to provide some control if needed, and awareness is needed of potential hazards downstream in the path of travel. If a victim is able to help himself or herself but cannot be approached safely, consider the use of a throw bag

**SAFETY TIP**

Holding the boat steady in current and intercepting a moving victim is very dangerous. When rescuers grab a victim in current from a boat, the victim's legs will naturally be drawn under the boat, putting them close to a spinning propeller. Losing the grip on the victim may cause the victim's legs to be pulled even farther back into the propeller danger.

from the boat, with the boat being kept in a safer area. The victim can then be brought to the boat and extricated in safer water.

## Watercraft Operation in Surf Conditions

NFPA considers surf conditions to be waves between 1 and 6 feet tall. Although 6 feet may not sound like much, in the wrong craft, such waves will make any water rescuer uneasy. Some boats are simply not meant to handle such conditions, and these conditions must be accounted for when selecting watercraft to tackle such a mission. Larger, cutter-style boats and true rough-water boats can handle surf with ease, assuming their operators are properly trained. Inflatables can handle much larger surf than johnboats can, but they are still susceptible to the effects of these bigger waves.

Operation of a boat in large surf should follow one general rule: Keep the bow pointed into the surf if possible. Most boats can handle waves much taller than themselves if the bow can be positioned to let the shape of the boat cut through the wave. A watercraft presenting parallel to large waves has a lot of tipping action, however, and the high center of gravity on some boats will increase this tipping action.

## Breaking Waves and Swells

Surf is generally made up of two types of waves: breaking waves and swells. **Breaking waves** are waves that roll over at the top and crash down to the water surface; they are found in shallower water (**FIGURE 3-16**). Swells, or rollers, are large waves, but are easier to operate in (**FIGURE 3-17**). They can still do damage, but the wave does not crest or break, so the angle of the wave usually is not as severe at the top. Many surf rescues will end up with a boat in the offshore areas where swells do not turn into breaking waves. While they may be much easier to contend with, swells must still be respected. A rolling swell or wave will travel across the water to shallow areas. Once the swell hits shallower water, the bottom will be taken from under it, but the inertia of the wave keeps it moving inshore. This explains why breaking waves are

**FIGURE 3-16** Breaking waves curl up and crash down, and can injure swimmers operating around them.
Courtesy of Steve Treinish.

**FIGURE 3-17** Rolling waves, or swells, are much easier to navigate than breaking waves, but are formidable and can become breaking waves in shallow waters.
Courtesy of Steve Treinish.

not present farther offshore. Rescues dealing with large surf are very skill dependent, and a qualified instructor or mariner must train rescuers in a craft that is capable of handling these conditions.

# Deploying and Retrieving Rescuers

Boat operations with rescuers deploying from them usually take the form of tending and deploying divers or rescue swimmers. There are a few key considerations when operating a boat around rescuers.

## Deploying Rescuers

Getting a technician-level rescuer in the water is generally easy, as going in the water is the focus of their training. For the sake of this discussion, we will assume the rescuers you are deploying are protected by proper personal protective equipment (PPE) for the hazards

present. Be aware of current or wave action, and recognize where it may push rescuers. Having rescuers deploy down current or downwind to a rescue is desirable, as it can keep the rescuers from being pushed under the boat, and the rescuers can keep their eyes on the victim. Preplanning the deployment in this fashion can be done even before the team arrives at the rescue site.

Operators should strive to use good spatial awareness in open water when any rescuer is in the water, as they may be hard to see, especially with wave action. Also maintain awareness of potential onlookers. A rescue boat operating its emergency lights will certainly attract other boats, whose operators may be distracted by "gawk" factors. Keeping an eye out on the bigger picture is a must.

## Retrieving Rescuers

A simple go rescue from a boat is not a big deal for most rescue technicians. Many times a rescue will occur in which a crew is dropped right in the water to perform a rescue, and the rescue is successfully completed with the motor turned off the entire time. Unfortunately, a common shortcut some rescuers will take when getting back on board is to use the lower unit as a step, especially the anti-cavitation plate located above the propeller. A slip in this area will almost certainly lacerate the rescuer's foot or leg. Even if the rescuer is not injured, the plate can certainly cut an expensive wetsuit, necessitating its replacement. Rescuers should be trained to use a boarding ladder or stirrup loop to return to the boat, or have the crew on board assist rescuers coming back on the vessel.

## In-Water Approaches

NFPA 1006 addresses COB events in three ways: as a rescuer in the water, as crew on board the vessel dealing with a COB, and as the operator of the vessel. Approaching any person in the water—whether a COB, rescuer, or victim—requires the same safety precautions and operation.

First and foremost, when approaching, you must be concerned about striking an in-water rescuer. As mentioned previously, a watercraft has no brakes. Slowing down before reaching the rescuers who need to be recovered is crucial. When slowing down, a boat will normally come off plane, and the bow will momentarily rise, which can occlude vision. Having the watercraft off plane at an idle speed is the best way to approach anyone or anything in the water. The use of a spotter in the bow area is highly recommended, and the operator must have communication with this spotter. Operators are well advised to keep the person in the water visible off one side. If the person is capable of swimming, having him or her swim a few feet to a stationary boat is much better than trying to motor the boat right up to the person to save a very short swim.

Another potential problem with operating watercraft close to in-water personnel is the risk of a prop strike. The easiest way to avoid this potential injury is to kill the motor, even if just by putting it in neutral, when a rescuer is ready to be picked up and is close enough to talk to. This strategy quickly eliminates the hazard from the equation. The in-water rescuer should be comfortable in the water, so he or she can swim a few feet to the boat instead of the operator trying to get closer with the motor running.

A well-trained operator can position a boat to pick up an unconscious person in the water, but the speed of approach, wind, waves, and currents must be carefully considered when doing so. Victims floating on top of the water do not present a large surface area to the wind, and the boat may be pushed across the water faster than the victim. Good operators gauge this movement and decide the optimal approach.

> ### LISTEN UP!
>
> Putting a water rescue mannequin in a PFD can make for a great practice session on windy days. Operators can quickly gain confidence in their ability to line up the boat, forecast movements, and control the throttle and steering, and the worst-case scenario results in just a beat-up dummy (**FIGURE 3-18**). Surface rescuers can also practice recovery efforts, as can technician-level rescuers. This kind of evolution can benefit many people who will be directly involved in a rescue.

Approaching an incapacitated person in the water should be done only if there are no other options, and only by an experienced operator.

A team might be leaning toward not allowing any motorized operations with rescuers in the water, but if the proper procedures are followed, these kinds of operations can be done safely (**FIGURE 3-19**). For example, a dive operation offshore may not bring divers up every time a boat approaches, because the boats and the dive crew should know before the dive begins what, where, and how things will happen. In a well-rehearsed team, the divers will know when the boats are approaching, and from what direction; the tenders will know where the divers are at all times; and the safety officer or dive supervisor will have direct contact with the boat crew. The team will have previously decided where the arriving boat will dock with the boat being used as the dive platform, and they will know which direction and procedures will be used when the boat leaves.

## Shutting Down and Securing Watercraft

The shutdown of watercraft is fairly simple and straightforward. The motor should first be placed in neutral, and then shut off. Fuel selectors and electric fuel pumps should be shut down or moved to the appropriate position. Any ancillary electronics or lights should be checked for proper shutdown so as not to drain battery power, especially if any of these electronics are connected directly to battery power. Some items, such as radios, may be wired directly to the battery to allow emergency use in case of motor failure. Bilge pumps must be operational if the boat is

**FIGURE 3-18** A water rescue mannequin can be a great training tool.
Courtesy of Steve Treinish.

**FIGURE 3-19** If the proper procedures are followed, motorized operations with rescuers in the water can be done safely.
Courtesy of Steve Treinish.

not covered with a mooring cover. Any gear that may be blown around or off the boat should be secured, cleaned, and dried as much as possible, and battery switches should be moved to the "off" position if applicable. The watercraft itself should be secured to the dock or mooring (if applicable) with the appropriate knots, bends, and hitches.

**KNOWLEDGE CHECK**

A rescuer in the water should avoid the use of which item to get back on board a vessel?

**a.** Assistance from the crew

**b.** Lower unit

**c.** Stirrup loop

**d.** Boarding ladder

*Access your Navigate eBook for more Knowledge Check questions and answers.*

# *After-Action* REVIEW

## IN SUMMARY

- New boaters should take and pass a boat operations class, regardless of their age.
- A watercraft technician is ultimately responsible for ensuring each system, both crucial and ancillary, is turned on and operating as desired before the watercraft departs for the rescue site.
- Operators should use common sense when boating in less than optimal visibility, and always remember the lag time that occurs between hazard recognition, the operator's reaction, and the boat's reaction.
- Crew resource management encourages crews to communicate more effectively by reducing the amount of pressure to *not* say something if someone recognizes a problem or concern, while maintaining the authority given to the person in charge.
- More advanced electronics—such as radar, weather reporting technology, and emergency lighting—are often found on larger boats, and can make operations safer and easier for the operator and crew.
- Generally, the less maneuverable boat has the right of way.
- Any watercraft towing another vessel has the right of way over other watercraft.
- Wind, currents, and tides can affect boat operations, especially at lower speeds, such as when docking or in a no-wake zone.
- Getting a technician-level rescuer in the water is generally easy. Having rescuers deploy down current or downwind to a rescue is desirable, as it can keep the rescuer from being pushed under the boat and allows the rescuer to keep his or her eyes on the victim.
- A simple go rescue from a boat is not difficult for most rescue technicians. Rescuers should use a boarding ladder or stirrup loop when returning to the boat, or have the crew on board assist rescuers coming back on the vessel.

- The actual act of shutting down watercraft is fairly simple and straightforward. The motor should first be placed in neutral, and then shut off. Fuel selectors and electric fuel pumps should be shut down or moved to the appropriate position. Any ancillary components should be checked for proper shutdown.

## KEY TERMS

*Access Navigate* for flashcards to test your key term knowledge.

**Breaking waves** Waves that curl up to the point that they collapse back onto the water with force.

**Crew resource management (CRM)** A communication system that encourages all crew members to verbalize their concerns, especially when dealing with safety, regardless of rank or authority.

**Downstream (down current)** The direction in which water is traveling.

**Downstream V** Current feature that forms a "V" shape, with the "V" pointing downstream; often formed when water flows between two solid objects, normally rocks.

**Eddies** River features that form when water curls around a solid object and creates a pool of calmer water flowing upstream; often found just downstream of rocks and bridge piers, but can also form behind trees, automobiles, and shore features.

**Eddy turn** Driving a watercraft bow first into an eddy, using the current differential to quickly turn the boat 180 degrees and then be held against the object forming the eddy.

**Ferrying** Use of the force of a current to drive a boat laterally from side to side, by changing the angle of the boat relative to the current.

**Peel-out** Use of current forces to quickly turn and enter a current; usually done when leaving an eddy.

**River left** When looking downstream, the land mass on the left.

**River right** When looking downstream, the land mass on the right.

**Rotational capsizing force** Current force that acts to overturn a watercraft that is perpendicular to the current when the watercraft is not moving with the current.

**Rotational turning force** Current force that is applied to the bow of a watercraft when entering current from another direction.

**Sterile cockpit** A reduction of noise or conversation to only mission- or operation-crucial communication during times of stress or critical evolutions.

**Strainer** Any object that lets water through it, but does not let larger objects pass.

**Swells** Waves that rise up, but do not crest or break; found offshore in deeper water.

**Trimming** Adjusting the angle of the engine or outdrive of a watercraft to maximize the applied thrust from the propulsion system.

**Upstream (up current)** The direction from which flowing water is coming.

**Upstream V** A current feature that forms a "V" shape, with the "V" pointing upstream; it often forms where water flows around a solid object, normally a rock.

# On Scene

Around 2245 hours, on a night you were hoping would stay quiet, you are called on a search and rescue mission of a different type. A neighboring department has a rescue boat that has gone missing; it has now been radio silent for more than 90 minutes. The boat was originally sent to assist victims of a watercraft accident on a remote bay, and the last report to the command post stated that the crew had one victim packaged and was heading back. This communication occurred at last light, but now the skies and horizon are pitch black.

Your department's rescue boat is very well equipped, with twin outboards, LED spotlights controlled from inside the cabin, a GPS/chartplotter, sonar, and full lighting, and it has been maintained several times over the season. You hook it up to your department's tow vehicle and respond to the launch site. There, you meet up with a very worried incident commander (IC) and several crews, all of whom seem a bit bewildered about what can be done. While the stricken boat has been on multiple calls this year, it is older, and this department does not have many water calls at this end of the lake.

You ask the IC which gear is on board the stricken rescue boat, but he is not entirely sure, other than confirming the presence of a marine radio, county fire dispatch radio, and spotlights. There were three rescuers aboard, all of

*(continued)*

# On Scene Continued

whom were confirmed to be wearing PFDs when they departed. Weather conditions are only fair, with patches of very dense fog and heavy clouds. While winds are calm, the temperature is dropping quickly toward 50°F. You decide to start toward the point last seen (PLS) of the stranded boat, with a crew of four rescuers including yourself.

**1.** Which extra gear should be taken along on this mission for added safety?

**A.** Extra PFDs, three throw bags, another Stokes basket, and some chemical light sticks

**B.** An extra battery for the portable radio, a paper map, a tool-box, and an extra propeller

**C.** A working cell phone, blankets or warm clothes, chemical light sticks, and an extra hand-held fire radio

**D.** A few snacks, extra drinking water, a towing line, and a cell phone

**2.** You decide that you should have additional help with navigation to the point last seen. The best navigation would be:

**A.** an electronic, helm-mounted GPS unit, backed up with a hand-held GPS unit or GPS-capable smartphone.

**B.** two smartphones carried by separate members of the crew.

**C.** a lighted compass.

**D.** a paper chart, with directions to the suspected location documented at the command post.

**3.** During the rescue, you encounter fog several times en route. The dense fog blocks almost any use of the spotlight, but you believe you are in the area where the boat last reported its position. Good crew resource management would include:

**A.** keeping the crew outside the cabin for good mental concentration.

**B.** using a member on each side of the bow as a lookout, one to help navigate and operate the GPS unit, one to monitor the radio, and asking for a sterile cockpit.

**C.** placing additional high-powered LED flashlights in service with the forward crew.

**D.** knowing where all safety gear is stowed, and each member knowing his or her duties should an emergency arise.

**4.** With a boat unable to communicate as the target and the weather conditions described in the scenario, what is the best way to safely continue the search for the stricken vessel?

**A.** Keep trying the marine and the fire dispatch radio, hoping to hear scratchy communications as the distance between the boats lessens.

**B.** Slow down to a no-wake speed, and use the spotlights in a sweeping, lateral motion at the edge of the beam's reach.

**C.** Slow to an idle speed and check in with command every 5 minutes for updates. Mark the GPS course with an updated position each time an update occurs.

**D.** Slow to the slowest speed possible, use the spotlight, position lookouts at the bow and each side if possible, and occasionally shut the motor off and listen for shouts and noise as a collective crew.

**5.** You kill the motor and hear screaming. You believe the crew are telling you their boat is sinking, and there is one person in the water. You approach the direction slowly, and are relieved to see the fog lifting enough to use the spotlight. Once you see the victim, what is the best approach?

**A.** Keep one person manning the light, keeping the beam on the victim at all times, and approach slowly, keeping the victim in sight and at the 10 or 2 o'clock of the bow.

**B.** Have the crew ready with extrication procedures, a throw bag, and extra flotation at the ready if needed.

**C.** Shut down the boat's engine if the victim is stable to eliminate the propeller danger.

**D.** All of the above.

*Access Navigate to find answers to this On Scene, along with other resources such as an audiobook and TestPrep.*

CHAPTER 4

## Operations Level

# Surface Water Rescue Operations

## KNOWLEDGE OBJECTIVES

After studying this chapter, you should be able to:

- Describe how to preplan water rescues. (**NFPA 1006: 16.2.1, 16.2.3, 16.2.4, 16.2.13**, pp. 103–106)
- Identify and describe water rescue equipment. (**NFPA 1006: 16.2.2, 16.2.3, 16.2.4, 16.2.5, 16.2.6, 16.2.7**, pp. 106–108)
- Explain the components of scene size-up at a surface rescue incident. (**NFPA 1006: 16.2.3, 16.2.4, 16.2.5, 16.2.6, 16.2.7**, pp. 108–110)
- Describe methods of rescuer communication and identify when to use them. (**NFPA 1006: 16.2.4, 16.2.13**, p. 108)
- Describe the components of scene operations and planning. (**NFPA 1006: 16.2.1, 16.2.4, 16.2.5, 16.2.8, 16.2.13**, p. 110)
- Explain the rescue options for a victim in sight. (**NFPA 1006: 16.2.6, 16.2.7, 16.2.13**, pp. 111, 113–119)
- Explain when and how watercraft may be used in surface rescue. (**NFPA 1006: 16.2.10**, p. 119)
- Explain the use of helicopters in surface rescue. (**NFPA 1006: 16.2.9**, pp. 119–120)
- Explain how to self-rescue from surface water. (**NFPA 1006: 16.2.2, 16.2.3, 16.2.11**, pp. 120–123)
- Describe how to extricate victims and rescuers from surface water incidents. (**NFPA 1006: 16.2.12**, p. 123)

- Describe basic rope systems used in surface rescue. (**NFPA 1006: 16.2.12**, pp. 123–128)
- Describe how to terminate a water rescue incident. (**NFPA 1006: 16.2.14** , pp. 128–131)

## SKILLS OBJECTIVES

After studying this chapter, you should be able to:

- Preplan and identify existing water hazards in your district or region. (**NFPA 1006: 16.2.1, 16.2.4**, pp. 103–106)
- Use hand signals to communicate with fellow rescuers. (**NFPA 1006: 16.2.2, 16.2.4**, p. 108)
- Talk a victim through self-rescue (p. 111)
- Use water rescue reach tools to perform a surface rescue. (**NFPA 1006: 16.2.6**, pp. 111, 113–114)
- Throw rope bags, coiled rope, and similar items to a surface victim. (**NFPA 1006: 16.2.7**, pp. 114–119)
- Self-rescue in surface water. (**NFPA 1006: 16.2.2, 16.2.11**, pp. 120–123)

# Technical Rescue Incident

"It hasn't rained like this in decades," you think to yourself. You have never seen streets and highways flooded so badly. You do not remember ever seeing water in locations to which responders are being dispatched, yet the calls are still coming in. Your next call is the typical "your sign doesn't apply to me" motorist who has driven around a barrier, certain his SUV could make it through the deep water. There is not much current visible, and the water is just below the vehicle's door handles.

Units are scrambling around the city on similar calls, and the water response teams are stretched thin. On your engine, you have two lifejackets, a section of search rope, and your turnout gear. Verbal contact is made with the driver, and he tells you that he is the only occupant of the SUV, and he thinks his car has simply stalled out. The driver seems very capable and fit, and says he is not injured in any way. It is approximately 125 feet from the closest pavement to his car, and you cannot figure out whether the water is rising, staying put, or receding. Other cars are visible, but only their roofs.

**1.** Is your basic cache of personal protective equipment enough to safely attempt a rescue?

**2.** How will you ascertain whether the other cars are occupied?

**3.** Which tactic may best fit this response?

**4.** What are some hazards associated with the conditions you face with this response?

**JONES & BARTLETT LEARNING**
**NAVIGATE 2** *Access Navigate for more practice activities.*

## Introduction

Surface rescue is a very broad discipline that encompasses many areas of water rescue. Surface water is water moving at a speed of less than 1 knot (1.15 miles per hour), which includes not only pools, beaches, ponds, lakes, and rivers, but also urban and rural floods, industrial accidents, and even flooded basements. Making conditions even more hazardous for everyone is the possibility of the water being physically cold. All of these environments may be the sites of incidents needing a rescuer response (**FIGURE 4-1**). Surface rescue skills are the basis for many other types of water rescue work.

The threat posed by surface water should not be underestimated. In most years, flooding causes more deaths and damage than any other type of severe weather, and in many years three-fourths of all federally declared disaster declarations are due, at least in part, to flooding. Nearly half of all flood deaths involve automobiles, and many of these deaths result from the driver intentionally driving the car into the water.

The surface rescue world is a very large and sometimes complicated space. Some surface rescues may take place in a crowded swimming pool and last just a few seconds, whereas others involve natural disasters

**FIGURE 4-1** Ponds, lakes, and rivers are only a few of the possible surface water environments.
Courtesy of Steve Treinish.

with extreme logistical and human resources needs, lasting for weeks. Surface rescue can take place in any water, at any time, in any conditions, and a wide range of specific training for these emergencies is available. Rescuers are encouraged to train and preplan their districts for the type of missions they might see, in the worst conditions expected. Rescue organizations at the operations level should meet the operations-level requirements in Chapter 16 of NFPA 1670, *Standard on Operations and Training for Technical Search and Rescue Incidents*.

# Preplanning

Site surveys of water areas can be readily performed by first doing some homework inside the station and then cruising the response area on a nice spring or summer day. Of course, swimming holes and pools are obvious target hazards, but other locations and conditions should also be considered long before a call comes in. Flood conditions will not only bring victim rescue into the mix, but may also require the rescuers to secure shelter for victims after the rescue. Many agencies—including the Emergency Management Agency (EMA), Red Cross, and National Guard—may be able to shelter and feed these victims. This task is not generally the job of water rescue crews, but they may initially be involved in this aspect of the emergency, especially in areas where help does not arrive quickly. Many victims in an evacuation will have no idea of where to go, how to get there, or what will happen next, and rescuers can greatly assist the community by providing guidance to these individuals.

## Operations Rescuer Roles

Operations-level rescuers must know how the technician-level rescuers will operate and why they use the equipment they do. Operations-level personnel will often be charged with tending tether lines, interviewing witnesses, and collecting information. More importantly, these personnel can perform plenty of shore- or boat-based rescue techniques. Each rescuer must know his or her role in the response, and this knowledge begins with preplanning. Bring in the technician-level rescuers or teams and figure out exactly what is expected from responders and how best to accomplish the tasks that may be faced. The preplanning stage is a great time to examine the risk–benefit tradeoffs and give all parties involved a solid foundation of knowledge for expected conditions. Larger areas of surface water may actually require an operations-based attempt, as technician rescuers can swim only so far before fatigue becomes a factor, or tether lengths become unmanageable.

## Weather Forecasts

Weather can play a big factor in surface rescue, both in causing the emergencies that necessitate water rescues and through continuation or worsening of conditions during the rescue. For example, a forecast of heavy rain, with frozen ground conditions, should certainly alert water rescue crews to the potential for flooding, as the frozen ground will not absorb rainfall but rather let it run off freely. This funnel effect may cause the water runoff to flood unexpected areas, which may trigger ongoing calls for increasingly more difficult flood or water rescues. A quick line of storms may cause a few isolated incidents, but several days of rain with temperatures barely above freezing can cause the funnel effect to wreak havoc on the local community.

The time of year and water temperature also play into preplanning. Classic "cabin fever" occurs when the air temperatures warm up in spring. Folks will rush outside to enjoy the nicer weather, and many people will be around the water with no thermal protection. Many of these people underestimate the effects that the cold water will have on them if they enter it. As rescuers, we are required to know about hypothermia and its dangers, so we need to dress and act appropriately.

## Scene Access

Just getting to a victim can be a key aspect of water rescues, and can be a frustrating problem in surface water rescues. Teams may very well have to blaze their own trails to visualize an area where a victim is stranded if roads, driveways, or normal banks are washed out or covered by water. Many times, victims or submersions will be spotted or witnessed from a bridge, but preplanned shore access and entry points become the key to ensuring the mission ends successfully. Teams can check out bicycle paths, railroad rights of way, boat ramps, and even the easiest places in which to hand-carry boats if this need arises. Many bicycle paths will accommodate smaller pickup trucks towing trailers, but are too small to facilitate the movement of larger apparatus (**FIGURE 4-2**).

Obtaining this information can be much easier than it sounds. Get out and look at your water district and hazards, in all seasons. Summer months may offer better weather for conducting such a survey, but be aware that wintertime weather can affect routes to the hazard. The bike paths used so frequently in summer may not be routinely cleared of snow during the winter season. Similarly, many lakes and reservoirs are drawn down at various times of the year for water supplies or flood control. When these bodies of water are at their lowest point, launch ramps may be unusable, requiring alternative plans to access them with a boat.

Go out, take pictures, make notes, and preplan an area during a major flood and then again during a normal rainfall. Are there differences? How pronounced? Areas with rapid urban or suburban growth often see flooding or water responses in places not affected in the past, due to the large amounts of roof, street, and parking lot surfaces constructed during their development. Retention ponds may also be used

**FIGURE 4-2** Consider access routes other than standard roadways when preplanning responses.
Courtesy of Steve Treinish.

for decoration and for water filtration to protect rivers and streams, but they present a possible hazard.

## Mapping

Topographic maps are a key resource when determining where rainfall and floodwaters may end up. Following the topographic contours on such a map at the kitchen table in the engine house may not actually let you picture conditions in a flood, but it may raise questions that other sources can answer, or raise a flag about an area that should be observed during periods when the water is higher.

Flood insurance mapping can also help forecast water events, especially larger events that occur only once or twice in long spans of time. The 100- and 500-year flood events are frequently underestimated by rescuers and authorities having jurisdiction (AHJs) alike, and the amount of ground covered by water during such events surprises many. Transferring this information to a satellite map can give a very realistic picture of the large rescue area to be addressed, with many potential problems being identified right at the firehouse.

Many agencies, including many county- and state-level EMAs, have a graphic informational system (GIS)—that is, a computer program showing hazards, terrain, and areas where water would go in certain flooding situations or dam failures. This information, and the ease with which it can be acquired, is a great asset to many agencies. The Army Corps of Engineers is a superb source of information on potential flood hazards, as many of the dams it manages have been built specifically for flood control purposes. No one knows better the effect of flooding around or downstream of these structures than this organization does.

## Rescuer Access

Scene access is a good place to start in any location, but getting equipment to the water can pose its own challenges. Boats, rope equipment, EMS supplies, and other resources may all be needed at the water's edge, so take a look at the shoreline (**FIGURE 4-3**). Do trees block the boat access? Will rescuers need to hack and chop their way through heavy underbrush to stage at the water's edge?

Recognize that conditions may change with different water levels or flows. In many high-use water

**FIGURE 4-3** Trees, brush, and steep banks can often challenge rescuers by restricting access to the water.
Courtesy of Steve Treinish.

areas, especially rivers, completely different places must be used to extricate victims during high-water events. Some bridges may allow very rapid victim extrication by use of aerial Stokes basket evolutions, but this possibility should be planned and integrated into the evolution in advance by well-trained crews.

## Heavy Use Clues

Another way to preplan a water area is to examine the things left behind by the people who frequent the area. The presence of trash, bait cans, beer and soda pop cans, cars, tires, and other items suggests a location might have a better than normal chance of being the site of a water rescue (**FIGURE 4-5**).

**FIGURE 4-4** The rescue team and the landowner worked together to create this river access for emergencies.

Courtesy of Steve Treinish.

### LISTEN UP!

The Columbus, Ohio, Fire Department and a large concrete business in the department's jurisdiction offer a great example of how preplanning and cooperation can work. The concrete business was developing a new quarry area near a major river hazard area. The fire department rescue crews approached the business managers about grading off a boat access ramp in the rear of the quarry to allow much easier access for rescuers and equipment. Not only did the company agree, but the quarry workers also laid stone down to make it an all-weather access point, and maintain it to this day. In less than 6 hours, a very nice access was created with a minimal effort from either agency (**FIGURE 4-4**). This ramp saves almost 30 minutes of travel time in certain areas of the river.

A

B

C

D

**FIGURE 4-5 A**. Some heavy use clues are hidden from roadways or common access. One might assume the picnic table at the water's edge is the only water access area. **B**. However, nearing the end of the pavement, a small entryway is seen. **C**. This entry leads to a pathway through the woods, completely unseen from the road. **D**. A very remote and heavily used fishing access is found at the end of the walking path.

Courtesy of Steve Treinish.

Sometimes, complications will be encountered when a victim is found in a location from which rescue seems impossible, at least as seen from shore. The victim may have originally accessed the water from another high-use area, only to wind up at the seemingly inaccessible location. No matter how rough the terrain is along a bank, the water can still place a victim there, or the victim may get there under his or her own power. Picture a fisherman wading far downstream of a riverbed, then breaking an ankle. He may have entered the water at a nice sloping shoreline, but may now be in a steep valley far away because he walked downstream many yards from where he started.

One fire department in the Midwest noted a huge increase in the ice fishing population on a large reservoir in its district, and was shocked to realize that the layoffs of hundreds of workers in the area were causing some people to literally fish for food. Homeless camps are often found around water, both for the bathing and personal hygiene the water offers, and because many remote water areas are thick with trees, keeping this population out of immediate sight. Knowing the location of these camps can allow the fire department to offer better protection to all the folks in a population, as well as reduce confusion during a response.

# Rescuer Equipment

Water rescue personnel should not just be familiar with personal protective equipment (PPE); the basic PPE should actually feel like part of the uniform (**FIGURE 4-6**). Other pieces of PPE that may be needed for surface water responses are water rescue helmets, thermal insulation, and emergency-signaling gear such as whistles, lights, and even cutting-edge

electronic gear resembling the personal alert safety system (PASS) device used in the fire service. Rescuers should be guided by the manufacturers' instructions and recommendations for PPE, as well as the standard operating procedures (SOPs) established by the AHJ for water responses. It is also advised that rescuers work in pairs and utilize a quick checklist to ensure personal flotation devices (PFDs) are buckled, emergency equipment is present, and all gear needed is properly fastened and ready to deploy.

# Personal Flotation Devices

Personal flotation devices are covered in Chapter 1, *Understanding and Managing Water Rescue Incidents*, but a quick review is presented here, and some uses for each PFD are examined.

Frontline rescuers on fire or EMS companies and police cruisers typically carry Type III vest-style PFDs. This kind of PFD provides proper buoyancy to most rescuers, may include pockets that can hold other gear, and is generally inexpensive. Some companies and most water rescue teams purchase and use Type V swiftwater rescue vests, but the higher cost of these PFDs makes them less common in most fire departments (**FIGURE 4-7**). Regardless of which type of PFD you or your team buys, the best PFD is the one you will actually wear.

Consider attaching a cutting tool or shears to your PFD. Becoming entangled in a line, whether in or around a boat or a bank, is a very bad position to be in, and rescuers must be able to cut themselves loose of such a predicament. Be aware of the entanglement hazard if rope slack or excess is not kept controlled. Reflective taping and whistles are other possible add-ons to PFDs to make them better suited for the user or the mission at hand.

**FIGURE 4-6** Water rescuers should consider PPEs part of their uniform.
Courtesy of Steve Treinish.

**FIGURE 4-7** Most water rescue teams use Type V PFDs, which are meant for swiftwater rescue.
Courtesy of Steve Treinish.

## Water Rescue Helmets

Detailed in Chapter 1, the basic water rescue helmet is lightweight, drains water from the head, and offers good trip, strike, and fall protection. Helmets can be fitted with flashlights and chemical marking sticks for visibility during nighttime operations, and are available in different colors, making company or rescuer identification or sorting very easy.

## Drysuits

Water rescue has followed the fire service trend of striving to reduce the amount of exposure to contaminants while rescuers are performing their work. Better education and research has prompted most AHJs to consider what is physically present in the water rescuers work in, although often the specific risk cannot be evaluated. Drysuits, although thought of as technician-level PPE, are becoming more widely used, even by boat crews. Thermal and chemical protection is much better when rescuers stay dry, and even the best boat crews run the risk of accidental immersion. Drysuits utilize a latex, silicone, or neoprene seal around the wrists and neck to seal water out, and a dry zipper to keep water out of the entry point. Most drysuits also require a boot or shoe to be worn over the latex sock. Heavy-duty boots should be avoided; specific water rescue boots are recommended.

### Donning a Drysuit

Care should be used when donning a drysuit. The latex seals are somewhat fragile and can easily be torn by jewelry, watches, or pendants. Both the neck and wrist seals should be pulled up with fingers, rather than pulled on like a T-shirt, to reduce the strain on

**FIGURE 4-8** Putting plastic grocery bags over latex socks will help you don water shoes or boots with less chance of damage to the socks.

Courtesy of Steve Treinish.

them. A helpful practice is to place plastic grocery bags over the latex socks when donning water shoes or boots (**FIGURE 4-8**). The latex socks are also fragile, and the slippery bags can help slide the boot on with less damage.

### Drysuit Care

Drysuits are not inexpensive, so proper care of them is a must. Each seal should be dried and dusted with pure talc powder after use. The talc neutralizes the current wearer's skin oils, but also allows an easier entry for the next user. Zippers should be lubricated with pure wax or zipper lube; use of a petroleum derivative will degrade the seals and the zipper. Manufacturers of drysuits supply detailed cleaning and maintenance information for their products, and this guidance should be followed closely.

## Emergency Signaling Equipment

In smaller areas of operation, rescuers can use a whistle to signal their location in the water, and can send specific signals by varying the number of whistle blasts. Signal mirrors and dye packs are also used to signal location in water, and can be tucked into a pocket.

### Electronic Signaling Equipment

Just as technology has increased effectiveness and safety in the fire service and aviation, so the marine industry is taking advantage of advances in the size and capabilities of electronic equipment. Man overboard (MOB) devices are commonly used in larger water areas; they can be manually or automatically triggered by the user if he or she falls in the water. Personal locator beacons (PLBs) transmit an electronic signal via satellite and signal authorities to the user's location in the water. Automatic information system (AIS) beacons transmit the user's location to any watercraft or shore operator equipped with the proper receiving equipment. On larger rescue watercraft, multipurpose display units are frequently installed that show mapping points, location, weather, and, using the AIS system, other watercraft around them. Thus, a rescuer who falls in the water can be tracked by multiple boats in the area, making recovery faster and safer.

## Team Communication

The rescue team must be able to communicate, both among its own members and with members of other teams at an incident. Everyone working together should be on the same page. Radio interoperability, radio frequencies, hand signals, whistle signals, and line signals should all be reviewed and settled long before a run begins, and adjustments made as needed.

One problem faced by many teams is a lack of common radio channels or use of different radios and frequencies by each jurisdiction or team. Some areas are trying to fix this problem by establishing common frequencies, sometimes on a statewide basis. This allows each agency to communicate with every other agency anywhere, at any time.

Another type of effective communications equipment that offers a large "bang for the buck" is hand-held marine band radios. Most of these radios are inherently waterproof, and they enable the user to communicate with other agencies such as marine patrols, the United States Coast Guard, and other boaters.

**LISTEN UP!**

On budget-minded teams, hand-held "talkabout" radios can be used very effectively on scenes where all rescuers are working in relatively close proximity. Each crew in a mission can be assigned a radio. While many of these radios are not high powered, they represent a quick and inexpensive way to get everyone on the same channel. Their use is also a great way to ease radio congestion.

## Rescuer Communication

Many well-trained rescuers can have a "conversation" using only hand or body signals. This kind of communication may be necessary when water and wind noise, distance, or radio failure do not permit voice communication. Some teams use hand signals that are unique to that particular team, but certain signals are typically used by many teams and agencies. More specific signals can be used for specific needs, such as signaling watercraft or crews on shore.

- *OK? /OK!* The rescuer makes a large "O" over the head with both arms, or uses one arm to form an "O" onto the head. This signal, which originated in the dive industry, can be used as a question or as a response (**FIGURE 4-9**).
- *Help!* The rescuer waves one or both arms back and forth to attract attention (**FIGURE 4-10**).
- *Stop!* The rescuer clenches one fist, with the arm held above the body (**FIGURE 4-11**).

**FIGURE 4-9** OK?/OK! signal.
Courtesy of Steve Treinish.

**FIGURE 4-10** Help signal.
Courtesy of Steve Treinish.

**FIGURE 4-11** Stop signal.
Courtesy of Steve Treinish.

Whatever the signals used, rescuers should know them by heart, and each team should at least be familiar with the signals of other teams. A lack of communication can severely hinder a rescue effort, or even cause injury to responders.

## Scene Size-Up

As discussed in Chapter 1, size-up can be a quick and easy mental note, a verbal plan, or a large investigation. During size-up, the rescuer weighs the information available against the tools available, and begins the development of the incident action plan. Size-up is not just done once and then forgotten, but rather should be dynamic and ongoing—that is, a continual process that helps a rescuer or crew determine the effectiveness of any evolution or rescue attempt. While Chapter 1 mainly deals with the awareness-level and incident management areas of an incident, the discussion of size-up presented here focuses on the rescuer's specific tasks when going to work around the water.

Two main types of searches are conducted before a rescue commences. **Active search measures** are performed by rescuers committing to a rescue event. Physically seeing a victim struggling in the water will require some sort of action to rescue the person, and rescuers can select the evolution that is best suited for the conditions and ensures the safety of the responding crew. **Passive search measures** are best suited for overdue boaters, missing watercraft, and incidents that have no obvious victims, scene location, or distress. Passive searches can turn into active searches at any time, so rescuers should still be ready to operate, with the anticipated resources being put in place.

## LISTEN UP!

With no obvious victim or witnesses, rescuers are encouraged to stage at a location that offers the ability to move quickly. Instead of stacking up all the pieces in one parking lot, a better move may be to have other companies investigate other areas or to stage in a location that enables responders to travel to more places quickly.

## Recognizing the Hazards

Given the many conditions that may lead to a surface water rescue, the specific hazards at the scene can change how a rescuer decides to attempt a rescue. Surface water incidents are dynamic scenes: Water conditions may change, the temperatures may affect operations greatly, and the capabilities of the victim, rescuers, and support crews must all be assessed.

## Water Conditions

Surface water is a liquid, static body of water. However, it may be very large, very cold, and very dark during night operations. Moving water presents a totally different scenario, so rescuers must be able to quickly decide whether the water is considered static or swift. Swiftwater is water that is traveling faster than 1 knot (1.15 miles per hour). Keep in mind that a victim moving in water, whether because of the water's flow or the victim's own movement, may end up in swiftwater in a different location, or static water may start flowing because of factors such as rainfall, inflow, or drainage. The power of swiftwater is often underestimated, and rescuers must respect it.

Wave or wind action may also affect the water conditions, making communication between rescuers or shore difficult. This action may also move the victim around in the water areas. A key question to answer when sizing up a surface water scene, therefore, is "Will any victims be subjected to movement by the water?" If so, at least consider the idea that the event may involve another form of water, such as surf or swiftwater.

## Water Temperatures

One of the fastest ways to glean information is by using your spatial awareness from the time of the call. A rescue involving surface water in Michigan in November will certainly require thermal protection, whereas an incident in the same type of water in Alabama in August may not require any insulation. Simply knowing whether you need thermal protection is a good start. Note, too, that the victim may not have any thermal protection, which may play into the choice of a specific rescue evolution. A lack of thermal insulation for rescuers may preclude immediately choosing to use a watercraft, or prompt a call for assistance from a team that uses cold-protection gear.

According to the National Center for Cold Water Safety, water temperature less than 70°F is considered dangerous for humans. Most humans experience severe hypothermia in freezing water in less than 1 hour. This knowledge helps incorporate a timeline into the size-up.

## Water Size

The size of the water area involved can greatly influence the size-up. A small retention pond may be visually cleared in seconds if conditions are right, and may result in a dive operation. Larger waters, such as large lakes or even oceans, may rule out immediate sighting of any victim or other scene, such as a crash site. This factor affects team safety a great deal, in that peripheral resources may be needed for larger areas, such as backup crews, scene lighting, and scene security.

## Weather Conditions

Rescues are significantly affected by weather conditions; indeed, both current and forecasted weather play into size-up. Increasing winds can make a calm lake very rough in just a matter of minutes, as can thunderstorms. Such a turn of the weather may mean the team keeps its small johnboat on the trailer and requests a larger, more stable watercraft. Excessive cold or hot air temperatures may also change the method of transportation over the water used during a search.

## Other Hazards

Other factors may also come into play during size-up of a surface water incident. Utilities such as electricity, gas and propane supplies, chemicals and contaminants, and even basic slip and fall hazards around the shoreline may all need to be addressed.

## Interviewing Witnesses

More intuitive information can be found as soon as a rescuer arrives at the scene. Are there folks present and pointing? If so, the following questions need to be answered:

- What happened? Slip and fall? Boat crash? Drinking or drugs?
- Where were the victims last seen? The specific location of the **point last seen (PLS)**, coupled with the water and wind conditions, may help rescuers predict the current victim location.
- How many people were seen in the water, and how many were seen coming out?
- Did anyone else attempt rescue? To truly call the water cleared, we need to ensure no other persons attempted rescue before arrival.
- What were the victims wearing? Thick clothes or jackets? PFDs?

Some of these questions can be answered very quickly without a witness. For example, suppose you arrive to find a car in a shallow retention pond with the driver sitting on the roof. She shouts that she was alone in her car, and you see she has no PFD.

All witnesses present should be questioned from the point where they watched the event take place. For more information on witness interviews, refer to Chapter 1.

## Maps

One key facet of surface rescue on larger bodies of water is mapping. Maps and charts can be used not only for tracking the search, but also for predicting where a missing person may be. For example, a boat overdue at a dock does not automatically mean there is a victim, but a good passive search would check the water areas as well as any attractions close by, such as bars or swimming or fishing areas. Mapping can be paper or electronic.

## Survivability Profiles

It is a tough call to make in surface rescue, but a survivability profile must be created for each incident, though many times it can be done quickly. Once again, the large, varying parameters of surface rescue make survivability of the water hazard a relative consideration. A person in cold water with no protection may die within 60 or 90 minutes of immersion, whereas a person in warm ocean water may float and travel with currents or tides for hours. Folks who were witnessed jumping from cliffs or from bridges may certainly die from the fall, but if their body is present, the response should favor the safety of the rescuer. A good way to look at victim timelines, then, is to give the benefit of the doubt when a question arises about saving lives, provided crews can operate safely.

## Additional Resources

The final step in the size-up process is deciding whether any additional resources need to be dispatched or to be on scene before a rescue attempt takes place. Scene lighting is a must at night, shore access may need to be considered, and even boat or auto traffic may need to be addressed to protect rescuers. Backup rescuers and teams should be at least as trained as the operating rescuers, and should be in place before the operation starts.

One important position that may be assigned is spotter. **Spotters** are crewmembers equipped with proper PPE, observation equipment such as binoculars, and a means of communicating with the incident or operations commander. Spotters can be utilized in passive searches by watching for potential victims who may be moving in or around the hazard or predicted locations, and in active rescues by reporting any potential hazards that rescuers may face. For example, spotters may be tasked with watching for bystanders or boat traffic that get too close to an operation and cause issues for rescuers or themselves.

# Scene Operations and Planning

Once on-scene witnesses have given your team something of a narrowed-down area—for example, a victim who was seen splashing around a few minutes prior, with no evidence that the victim exited the water under her own power—the scene operations change direction, focusing on a more refined search and rescue/recovery effort.

# Victim in Sight

Rescuers may not get many chances at the true working rescue of a distressed person in the water, especially one with no flotation. Typically, by the time rescuers are called, the victim has already submerged or exited the water by some other means. Nevertheless, especially in districts with high-use water areas, it may be the case that a victim has not yet submerged. People (sometimes even rescuers!) may find themselves in positions needing rescue, such as in sheer drop-offs, while floating on debris, or even when wearing flotation but suffering from injury.

Most rescuers know the old training slogan of "reach, throw, row, and go," but there are other options for some victims and rescuers. "Yo" self-rescue is simply verbally directing a victim to safety or a safer location, and helicopter rescue is commonly used in larger metropolitan, coastal, or rugged terrain. Regardless of the equipment available, these adages stack the rescue type against the potential risk to quickly evaluate the safest evolution to perform a rescue. Experience and scene size-up can quickly take some evolutions off the table, such as a victim too far from shore to benefit from throw bags.

A review of the most common rescue techniques follows, with each successive type of rescue attempt presented being more complex and potentially dangerous to rescuers. Rescue crews often arrive at a rescue method decision in seconds, based on experience, preplanning, and scene size-up.

## Verbal Self-Rescue

If a victim is stable in the water and can help himself or herself, simply making verbal contact with the victim may mitigate the situation. Perhaps the victim cannot see the optimal place to exit the water or another rescuer approaching. A person who has fallen through ice could quite possibly be talked into a proper self-rescue if close enough to shore. Sometimes a person may not need immediate rescue at all if he or she is in a safe place and rescuers can keep him or her there until a better means of evacuation arrives.

The hazard may also dissipate. Consider a victim in a tree in a flash flood, with no obvious injury. The weather forecast is for clearing skies, and the flood is receding quickly. If that victim can stay put and not enter the water, the water itself may recede, or time can be bought to set up and perform a safer rescue. The tree may not offer the most comfortable perch, but staying there beats losing a life. Simply put, some problems can be solved verbally, and sometimes the best action is delayed action or even planned *in*action.

# Reaching Out for Rescue

The safest, most direct way to the victim without actually entering the water is to reach for him or her. Most fire trucks carry tools that can be used for this purpose, or makeshift aids can be utilized on scene. Pike poles obviously work well, but are somewhat heavy, especially when extended fully to the victim. Trying to hold a 16-foot pike out to someone in the water by grasping just the butt end is very hard. Shepherd's crooks are lighter in weight, are found at many pools and swimming sites, and work very well for reaching visible victims (**FIGURE 4-12**). The large hook is used to actually sweep and hold the victim as he or she is pulled toward shore.

Modified tools can include extension ladders with floats attached, or even a PFD strapped to the end of the pike to provide some flotation assistance to the victim. Hose inflator systems are primarily considered an ice rescue tool but can also be used successfully in certain swiftwater and surface rescue work. Items on scene that can be used in a pinch are limited only by the rescuer's imagination and safety considerations (**FIGURE 4-13**). Tree limbs, paddles or oars, and even long-handled tools can all be used.

The important thing to remember when using such a tool is that the victim will pull back. Hang on, but be ready to release. The victim will sometimes pull so hard that the rescuer will end up in the water, too. Reaching rescue is most likely the method that civilians will use unless they have some specialized equipment available to them at the water's edge.

Do not discount the possibility of reaching out with a ladder truck or aerial platform. Even if the ladder itself does not reach the victim, it may buy some extra distance in which to then throw or reach, keeping the rescuers in a safe zone.

**FIGURE 4-12** Shepherd's crooks, found at many pools and swimming sites, work well for reaching visible victims.

Courtesy of Steve Treinish.

# Voice of Experience

Being a well-equipped team and maintaining a high level of competence in the varied roles of a water response unit are the keys to a safe and effective surface rescue. I can think of a particular call when I was in command of our fire boat. A group of young children, ages five to nine, were spending their summer afternoon in a local sailing camp. Miami is known for its fantastic summer days, but less known for its sometimes violent afternoon thunderstorms. These storms can move in quickly and make a dramatic change in the conditions of the waters that surround our land and assorted islands.

It was this very weather that caught this group of young campers and their adult leader by surprise. There were nine sailors in total, each in a small sailboat no more than six feet in length. The sailboats were tethered to each other, bow to stern, when a storm rapidly presented itself, capsizing every camper and their adult leader. There was no motorized vessel nearby to make a rescue. Their only radio communication was a portable VHF marine radio held by the camp leader, which was lost when his boat capsized. Despite this unforeseen tragedy there was one fortunate occurrence: A witness from a shoreline condominium witnessed the scene and called 911.

At the moment of dispatch we knew that this was no ordinary call. Our response would be short, but the visibility in the storm was less than 50 meters. Although our fire boat is equipped with radar, through our dispatcher we kept in constant communication with our land-based witness. The witness could no longer see our small, capsized victims, but she was able to direct us to their last known location, thus placing us near the correct position to make a rescue.

Once we had our victims in sight, we recognized the gravity of the situation and our need to act swiftly. We had minutes to make a rescue, as each child was either holding onto another child or holding a short rope floating from their capsized boat. The storm was at its peak, and in the rough weather the campers, along with their leader, had clear looks of panic on their faces. Threatening to make us unable to perform a rescue, and further complicating our scene, were sandbars that were shifting around our rescue vessel during the storm. We had some decisions to make. Do we use throw bags? We weren't confident with this choice because of the young age of our victims and their ability to grab the line in a rough sea. Would we make matters worse? Should we place a rescuer in the water? This operation appeared unrealistic as we were only a four-person crew and yet still had to operate a 48-foot vessel and make nine rescues. We decided to reach out for our victims one by one, carefully placing the bow of the vessel in position with a rescuer tethered and hanging over the edge. This operation required effective team communication as we relied on hand signals to direct our fire boat operator, who was inside the cabin.

I am proud to say that each camper and their adult leader were rescued that day and were safely returned to shore. Beginning with our attentive witness, I now recognize all of the integral pieces that had to go just right in order to have been successful. We can rely not only on our rescue equipment but our ability to gather information, assess the scene, and make operational decisions that are appropriate and realistic for the situation at hand.

**Robert Hevia, CFO**
District Chief
City of Miami Fire-Rescue
Miami, Florida

**FIGURE 4-13** Many items found at a scene can be used to facilitate a surface rescue.
Courtesy of Steve Treinish.

**FIGURE 4-14** Recovery jaws use hydraulic force to open and close the tool's large, curved jaws.
Courtesy of Steve Treinish.

# Reaching Underwater

Very often, surface rescues result in underwater recoveries or floating deceased victims. In worst-case scenarios, a victim is never found or recovered. Many fire- or EMS-based rescuers will never face the typical drowning in process, because often the victim is already submerged before rescuers arrive. If the water is shallow enough that you can reach the bottom with rescue tools, some search techniques can be used immediately. Even with more technical resources such as dive teams en route, the name of the game is still rescue. With this point in mind, surface rescuers are somewhat limited in the scope of their operations, but some very effective, tried-and-true measures can still be employed on a call with no victim in sight, but a good PLS.

## Probing Underwater

Pike poles or similar probing tools can be very effective in some scenarios, especially in water deeper than the rescuers can stand in and when the probing is performed from a boat. Using pikes or probe poles underwater can be tedious, and the depths that can be probed are limited by the length of the tool, but sometimes they are the only tools available. Probing works better when the bottom is fairly clear, but knowing the makeup of the bottom material can help rescuers discern what they are feeling.

A body is a relatively large object that is easily felt underwater when circling in soft mud or sand. Consider using the butt end of the pike pole to tap or make small circles on the bottom. This keeps the pole in contact with the bottom, making it easier to operate and giving the rescuer a better feel for the bottom material or objects. Heavy vegetation may hinder the pole, but a recovery can be made—it just takes a lighter "feel" and is much slower.

Due to the relatively light weight of a submerged body, a better way to investigate possible "hits" is to keep the first pole on the object underwater, and then let a second rescuer use the hook end of a second pole to recover the object. When pulling up a victim on a probe pole or pike pole, go slowly but steadily, and be ready to grab the victim. Victims can sink much faster than many rescuers realize. Often, the rescuer operating the pole will drop it quickly to get a hand on the victim, but it is better to keep upward pressure on the victim and let another rescuer make the grab.

## Recovery Jaws

In depths less than approximately 20 feet, **water-type recovery jaws** can be used to pick up victims off the bottom. With these tools, hydraulic force is used to open the large, curved jaws, and then to close the jaws as they are brought back up (**FIGURE 4-14**). The victim is immediately felt as a heavier pull, but not so much pull as to be a snag. Keep the jaws coming up steadily, which will keep water flowing over the tabs, locking the victim inside the two hooks. When the victim is in sight, be ready to grab him or her. Do not simply set the jaws back down; the two jaws will relax and quite possibly drop the victim before rescuers can get a hand on him or her. As when working with probe poles, keeping the victim attached until another rescuer makes positive victim contact is the better choice.

**LISTEN UP!**

A great probe pole can be fashioned from a piece of ¾-inch or 1-inch aluminum pipe, with a small grappling hook attached on one end, usually with pipe thread. The butt end is used to probe, and the hook end is used to snag victims. Using the hook end underwater will result in many snags and slow progress. A little practice with an underwater mannequin will go a long way toward increasing skill with this probe pole.

## Sonar

Although they can take a bit longer to set up if not installed directly on a boat, **sonar units** have revolutionized the rescue business, especially in dive operations. They are very effective at reading the bottom of large areas, and are especially useful in missions involving missing boaters or floating watercraft with no witnessed submersions. A well-trained sonar operator working over a clear, unobstructed bottom can cover an area of water very quickly (**FIGURE 4-15**). Sonar units are a great reminder of the value of pre-planning and determining the resources available to the team. If your team does not have access to sonar units, finding out how they can be obtained might be an important part of your team's preplanning efforts. More information on using sonar units is found in Chapter 14, *Dive Rescue Operations*.

## Drag Hooks

**Drag hooks** are used less frequently today (sonar is preferable to this option), though they are still found on many boats. With this type of underwater probing, a large bar equipped with small hooks is dragged through the water in hopes that it will snag the victim

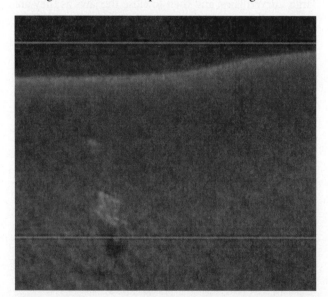

**FIGURE 4-15** Sonar units can be used to search an area of water quickly. In this image, the victim is lying face up, with legs slightly bent.
Courtesy of Steve Treinish.

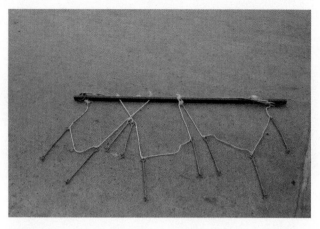

**FIGURE 4-16** Drag hooks can be used to snag a victim.
Courtesy of Steve Treinish.

(**FIGURE 4-16**). Drag hooks can be successful, but two problems are often encountered. First, rescuers may not realize how light a body feels on a set of drag hooks or, even worse, crews may tie the drag hooks directly to the boat towing them. Rescuers may then lose the body of the victim when making turns or by bumping the body along the bottom. Second, the PLS is a critical piece of information, especially in dive operations. Many times the PLS is very good, but drag hooks take the body off that point, so that the recovery takes longer.

## Throw Rescue

Although throw bags are probably the most popular throwing rescue tools, other effective throwing tools are available, including line guns, rescue rockets, ring buoys, and even Frisbee-style devices. Do not discount the value in dropping or lowering things to victims if that is the only means to supply flotation. A rescuer on a bridge with no rope would certainly be justified in throwing a PFD to a victim in trouble in water below the bridge. We can always get more PFDs; the victim cannot wait. Be wary, though, of tempting a victim to leave a safe area in an effort to reach the flotation. If the victim is found on an abutment and is stable, perhaps quickly lowering a PFD via a rope, with instructions on how to don it, would keep the victim safer by reducing the urge to jump or reach out to a thrown rescue device. A bad jump or loss of balance might put the victim into the water, exposing the individual to a much greater hazard.

## Throw Bags

Throw bags are probably the most common water rescue tools found in the fire service. They are found on many watercraft, rescue trucks, dive rigs, and frontline fire apparatus almost everywhere in the United States. A throw bag consists of a length of floating water

**FIGURE 4-17** Throw bags are commonly used in water rescue during primary rescue attempts.
Courtesy of Steve Treinish.

rescue rope, ranging from 7 mm to ½ inch in diameter, that is stuffed into a bag; this bag is thrown into the water near the victim (**FIGURE 4-17**). Most bags are equipped with a small float in the bottom of the bag that allows the bag to stay at the surface.

Throw bags can also be pressed into service as dive tethers, tag lines, or even anchor lines in a pinch, but they should not be used for life-safety loads hanging humans, unless it is truly a loss of life situation. The smaller-diameter lines deploy better through the air than thicker ropes do and let the user throw the bag over a longer distance. The larger-diameter ropes create a lot of drag through the air, which limits the distance over which the bag can be thrown.

**Deploying the Throw Bag.** Seasoned rescuers may be familiar with the use of a throw bag, but younger members who have minimal fire training at the beginning of their careers may be less well versed in their use. A throw bag is not thrown like a football, but if you tell an untrained person to "throw the victim a throw bag" you may get exactly that result. Throwing the entire bag to the victim is most definitely *not* how it is done. Instead, the rescuer opens the bag, grabs and holds the end of the line, and throws the bag with the remaining rope on board. As the bag travels through the air, the rope deploys from the bag, and ideally will fall close enough to the victim to be grabbed.

### SAFETY TIP

Many throw bags are stored with a loop tied in the end the rescuer holds onto. This loop should not be placed over the wrist, as that would leave the rescuer attached to a rope from which he or she may not be able to escape, causing the rescuer to end up in the water with the victim. A knot may be used as a stop for the hand, to provide some pull, but using the bag in this fashion allows the line to quickly be dropped if needed.

Rescuers should always yell "Rope!" when throwing, as the victim may hear this message and know to look for the rope. Rescuers inadvertently in the water who hear this call will also know they are in sight of team members, and recognize that help might be landing on or beside them.

With some practice, throw bags can be rapidly redeployed, or the rope coiled and rethrown in case of a missed throw. Follow the steps in **SKILL DRILL 4-1** to deploy a throw bag.

### SAFETY TIP

Do *not* place rocks or other heavy objects in the bag to provide more throwing weight. Water works well for this purpose, and will not injure the victim further.

**Throwing Styles.** While the underhand throw shown in Skill Drill 4-1 is the typical practice throw, many rescuers find that when throwing from a steep bank, from a bouncing boat, or under a thick canopy of brush or trees along a river, other methods must be used. Sidearm, overhand, or grenade-style throws can all be used. Rescuers can also expect to throw from their knees, or perched in awkward ways. In boats in turbulent water, staying on the knees and keeping your center of gravity down is much safer. Grenade throws are made by staying low and lobbing the bag in with a sidearm motion, as described in **SKILL DRILL 4-2**.

**Reloading Throw Bags.** Throw bags are designed so that if the rope throw is short or off target, the rope can be quickly "reloaded" for another try. Instead of re-packing the entire bag, the bag itself can hold some water to provide the weight needed to throw it again. The loose rope can be coiled on the nonthrowing hand, and the weighted bag thrown again. With some practice, throw bags can be reloaded and thrown again very quickly.

**Repacking Throw Bags.** Throw bags should be repacked after use or cleaning, and must be repacked properly so that they will deploy smoothly the next time they are needed. Here is one quick way to repack a throw bag:

1. Open the bag as far as possible, and grip the bag with your middle, ring, and pinkie fingers.
2. Run the rope through the "OK" sign made with your thumb and index finger.
3. Use your other hand to pull the rope through the OK sign and into the bag. The rope should be stuffed in the bag, not coiled. This packing lets it fly through the air better during throws (**FIGURE 4-18**).

# SKILL DRILL 4-1
## Deploying a Throw Bag NFPA 1006: 16.2.7

**1** The rescuer loosens the drawstring on the top of the bag. This lets the rope deploy cleanly, without getting tangled or bound up in the bag. Do not place the loop over the wrist: The rescuer may not be able to remove the loop and may be pulled into the water.

**2** Hold the end of the rope and throw the entire bag (in this case, underhand). The point of aim should be over the victim's shoulders. Aiming directly at the victim may result in a short throw.

**3** Loudly yell "ROPE!" when throwing, to get the victim's attention.

**4** If the bag is overthrown, the victim can grab either the rope or the bag as it is pulled to shore. Once the victim has the rope secured, the rescuer can simply back away from the water. This technique keeps the rescuer away from the water's edge and reduces risk.

Courtesy of Steve Treinish.

# SKILL DRILL 4-2
## Throwing Grenade-Style NFPA 1006: 16.2.7

**1** With the end loop in one hand, use the other hand to pull the bag to the side and behind.

**2** Swing the bag forward and release it, while holding onto the end loop.

Courtesy of Steve Treinish.

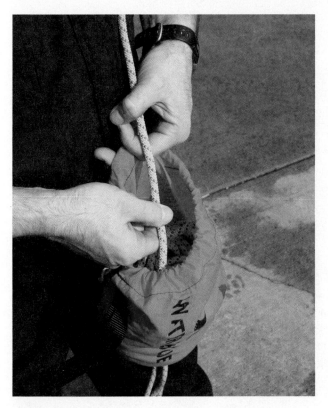

**FIGURE 4-18** Repacking a throw bag.
Courtesy of Steve Treinish.

## Coiled Lines

The ability to throw a hand-coiled line is important for two reasons. First, throw bags can miss their target.

Rather than restuffing the entire bag, the rope can be stacked on the rescuer's hands and thrown again. Second, sometimes another type of rope may be the only option available. Ski-ropes, anchor lines, and even utility rope found in many vehicles can be thrown to victims in the water, but only after some practice.

When throwing coiled ropes, the rope should be divided into two sections. The first section consists of five to seven coils of rope that lie on the nonthrowing hand. The second section lies on the throwing hand, and the weight of these coils pull the nonthrowing hand coils off the hand when it is thrown. The coils that are thrown stretch out and deploy from the weight behind them. Follow the steps in **SKILL DRILL 4-3** to coil and throw loose rope.

### LISTEN UP!

Many rescuers pull out a throw bag and practice with it occasionally, with a partner playing a victim a few dozen feet away. The "victim" usually puts the hands in the air and stands still, often on the back ramp with no blocking objects or other issues to consider. Consider having the person playing the victim jog across the area, and rescuers throw from between trucks or from under trees. These scenarios add fun to the exercise, are very realistic, and can be used to add more rescuers (more confusion and ropes) and even more victims in the mix. Simulating a tipped canoe with three victims will require a lot of quick coordination from the rescuers practicing with throw bags.

# SKILL DRILL 4-3
## Coiling and Throwing Loose Rope NFPA 1006: 16.2.7

**1** Lay the end of the rope on the nonthrowing hand, with the palm up. The number of coils depends on the length of the rope. Each coil should be small enough to easily handle without tangling. Too-long coils may get tangled in the feet or other ground objects. Place the coils flat on the palm of the nonthrowing hand, with each coil positioned side by side, getting closer to the fingers. Load the throwing hand the same way. If possible, place all the coils on the nonthrowing hand, as it will allow the rescuer to pick about half the coils for the throwing hand in one smooth motion.

**2** When throwing, extend the off hand out, so that the rope is easily pulled off. Throw the coils on the throwing hand together.

**3** The weight of the coils on the throwing hand will pull the coiled rope off the nonthrowing hand, while flaking apart in the air. Point the nonthrowing hand toward the victim, and get the thumb out of the way of the rope.

**4** Aim for the same area as with the throw bag toss: just over the shoulder and past the victim. It does take some practice, but a throw of 30–40 feet is not hard to accomplish and is identified in NFPA 1006 as a goal. A double throw of a bag and a coiled line should be done within 40 seconds to a target 40 feet away.

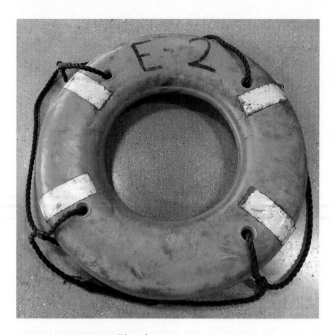

**FIGURE 4-19** Ring buoy.

Courtesy of Steve Treinish.

**FIGURE 4-20** Practice the sidearm lob method of throwing so that it comes more easily at rescue incidents.

Courtesy of Steve Treinish.

## Ring Buoys

Ring buoys can be thrown alone when a victim is separated from rescuers by a longer distance, such as a bridge or ship deck, or a speeding boat (**FIGURE 4-19**). They can also be attached to throw bags or lines and used in a throw rescue to provide the rescuer with a method to pull the victim in.

Ring buoys are bulkier and heavier than a throw bag, but they can cover good distance in a throw and provide more flotation for a victim. The sidearm lob is typically used to throw the ring buoy correctly and accurately. With this technique, the rescuer swings the ring back and around, holding it by the inside of the ring or the rope attached to the sides. Swinging your entire body and releasing at the correct time will let the ring travel straight and far; achieving this kind of performance, as with throw bags and coiled rope, takes practice (**FIGURE 4-20**).

## Watercraft in Surface Rescue

Watercraft are the "go to" tool for surface rescue of victims who are beyond the scope of reaching tools and throw bags. Short of surface rescue technicians attempting a swimming rescue, the quickest and safest way to reach a victim or victims is in a boat. Rescuer safety is increased in a number of ways when watercraft are used, with the most obvious advantages being flotation and ease of travel. Swimmers do not have to swim when a boat is available;

they merely ride, and usually ride much faster than swimmers can make progress through the water. Hypothermia, chemical exposure, debris, and currents can all injure or kill improperly equipped rescuers, but watercraft can allow crews to deal safely with all these hazards. Lighting can be carried and utilized on the water with watercraft. Even more importantly, patient care can begin more quickly on a watercraft, even to the point of advanced life support (ALS) measures being administered while en route back to shore. Watercraft are so important in the water rescue world that NFPA 1006 specifically mentions the need for operations-level rescuers to be versed in the operation, equipment, use, and safety procedures of watercraft. For more information or to review watercraft, see Chapter 2, *Watercraft Rescue Operations*.

## Helicopter Use

Many people have seen footage of U.S. Coast Guard rescue swimmers being lowered into rough-and-tumble seas to rescue victims. Although this is a very real capability and resource for some agencies, it is also a very specific, highly expensive, and training-intensive capability. Many fire and rescue agencies work hand in hand with helicopter-based crews, but few can maintain this equipment. Even fewer actually use or have helicopters for deployment in this way.

If helicopters are available for water rescues, they can offer many capabilities to crews during a surface rescue event. A surface rescue may not always be a specific, single-victim mission; on some occasions, it may be a very large response to a flood or natural disaster. Hurricane Katrina is an example of a natural disaster in which helicopters were used for surface rescue, transport, and reconnaissance. Personnel in helicopters can take aerial photographs that assist in planning routes, evacuations, shelters, and other operational needs. A helicopter equipped with infrared cameras and night vision can be deployed on actual rescue missions. The technology and altitude that helicopters can provide may prove a great asset. In addition, nighttime scene lighting from helicopters is fast, truly portable, and super-bright. Scenes can even be well documented using the video capabilities on these aircraft. Helicopters may be used when rescue crews need transportation to sites or staging areas that are inaccessible by motor vehicle or boat, and supplies can be airlifted and dropped to victims or rescue crews at these sites. Similarly, helicopters can aid mass evacuations.

Approaching helicopters should be done carefully, owing to the danger zones associated with these aircraft. Chapter 16, *Helicopter Rescue Support* takes a closer look at operating with and utilizing helicopters in water rescues.

# Self-Rescue

While many rescue personnel may be called to the scene of a surface rescue, not all of them may be good swimmers. Although the use of PFDs is required in this line of work, accidental immersion could happen to any rescuer—and every rescuer should be prepared for this possibility. "An ounce of prevention is worth a pound of cure" is an old adage, but it certainly rings true in the water rescue business. Regardless of the type of water you face in your district, your preplanning should include each of the water types that could be faced. You are not prepared to enter the water in a flash flood when the temperature is 40°F if you have only trained and learned to float in a swimming pool filled with 80°F water. Plan for the worst!

The first way to stay out of trouble is to assess your own swimming skills. Then consider the fact that even good swimmers cannot stay face up while unconscious. Simply placing yourself in a safe area on the shore can mean the difference between life and death.

As an example, consider an emergency at a low-head dam. The area immediately downstream from the dam will recirculate the river water, along with anything in it. This area of recirculating water is called a boil line. How many rescuers would make the heads-up decision to stay downstream of the boil line unless it was absolutely necessary to go upstream? If you should fall into the river, falling in downstream of the boil line will likely keep a rescuer from being sucked into it, whereas falling in beside or upstream of the boil line might be a death sentence. Stay away from these kinds of traps, and wear your PPE.

Before getting anywhere near the water, it is important to practice self-rescue skills.

## Staying Afloat

The first order of business in an unplanned submersion is to stay afloat. Weak swimmers might be better off focusing all their energy on simply staying afloat rather than trying to make it to shore. Once afloat, they can wait in the water for rescue to come to them.

All rescuers without a PFD in water deeper than their neck, or in water flowing with enough force to knock them off their feet, must make some sort of effort to keep their head above water. Less-experienced, fearful, and out-of-shape rescuers will quickly find that fatigue becomes a major factor in achieving this goal. If rescuers jump in deeper water while wearing heavy-duty boots, they will find that, in addition to the boots creating a problem, treading water with water-laden boots wears out the muscles surprisingly fast. This kind of fatigue can lead to panic and a reduction in the effectiveness of motor skills and proper technique. In turn, these effects can lead to even more panic, creating a vicious cycle that may not be easy to break.

## Floating

The easiest way to stay above water is to float. Deep inhalations with the lungs will help provide buoyancy, as will positioning the body face up and arching the back slightly (**FIGURE 4-21**). Staying face up helps keep the airway dry. Any heavy objects

**FIGURE 4-21** Floating is most easily accomplished by taking deep inhalations, positioning the body face up, and arching the back slightly.

A

B

**FIGURE 4-22 A.** Drownproof floating. **B.** Drownproof breathing.

Courtesy of Steve Treinish.

(e.g., multiuse tools, large knives, radios) should be dropped. It makes no sense to try to float to save your life, while keeping a few pounds of tools on the belt. Relaxation, while probably not at the forefront of reasoning at this point, will do much to help flotation. Easy kicking or treading of the legs can also be done, and may even start some movement toward safety.

## Drownproofing

The natural weight of one's arms and legs can contribute to difficulty in staying vertical in the water. The human head is also much heavier than most people realize, and it can take quite a bit of strength to keep it upright for long periods of time. The method of survival floating called **drownproofing** takes this reality into account. Drownproofing is done by lying face down in the water, without moving, and letting the head, arms, and legs hang down limply (**FIGURE 4-22A**). The breath is held, and when the person needs to breathe, the head is raised just enough to exchange air; then the breath is held again for a few seconds (**FIGURE 4-22B**). Again, relaxation is the key to success with this technique. The lack of movement also helps aerobically, as not much physical work is done while drownproofing.

## Other Flotation Aids

Anything that floats may help a rescuer who has fallen into the water. Logs, buoys, and even basketballs and soccer balls will float and offer some buoyancy. By using any makeshift aid, the flotation problem is at least partially solved, and egress to shore can be considered. Some flotation aids, such as boogie boards, allow the rescuer to lie on his or her stomach area; the

legs can then be kicked for forward motion, and the head positioned to scan ahead for decision making. Other flotation aids, such as the generic collar-style PFD, will be easier to manage with the rescuer lying on his or her back, keeping the head and airway protected from submersion.

## Emergency Swimming

Once adequate flotation is secured, and the rescuer can keep the airway above water, some kind of effort might be made to get to safety. Ideally, the other rescuers will have noticed someone in the water or missing, but in a worst-case scenario, you can save yourself.

## Swimming Strokes

The most commonly used swim stroke is the forward crawl; however, a person not used to swimming this stroke can tire *really* quickly (**FIGURE 4-23**). Also, weak or nervous swimmers may not want to place their face in the water.

**FIGURE 4-23** The forward crawl is an effective swimming stroke, but can be tiring for swimmers not used to it.

Courtesy of Steve Treinish.

The sidestroke is done by lying on one's side, keeping one's head above water, reaching out with the arms, and pulling the water toward oneself while at the same time spreading the legs and kicking them in, similar to a frog. This will propel the swimmer forward, and by holding the arms and legs tucked in, the forward glide can be extended a good amount. Because the head remains above water, the sidestroke keeps the airway dry and allows the rescuer to maintain eye contact with the shore or watch for rescuers.

The backstroke is another option, but has the drawback of the swimmer not being able to see where he or she is heading. The swimmer lies on his or her back, or leans back slightly from vertical, and kicks the legs while reaching back with the arms, then sweeping them down toward the body. This stroke is easily done from a water tread, and allows the swimmer to monitor the area with the eyes and ears.

## Determining the Best Destination for Self-Rescue

Water and shore conditions can affect where swimmers must travel to find a place for self-extrication or rescue. This decision largely depends on the fall into the water itself and where in the water it has placed the rescuer. The quickest way to shore may not be the best choice if it sends the rescuer to a steep cliff wall. Getting to shore is half the battle, but getting onshore can be the other half: Swimmers without flotation who cannot complete a water exit may end up suffering from fatigue and drown before being rescued. Chapter 11, *Surf Rescue Operations*, explains that the best place to go may be, in fact, farther offshore.

## Rip Currents

**Rip tides**, also known as rip currents, are a hazard found in surf or beach areas. A "rip" is an area where all the water piled up onshore from waves or surf drains back into the ocean (**FIGURE 4-24**). Rip tides are similar to swiftwater currents in that they are hard to beat, and they often claim the lives of people who fight them the wrong way. If rescuers find themselves in a rip current, the best plan is to swim parallel to shore instead of directly to it. Ideally, the rescuer will swim out of the rip current and be able to start back to shore without fighting it, or perhaps other help will arrive.

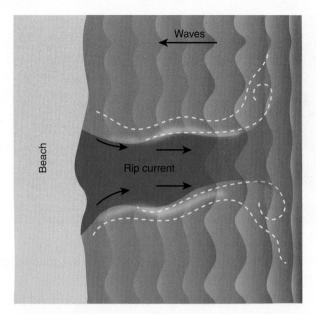

**FIGURE 4-24** Rip tides occur when surf action piles water into one particular area, and that water drains back out to open water.
© Jones & Bartlett Learning.

## Defensive Swimming Position

If currents are an issue, assume a **defensive swimming position**. Your feet should face downstream so that visibility is maintained, and your knees and legs can be used as shock absorbers if something is struck underwater (**FIGURE 4-25**). Use your arms to paddle and move your body from side to side. If you are facing downstream, use your arms to point your head in the direction you want to travel, then use your arms to paddle backward. This changes the body angle presented to the flowing water, and will help you get to shore or around objects. The general idea in defensive swimming is to use visibility and your arms to steer around hazards, and let the river win for the time being. Once in safer or calmer waters, the forward crawl stroke can be used for better speed and power. More swimming information can be found in Chapter 6, *Swiftwater Rescue Operations*.

**FIGURE 4-25** In the defensive swimming position, the feet face downstream so that visibility is maintained, and the knees and legs are used as shock absorbers if something is struck underwater.
Courtesy of Steve Treinish.

# Extricating Victims (and Rescuers)

A water rescue is not complete until the victim is in a safe place, such as in the hands of EMS personnel, or a position where he or she will not enter the water again. Many rescuers think of a mission ending at the water's edge, but the entire team is responsible for getting the victim completely removed from the situation and ensuring all rescuers are out safely. For more information on in-water medical treatment, see Chapter 5, *Surface Water Rescue Technician*.

## SAFETY TIP

An annoyance for many around the water is rescuers tripping and falling in the shallow water right at the edge while they are trying to get out through mud or large banks of rocks. Slips and falls can lead to many different preventable injuries, especially facial and ankle injuries. Dive rescuers are especially susceptible to these types of injuries due to the amount of gear that they may be wearing. It is not uncommon to witness rescuers struggling to exit the water, while onshore personnel watch without even offering an arm or hand. Many technical rescuers exiting the water greatly appreciate a rope grab for them to use. In extreme cases, ground ladders can be used to enter and exit the water safely.

## Removal Up Banks or Slopes

The water's edge is often sloped or even vertical. Although only rescuers trained as rope technicians should do vertical lifting, a very simple rope system can be established to handle gentler slopes, or to keep rescuers from slipping in mud while trying to get the victim out. Stokes baskets are the most common means of

**FIGURE 4-26** Stokes baskets are often used to transport patients up a slope.
Courtesy of Steve Treinish.

securing the patient for the ride (**FIGURE 4-26**), but other equipment may be used, such as "skeds" or even smaller boats. A sked is a large piece of plastic used to wrap the victim, with the plastic then being slid, carried, or hoisted. Stokes baskets or litters built with chicken wire–type floors and sides will let much of the water weight drain off, while giving rescuers an easy-to-grab system; unfortunately, they do not slide across certain surfaces—such as muddy banks or slushy ice—as easily as the plastic skeds.

# Basic Rope Systems

Although NFPA 1006 does not specify which rope skills a water rescuer should have, it does state that surface rescuers at the operations level must be able to identify the systems best used by the AHJ for what may be expected. This section describes only direct and simple mechanical advantage systems. For the purposes of this discussion, assume that a crew is working to bring a litter up a shallow-angled, muddy slope.

## Common Positions

A simple mechanical advantage system is a very basic evolution, without much complexity. Nevertheless, it does have some critical elements. One person must assume the role of leader at the top of the slope, especially when giving commands to move, hold, or lower a load. This person must be able to see the load movement, hear the rescuers carrying the litter, and relay commands to the crewmembers pulling the rope used. The leader of the haul system should also have direct contact with the rescuer at the front of the litter, as this person has a view of the terrain immediately in front of the path to be traveled, and may stop or continue the pull as needed.

## Common Commands

Some common terminology should be used to prevent misunderstandings and increase safety and effectiveness when using rope systems. Crews carrying litters or Stokes baskets should coordinate their efforts by giving the following commands:

- "Ready to lift." This command places the rescuers into place and prepares them to pick up the litter.
- "Lift." All rescuers lift the litter to a standing position, but remain stationary.
- "Haul." The rope crew at the top pulls the rope, creating forward motion.
- "Stop." All movement ceases. The term "whoa" is discouraged, as it sounds like "go."
- "Lower." The rope crew at the top allows the litter to move downslope.
- "Ready to sit down." The rescuers prepare to gently set the litter down.
- "Sit down." All rescuers gently lower the litter to the ground.

## Safety Checks

Any evolution involving rope systems should be checked for proper rigging and operation—including the safety devices used to prevent the load from becoming uncontrolled—before any movement occurs. Having a second or even third person check the system and confirm the system is ready for use is recommended.

## Rope Belays

Any rope system used for pulling, lifting, or lowering a victim should be equipped with some kind of safety belay device. If rescuers lose their grip on the rope, the safety device will do two things: It will keep the victim from falling back to wherever he or she has been pulled from, and it will let rescuers recover and continue the pull without losing too much ground. Very complex safety systems can be installed, and true vertical, live-load rope systems require much more work to set up and operate. Water rescue personnel who attempt this kind of rope use should be trained to the technician level in rope rescue as well.

### Prusik Hitch Belays

Prusik loops are smaller-diameter cords (normally 6 or 7 mm) tied in a 2- to 4-foot-long loop with a Prusik knot. The Prusik line becomes a "grab" on the main line if the pull from the rescuers slip. It can be installed in conjunction with a Prusik minding pulley, or it can be installed as a stand-alone safety measure, in which case it must be anchored to something substantial enough to hold the entire load. Remember, rescuers may also be hanging onto the Stokes litter for support, and their additional weight needs to be considered during a belay.

To install a Prusik loop on a rope, follow the steps in **SKILL DRILL 4-4**. The Prusik loop is draped over the main line, then wrapped around the main line three times, back through the first loop that was created when it was laid over the main line. The knot is "cleaned up," and each loop tightens into a row of Prusik cords wrapped around the main line. If one of these loops is stacked onto the other, and not lying flat, the Prusik hitch will slip. Make sure the Prusik knot does not interfere with the friction loops provided by the Prusik hitch. These loops, when tightened down, will bind and clamp the rope, allowing the Prusik hitch to assume the load.

To ensure the safety of the main line, the Prusik hitch must be kept tight against the anchor, and the main-line rope fed through it. To do this, simply twist and loosen the wraps slightly. The Prusik hitch will not slide easily over the rope without relieving some of the tension on it if it has been pulled tight. The Prusik loop is then attached to the anchor, commonly with webbing and carabiners. Next, the Prusik hitch is pushed as far toward the load as possible to minimize the amount of backward travel the load is subjected to if it slips. The Prusik hitch is now ready to assume the load if something slips, and must also be minded or slid over the traveling main line by a rescuer. If something or someone slips, the rescuer should instantly release the Prusik hitch, and the load will transfer to the Prusik rope.

## Simple Pulls

With the large numbers of responders who often manage to find their way to a water incident, a simple, straight pull up the bank is possible and tempting, but a safety line must be installed and manned as the rope feeds through it. Another possibility, especially with minimal rescuers, is a **simple 2:1 advantage** (**FIGURE 4-27**). In this case, the rope is deployed as a large bight to the load, with the running and working ends of the rope kept at the higher flat ground, or the safe point. A pulley is then installed on

# SKILL DRILL 4-4
## Attaching a Prusik Loop to a Rope

**1** Stretch out the Prusik loop between your two hands, with the connecting knot around midpoint. Hold the loop against the rope on the side of the rope facing you. Have a smaller portion of the loop (about 6 inches [15.2 cm]) off to the right of the loop.

**2** Bring the larger side of the loop around the main-line rope toward you and pull it through the smaller side of the loop.

**3** Make sure the grapevine knot passes well through and that the coils formed around the main rope are even.

**4** Bring the larger side of the loop through the same path as before.

*(continued)*

# SKILL DRILL 4-4 Continued
## Attaching a Prusik Loop to a Rope

**5** Make sure the Prusik hitch coils are even and parallel.

**6** Tighten the Prusik hitch around the main rope.

**7** Inspect the hitch to confirm that it is tied correctly.

Courtesy of Steve Hudson.

the load, and the bight installed through it. The working end of the rope is anchored to a suitable point at the destination or safe point, and the running end is pulled, again with a belay device installed and working. The rescuers at the safe point provide the pull, but their efforts are doubled because the pulley on the load is moving uphill, too. Again, rope length must be considered, as well as a place or direction from which

to pull: The rescuers will need to pull 2 feet of line for every 1 foot the victim moves.

## Change of Direction

Rather than rescuers pulling ropes hand over hand, it is much preferred to have the rescuers walk a greater distance with the rope in a firm grasp. This approach keeps more hands on the rope pulling, decreasing the

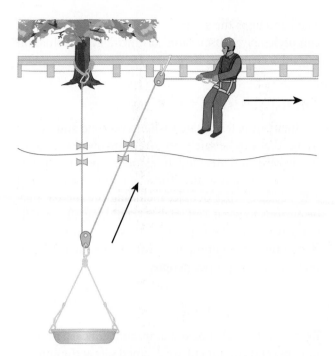

**FIGURE 4-27** A 2:1 mechanical advantage system.
© Jones & Bartlett Learning.

possibility that the rope may slip. Placing a simple pulley in a strategic location can allow the rescuers to keep walking with the rope, and can often place the pulling crews on flatter or drier ground. For more information on rope systems and hardware, refer to Appendix A.

## Removing the Rescuers

Rescuers need to be removed from the water, too, even though most, if not all, the focus goes to the victim at the removal point. Simply outfitting one or two trained personnel with hip waders or a drysuit and PFD will enable them to stay dry and warm but enter the water to a shallow depth to assist the rescuers out. Deep silt, mud, and unseen rocks or debris may lead to all of these personnel getting hurt or even stuck.

If rescuers are tethered, these ropes can be used to extricate them from grades and steeper banks (**FIGURE 4-28**). Use the tether to support the rescuer as he or she climbs the bank or up and out of the water. Consider tying off a rope at chest height between the safe zone and the shoreline to give the rescuers a grab line as they climb up the bank.

## Aerial Extrication

An **aerial extrication evolution** uses an aerial ladder or platform as a lifting point. Aerial ladder or platform ladder apparatus can offer an easy way to extricate victims, especially straight up. Good road access is a must for the ladder apparatus to be set up properly. The tip of the ladder is used to provide a lift point, similar to a

**FIGURE 4-28** Simple rope and belay systems can be used to extricate rescuers from grades and steeper banks.
Courtesy of Steve Treinish.

**FIGURE 4-29** When good road access is available, an aerial ladder or platform ladder apparatus can offer an easy way to extricate victims, especially straight up.
Courtesy of Steve Treinish.

crane. The line or lines are fed under the aerial ladder so as to provide lift, lower, and safety capabilities from the street (**FIGURE 4-29**). The aerial tip or bucket can be positioned over the victim and/or rescuers, and then the rope system or the ladder itself can be used to lower the Stokes basket or litter down to the crews. Once the victim is packaged, the litter can be raised to safety or waiting EMS crews.

The aerial extrication evolution is *not* appropriate for untrained or even rusty crews. Both the rescue crews and the ladder operator must practice it often. Failure to set up something properly or use of the wrong area of the ladder itself could result in a catastrophic failure, or even death.

# Terminating an Incident

Although awareness-level personnel can handle many incident termination tasks, operations- and technician-level personnel are responsible for ensuring all termination activities have been performed and all necessary safety precautions taken. Such activities include the following:

- Retrieving all PPE, tools, and other gear brought to the scene and evaluating it for wear
- Accounting for all personnel and bystanders at the scene
- Notifying responsible parties of any changes or damages to property during the incident response and any potential lingering hazards
- Documenting all components of the scene, including injury, loss, and damage
- Transferring control of the scene to a responsible party
- Conducting postincident debriefing, analysis, and critique as needed
- Officially terminating command per local protocols

The event is not over until everyone is safe and proper documentation and debriefing have taken place. Properly ending a mission may require the efforts of all levels of rescuers to safely return all gear and all rescuers to service, and to record all of the scene data needed for future reference. Because surface water rescue is the simplest water rescue discipline, a comprehensive discussion of incident termination activities is included here.

## Termination Defined

An incident is terminated when everyone and everything (that can be safely removed) involved is clear of the hazard and deemed in good shape—medically, mentally, and physically. The scene must be rendered safe because other people, including the media, may be curious to see it and stick around long after rescuers leave. Any damage done to surrounding areas, structures, or equipment must be documented and reported to the proper people.

## Safety at Termination

Rescuer safety must be of paramount concern until all rescuers and equipment are deemed safe at the conclusion of an incident response. Quite often, rescuers, commanders, support crews, and even media reporters will want to offer kudos or "talk shop" with the rescuers directly involved almost immediately after they exit the water. This distraction can occur even when other rescuers are still working in the hazard.

Rescuer fatigue is often a factor during the later stages of a water rescue, so rotating crews into the mix to clean up will let the previous workers take a break and help the new crews concentrate on their task of safely breaking down gear and systems. In extreme cases, rescue equipment may be abandoned or written off completely while conditions improve around it. No equipment is worth a life; equipment can always be replaced or retrieved later.

**LISTEN UP!**

Most rescuers who have brought a rescue to an end will relate to the "big breath of relief" felt after a successful mission. Concentration can erode into conversation, with rescuers sharing excited and happy details about how well things went. Conversely, unsuccessful rescue attempts may cause rescuers to retreat mentally, or replay roles or tactics. If a rescue turns into a recovery, it may be better to suspend gear retrieval or to bring in other personnel who have not been a part of the rescue attempt or recovery to retrieve the equipment.

## Gear Retrieval, Decontamination, and Inspection

Water rescue gear and the tools to support it after an incident are often carried back to the area from which they originated, typically the apparatus. This is a good way to initially sort the gear and tools and to ensure each piece gets back to where it is carried. Any gear should, at a minimum, be grossly decontaminated with copious amounts of clean, fresh water. If available, a cleaning solution containing a mild soap and brushes can be used to help wash off dirt, mud, and other common contaminants. If possible, washing gear on concrete or pavement helps keep dirt and mud off gear that has been freshly rinsed.

All gear needs to be examined for proper use and packaging, preferably before it goes back to the station. Additional calls for rescue may lead to the crew being dispatched immediately after returning to service, especially in flooding or high-risk weather. If the crew is sent to another call two blocks away, it will not have enough time to sort through a huge pile of equipment in the cab of the engine. While some gear may need more attention at the station, at least consider having the basics ready in case the crew receives a new call on the way home.

Any gear that may be used in subsequent calls must be inspected to ensure its proper operation or use if needed. PFDs should be ready to respond, and if they are typically equipped with knives, whistles, or other safety devices, these items should be accounted for and at the ready. If a PFD or other safety gear is unable to be used on another incident, it may be better to place that gear out of service until it can be fixed or replaced.

///////////////////////////

### SAFETY TIP

One common problem seen in water rescues is postincident cleanup in which rescuers are not protected. Hot weather often prompts rescuers to doff PPE immediately after the rescue, take a break, and only then clean up. One way to ensure safety in such a scenario is for all rescuers to carry all equipment used during a rescue to an area safely away from the water before they doff their PPE. This practice keeps rescuers protected when close to the water, and it keeps rescuers prepared if accidents happen during incident termination.

It is also a good idea to keep extra sets of basic PPE and throw bags close by and readily accessible in case of an unforeseen emergency during incident termination. This equipment should be the last thing placed on the truck before leaving a scene.

### Rescuer Decontamination

Not only should the gear be cleaned, but efforts to decontaminate the rescuers should be considered as well. A simple wiping of the eyes or nose on a hot summer day can transfer contaminants from the wet hands onto the face of a rescuer, and can easily transmit chemicals or bacteria into the eyes, ears, and nose. Many rescuers think nothing of eating lunch and handling food with wet or dirty hands. Washing the hands immediately after handling wet gear of any kind is the best defense to prevent exposure. Alcohol sanitizers and old-fashioned hand soap are great weapons to keep crews clean. Tenders can also benefit from using the medical gloves found in most fire apparatus.

## Documentation

Incident documentation is important to enable teams or rescuers to apply for possible reimbursement or replacement for equipment or other costs such as fuel, food, and mileage traveled. This kind of documentation covers the entire operation, as opposed to operational documentation, which is done while the incident is ongoing and used to determine the PLS, progress, and hazards. A lack of incident documentation can greatly affect later care of injuries suffered in the line of duty when insurance or worker's compensation claims are processed. Pictures and a detailed narrative of any unusual events, such as injuries or unexpected breakage of gear, may help with explanations. In the case of a fatality in which a victim was touched or moved by rescuers, a written statement by the rescuers closely involved will assist the subsequent investigation done by law enforcement.

Most agencies that use online reporting require an injury report be filed as part of the incident report, and deadlines for most of these reports are just a few days after incident termination. Any injury should be noted, even those that seem nothing more than a very slight sprain or strain. After the adrenaline wears off, pain from an injury can become worse.

Documentation may be needed for any property or equipment damaged in the course of rescue. Some rescue equipment may be routinely damaged during the course of an incident or training: PFD buckles can break, especially in cold weather; suits, gloves, and boots can be torn or punctured; and gear is lost underwater. The department administration should be informed whenever damage or loss occurs so that they can facilitate repair or replacement.

Homeowners or property owners should also be informed about damage to their property. Many times gates or locks are cut or forced open or trees or shrubbery destroyed to obtain access to the water area or to evacuate victims. When property owners carry homeowner's insurance, their insurance adjusters will appreciate a detailed explanation of damage that may have occurred in the course of a rescue. Government help may sometimes be made available to teams and civilians in the form of costs and

**FIGURE 4-30** Postincident debriefing allows rescuers to review the details of the incident and evaluate both what worked well in their response and how they could improve.
Courtesy of Steve Treinish.

recovery assistance, especially once disaster declarations are announced at the state or federal level. Many emergency managers will provide help in navigating this cost recovery process.

## Critical Stress Observation

Rescuer safety also includes managing critical incident stress. Critical incident stress (CIS) is a normal human reaction to an abnormal situation. It can vary from rescuer to rescuer, and different stressors affect rescuers differently. For example, the first recovery of a deceased child that a diver makes will certainly elicit some stress response, which may differ from the stress experienced by someone who has performed this task dozens of times. Even the veteran may be affected, albeit perhaps to a lesser extent, having been desensitized to such scenes.

Critical incident stress management (CISM) can start on scene with defusing—that is, a brief reminder of the signs of CIS, what to expect in the next few hours, how to manage it, and how to get help. Defusing is often followed by a critical incident stress debriefing (CISD), which is more structured and is intended to help all responders who choose to participate express and deal with mental factors that are still causing problems. Rescuer suicide is a topic many prefer to avoid, but it is on the rise, and we need to ensure our teams are mentally safe as well as physically safe.

## Postincident Debriefing

After everyone is safe and all gear secured, a quick debriefing is always suggested. In its most basic form, a postincident debriefing is a quick recap of the mission, challenges,

and results that can be done within view of the scene (**FIGURE 4-30**). It puts everyone on the same page regarding the events and can provide details as to why something was ordered or where something occurred. Some teams prefer to debrief with only members of that team or other qualified teams present. This way, members can debrief quickly and more thoroughly by "speaking the same language." Debriefing within a team can also foster a better environment for talking about difficult issues or managing problems, as discretion can be used and the audience is limited to those involved.

## Postincident Analysis

A good **postincident analysis**, or critique, can be done after any water rescue, and is usually performed within a few days after the event ends. Often, once rescuers have rested and calmed down, and are placed in an environment better suited to meetings, better analysis occurs.

Two distinct types of critiques may be performed. The first is a **mission analysis**, which includes all responders, regardless of training level or agency affiliation, in the discussion. The more perspectives included, the more complete the analysis and the greater the possibility of a better operation next time. Rather than placing blame and determining what went wrong, participants should concentrate on listening to everyone involved and ensuring future responses are safer and more efficient. Dispatchers often have a good perspective on a run, and including audio recordings of operations in the mission analysis can be a great move. It takes a very secure team to be able to examine its actions and admit shortcomings or mistakes—but it also saves lives. A team that admits issues and solicits solutions takes a huge step toward preventing those issues in the future. Going public with the mission analysis's results by sharing them with other teams may save more lives by triggering more wide-scale change in the discipline.

The second type of analysis is a company- or team-level critique. It can be very specific, dealing with procedures or equipment used in the mission. Keeping the number of participants relatively small often encourages honesty, and specialty rescuers can understand and speak the language without interruption. Quite often, recreating problems identifies better solutions. Giving rescuers the ability to identify and solve problems also encourages "buy-in" to the team.

### The FAILURE Acronym

Many teams use the acronym FAILURE to describe how rescues fail. This mnemonic is a great way to identify what happened and which improvements are needed, whether in terms of awareness-level training for more personnel or the need for a specific piece of gear. No one wants to be a part of any failure, but it is everyone's responsibility to learn from the experience. There is no perfect response, and there is no shame in acknowledging that.

The FAILURE acronym can relate to water rescue as follows:

**F**—*Failure to understand the water environment.* Perhaps the water moved faster than expected or unexpectedly spilled into another area. Perhaps the water made travel harder, or tired crews faster, or crews had never trained in the conditions faced.

**A**—*Additional medical concerns not considered.* Hypothermia, shock, and other injuries can all affect a victim's ability to self-rescue or even verbalize thoughts, and rescuers must allow for a victim who cannot contribute to his or her own rescue.

**I**—*Inadequate rescue skills.* This is huge! Many skills in the fire service are erodible, meaning they will dull or be forgotten over time. This kind of deterioration is especially likely in the dive world. Constant, realistic training is a must, in conditions expected, in the same PPE for the rescue, with the worst stressors expected.

**L**—*Lack of teamwork and/or experience.* Many rescuers struggle with saying, "I don't know." Some trucks and departments hesitate or even refuse to call for help or additional teams. Leave the ego at the station.

**U**—*Understanding logistics of an operation.* Mission logistics can range from needing a longer rope to needing to feed and house dozens of rescuers on multiday missions. Preplanning can greatly reduce logistical issues, as can teams quickly forecasting needs before the needs actually emerge.

**R**—*Rescue versus recovery not considered.* Dead is dead, and no rescuer or team should be expected to risk their own lives for someone already gone. Any line-of-duty death changes lives forever, but imagine the loss of a rescuer trying to save a victim who has been deceased for hours or one who does not even exist.

**E**—*Equipment not mastered.* Just like the erodible skills, equipment such as boats, rope gear, and dive gear must be used often enough that rescuers have 100 percent confidence in their use and are able to troubleshoot and fix problems in a timely manner.

## After-Action REVIEW

### IN SUMMARY

- Surface rescue is a very broad discipline that encompasses many areas of water rescue; it can take place in any water, at any time, in any conditions.
- A wide range of specific training for surface rescues is available. Rescuers are encouraged to train and preplan their districts for the type of missions they might see, in the worst conditions expected.
- Water rescue PPE should feel like part of the uniform.
- No matter which hand or body signals the team uses, rescuers should know them by heart, and each team should at least be familiar with the signals of other teams.
- Surface rescuers are somewhat limited in victim search operations, but some very effective measures can still be employed on a mission with no victim in sight, but with a good point last seen.
- Rescuers may not get many chances at the true working rescue of a distressed person in the water, especially one with no flotation. When one of these situations does happen, various rescue techniques can be used to rescue the victim.
- The use of watercraft in rescue offers flotation as well as protects the rescuer from hypothermia, fatigue, chemical exposure, and debris.
- If helicopters are available for water rescues, they may be used a variety of ways to assist in a surface rescue event—for example, by providing aerial photography, infrared cameras and night vision, scene lighting, supply delivery, and victim extrication.
- Although the use of PFDs is required in the surface water business, accidental immersion could happen to anyone. For this reason, all rescuers should be well trained in self-rescue skills.

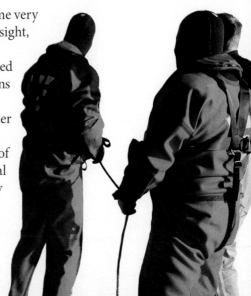

## KEY TERMS

*Access Navigate* **for flashcards to test your key term knowledge.**

**Active search measures** This phase of search measures includes those that are formalized and coordinated with other agencies. (NFPA 1006)

**Aerial extrication evolution** The use of an aerial ladder or platform to provide a lift point to recover victims; it uses a rope system in conjunction with the ladder itself.

**Defensive swimming position** A position used by swimmers in fast current; the body is positioned to protect from injury in the water with the back down, head in the rear, feet forward, and knees bent to absorb shock.

**Drag hooks** A set of small, clustered sharpened hooks used to snag a body on the bottom of a water area.

**Drownproofing** Letting the body float limply on the water, with the head lying in the water to save strength; breathing is done by raising the head.

**Mission analysis** A type of postincident analysis that includes all responders, regardless of training level or agency affiliation.

**Passive search measures** Search efforts that do not require active searching by the rescuers. (NFPA 1006)

**Point last seen (PLS)** The last geographical location where the victim was physically witnessed to be; it can be on the water or the shore.

**Postincident analysis** A critique that can be done after any water rescue; it is usually performed within a few days after the event ends.

**Rip tides** The water flowing back out to open water past a surf line, after being pushed onshore; also called *rip currents.*

**Simple 2:1 advantage** A basic rope system using a movable pulley on the load.

**Sked** A litter-like hard plastic sheet used to package victims and provide immobilization, carrying points, and sliding and hoisting ability.

**Sonar units** Electronic devices used to examine the bottom contours, depth, and density, depending on the unit chosen; useful for searching high-hazard areas, or for covering larger areas more quickly than divers can.

**Spotters** Personnel positioned to scout for or search for missing victims, watercraft, and other issues.

**Water-type recovery jaws** A long set of hydraulically operated hooks, which are spread out as they are forced to the bottom and closed as they are brought up; they are used to hook and retrieve victims in deeper water, usually 8–20 feet.

# On Scene

Your department has been looking for a missing person for more than 2 hours. The person was on the phone with your dispatch center, and was roaming the bank of a large lake looking for mushrooms. He had walked about an hour when he suddenly slipped and fell down the bank to the river; he ended up lying partially in the water. The victim is unable to move his legs, according to the dispatchers, and they think he is starting to fade.

Your company hears a very weak cry for help, down a fairly steep bank. Through the trees, you spot the victim lying in approximately 2 feet of water, barely able to keep his head up. Late in the season, the ground is very slippery and void of vegetation, and conditions are hard to work in. The youngest member of your company has a PFD on and proceeds down to the water to ensure the victim can keep his head above water. He plans to advise what he has and needs verbally. You cannot hear the rescuer very well, so you start down the bank a bit to improve communications.

**1.** You find yourself in the water after a slide down the muddy bank. You are wearing your duty uniform, but did not think you would be this close to the water. You are certainly not a strong swimmer. Which actions should you take?

**A.** Thrash in the water, making lots of noise and disturbances so your crew will hear you.

**B.** Take big breaths, lean back, and float; once stabilized, begin a slow, easy swim to shore or shallow water.

**C.** Make noise to alert your crew, and be ready to grab whatever flotation aids they can get to you.

**D.** Both B and C.

**2.** While the rescue was easy, the extrication from the riverbank is not. The grade is approximately 15 percent, but the bank consists of very muddy, slippery clay. Which rope systems would be easiest to use and would ensure safe transit up the slope for rescuers and the victim?

**A.** Simple pulls, using rope grab hardware to prevent a slide or fall back

**B.** 2:1 mechanical advantage, using a pulley at the load

**C.** The winch on board the department's utility pickup truck

**D.** A prerigged 4:1 system, with a Stokes litter to haul the victim up

**3.** The air temperature is 41°F. What is the basic treatment for hypothermia?

**A.** Remove wet clothes, place warming blankets on the victim, and place the victim in a warm ambulance.

**B.** Remove wet clothes and pack heat packs into the core area.

**C.** Provide two large-bore intravenous lines with heated fluids, and warm liquids by mouth.

**D.** Leave wet clothes in place, and cover the victim with warming blankets.

*Access Navigate to find answers to this On Scene, along with other resources such as an audiobook and TestPrep.*

# Technician Level

# Surface Water Rescue Technician

## KNOWLEDGE OBJECTIVES

After studying this chapter, you should be able to:

- Explain the components of a swim test for water rescuers. (**NFPA 1006: 16.3.1**, pp. 135–137)

- Describe how to perform a swimming rescue of a victim in liquid, nonmoving water. (**NFPA 1006: 16.3.2**, pp. 136–138)

- Identify defensive swimming tactics, given a panic or hostile situation in the water. (**NFPA 1006: 16.3.3**, pp. 138–143)

- Identify and describe rescue tools that can assist water rescue technicians. (**NFPA 1006: 16.3.2**, pp. 143, 145–146)

- Describe how to conduct tethered rescue operations. (**NFPA 1006: 16.3.2**, pp. 146–147)

- Describe medical treatment that can be performed while the victim is still in the water. (**NFPA 1006: 16.3.2**, pp. 147–150)

- Identify and describe in-water tools that can assist in victim packaging. (**NFPA 1006: 16.3.2**, pp. 149–153)

- Discuss some leadership skills needed to deploy, supervise, and protect other personnel involved in a surface rescue. (**NFPA 1006: 16.3.4**, pp. 153–154)

## SKILLS OBJECTIVES

After studying this chapter, you should be able to:

- Make a head-up running leap entry into water. (**NFPA 1006: 16.3.2**, p. 140)

- Safely flip an unconscious patient. (**NFPA 1006: 16.3.2**, p. 150)

- Use litter flotation. (**NFPA 1006: 16.3.2**, p. 152)

# Technical Rescue Incident

It is a superb summer evening, a scene right out of a magazine—warm weather, sunshine, and your crew halfway through a great shift. While you are riding in charge of the engine today, your crew receives a call for a person in the water at a newer, high-value condominium complex on the outskirts of the district. Your truck carries the basics needed for water rescue, including some Type III life vests, a hose inflator, and some throw bags. Your crew is comfortable around the water, and you are not overly concerned because the ponds at the condominium complex are fairly small. When you get in the truck, the mobile data terminal (MDT) comments include that the victim is refusing to come out of the water, and is screaming at things.

You arrive to find about a dozen or so bystanders watching a person cling to a fountain running in the center of the pond. One bystander approaches you and states that the victim has been swimming for the last 15 minutes, and is screaming at seemingly imaginary people. According to this bystander, it is rumored that the victim uses methamphetamine. You watch as the victim pushes off the fountain, swims a few yards, and then swims back to the fountain and continues screaming. Although he is acting very strange and aggressively, he seems quite the capable swimmer when he releases the fountain. You have not seen one of his hands at all, however, and even after multiple attempts, he does not respond to verbal commands or even acknowledge the presence of anyone on the bank.

**1.** Is this a rescue attempt that should be operations or technician based?

**2.** Which rescue evolutions can keep a safe distance between rescuers and the victim?

**3.** Would it be possible to let the local law enforcement officers use your equipment to take this person into custody?

**4.** What will be your initial incident action plan?

Access Navigate for more practice activities.

## Introduction

Perhaps the water rescue discipline that encompasses the widest range of water rescue missions is surface rescue. An incident that occurs in any water that is standing but not frozen, crashing with waves, or running at a current faster than 1 knot is fair game to be grouped in this category.

Rescue technicians in the rescue disciplines are typically tasked with actually entering or exposing themselves to the same hazard in which the victim is trapped. Usually, a good risk assessment will determine the least hazardous rescue methods to use, but sometimes it boils down to where a rescuer simply has to "go." When rescuers subject themselves to the hazard entrapping the victim, they need to thoroughly understand it, utilize the proper protection from it, and know how to remove the victim, themselves, and their team from it should the need arise. Many different water conditions can occur in surface rescue, and many different rescue scenarios are possible. Often, an evolution based on operations-level training may be used to start a rescue, with an entry rescue being done concurrently. It does not matter who gets the "grab" first; all that matters is that the grab is made.

## Swim Testing

Before water rescuers enter the environment, they must prove themselves—not by just jumping in with a personal flotation device (PFD) and hanging out in the pool for a bit, but through an actual swim test (**FIGURE 5-1**). Such a test proves they have the skills, endurance, and comfort in the water necessary to do the job of a water rescue technician. Not only do rescuers need to demonstrate these skills in a controlled environment, but they can also gain a great deal of confidence by proving to themselves that they can swim and operate effectively and safely. Instructors quickly figure out who are the stronger swimmers, who can handle water and who truly loves it, and how the overall group dynamic feels. Practice tests can also

**FIGURE 5-1** A swim test is necessary before a rescuer can be allowed in the water at a rescue.
Courtesy of Steve Treinish.

be given to acclimate the rescuer to exactly what the swim test will reflect, and which skills may need to be polished or practiced. Some agencies now require a basic swim test for operations-level rescuers, and a harder, more physical test for technician training.

## Watermanship Skills

NFPA 1006, *Standard for Technical Rescue Personnel Professional Qualifications,* actually mentions by name a sample swim test that can be used for testing swim rescuers—a dive master-level swim test, developed by the International Association of Dive Rescue Specialists (IADRS) (**FIGURE 5-2**). The IADRS Watermanship Test includes the skills, ability, and stamina needed to start a training class, and can be manipulated by the authority having jurisdiction (AHJ) as needed. It has been adopted by many AHJs and training agencies as a prerequisite before the actual water rescue training starts, and can also be used on a pass/fail basis or to rank members applying for a position on a water rescue team. It can be modified to reflect the equipment worn or needed in the AHJ's hazard areas.

The IADRS Watermanship Test consists of at least the following tasks:

- A 500-yard, nonstop forward crawl swim, with no PFD
- An 800-yard snorkel course, with the face in the water at all times, and no use of the hands
- A period of 15 minutes of treading water, drownproofing, bobbing, or floating, with the hands out of the water during the final 2 minutes
- A 100-yard tired-diver tow or push (the 100-yard tow can be modified by having surface rescue students wear whatever personal protective equipment [PPE] is deemed necessary, most likely at least a PFD)

The test is a pass/fail test, and is graded with a point system. The more quickly and more satisfactorily a candidate performs, the more points he or she receives for each task. Failure to complete any task results in a disqualification, and the student must retake the test at a later date. The IADRS Watermanship Test can be a good way to select and rank candidates for training, and can show some students that they need more physical preparation before starting the class.

Consider using swiftwater gear for swiftwater classes or full public safety dive gear for dive candidates during the test. The AHJ should tailor the test to reflect the actual PPE used in the jurisdiction and the conditions rescuers would be expected to face.

## Yearly Requalification Testing

Many agencies and teams also require a swim test as a yearly recertification for members to remain active on the team. This test is just a baseline, and the AHJ is responsible for testing students realistically and thoroughly. As with any training, safety is paramount, especially when testing swimmers for the first time. Many swimmers overestimate their physical endurance, or fail to realize that a too-fast pace may cause problems. At a minimum, lifeguards with equipment should be stationed around the pool during the test. Ideally, an emergency medical services (EMS) crew and supplies should be available as well.

**LISTEN UP!**

The initial swim test is a good way to encourage teamwork before the meat and potatoes of a water rescue class begins. Many students naturally want to help the others succeed, and some will coach and verbally support other slower swimmers. This type of assistance is great, and could be the sign of an outstanding student later on.

## Submerged Victims

Most rescuers think of a submersion as a dive rescue event, but especially around beaches or shallow ponds, rivers, or other water, there are ways to search areas without diving. Any responding resource can be canceled if a victim is found, and many successful rescues are made by nondiving rescue crews.

## Human Chains

The human chain technique is simple to use in shallow and relatively safe areas. Each rescuer joins hands or locks elbows with the next, and the entire line of

## I.A.D.R.S ANNUAL WATERMANSHIP TEST

### Evaluation parameters
There are five exercises that evaluate stamina and comfort in the water, each rated by points.
The diver must successfully complete all stations and score a minimum of 12 points to pass the test.
The test should be completed with not more than 15 minutes between exercises.

### Exercise 1: 500 yard swim
The diver must swim 500 yards without stopping using a forward stroke and without using any swim aids such as a dive mask, fins, snorkel, or flotation device. Stopping or standing up in the shallow end of the pool at any point during this exercise will consitute a failure of this evaluation station.

| Time to complete | Points Awarded |
| --- | --- |
| Under 10 minutes | 5 |
| 10-13 minutes | 4 |
| 13-16 minutes | 3 |
| 16-19 minutes | 2 |
| More than 19 minutes | 1 |
| Stopped or incomplete | Incomplete |

### Exercise 2: 15 minute tread
Using no swim aids and wearing only a swimsuit the diver will stay afloat by treading water, drown proofing, bobbing, or floating for 15 minutes with hands only out of the water for the last 2 minutes.

| Performance criteria | Points Awarded |
| --- | --- |
| Performed satisfactorily | 5 |
| Stayed afloat, hands not out of water for 2 minutes | 3 |
| Used side or bottom for support at any time | 1 |
| Used side or bottom for support > twice | Incomplete |

### Exercise 3: 800 yard snorkel swim
Using a dive mask, fins, snorkel, and a swimsuit (no BCD or other flotation aid) and swimming the entire time with the face in the water, the diver must swim nonstop for 800 yards. The diver must not use arms to swim at any time.

| Performance criteria | Points Awarded |
| --- | --- |
| Under 15 minutes | 5 |
| 15-17 minutes | 4 |
| 17-19 minutes | 3 |
| 19-21 minutes | 2 |
| More than 21 minutes | 1 |
| Stopped at any time | Incomplete |

### Exercise 4: 100 yard inert rescue tow
The swimmer must push or tow an inert victim wearing appropriate PPE on the surface 100 yards nonstop and without assistance.

| Performance criteria | Points Awarded |
| --- | --- |
| Under 2 minutes | 5 |
| 2-3 minutes | 4 |
| 3-4 minutes | 3 |
| 4-5 minutes | 2 |
| More than 5 minutes | 1 |
| Stopped at any time | Incomplete |

### Exercise 5: Free dive to a depth of nine feet and retrieve an object.

| Performance criteria | Points Awarded |
| --- | --- |
| Performed satisfactorily | Pass |
| Stopped or incomplete | Incomptete |

Additional copies aviailable at no charge via the International Association of Dive Rescue Specialists webpage. Visit www.IADRS.org

**FIGURE 5-2** IADRS Watermanship Test.

Reprinted with permission of International Association of Dive Rescue Specialists. www.IADRS.org.

rescuers shuffles their way along the search area, often literally stumbling over the victim (**FIGURE 5-3**). A human chain is a very effective way to quickly search shallow areas such as beaches, ponds, and even swimming pools. Swimming pools, with no chemical treatment, will cloud to the point of no visibility, and each year missing children are found at the bottom of such pools. Many fire and EMS companies carry only basic water rescue gear (i.e., a PFD for each member and a

few throw bags), and this evolution can be done with such gear, provided the environment does not directly require more protection.

The major issue with human chains is the need to protect the rescuers, especially those "drafted" on scene. PFDs and thermal protection must be maintained for each and every rescuer. The use of a pike pole or a probe pole to sound the area in front of the rescuers will help identify drop-offs, holes, entanglements, and possibly

**FIGURE 5-3** A human chain is a very effective way to search fairly shallow areas such as beaches, ponds, and even swimming pools.
Courtesy of Steve Treinish.

**FIGURE 5-4** A pike pole or probe pole can be used to sound the area in front of the rescuers.
Courtesy of Steve Treinish.

even the victim (**FIGURE 5-4**). Many dive crews use a human chain to rule out the shallow areas presented on dive missions, making it easier for the diver to swim patterns in the deeper water left unsearched. A potential problem with human chains is the possibility of destroying evidence immediately on shore at the mission area, but overall, human chains can be a highly effective search tool when the proper technique is used.

# Swimming Rescue

In addition to mastering the tried-and-true methods of reach, throw, and row, technician-level rescuers are trained to enter the victim's environment to perform rescue. Of course, rescuer safety remains the first priority. Rescuers need to protect themselves from both the conditions and the victim, who may be capable of injuring himself or herself as well as a rescuer. A well-meaning rescuer may be injured by a wild grab or swing from the victim, but the rescuer

may be so focused on the victim that other things—such as other victims, boats, or seeing or hearing other rescuers offering assistance—might be missed. For all rescuers, being aware of the surroundings is a must.

# Getting to the Water

Although many rescuers are great swimmers and can stay afloat for hours with proper PPE, the reality is that humans are not very fast in the water. Most of us can run faster than we can swim, so using the shore to get as close to the victim as possible will go a long way toward reducing the time it takes to swim to the victim, while keeping the victim in sight longer. Jumping into the water as soon as rescuers get there might be an instinct that needs to be controlled. Once a swimmer commits to the water level, line of sight is often reduced due to wave action or surface agitation, and rescuers must try to keep eyes on the victim as long as possible.

Docks, pilings, seawalls, and other access can be used to approach the victim as closely as possible, as can bike paths, streets, and bridges. If there is access to a boat, that might be the way to go. This possibility must be weighed against the advantages of a direct swim, and you will need to consider issues such as how long it will take to get the boat operational and how it is propelled against how the rescue swimmer can proceed and how far offshore the victim is. Consider committing both resources, if you have the proper setups. It does not matter who gets there first; it is the victim who matters.

# Getting in the Water

Many water rescuers are amazed at the number of people, both civilians and rescuers, who will blindly dive into unknown waters that offer little or no visibility. Even more unexplainable is the fact that many of these rescuers know how risky jumping or diving can be. It is very possible to get injured, impaled, or even killed by leaping blindly into low-visibility water (**FIGURE 5-5**).

### SAFETY TIP

One preplanning trick when the water is very low is to check out bottom conditions. Just walking the bottom can open up rescuers' eyes as to what might be in shallower waterways. Rebar, construction debris, trees, concrete, rocks, and assorted other hazards are all often present. The realization that they might strike these things jumping in blind teaches a good lesson to many rescuers.

**FIGURE 5-5** Assess for the presence of debris and other physical dangers before making a conservative water entry. A dam on this river was removed, showing the debris that was previously 4–5 feet underwater.
Courtesy of Steve Treinish.

A feet-first entry into the water is usually desired. Although the urge to take a running dive off the shore to get some quick distance can be strong, that might not be a good idea for safety reasons. The process might seem slow, but getting into the water more cautiously is safer for rescuers and ultimately better for victims.

Even if water is known to be normally clear of debris, or deep enough to safely jump into, rescuers still need to keep the victim in their sights. To do this, a **head-up running leap** entry can be performed. The floating action of the PFD coupled with a sweeping kick of the legs when entering the water can keep the rescuer's head above water. When entering the water, kick both legs as if treading water, and sweep both arms from the back to the front the same way. To make a head-up running leap entry, follow the steps in **SKILL DRILL 5-1**.

## Approaching In-Water Victims

Many rescuers have been injured by approaching a victim straight on, or by underestimating the measures the victim may take to save himself or herself. People who are truly panicking have the power to perform almost superhuman feats, and rescuers need to be aware of the risks they pose. Victims may strip rescuers of their flotation; punch, kick, or scratch rescuers; or even try to use a rescuer as a flotation aid (**FIGURE 5-6**).

**Defensive swimming** and tactics are mentioned specifically in NFPA standards for surface rescue, but should also be in rescuers' minds on any rescue, at any time. Most swimming approaches give the rescuer some time to get a feel for the victim, including the individual's mental status. Some victims will be well established in the fight mode of panic, and are considered to be in the **active drowning stage**. In this stage,

**FIGURE 5-6** Panicked victims may try to use the rescuer as flotation aid.
Courtesy of Steve Treinish.

which is guided by instinct, the hands are out, pushing or stroking down, and the head is tilted up to protect the airway. A frontal approach to such a victim is possible, but be ready for a battle—and recognize that panic is a formidable foe.

## Impaired Victims

A victim who has given up may not have a very active fight mode in place to offer much resistance; this condition is called the **passive drowning stage**. Shock, hypothermia, emotional overload, or fatigue can all result in a victim not really doing anything. Direct approaches to the victim are possible here, but remember the adrenaline response. Once the victim realizes help is there, he or she may "wake up" and need to be controlled.

**Hypothermic Victims.** It is extremely important to understand the effects of hypothermia on water rescue victims. The importance of providing flotation assistance to these people cannot be overstated. They may be incapable of aiding in their own rescue, and rescuers should be cognizant of protecting their airway and facial areas.

Dr. Gordan Geisbrecht, a hypothermia researcher, has created several training videos and offers a wealth of information to the public and rescue teams on hypothermia. His "1-10-1" rule offers the following advice:

- 1 minute: Victims entering freezing water have roughly 1 minute to control their breathing. A typical mammalian reflex will cause deep gasping as the body tries to hyperventilate. A risk of this reaction in water is ingestion into the lungs.
- 10 minutes: After the first minute, victims have approximately 10 minutes of meaningful movement to self-rescue before their extremities start

# SKILL DRILL 5-1
## Making a Head-Up Running Leap Entry NFPA 1006: 16.3.2

**1** The objective when leaping into the water in this fashion is to keep the head up to maintain sight of the victim. Start into the water as low as possible to the ground.

**2** Hold the rear leg back and the arms behind the shoulders, straight out from the body, while leaping into the water. Keep the head looking forward at the victim.

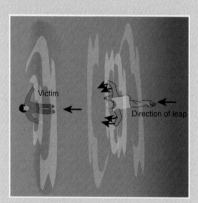

**3** Sweep the legs together in a scissors kick, and simultaneously sweep the arms forward, similar to the movements when treading water. This will provide maximum upward thrust.

**4** Expect a large forward splash when the body hits the water. Water may enter the eyes and mouth if the timing of the sweep is off.

to fail. After this point, the arms and legs do not function well, and without a PFD, treading water or swimming can become impossible. Victims not wearing PFDs will flush colder water over their bodies, resulting in a faster decline in body temperature, and are likely to drown.

- 1 hour: With proper flotation, victims have approximately 1 hour before hypothermia results in unconsciousness. This allows a good deal of time to be noticed or rescued.

**Drug and Alcohol Use.** In some water rescues, victims may be under the influence of drugs and/or alcohol or may have mental health issues. People suffering from suicidal thoughts or mental illnesses can harm rescuers, especially if these victims are armed. A great preplanning activity is to invite local law enforcement officers to meet with the team and discuss exactly what will happen if this kind of run occurs. Will officers follow in a boat? Do they have PPE in their cruisers? The same risks can arise when victims are under the influence of drugs or alcohol; that is, they may be in harm's way but not want to be rescued at all.

Unfortunately, alcohol is frequently found in many of the same activities that present water rescues. A long day of boating and drinking can result in a crash and present with multiple victims, all under the influence. Incidents in which victims are fueled by drugs or alcohol can turn hazardous for rescuers in the blink of an eye. Remember, if the victim is swimming, no matter how erratic or out of touch with reality that person is, it may pay to keep your distance and try a rescue from a boat.

## Flotation First

To keep the victim from injuring rescuers, even unintentionally, consider taking a flotation aid to the victim. The victim can then grab for the flotation you provide, and putting something that floats in the victim's hands keeps those hands away from you, making the entire process safer and allowing for more rescuer movement if needed (**FIGURE 5-7**). Extra flotation devices such as a rescue tube or an extra PFD can add minor drag in the water, but will go a long way toward keeping the victim on the surface, regardless of his or her condition. Victims

**FIGURE 5-7** Offering the victim a flotation device can keep the victim from using the rescuer as a flotation device.
Courtesy of Steve Treinish.

who have spent a good deal of time trying to stay afloat or swim for distance will likely be exhausted and, similar to hypothermic victims, may not be able to help themselves or their rescuers.

Once the victim gets the flotation aid, some coaching may be required to get the victim to settle down. This can be done from a short distance away, and the rescuer may approach after a few seconds to assist. Once the victim is settled, the next step in the rescue can be decided.

## Self-Defense

Sometimes a victim will wake up or change focus, and rescuers will need to retreat to stay above the water. Defensive swimming tactics let rescue swimmers back out of this kind of situation, and regroup for another attempt at rescue. All rescuers should know and practice these tactics with realistic "victims," not class buddies just going through the motions. Whatever you have to do to protect yourself on the water, do it. You may be faced with the undesirable, but essential, need to strike a victim to get away and ensure your own safety. In a true "him-or-me" situation, you must take the "me" choice and protect yourself.

## The Approach

Obviously, rescuers will be swimming for a victim because that person is in distress and cannot keep himself or herself on the surface forever. Rescuers should keep a few things in mind when making initial contact with someone in this position.

**FIGURE 5-8** Reaching for the left arm of a victim by a right-handed rescuer will likely let the victim "swarm" all over the rescuer.

Courtesy of Steve Treinish.

**FIGURE 5-9** Reaching across a victim for his or her arm keeps the rescuer's arm and elbow in a blocking position and maintains some initial distance between the rescuer and the victim.

Courtesy of Steve Treinish.

Reaching for the left arm of a victim by a right-handed rescuer will likely let the victim "swarm" all over the rescuer, and escape from this position may not be easy (**FIGURE 5-8**). A better method is to reach across the victim for his or her arm. This technique keeps the rescuer's arm and elbow in a blocking position and maintains some distance between the rescuer and the victim for a few seconds until the rescuer can get a read on a victim's condition and self-rescue ability (**FIGURE 5-9**). A rescuer can then swim or push backward away from a panicked victim. In extreme cases, a rescuer may need to put a foot in the chest of an attacking victim and physically kick the victim away. Once this happens, a firm backstroke will likely put some space between the rescuer and the victim.

Another method is to approach the victim from the rear. This strategy offers two advantages for the rescue swimmer. First, the victim cannot reach

**FIGURE 5-10** Approaching the victim from the rear allows the rescuer to grab the victim under the arms, lean back, and support the victim.

Courtesy of Steve Treinish.

behind himself or herself very well, so the rescuer can get a good grip and control on the victim. The rescuer also has the ability to push off from the victim if things get out of control and place a few feet between them before the victim can turn, if the person can turn at all. Second, the rescuer can grab the victim under the arms, lean back, and support the victim on the rescuer's body (**FIGURE 5-10**). This step often calms the victim enough to let rational thought take over.

One final thought about approaching panicky victims in the water: The last place they want to be is underwater. Instead of pushing straight backward from an overly aggressive victim, consider submerging and backing off underwater. This often will cause the victim to release his or her grasp very quickly. Beware of kicking legs and flailing arms if you perform this maneuver, though.

**LISTEN UP!**

As rescuers, we are trained and equipped to handle this kind of scenario. If we get a face full of water, that is simply part of the job. The victim may not hurt us, but we should expect to get wet and handle a face full of water for our customers.

## Assuming Control

Once you have made physical contact with a victim, flotation must be your single most important concern. Grabbing around the victim's body from the rear puts the rescuer in a good position of control, keeps flailing arms and legs clear, and lets the rescuer talk or shout instructions. It may take quite a shout to jolt the victim enough to follow instructions. Leaning the victim back onto your own chest places the victim's head and mouth far away from the water, and also helps to keep the individual's weight from pulling you forward and farther down in the water. Some victims will settle to the point the rescue can then proceed with minimal problems, but others may continue to panic until they are physically removed from the water. Either way, the rescuer is to ensure they stay above the water.

> **SAFETY TIP**
>
> One common trait that rescuers share is the desire to go directly to the victim and help. Too many rescuers on the side of a watercraft pulling a victim on board can overturn the boat. Consider placing some rescuers on the high side of the boat to balance it.

# Other Rescue Tools

Items other than PFDs can also be used in conjunction with surface rescue swimming. Some of these tools are made especially for water rescue and can increase the speed and efficiency with which rescuers respond.

## Swim Fins

**Swim fins** are known to most people as the fins that divers using self-contained underwater breathing apparatus (SCUBA) or snorkelers wear. These flat rubber or plastic blades can be strapped on the feet and kicked, providing great propulsion (**FIGURE 5-11**). Standard dive fins can be worn during water rescue, but shorter fins are also available that are specific to rescue applications. Most are worn over the feet or wet boots and secured with adjustable straps.

A tradeoff when using fins is the lack of mobility on shore. It is very difficult to walk while wearing swim fins—and don't even think about running unless you want to end up face first on the ground. Carry the swim fins to the point of entry, and then put them on there.

## Swim Mask and Snorkel

Even without training as a diver, a surface rescue technician can use a **dive mask** and **snorkel** effectively (**FIGURE 5-12**). Dive masks are constructed of tempered glass lenses held in a silicone or rubber skirt and strap. They provide a means to seal the nose and eyes from the water and preserve visibility underwater. A snorkel is a breathing tube that the user inserts in the mouth, drawing in air from a point higher on the head, which in theory should be drier. Many rescuers prefer to

use a mask and snorkel because the face can be splashed with water and the eyes and the airway still stay dry.

> **SAFETY TIP**
>
> Until proper training is obtained, no rescuer should try to free dive or SCUBA dive. Injury or death can occur if this evolution is done improperly. To read more about free diving, refer to Chapter 12, *Surf Rescue Technician*.

**FIGURE 5-11** Swim fins.
Courtesy of Steve Treinish.

**FIGURE 5-12** A surface rescue technician can use a dive mask and snorkel to keep the eyes and airway dry.
Courtesy of Steve Treinish.

# Voice of Experience

On June 23, 2016, I was the Kanawha County (WV) Fire Coordinator and Deputy Director of Emergency Management. Earlier that day, the National Weather Service (NWS) in Charleston, WV, had given us a heads-up forecast that we might get flash flooding. Flash flooding is our number-one weather hazard, and we take it seriously. Our local fire department swiftwater rescue teams were asked to go on standby, and all response agencies were notified of the forecast.

By mid-afternoon, we had dealt with some localized flooding, but it was not severe and no life-threatening events had occurred. However, other parts of the state had a lot of rain that caused serious flooding, and local response teams there were becoming overwhelmed and asked the state to send help. The West Virginia Department of Homeland Security and Emergency Management called me and asked for some of the local response water rescue teams to deploy to other counties.

Kanawha County has been blessed by response agencies that had prepared for this. After some failed rescue attempts that resulted in the loss of life, many of the fire departments and the Kanawha County Ambulance Authority had trained personnel in water rescue. We were able to send two fire department water rescue teams to another county, where they made several water rescues.

While the two teams were out of the county, severe weather occurred over parts of our county. The NWS characterized it as a thousand-year rain event that caused severe flooding. Some areas had flooding that exceeded anything of record. In our county, the most serious flooding occurred in the immediate response area of one of the fire departments that had responded to another county.

Luckily, other agencies were able to respond, including fire departments in the adjoining jurisdictions with strong mutual ties. Many other fire department-based water rescue units were able to respond with boats and equipment. EMS, emergency management personnel, and sheriff's deputies assisted the water rescue teams from the other parts of the county, and countless lives were saved. Unfortunately, we still lost six people in the flooding, but many more would have been lost if we did not have redundancy in our swiftwater capabilities.

Rescuers evacuated nursing homes that ended up under several feet of water. They evacuated apartment buildings and moved residents to higher ground and sometimes to upper floors. In a few instances, after moving people to higher floors in apartment buildings, the rescuers stayed with them all night until a safe evacuation could be made. All of this was possible due to having trained personnel, a familiarity and working agreements with each other, and a knowledge of the area. This is not by chance. They work together, train together, and communicate with each other.

**C.W. Sigman**
Director
Kanawha County Emergency Management
Charleston, West Virginia

## Rescue Buoys

Remember the television show *Baywatch*, and the lifeguards running down the beach to make a Hollywood save? The red plastic jug they are carrying is a **rescue buoy**. A rescue buoy, or "can," is a hard, plastic flotation device that is carried on the shore, towed through the water, and held by conscious victims in the water (**FIGURE 5-13**). It consists of a shoulder harness, usually a quick-release type fastened by Velcro; a short line used to connect the can to the user; and the buoy itself. The buoy has handles molded into it that make it easier to grasp. While much more common on beaches and used by lifeguards, rescue buoys can be successfully used by surface swimmers when needed.

To deploy a rescue buoy, the rescuer must throw the shoulder harness over the shoulder, but should not drag it on shore. Instead, the rescuer should carry the rescue buoy as close as possible to the entry zone to minimize the chances of tripping. When the rescuer is ready to swim, he or she drops it into the water and tows it to the victim.

The length of the tether can determine how effectively the rescuer uses the rescue buoy. Too long a leash, and the rope can trip a running rescuer and become an entanglement risk in the water. Too short a leash, and the buoy may hit the rescuer's feet when swimming.

## Rescue Tubes

Another device that is similar to the rescue buoy is the **rescue tube** or belt. A rescue tube is a long piece of foam rubber coated with vinyl that wraps around the victim like a belt, except that the tube goes around the chest (See Figure 5-13). The tube has a quickrelease shoulder harness, a tether line, and the same type of deployment as the rescue buoy. Compared to a rescue buoy, it offers less drag through the water due to its slimmer profile. Rescue tubes cannot handle as much wear and tear as a buoy can, however, because the vinyl can rip or split.

Notably, the rescue tube offers a major advantage in the water with victims, especially with calm victims. Specifically, the tube can provide adequate flotation just by hanging onto it, but the swimmer can also get behind the victim and fasten it using the webbing and snap. This puts the flotation all the way around the victim and will keep him or her safely enclosed in the tube, even if both arms fall down.

### Assisting Victims with Rescue Buoys and Tubes

When approaching the victim with a rescue buoy or tube, use the device's tether line to your advantage, and put some distance between you and the victim.

**FIGURE 5-13** Rescue buoy and rescue tube.
Courtesy of Steve Treinish.

**FIGURE 5-14** The rescue tube may be pushed into or handed to a person in trouble, enabling the rescuer to avoid direct contact with a potentially panicky victim.
Courtesy of Steve Treinish.

Either the tube or the buoy may be pushed into or handed to a person in trouble (**FIGURE 5-14**). This is a great way to stay out of a fight on the water, and lets the victim grab for the flotation and keep his or her hands on it. Most victims will not let go of the flotation aid, and if they do not have to release that hold, the better they will cope. Most people will calm down enough to follow further instructions at this point. Even if the victim becomes aggressive, you can use the flotation device to block his or her advance. Once the victim has the rescue buoy or tube in his or her grasp, he or she may proceed to the safe zone, whether that is back to shore or to a boat, or even just remain in place and await more help.

What if the victim cannot help? No problem. Simply slide your arms under the victim's arms from behind and grab the rescue buoy or tube, holding it to the victim's chest by the handles (**FIGURE 5-15**). This places you in a helpful position, on your back and able to use your legs to swim. It also pins the victim between the two flotation devices available, the buoy and you. This is

**FIGURE 5-15** If the victim cannot help, the rescuer should slide his or her arms under the victim's arms from behind and grab the rescue buoy, holding it to the victim's chest by the handles.
Courtesy of Steve Treinish.

**FIGURE 5-16** Once a rescue tube is attached to the victim, the rescuer can use both arms and legs to swim to safety.
Courtesy of Steve Treinish.

**FIGURE 5-17** Rescuers can lie on swim boards and use their legs to kick. Owing to the higher ride in the water, they can maintain visibility of the victim throughout.
Courtesy of Steve Treinish.

one drawback of the rescue buoy: If the victim cannot hold onto it, the rescuer must devote some energy to keeping the victim afloat all the way to the safe zone.

Buoys offer enough buoyancy that two or three people can use it for partial flotation if they cooperate. That kind of sharing may be the right decision, especially if the rescuer decides it is better to "defend in place" (to borrow a term from firefighting).

Once a rescue tube is attached to the victim, the rescuer can swim to the safe zone. The rescuer still needs to monitor the victim, but can use both arms and legs to swim (**FIGURE 5-16**).

Tubes also make superb tools to help float people, backboards, or Stokes litters if needed. Two tubes placed under a victim will go a long way toward keeping the victim's body afloat while the rescuer controls the airway or even provides cervical spine immobilization. The downsides to rescue tubes are that while they can be used to support multiple people, they are not as buoyant as buoys, and if a victim puts up a fight, tubes can easily be pushed aside, giving the victim access to the rescuer.

## Swim Boards

**Swim boards**, boogie boards, and rescue boards are generic terms for any shorter, floating board used in rescue (**FIGURE 5-17**). Rescuers can lie on these boards, use the legs to kick or forward crawl swim, and, owing to the higher ride in the water, still maintain visibility of the victim. Surfboards are longer but can be used similarly to rescue boards, although surfboards by design are propelled with the hands and arms, not kicked. Any of these tools will provide the rescuer with some flotation, allow a quicker response, and give the victim something to lie or lean on so as to stay on the surface.

## Tethered Operations

**Tethered line rescues**—also known as "live bait" rescues—are possible in all water rescues. Even dive operations normally entail a tether. Any technician-level rescuer

---

**LISTEN UP!**

A large boat crash on open water is still considered a surface rescue—and it may be a common scenario encountered by many agencies. Sometimes there may be 6 to 10 people (or more) in the water when you arrive, and some may not be wearing PFDs. First-arriving responders, especially if they are in smaller watercraft, might be well advised to let the victims float there using PFDs or rescue buoys while the scene is triaged and they are awaiting the arrival of additional resources. Having a few rescue buoys or tubes to distribute initially can go a long way toward calming the victims in this type of response.

may encounter a situation that requires being hooked up to a tethered line to complete the mission. Surface rescue is no different. In fact, tethering can make things much easier for rescuers.

In a tethered operation, the swimmer attaches a line to his or her PFD, if it has the proper attachments, and swims to the victim (**FIGURE 5-18**). The rescue crew on shore plays out and controls the line and, when the signal is given to retrieve, pulls the rescuer and the victim into shore. This strategy lets the rescuer focus on keeping the airway clear and dry and keeping the victim afloat—something that is much easier to do when someone else supplies the propulsion back to shore. Three elements that can have a major effect on this operation are speed of retrieval, choice of rope, and entanglement concerns.

## Speed of Retrieval

The speed of retrieval is the speed at which the rescuers pull the victim and the rescuer as a package to shore. Most people are familiar with fishing bobbers and know what will happen if one is pulled back to the user too quickly. While the fishing bobber is stationary, it floats. With a slow, controlled pull, it still floats and moves over the water very easily. Pull hard and fast, though, and the bobber is pulled underwater and stays there until the pull is reduced or it reaches shore. Likewise, during a surface rescue, if the crew on shore gets excited and pulls the line in too quickly, the rescuer–victim package will be submerged. A slow, controlled pull can avoid this outcome. The easiest way to monitor for this condition is to pay attention to the rescuer and not get excited when the signal to retrieve is given.

**FIGURE 5-18** In tethered rescue, the rescuer attaches a line to his or her PFD and swims to the victim while a second rescuer controls the line from the shore.
Courtesy of Steve Treinish.

## Choice of Rope

Another major problem that could arise relates to the choice of rope. Most kernmantle rescue rope is heavy by nature. It is manufactured to be strong and hold thousands of pounds. It also can hold a lot of water in the fibers, making it even heavier. This presents two problems: (1) It may be hard to swim anywhere when attached to kernmantle rope, and (2) if the rope sinks to the bottom and snags or gets tangled, it may pull the swimmer underwater. Some companies use nylon rope for this purpose, but it still may sink and present the same hazard.

A good-quality water rescue–specific rope is made from polypropylene and floats. It also does not absorb water, so its weight will not increase during the rescue. Water rescue rope also has good breaking strength, usually around 3000 pounds, so it is an excellent choice for any water mission, including for static lines. Smaller-diameter water rescue ropes can be used in throw bags; because this type of rope is smaller, 75-foot-long bags are not uncommon. If necessary, water rescue ropes can be tied together to form longer lines.

**SAFETY TIP**

If water rescue rope is placed in service, it must be marked as such, and not used for life safety. The differences in durability and breaking strengths for ropes can create unique and potentially dangerous problems if a rope lifting operation and a water rescue operation mix.

## Entanglement Concerns

Whenever rescuers are working around rope, they should use extreme care to control slack, loops, and bights of rope that may entangle them. Water current, ice, heavy wave or surf action, or a combination of all of these conditions can cause problems by entangling rescuers or by fouling in or around objects found in the water. Surface tenders can control the amount of rope in the water with the rescuer, but there is a fine line between holding tension and preventing forward movement. A worst-case scenario is a rescuer who is pulled underwater or a rescuer with a loop of rope around the neck. Technician-level rescuers should always carry a one-handed cutting tool with them in case they need to escape a rope entanglement.

## Medical Treatment

Many victims, including competent swimmers, find themselves in a whole new ballgame when they are injured in the water. Just as unconscious people cannot swim, badly injured swimmers may find themselves unable to do so. Many of these swimmers are victims of cervical spine injuries, which may be experienced both in open waters and in swimming pools. Water skiing, falling overboard from a boat, or jumping into shallow water headfirst can all result in spinal injury or major and permanent paralysis.

Falling off a dock and striking the head on something underwater may be enough to kill someone. Even if rescuers get to the victim and provide positive flotation, they still must properly treat the victim. Most very basic emergency medical care can be provided on the water, but cervical spine injuries are especially tricky if the victim is still in the water. This environment increases the possibility of the head or neck moving and resulting in further injury, especially in water rough with waves or chop.

## Cervical Spine Injuries

How victims injured themselves is really not pertinent during the water rescue. Most rescuers can recognize cervical spine injuries and understand how they occurred. If the victim is removed from the water incorrectly or too quickly, however, the rescue effort may exacerbate the original injury. Although most EMS care is easier to do on terra firma, most victims would like their rescuers to do as much as they can, where they can. While some conditions or hazards, such as hypothermia, impaired airway, or cardiac arrest, may call for a quick extrication from the water, cervical spine (c-spine) precautions can still be applied in many cases. In water that is not directly hazardous (e.g., the local municipal swimming pool), rescuers can still protect the victim's head and spine, thereby providing the best medical care possible for the individual.

**C-spine treatment** involves bracing or immobilizing the victim's head so that the chin is in line with the body while maintaining the head tilted in a neutral position. A **backboard** is used to lay the victim flat, with the victim's body being firmly attached to the board using straps. The head is then braced with foam blocks, rolled-up blankets, tape, or a combination of all three. Different methods are used by different agencies, but they generally tie into this common idea.

Being in the water may result in more movement to the patient, but rescuers still need to do what they can. A rescuer reaching the victim before the backboard does can position the patient in ways that provide some cervical spine immobilization.

## Vise Grip Hold

One of the quickest ways to initially take care of the victim's head and neck is to simply extend the individual's arms up over the head and hold them together (**FIGURE 5-19**). While not exactly optimal emergency treatment, this **vise grip hold** helps keep the head from rotating. Of course, this position is possible only if the victim has enough flotation to stay on the surface. Most PFDs will bunch up at the shoulders and provide even more support in this position.

Applying a vise grip hold sounds easy, but for a rescuer alone with a victim with no flotation in deeper water, it is an extremely heavy task load. Remember, do what you can with what you have—including any extra flotation or rescue tube you brought along. Placing the victim on his or her back, floating the victim, and protecting the airway may be the only care possible for a victim with a cervical spine injury while in the water. A victim wearing a PFD by chance, such as water skiers or some swimmers, makes this entire scenario much easier.

## Face-Down and Unconscious Victims

Face-down victims need to be turned over before further treatment can be administered. Grabbing an arm or a shirt to flip the victim without managing the neck and head will result in the head rotating a great deal, which is not a good thing. These victims

**FIGURE 5-19** Vise grip hold.
Courtesy of Steve Treinish.

can be rotated into a supine position while maintaining at least some cervical protection, but rescuers can and will submerge for a few seconds during this process. To flip an unconscious victim, follow the steps in **SKILL DRILL 5-2**.

## In-Water Aids

Even if rescuers are in the water alone, or if the proper equipment has not arrived, they cannot release the cervical spine hold. Letting go might allow the spinal column to move or let the head and neck move, which may injure

**FIGURE 5-20** Many cervical collars used in the field are the folding style, which lie very flat for easy storage.

Courtesy of Steve Treinish.

the victim further. But what if you are floating in water 10 feet deep? Consider slipping rescue tubes, other PFDs, or even ring buoys under the victim's back to provide extra flotation. This flotation will keep the airway as high as possible and will make it easier for the rescuers to manage the victim and still protect the airway.

If the rescuer has no equipment, he or she can lean back and place the victim on the rescuer's own chest, letting the PFD float them both. The victim's legs will droop, but this assistance is better than nothing.

## Backboard Use

Backboards are a crucial part of c-spine immobilizations, and they can be used in the water. Some backboards do float, but the majority of them do not supply enough flotation for the victim to fully let go of it.

Backboards do not cradle the patient very well, and they can get very tipsy in the water. The victim presents a very high center of gravity on the board and can easily slide off one side or the other. In wave conditions, this risk is even more pronounced.

It takes several rescuers to package the patient on a backboard. During this process, if possible, one rescuer should be assigned the victim's head and neck area. This rescuer also must make sure the head is not submerged. One rescuer must take control and provide traction and manual immobilization for the head, not leaving unless someone else assumes this responsibility (**FIGURE 5-21**). This rescuer also usually provides the count or the commands to do the other work, ensuring that things are done in sync.

The other rescuers assemble on each side of the victim, providing flotation and support and trying to keep the patient's body as straight as possible, in line from head to toe (**FIGURE 5-22**). Other rescuers can then move or float the backboard into place and position it under the victim. Moving the board into place from the feet keeps the rescuers in place on each side of the victim, resulting in better flotation.

You might imagine trying to put a board in place from the side of a victim while trying to keep him or her above water. Trying to place the board under the victim by sinking it from behind the victim's head is not recommended, because the backboard can strike the head and this strategy places the rescuer immobilizing the head out of position.

PFDs, ring buoys, or rescue tubes can all be used to help float the backboard. The tubes are especially helpful because they can be easily slid or forced under the backboard, and because they create a natural sling effect. It is also fairly easy for a rescuer to quickly pull the tube from under the victim and immediately place

# SKILL DRILL 5-2
## Flipping an Unconscious Victim NFPA 1006: 16.3.2

**1** If approaching the victim's right side, lay the left arm on the victim's back, with the left hand holding the back of the victim's head and neck.

**2** Hold the right arm under the victim's chest, with the right hand holding the head and jaw area. This stabilizes the head somewhat inline with the body. Submerge yourself under the victim, and with one smooth move, rotate under the patient 180 degrees.

**3** Come back up on the other side, moving around the victim underwater as the victim is turned.

**4** The victim should be face up, and the rescuer should be back up beside the victim's head. A manual c-spine position can be held until further help or treatment arrives. This is tricky but works very well and does a good job at maintaining the victim's head in a neutral position.

Courtesy of Steve Treinish.

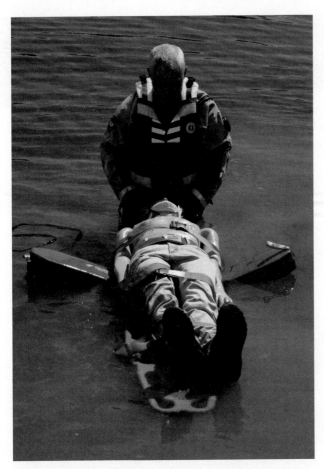

**FIGURE 5-21** While packaging a victim on a backboard, one rescuer must maintain traction and manual immobilization for the victim's head.
Courtesy of Steve Treinish.

it under the backboard once in place. The tube will take on a crescent shape while floating, which cradles the board. Ring buoys can be used, but the flotation does not extend as well past the sides of the board, so the board will have a tendency to tip or roll over in the water. PFDs are better than nothing, but present a certain amount of floppiness that must be controlled.

## Stokes-Type Litters

A piece of equipment that most rescue companies carry, or that is sometimes found on ladder apparatus or even on the shore somewhere, is the litter, also known as a

**FIGURE 5-22** While packaging a victim on a backboard, rescuers assemble on each side of the victim, providing flotation and support and trying to keep the patient's body as straight as possible.
Courtesy of Steve Treinish.

Stokes basket (**FIGURE 5-23**). Most Stokes baskets are constructed from plastic, but some are made from chicken-type wire. The wire litters let water drain out, but extra care must be used to prevent evidence being lost. Litters offer great victim packaging, as the victim sits much deeper in the cradle. Litters also offer good handholds, durability, and the ability to place a fully immobilized victim in the litter while on a backboard.

## Litter Flotation

The use of a flotation kit on litters makes it easier to package any victim. The flotation is attached around the sides of the litter and works to the rescuer's advantage in deeper water, since the victim lies in the litter, not on it. The litter is used and maneuvered just like the backboard, in that one rescuer must control the head and neck, and the other rescuers bring the basket to the feet of the victim, partially sink it, and move it underwater to a point under the victim. Releasing the pressure needed to sink it will float the litter into position and cradle the victim. To use litter flotation, follow the steps in **SKILL DRILL 5-3**.

A big concern is the flotation of the litter itself. Ready-made factory kits are available to provide flotation to the basket, but many agencies have adapted their litters themselves by strapping or tying foam tubes or sealed plastic pipes to the sides of the basket. This adaptation can work very well, but it must be tested in the water, with sufficient weight or perhaps a mannequin, before placing a victim on it. After the victim is loaded into a newly modified Stokes basket is no time to find out that the pipe glue did not seal, or that a few more pounds of flotation were needed. A good rule of thumb is to secure and use extra flotation if possible, and keep it there until the litter reaches dry land.

# SKILL DRILL 5-3
## Using Litter Flotation NFPA 1006: 16.3.2

**1** Loading a victim onto a Stokes litter in the water requires teamwork from the rescue crew. One rescuer must maintain c-spine immobilization at all times.

**2** While the head and neck are managed, the other rescuers position the Stokes basket at the victim's feet.

**3** It is easier to move the Stokes basket while keeping the victim as still as possible. Rescuers push the Stokes basket below the victim and move it up toward the head.

**4** Release the litter to let it float up and under the victim. One rescuer continues to maintain c-spine immobilization until the victim is packaged per the AHJ's protocol.

**FIGURE 5-23** Litters can be very helpful in victim rescue.
Courtesy of Steve Treinish.

**FIGURE 5-24** Once the victim is packaged and placed in the litter, slide or lift the litter to the bank, boat, or other stable surface.
Courtesy of Steve Treinish.

Once the victim is packaged and placed in the litter, extrication can take place by sliding or lifting the litter to the bank, cot, or boat (**FIGURE 5-24**). Some larger watercraft are equipped with small cranes that enable victims to be lifted onto the deck. Standard patient care can then be given per the EMS protocol under which the rescuers are operating. That care may include removing the backboard from the Stokes basket or placing a backboard on a victim if one was not used in the water. In many AHJs, once a victim is immobilized and placed in a litter, the entire assembly goes to the emergency room and is kept as a unit until cleared by a physician.

### SAFETY TIP

Especially in water deeper than the chest, the litter or backboard may potentially be used for flotation support. If a rescuer sinks in the water and is holding onto the board or litter, the victim sinks, too—but the victim is strapped to the board and cannot do anything to help himself or herself. When multiple rescuers try to package someone in deep water, it is much easier to dunk the victim's head than most rescuers might imagine, especially if the water is choppy. The victim will be unable to make any noise to bring attention to this problem, and it is not uncommon for victims to have their heads inadvertently dipped or held underwater. A training class may package a student or two and give a good lesson on how bad this experience feels to the victim.

### LISTEN UP!

One thing that can make life much easier for rescuers, especially those in watercraft, is to snap a few longer straps to the foot area of the board or litter (**FIGURE 5-25**). The crew can then pull the entire package up and onto the deck with one smooth pull, rather than having to stop and reach down farther when the patient is halfway out of the water. In addition, more rescuers can help with the lift, instead of a few rescuers bunching up at the side on which the victim is being brought up.

## Supervisory Skills

According to NFPA 1006, a technician rescuer needs to be able to supervise, coordinate, and lead rescue crews during operations. These skills are already known to rescuers who are well versed in conducting size-ups and evaluating incidents. Having crews on the same page is crucial, and the best way to achieve this goal is to ensure information is given and received by all involved. As a rescue leader, your job is to supervise, not perform. It is easy to become engrossed in the actual activity of an incident, but if sufficient human resources are available to perform a task, you should operate in a much more hands-off manner. Let your crew perform the work while you manage the operation.

### Pre-entry Briefing

A quick briefing before the rescue begins may take the form of a quick shout to the chief standing on the dock.

**FIGURE 5-25** Straps snapped to the foot area of a board or litter help the crew lift the package with one smooth pull.
Courtesy of Steve Treinish.

A longer, more detailed briefing may be needed if the incident involves larger areas or more complex evolutions, such as multiple watercraft all working a crash on the water. This briefing offers an opportunity to ensure each rescuer tasked with involvement in the incident is qualified and adept at the work to be performed. It is much better to question rescuers' qualifications and ability quickly on shore, rather than in the midst of a potential hazard. Preoperation checklists can also be used before committing rescuers to an operation.

## Safety

Safety is an ongoing, continual evaluative process that does not end until the last victim, rescuer, and piece of equipment are back in a safe area with no other rescue work to do. Many rescuers are injured after the actual rescue is done, while rescuers and equipment are being removed from the hazard. Concentration lapses, fatigue, and inattention to detail can all cause problems after the rescue itself but before the incident is officially concluded.

Crew leaders must be able to operate while monitoring the conditions around themselves and their crews. As mentioned in Chapter 1, *Understanding and Managing Water Rescue Incidents*, one of the best ways to mentally test the safety of an environment, a system, or an evolution is to ask, "What happens if things fail catastrophically?"

The appointment of a safety officer who is not directly tied to a task is invaluable and provides another level of spatial awareness. This safety officer should be trained to at least the highest level of response being used during an incident, and should have direct communications with the incident or operations commander if it becomes necessary to halt a dangerous operation. More information on safety can be found in Chapter 1.

## Accountability

Company- or crew-level accountability should be maintained at all times. This does not mean the crew must be on a leash and kept so close that tasks are impossible to perform. Rather, the leader needs to know where they are and how to get them help if needed. The most effective span of control is generally thought to be five to seven members, and adhering to this rule will help track the entire crew under your command.

## Postincident Briefing

After the incident ends, all members should be accounted for; results passed on; equipment checked and returned to service; and the operation examined for effectiveness, clarity, and overall success. This briefing does not have to be of long duration, but it should give responders a chance to ask questions and pose concerns. Problems and concerns should be noted and efforts made to reduce or eliminate them in future trainings or evolutions. In addition, the findings from the postincident briefing represent a good starting point for another session of preincident planning.

---

**LISTEN UP!**

Use training evolutions to prepare other team members for leading rescue evolutions. Place other rescuers in the boss's seat and let them call the plays, as long as it is safe. Especially in volunteer agencies, it could be you placed into this position, even if only temporarily.

---

**KNOWLEDGE CHECK**

The question, "What happens if things fail catastrophically?" is used to test:

a. strategic goals.
b. qualifications.
c. tactics.
d. safety.

*Access your Navigate eBook for more Knowledge Check questions and answers.*

# *After-Action* REVIEW

## IN SUMMARY

- Surface rescue encompasses many types of rescues and conditions, and rescuers must train for the water they expect to find in their areas.
- Swim testing is essential to find out how comfortable potential trainees are in the water. It is a good idea to have an annual test for trained rescuers, and the test should be adapted to the types and conditions of water found in the areas served by the team.
- Most rescuers think of a submersion as a dive rescue event, but especially around beaches or shallow ponds, rivers, or other water, there may be other ways to search areas without diving.
- Most people can run faster than they can swim, so using the shore to get as close to the victim as possible will reduce the time it takes to reach the victim, while keeping the victim in sight longer.
- A variety of tools—such as swim fins, swim masks, snorkels, rescue buoys, rescue tubes, and swim boards—can increase the speed and efficiency of surface water rescues.
- Tethered rescues can be done in all water rescues. Any technician-level rescuer may get into a situation, including during a surface rescue, that requires being hooked up to a tethered line to complete a mission.
- Most very basic emergency medical care can be provided on the water, but caring for victims with cervical spine injuries is especially tricky if the victim is still in the water. This environment increases the possibility of the victim's head or neck moving and resulting in further injury, especially in water rough with waves or chop.

- Once a victim's cervical spine is protected, consider slipping rescue tubes, other PFDs, or even ring buoys under the victim's back to provide extra flotation. This flotation will keep the airway as high as possible and make it easier for the rescuers to manage the victim's care.
- Technician-level water rescuers need to be able to supervise, coordinate, and lead rescue crews during operations.

## KEY TERMS

*Access Navigate* **for flashcards to test your key term knowledge.**

**Active drowning stage** The drowning stage in which the victim is still fighting to keep the head above water by doing anything necessary.

**Backboard** A hard, flat piece of wood or plastic used to provide a firm platform to secure patients suspected of having cervical spine injuries.

**C-spine treatment** Cervical spine injury management, which relies on the general practice of keeping the head, neck, and body immobilized to prevent further injury. The head is kept in a neutral, centered position.

**Defensive swimming** Self-preservation by using blocks, thrusts, and swimming maneuvers in the water to ensure rescuer safety.

**Dive mask** A device consisting of a silicone skirt and glass lenses that attaches around the head, keeping water from the wearer's eyes and nose.

**Head-up running leap** An entry into surface water in which the rescuer kicks the legs together and sweeps with the arms in an effort to keep the head above water and maintain sight of the victim.

**Passive drowning stage** The stage of drowning in which the victim is no longer fighting to stay above the water.

**Rescue buoy** A hard plastic float, resembling a football, that is equipped with molded handles and a tether for the rescuer.

**Rescue tube** A needle-shaped tube made of foam rubber, which is towed by the rescuer and used for victim flotation; it may also be used as an accessory float due to its shape and ability to bend.

**Snorkel** A plastic or rubber breathing tube through which the wearer draws breath from above the head.

**Swim boards** Small, flat flotation aids used for recreational swimming, flotation, or swiftwater rescue swimming.

**Swim fins** Rubber or plastic blade-like propulsion, worn and kicked by the feet.

**Swim test** An in-water exam to test needed skills, comfort level, and stamina to start a water rescue training class.

**Tethered line rescues** Rescue evolutions in which the swimmer is physically attached to a rope or line and controlled from shore or a boat.

**Vise grip hold** Using the victim's arms as an impromptu cervical spine brace by raising both arms up above the head and holding them in place.

# On Scene

You have been called to a boating accident at a local recreational lake. You are placed in charge of operations and have a decent view of the crash site from the response point in a local marina. You observe one boat still floating, with two small children screaming on board, and one personal watercraft barely floating, with only the nose showing from the water. It is a busy weekend day, and the lake is very choppy. Numerous boats are driving by, and many are stopping to watch. Using binoculars, you make out three people in the water, just one of whom is wearing a PFD.

Your department is responding with a rescue boat, three surface rescue technicians, three operations-level trained fire fighters, and two brand-new fire fighters who have only the basic skills learned at the training academy. The rescue boat has six PFDs, a backboard, a floating litter, and numerous throw bags. Your mutual aid department has another inflatable rescue boat, but it was not dispatched on the initial assignment. Law enforcement does have a large center-console boat on the water for patrol. You cannot see it anywhere yet, but you do have communications with it. Several civilian watercraft are docked nearby, with owners on board, including two pontoon-style boats.

# On Scene Continued

**1.** To get an initial "read" on this incident, which information do you need from the first-arriving crews?

**A.** The number of boats, the total number of people and how many are accounted for, needs for quick triage, and any additional needs

**B.** Which dive team must be called in, which towing company can tow boats, and whether any businesses have extra watercraft to use

**C.** Whether the crews on the first-due boat have enough EMS supplies

**D.** Where the crash site is drifting in the wind and waves

**2.** Which scene accountability must be in place?

**A.** Each boat's crewmembers must be identified with a name and qualification, the boat to which they are assigned, their task, and their location.

**B.** Every member operating around or on the water must be tracked, have the proper PPE, and be able to call for help if needed.

**C.** Passports must be collected on the water before committing rescuers into the rescue evolution.

**D.** Both A and B

**3.** How will you provide care to all the people in the water?

**A.** Get them flotation, provide quick in-water triage, and place walking wounded in one boat for further treatment or transport.

**B.** Call in the mutual aid department, assigning this crew a specific victim before they launch.

**C.** Load everyone up on the law enforcement boat, and provide a triage and care on the way back to the command post.

**D.** Triangulate the scene or leave the law enforcement boat to guard and mark the scene, and use the civilian pontoon boats to shuttle wounded victims to the extrication point.

**4.** What is the best way to preserve the scene and evidence for law enforcement investigators?

**A.** Triangulate the crash when initially on scene, using one of the untrained fire fighters to take pictures with prominent landmarks, if possible. Try to set up a perimeter on the water, using civilian boats if needed.

**B.** Tow the boats left on the water back to the closest shore.

**C.** It will be nearly impossible to preserve any type of scene because of the waves and wind.

**D.** Describe in detail to the dispatchers exactly what you are seeing from the command post.

*Access Navigate to find answers to this On Scene, along with other resources such as an audiobook and TestPrep.*

# 6

# Swiftwater Rescue Operations

## KNOWLEDGE OBJECTIVES

After studying this chapter, you should be able to:

- Identify and describe the best conditions for swiftwater rescue training. (NFPA 1006: 17.2.2, pp. 160–161)
- Discuss the effect of current speeds and current forces on swiftwater rescue. (NFPA 1006: 17.2.2, 17.2.3, pp. 161–163)
- Identify and describe the personal protective equipment necessary for swiftwater rescue. (NFPA 1006: 17.2.1, 17.2.2, 17.2.4, pp. 163–165)
- Identify and describe the features of swiftwater. (NFPA 1006: 17.2.3, pp. 166–170)
- Discuss the importance of defensive swimming in swiftwater rescue. (pp. 170–173)
- Identify and describe swiftwater rescue roles. (NFPA 1006: 17.2.1, 17.2.2, pp. 173–175)
- Discuss the use of ropes and rope systems in swiftwater rescue. (NFPA 1006: 17.2.1, pp. 175, 177–194)
- Explain how to terminate a swiftwater rescue incident. (NFPA 1006: 17.2.5, p. 194)

## SKILLS OBJECTIVES

After studying this chapter, you should be able to:

- Get over a strainer. (p. 172)
- Deploy a droop line. (NFPA 1006: 17.2.1, p. 180)
- Peel out. (p. 192)

# Technical Rescue Incident

You and your partner are on your 13th run on the medic unit. You are exhausted and ready for the rewarmed dinner and coffee you frequently see at night during your shift. On the way back to the station, your department is dispatched for another call: a truck in the water at a well-known river that parallels the road on which you are traveling. It takes you only a few minutes to get there.

From the street a few hundred yards away, you can see the cars stopped along the road. You can also see an object in the river, approximately 45 feet offshore. The river is flowing high, but not very fast—you have certainly seen it moving faster. You and your partner don the personal flotation devices (PFDs) that are always kept in the medic unit and grab an extra PFD, plus some throw bags.

A short scramble down the riverbank places you on a small gravel bar, directly inshore from the victim. The victim is a 56-year-old man who had worked on his truck's brakes earlier in the day. When they failed while he was driving, he ended up in the river but was able to climb up onto the truck's roof. The victim is stable, but not a strong swimmer. You hear sirens in the distance, so help is coming.

**1.** What can you do with the equipment you have on scene?

**2.** Should anything be done at this point, or does that place the victim in greater danger?

**3.** Which method of rescue can be attempted that provides the greatest safety for responders?

 JONES & BARTLETT LEARNING NAVIGATE 2 *Access Navigate for more practice activities.*

# Introduction

Despite how minimal the swiftwater operations-level requirements appear to be in NFPA 1006, *Standard for Technical Rescue Personnel Professional Qualifications*, they encompass a very large world in water rescue. Flood events are occurring with greater frequency, and both loss of life and loss of property (measured in dollars) are increasing. Humans are changing the landscape by developing millions of acres each year, which in turn changes the natural drainage and absorption of the soil. The addition of roads, parking lots, and the like allow for more runoff at a faster rate. When mass development occurs, most small creeks and streams cannot keep up with the excess runoff from the altered landscape. When water is introduced in this picture, such as from spring rains, storms, or snowmelt, the ground simply cannot soak it up. The runoff has to go somewhere, and it may get there much faster than it did in past years. Swiftwater events can occur during flooding, recreational use, and accidental immersion into moving water of any kind, and the body of water involved can be as large as the Mississippi River or as small as a creek that, under nonflood conditions, is only inches deep and a few feet wide (**FIGURE 6-1**).

The National Fire Protection Association (NFPA) has streamlined some standards related to water rescues by referring to the prerequisites listed. For example, swiftwater operations-level students are expected to have met the surface operations-level requirements, one of which involves the use of water rescue personal protective equipment (PPE). In NFPA 1006, **swiftwater** is defined as any water moving at a speed faster than one knot, or 1.15 miles per hour.

**FIGURE 6-1** With climate changes and increasing development and manipulation of wetlands and waterways, urban flooding will continue to be a problem.
Courtesy of Steve Treinish.

# Training Conditions

One important facet of fire service training is the balance of realistic training, availability of training props and environment, and safety. No other fire service subdiscipline presents quite as much of a challenge in terms of mimicking the expected environments for students as swiftwater rescue does. Departments often have scheduling issues when dealing with vacations, sick leave, other department business, and human resources costs that complicate planning for training sessions. Putting training together within a few hours can be problematic when the water is flowing naturally, so some resulting swiftwater training gets dictated by time, not realism. Rescuers can get caught in the "This has been scheduled for four months, and it will happen!" issue, even if the area has seen a record drought rather than normal to increased precipitation.

Swiftwater training has a unique requirement in that the "classroom" should ultimately reflect the environment expected during extreme events. Rescuers may be placed in danger when they are forced to operate in water that is more forceful or faster than they have ever seen in training, and they may wrongly assume more risk in a life-and-death situation. Instructors are encouraged to use slower water to build confidence and skills, and gradually increase the current speed as training progresses. Performing swiftwater training in slow-moving water only may foster a false security in rescuers and expose them to injuries (or worse) when faster water requires action.

Many agencies perceive a difference in operations-level swiftwater training and technician-level flow rates, but as rescuers learn, experiencing faster and faster currents can help foster respect for moving water and keep the rescuer safe. Moving water presents the same hazards regardless of the training level. The difference in swiftwater technician and operations training lies only in the techniques learned, not the water flow.

## Realism

One issue that agencies, instructors, and students must understand is the idea that the speed of water is relative. A large river may be flowing over the 1 knot requirement, but in actuality be very docile. In that large river, boating conditions will likely be much more like those found in a lake than in a river. This kind of venue may be a great starting point for students, but for experienced rescuers or boaters there will just not be much stress present.

**FIGURE 6-2** Swiftwater training must establish a base of skills and competence in different conditions, but ultimately provide an environment that matches the expected incidents.
Courtesy of Steve Treinish.

Put those same boats in a tight stream moving at 4 or 5 knots or even faster, with trees, strainers, and other hazards to manage, and the same setting presents a totally different, but realistic scenario (**FIGURE 6-2**). If the students can expect these conditions in their districts, even if they are not the norm, then students should be trained in these water conditions to create respect and familiarity.

**LISTEN UP!**

Rope rescue training can face a similar problem. A student hanging from a ledge only 10 feet tall does not feel anything like a rescuer would if hung from a water tower at 140 feet, even if the skills and the work done are the same. Some "ill-at-ease" feelings may be required to show rescuers which conditions they may really be working in.

**SAFETY TIP**

Under no circumstances should students or inexperienced rescuers be placed in conditions for which they are unprepared. Starting slow and progressing into faster water as skills are mastered is a much better path, and will help build the confidence needed to tackle this environment.

## Dam Releases

Controlled releases from impoundment dams are a great way to provide adequate water for a controlled river flow. Flows can be tailored as needed, or even shut off completely in case of emergency (**FIGURE 6-3**). Forming relationships with the agency controlling the

**FIGURE 6-3** Controlled releases from impoundment dams offer great opportunities to train in a controlled river flow.
Courtesy of Steve Treinish.

flow through the dam can also lead to some great pre-planning. These personnel can often provide details on the planned flow, resulting water depths, and time factors. The ability to call for an emergency shut-off at the dam discharge is always a good option in extreme problems, and having the ability to do so within seconds increases safety. There will still be residual water flowing downstream, but dam personnel will know how long it should take to dissipate.

> **LISTEN UP!**
>
> Maintaining ongoing communications with dam personnel while rescuers are working in or around the water will help ensure that water levels match the expected levels, that flows can be adjusted, and that emergency shutdowns can occur if needed. An "all clear" signal after rescuers have left the water should also be relayed to the dam staff.

## Seasonal Training

Precipitation levels vary around the world. In the United States, the summer and fall seasons generally do not lend themselves to planned swiftwater training, as many U.S. reservoirs are used by wildlife and for recreation. Most releases from reservoirs depend on obtaining approval just before the planned date of release and factor in the amount of water in the reservoir, conditions downstream, and the impacts on wildlife and fish habitats. Springtime is usually a much better season in which to train, as springtime flood control and runoff often create a need to flow water—and with spring rains, there may be plenty of

water to spare. Although fall is normally dry in many U.S. regions, it may be possible to get water for training through the winter drawdown of many lakes, which is intended to provide flood control the next spring. You may not get the precise water conditions you want, but the flow may be enough to do some basic evolutions.

## Training Facilities

Some facilities are now using watersports training locations as full-time paddling and swiftwater training sites. These areas have been created to form the same river features that canoes and kayakers face in competition. While not everyone may be able to access these sites, they are truly state-of-the-art facilities for honing swiftwater rescue skills.

# Swiftwater Savvy

Swiftwater environments are a very dynamic field, in that conditions can change dramatically from mission to mission, or even within a short distance of water during the same mission. The seemingly increased sense of urgency may be greater if the victim is actually moving with the water. The confusion that victims and rescuers may experience as water finds its way into areas not normally wet contributes to this sense as well. Swiftwater can be an extreme danger to untrained rescuers, and we face more of it every year. Fortunately, moving water behaves in ways that are predictable and manageable, and we must learn to take advantage of this behavior in swiftwater operations.

Rescuers must remember that the water will never stop. This fact alone seems to put all kinds of people in jeopardy, in that they drastically underestimate the amount of strength the current will present. Operating in moving water is especially dangerous because, even as the rescuer's strength is ebbing, the water continues to flow.

## Current Speeds

The actual speed of the water is a major influence on how much energy the water carries, and is crucial knowledge during a search to physically locate a victim if that person is traveling with the flow. A quick estimate of water flow rate can be used to determine the distance over which a victim may travel. Obtaining an estimated distance traveled will let rescuers utilize features such as access sites and bridges downstream to identify spotter locations, potential rescue sites or access sites, and locations where they can stage, deploy, and extricate resources. Water speeds of up to 25 to 30 miles per hour are possible in the drainage ditches in urban areas, and humans are simply no match for this kind of flow.

One quick way to estimate the speed of the current is to assume that a floating object traveling 100 feet in 1 minute equals 1 knot. A distance of 100 feet can be measured or estimated, and then a float of some kind can be thrown and the time the float takes to travel that distance can be recorded. This time may then be compared to a current chart to estimate the water's speed (**TABLE 6-1**). Note that faster currents are gauged with a 300-foot measure (i.e., an object traveling 300 feet in approximately 18 seconds is traveling at 10 knots) and might come in handy when assessing water speeds in rivers or flood channels. A speed of 25 knots is very rare, but has been seen.

Another way to quickly get a distance scale to use is to use a throw bag. Many throw bags are 50 or 75 feet long. Deploy the bag into or along the river and it will stretch itself out, giving a known length on which to calculate an approximate speed.

## Current Forces

Although operations-level rescuers are not entering the water directly, they should have a good understanding of the power and dynamics of swiftwater they will face working on shore, or if they accidentally fall in. The larger the object in the water, the more force that will be applied to it, since larger objects have a larger surface area that will be subjected to the flowing water. **TABLE 6-2** is a current force chart. As seen in this table, if the current speed doubles, the force from it quadruples.

Current velocity also plays into the force of the current. Specifically, if the speed of the current doubles, the force applied by that current is quadrupled. In a river that averages 50 feet wide and a current of 2 mph, the water speed will increase to 4 mph if the river bank narrows to 25 feet and the depth stays the same. Needless to say, if you get pinned against something in fast water, you may not be able to simply wiggle off of it to safety. At water speeds around 3 mph, as little as 6 inches of moving water can take the legs out from under a rescuer, and most automobiles can float off the roadway in

**TABLE 6-1** Current Speed Chart

| Speed | | Distance (Feet) | | | |
|---|---|---|---|---|---|
| Miles/Hour (Knots) | Feet/Second | 25 | 50 | 100 | 200 |
| 1.2 (1) | 1.7 | 14.7 s | 29.4 s | 58.8 s | 117.6 s |
| 2.3 (2) | 3.4 | 7.4 s | 14.7 s | 29.4 s | 58.8 s |
| 3.5 (3) | 5.1 | 4.9 s | 9.8 s | 19.6 s | 39.2 s |
| 4.6 (4) | 6.8 | 3.7 s | 7.4 s | 14.7 s | 29.4 s |
| 5.8 (5) | 8.4 | 3.0 s | 6.0 s | 11.9 s | 23.8 s |
| 11.5 (10) | 16.9 | 1.5 s | 3.0 s | 5.9 s | 11.8 s |
| 23.0 (20) | 33.8 | 0.7 s | 1.5 s | 3.0 s | 5.9 s |

**TABLE 6-2** Current Force Chart

| Current Velocity | Average Total Force of the Water | | |
|---|---|---|---|
| Miles per Hour | On Legs | Body | Swamped Boat |
| 3 | 16.8 lbs. | 33.6 | 168 |
| 6 | 67.2 | 134.0 | 672 |
| 9 | 151.0 | 302.0 | 1512 |
| 12 | 269.0 | 538.0 | 2688 |

Ohio Department of Natural Resources.

just 24 inches of water. Many rescuers find themselves on incident responses because drivers failed to heed the "High Water" sign posted for a road. Once the water gets a person off his or her feet or a car moving, both will be at the mercy of the water flow.

## Search and Assessment

As with surface rescue, a size-up of the rescue call should be done upon arrival. Several factors may affect the team's ability to locate any victims, including noise in the environment as well as brush and trees blocking views both in and out of the area. Current speed itself will place the victim downstream from where the person entered the water. River features such as inside bends, eddies, slower water, and self-extrication possibilities should all be considered. Witnesses may not be able to describe to dispatchers exactly where they are, so landmarks should be noted. In extreme cases, vehicle sirens may be sounded and callers questioned about whether they hear these signals.

Once the victim or victims are located, another size-up of the actual rescue environment should be done. The banks may be steep enough to require rope systems to lower personnel and equipment, or they may be so packed with brush and trees that access is difficult. The physical condition of the victim must also be considered, especially in terms of hypothermia, exhaustion, and other injuries before or during the immersion. Water flows and approximate depth can be estimated, and rescue types selected based on what will work best—for example, boats with motors, boats on a rope system, or walking or wading rescue.

Once all of this information is obtained, all responders involved should be notified via radio or verbally to ensure that everyone is on the same page.

## Swiftwater PPE

PPE for swiftwater rescue is not much different than PPE for any of the other water rescue disciplines. Some of the PPE is modified to fit specific evolutions, but the same protection usually applies. This section provides a quick review of this equipment.

## Personal Flotation Devices

As always, personal flotation devices (PFDs) are absolutely required for anyone within 15 feet of the water's edge, or for any responder at risk of accidental immersion, such as one operating on a bridge or steep bank. For operations-level rescuers, entry into the environment is discouraged but possible in case of accident, so the typical Type III PFD amply protects most rescuers. The trick is to wear the PFD—not to disregard or forget it. Operations-level swiftwater rescuers are also expected to assist technicians who may be doing tethered rescues, so the technician gear must be examined and understood.

Type V PFDs include swiftwater-specific vests, most of which are equipped with **tri-glide** buckles on the front and releasable rings on the back (**FIGURE 6-4**). During a tethered rescuer operation, the rescuer normally attaches the tether at the rear of the PFD. If the person slips and loses footing, the force of the water often quickly pulls the rescuer's arms and legs downstream, and the water hitting the back of the head can force it underwater. To protect against this possibility, the back of a Type V PFD usually includes a steel ring on a strap that is held in place by Velcro tabs. This strap wraps around the waist of the PFD and finishes through a small buckle; then that tail is held tightly by another folding friction buckle. On the friction buckle,

**A**

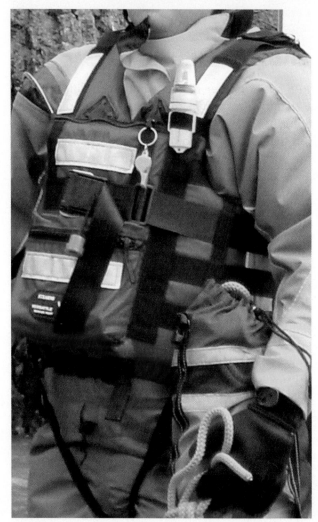

**B**

**FIGURE 6-4** Most Type V PFDs are equipped with releasable rings on the back (**A**) and tri-glide buckles on the front (**B**).
Courtesy of Steve Treinish.

a small, red ball can be pulled to release the friction. When the rescuer pulls it, the entire buckle assembly unlaces itself, getting assistance from the flow pushing on the rescuer in the water. This allows the steel ring to remain on the rope but releases the rescuer. The rescuer can then tether from the back, but is still able to obtain a quick release in case of trouble. Thus, the rescuer can release himself or herself from a tensioned rope, even under heavy pull. It is much better to swim for safety by moving with the water than to fight the flow (and likely lose) while remaining stationary.

## Swiftwater Helmets

**Swiftwater helmets** protect the rescuer's head in case it should strike anything, including the rocks or debris that may be encountered at higher speeds while moving with the current. One potential source of danger is the rescuer's total loss of control in moving water, in which case objects might be struck with the head before the rescuer is able to raise arms or hands to block the collision. The helmet should not include a large visor on it, so as to reduce the amount of flexion to which the neck is subjected if rescuers jump or fall in. A *small* visor on the helmet, however, may assist the rescuer in maintaining a small air pocket from which to breathe when water is moving over the rescuer's back and head. A rescuer may be placed in this position when being lowered intentionally in certain rescue procedures, or if trapped on a taut rope.

## Thermal Protection

Rescuers can underestimate the importance of thermal protection in moving water, but they should recognize that water wicks heat away from a body 25 times faster than air does. Wetsuits can be used in place of drysuits, but water constantly flushing through the former type of suit quickly draws heat away from the rescuer's body. Many drysuits are made from a tri-laminate material that, while very strong and durable, is very thin. Once that thin material is crushed against the skin by the water pressure when submerged, the sensation is akin to having the water itself pressing on the skin or the undergarments.

To ensure that they have adequate thermal protection, rescuers often tailor the insulation to the mission. Anything from duty fatigue wear to heavy synthetic

### SAFETY TIP

While many departments do not have specific drysuits for surface water or swiftwater work, the use of ice rescue suits is not recommended. Ice rescue suits are not sealed at the neck, and they may flood if a rescuer stumbles or falls into the water.

**FIGURE 6-5** Thermal layers should wick sweat and moisture away from the skin.
Courtesy of Steve Treinish.

**FIGURE 6-6** Swiftwater cutting tools, such as this one, should be worn by every rescuer working on or around the water.
Courtesy of Steve Treinish.

fleece may be chosen as the layer worn under the drysuit (**FIGURE 6-5**).

Neoprene or dry gloves also offer thermal protection, with dry gloves providing basic contamination protection. Heat loss from the head area can be mitigated by using a **skull cap**, which is a tight-fitting hat, commonly made of neoprene, worn under the helmet. Even a thin layer of neoprene can greatly reduce heat loss and increase comfort. Remember, dress to the water's temperature, not the air temperature.

**LISTEN UP!**

One material that rescuers wearing drysuits should avoid is cotton. When sweat or water soaks the skin of a rescuer, the rescuer may become chilled from wet underclothes lying on skin. Moisture-wicking fabrics allow for much greater comfort.

## Cutting Tools

Every swiftwater rescuer and shore crew worker should have at least one cutting tool (normally a knife) readily available that can be drawn and used with one hand

(**FIGURE 6-6**). Blunt-tip blades are better because the lack of a sharp point means less potential damage to the equipment and the user. Also, rescuers rarely need a sharpened point.

## Whistle

As moving water often creates a loud atmosphere to work in or around, every rescuer should have a whistle, ideally a pea-less whistle. A pea-less whistle does not use a small ball of cork to make sound, and it will work even when wet.

Whistle blasts are frequently used to signal basic communications, and can be used in conjunction with hand signals to direct movement, attention, or emergencies. A typical set of whistle signals follows:

- One blast of whistle: Look at me!
- Two blasts: Look upstream!
- Three blasts: Look downstream!
- Four blasts: I need help!

Hand signals are determined by the authority having jurisdiction (AHJ), but a general guideline might be as follows:

- OK? or OK!: This signal can be used as a question or response, and is given by forming a large "O" over the head with one or both arms.
- Help!: Wave one or both arms overhead in a back-and-forth motion.
- Pointing with the entire arm: This signal indicates the direction in which the rescuer or victim should travel.

Regardless of the AHJ's signal selection, all of the signals must be known before anyone involved needs them.

## KNOWLEDGE CHECK

Which of the following components makes the difference between swiftwater technician and swiftwater operations training?

**a.** Rate of water flow
**b.** Depth of water
**c.** Techniques learned
**d.** Safety requirements

*Access your Navigate eBook for more Knowledge Check questions and answers.*

# Swiftwater Features

Fortunately for rescuers, flowing water is relatively predictable in its shape and form when moving. Some of these features can be deadly, whereas others can save lives. This section reviews a few items that all swiftwater rescuers should be aware of. See Chapter 2, *Watercraft Rescue Operations* for additional information.

## River Flows

Water flows are normally faster on the surface and in the middle of a river or stream running in a straight line. **Helical flow** is water that tends to circle into the middle of the flow, due to the friction of the flowing water against the bottom surfaces of the stream or river. **Laminar flow** is water flowing faster at the surface than the bottom, due to the same friction (**FIGURE 6-7**). With this type of flow, a person in the water who merely floats with no swim effort will likely be drawn to and stay in the middle of the flow; some effort must be made to move to one side or the other.

In rivers, water is also subjected to centrifugal force around corners or bends, which increases the water speed at the outside of the corner. Inside the corner, water often slows, so sand or gravel bars are frequently found in these areas (**FIGURE 6-8**). Such structures make good places of safety.

**Turbulent flow** is water flowing in irregular fluctuations, or mixing, creating whitewater that consists of 50 to 60 percent aerated water (**FIGURE 6-9**). Some smaller PFDs do not offer enough flotation to properly float the user in turbulent water.

## River Directions

Imagine how confusing directions or location identification could get if teams did not use common terminology when dealing with flowing water, especially at larger scenes or when rescuers do not normally work together: Was it the right (correct) bank or the

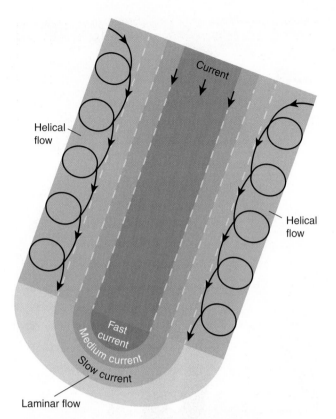

**FIGURE 6-7** Helical flow and laminar flow.
© Jones & Bartlett Learning.

**FIGURE 6-8** Water flows often slow on the inside of a curve, sometimes creating sand or gravel bars there that can provide refuge for victims and rescuers.
Courtesy of Steve Treinish.

### SAFETY TIP

When faster water flows around a corner, it can create an undercut—an area where the soil at the water level erodes out and the top of the bank can fall into the water (**FIGURE 6-10**). Rescuers operating near an undercut should be aware of the overhead hazard, and rescuers on the shore should be aware that the ground they are standing on might give way and collapse into the water below. Trees on undercut shores frequently fall across the water, creating strainers.

**FIGURE 6-9** Turbulent water has a great deal of air mixed in and may behave differently than denser water.

Courtesy of Steve Treinish.

**FIGURE 6-10** Faster water flowing around a corner can create an undercut, where the soil at the water level erodes out and the top of the bank can fall into the water.

Courtesy of Fred Jackson.

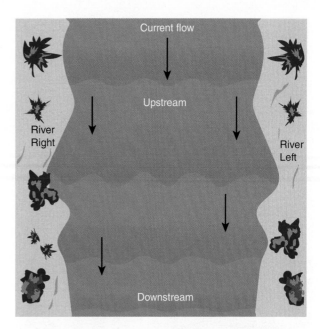

**FIGURE 6-11** River directions are determined by looking downstream

© Jones & Bartlett Learning.

**LISTEN UP!**

Some agencies or departments refer to "upstream" and "downstream" as "up current" and "down current," respectively, as the water can move in different directions on the river. All the terms mean water direction as a whole, but "up current" and "down current" may be better terminology in certain conditions. See the discussion of eddies to see why.

right (side) bank? To avoid this kind of confusion, use of the terms "upstream," "downstream," "river left," and "river right" has become universal in water rescue (**FIGURE 6-11**):

- Upstream—Where the water is coming from, collectively
- Downstream—Where the water is going to, collectively
- River left—Looking downstream, the left-hand shore
- River right—Looking downstream, the right-hand shore

Some agencies have adapted the terms *command side* and *away side* for use in conjunction with the preceding terms. Since most rescues will have one side with more staged resources and access, it can be logical to make that the command or operations side, with the opposite bank being the away side.

## Eddies

The safest places in flowing water are eddies. An eddy is a relatively calmer pool of water formed when moving water flows around a large object (**FIGURE 6-12**). The water that curls back upstream will still be moving, but usually at a much slower rate. This gives the swimming rescuer a place to rest and decide the next move or to take a victim until further help arrives. A rescuer who inadvertently ends up in the water can try to get to an eddy as a safe haven, or a boat crew may use an eddy to hold the boat steady without a lot of throttle and steering input. If you think of the river as a freeway, eddies are the rest areas.

The water that curls back up toward the obstruction can create a great deal of confusion for swiftwater operations and terminology. The general water movement is downstream past the eddy, but in the eddy itself the water is moving in the opposite direction. When coming in with a boat, rescuers need to lean down

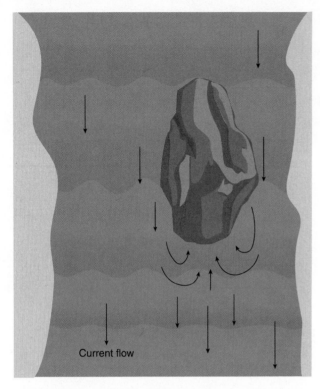

**FIGURE 6-12** An eddy forms when water flows around an object, causing a small area of water to curl back up current. Eddies make for great resting or staging areas for rescuers.
© Jones & Bartlett Learning.

**FIGURE 6-13** Strainers allow water to pass through, but trap other objects, including humans.
Courtesy of Steve Treinish.

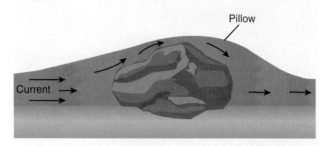

**FIGURE 6-14** A pillow forms when water flows up and onto an object, creating a visible bump in the water.
© Jones & Bartlett Learning.

current of the eddy but are leaning upstream when the river is viewed as a whole. In this text, downstream and upstream are determined based on the water movement as a whole, while up current and down current positioning is considered when dealing with flow, hazard, and obstacle characteristics.

## Strainers

As mentioned in Chapter 2, water moving through, but not around, an object creates a strainer. Trees, cars, bridge pilings, and some other obstacles can all trap objects against them but will let water through (**FIGURE 6-13**). Both rescuers and victims can end up in strainers—sometimes under them and sometimes on top of them. Either way, the force of the water flowing through the strainer can very easily trap someone underwater and pin the person there. A common scenario for rescuers is finding a canoe that has been pushed into a strainer with a trapped occupant either pinned against the strainer or on top of the strainer.

## Pillows

Pillows are formed on the water when water piles up against a submerged object, leading to a raised area on the water (**FIGURE 6-14**). A pillow may signal the presence of a shallow object that can injure or damage someone or something striking it.

## Upstream and Downstream *Vs*

Upstream *Vs* indicate a stationary object around which water is flowing, with wake disturbances found on each side. The *V* points upstream and marks the hazard. Downstream *Vs* usually signal a path between two objects, with the *V* pointing downstream (**FIGURE 6-15**). Swimming rescuers or boat crews may try to stay in a downstream *V* while navigating for better depth clearance.

## Hydraulics

A **hydraulic**—the ultimate river hazard—forms when enough water drops on a vertical plane to cause a backflow of the water immediately downstream. This water rolls back upstream over and over, only to be forced to the bottom by the water flowing over the obstruction. This creates a hazard that not only will suck things (including rescuers and victims) into it, but also keep them there, submerging them again and again (**FIGURE 6-16**).

Most **low-head dams**, which are vertical or near-vertical concrete walls spanning the entire river, create

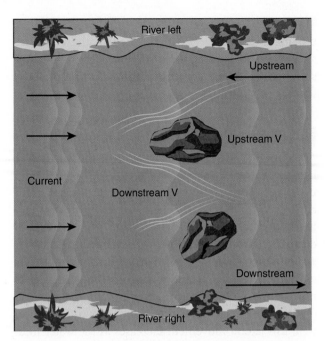

**FIGURE 6-15** Upstream *V*s are created when water flows around an object; downstream *V*s are created when water funnels between two objects. Stay away from upstream *V*s, and go through downstream *V*s.

© Jones & Bartlett Learning.

**FIGURE 6-17** High enough water can eliminate a vertical drop over the dam, reducing the hydraulic or hiding the dam completely.

Courtesy of Steve Treinish.

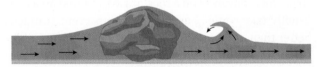

**FIGURE 6-18** Standing waves form when water passes over an object, is forced up, and then falls back on itself while trying to refill the depression formed. The current continues to pass downstream under the wave.

© Jones & Bartlett Learning.

**FIGURE 6-16** A hydraulic forms when water falls over an object, and is sucked back into the falling water. Once a hydraulic "grabs" an object, it can be kept there long periods of time.

Courtesy of Steve Treinish.

some kind of hydraulic, although the amount of water affects the amount of "hold" the dam has. Higher water flows increase the holding power of the hydraulic, but high enough water can eliminate a vertical drop over the dam, reducing the hydraulic or hiding the dam completely (**FIGURE 6-17**). Nevertheless, as the water drops, the hydraulic will return. Low-head dams are called "drowning machines" for good reason.

Other objects can also create hydraulics in the water, including fallen trees and natural rock formations. Many dams are shaped with slight arcs: Arcs that are built into the upstream create significant hazards because swimmers will be continually pushed toward the center of the current.

## Standing Waves

**Standing waves** form when water moves along a more horizontal plane over an object, and the water then rushes back in to fill the resulting depression. A stationary wave is formed in the flow and will always face upstream toward the falling water (**FIGURE 6-18**). Victims will rarely end up in standing waves, but technician-level rescuers use them to navigate the river when swimming. It can be difficult for the untrained eye to differentiate between a standing wave and a hydraulic. The most important feature of a standing wave is the fact that objects pass through it, and it does not hold objects as a hydraulic does.

# Defensive Swimming

Even if not trained to the technician level, all personnel working on or around swiftwater should know how to save themselves if they end up in it (**FIGURE 6-19**). The most important point to remember when swimming in swiftwater is that you will not beat the river, so you must use it. That is, use the current to direct yourself into a safer spot, whether it is an eddy, or a section of bank, or even an island.

## Defensive Swim Position

As mentioned in Chapter 4, *Surface Water Rescue Operations*, the defensive swimming position is to lie on the back, facing downstream. In this position, the head remains up for visibility and protection, the feet are used as bumpers, and the knees are used as shock absorbers (**FIGURE 6-20**). Yes, swimmers might strike their buttocks on a rock or two, but that area can usually take some abuse, and such an event is much better than striking a rock with the head. When in the defensive swimming position, ride the river using the increased visibility to work around hazards or obstacles such as rocks or strainers. When the time comes to swim for positioning, go for it, and go hard. Flip your body forward and swim using a forward stroke.

**FIGURE 6-19** Anyone can experience an accidental immersion and must know the proper reaction to such an event.

Courtesy of Steve Treinish.

Try to keep your head up and monitor your progress, but make it count. When reclined in the defensive swim position, you can also somewhat swim with the shoulders, but that technique generally works better for ferry swimming.

## Ferry Swimming

Adjusting your **ferry angle** involves using the river to help push yourself or an object to one side of the river or the other. Turning your body to angle it against the current will guide you to the shore at which you are aiming (**FIGURE 6-21**). This ferry angle can increase or decrease your movement laterally in the water. There will still be down-current movement in faster currents, but swimmers, boats, and victims can all use ferry angles to move laterally in the water. As the flowing water pushes on an angled object, the water will be moved laterally as well; this lateral movement increases as the water force increases. While the ferry angle can be adjusted in the water, increasing this angle past 45 degrees will result in a much quicker downstream movement.

> **LISTEN UP!**
>
> Once rescuers truly understand the concept of ferry angle—whether swimming, paddling, or boating—operating in current becomes much easier. Working with ferry angles truly lets rescuers use the river to their advantage.

# Managing Strainers

Even the best river swimmers may find themselves pushed into a strainer. Trees that have fallen across the water are probably the best examples of strainers, and they are very hazardous if they span the entire channel. If rescuers find themselves about to tangle with a strainer, the best plan is to get up and over it, rather than

**FIGURE 6-20** In the defensive swim position, the person lies on his or her back, facing downstream, and with head above water. The knees can be used to push off objects, and a backstroke of the arms can be used to move laterally in the flow.

Courtesy of Steve Treinish.

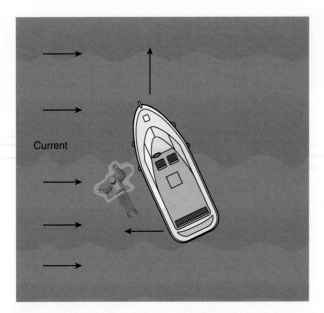

**FIGURE 6-21** Ferry angle is a diagonal push from the current used to move laterally across the river. Free-floating objects will travel downstream as the current pushes on them. Boats on static lines will simply be pushed across the water. The ferry angle can be adjusted by using different lengths of lines at the boat's bow and stern.
© Jones & Bartlett Learning.

to fight a losing battle by trying to avoid the obstacle. Follow the steps in **SKILL DRILL 6-1** to manage strainers while swimming currents.

## Escaping Hydraulics

Perhaps the most dangerous feature in rivers today is the human-created low-head dam. While some natural low-head dams exist, most are actually human-made structures intended to provide flood control or to form recreational water areas. Some of these dams are relatively hidden and drop only a few inches, whereas others are much taller and very prominent. Bigger low-head dams can drop as much as 10 to 15 feet and have widths of more than 400 feet. Another feature of such dams is their shape: A dam that curves so that its middle section is located farther upstream will let water flow back into it from the banks, making escape from the water more difficult. In such a case, not only is a victim now spinning in the hydraulic action, but he or she is also being washed back to the center of the stream. This can make the rescue evolution more difficult for rescuers, since they may have to operate a long distance from the bank.

The safest way to work around low-head dams is to respect them. A lot of water rescue training is done when water flows are reduced, either from lack of water or for safety. When the water is flowing hard, however, the situation will be very different, no matter how many times slow-flow evolutions have been

**FIGURE 6-22** With most boils, a slight slope is created between the boil line and the face of the dam where the water is being sucked back into the dam. Gravity and this slope will help push you back in.
Courtesy of Steve Treinish.

practiced. Simple positioning on the shore can play into safety, no matter the activity. On shore, put yourself downstream of the boil line, unless it is necessary to do otherwise. Many people are drawn to the power of water action, but that is a very bad place to fall in.

If you ever fall or are sucked into the hydraulic, you will not have a lot of options. Indeed, escaping any hydraulic will likely involve some luck. The best course of action is to try to get downstream of the boil line. Even the best PFDs may not provide enough flotation to keep rescuers on top of the water. It is very likely you will be sucked into the face of the dam and pushed back underwater. Compounding the water movement is the fact that most boils have a slight slope created between the boil line and the face of the dam where the water is being sucked back into the dam. Gravity and this slope will now help push you back in (**FIGURE 6-22**).

### SAFETY TIP

Rescuers have been killed while trying to get victims out of a hydraulic because they did not realize how close they were to the dam or the boil until it was too late. It is very easy to lose depth perception when approaching a dam on the water, especially when multitasking. Engage shore personnel to help you judge a safe distance from the boil, and be very aware of the direction in which the water is flowing when you operate around it. Operating around a low-head boil in higher-flow conditions should make most rescuers a bit nervous when they get close to the boil.

Another hazard that arises with low-head boils is the other debris that is rolling around in the water. Huge trees, boats, docks, and a host of other trash can be trapped in the boil for days. Put yourself into this mix, and there is a high likelihood of getting hooked or hammered by something. Remember, the river never rests, and the best defense is to maintain a very healthy respect for this type of hazard.

# SKILL DRILL 6-1
## Getting Over a Strainer

**1** Head toward the strainer on your belly, face first, using a very aggressive forward crawl stroke. Develop as much speed as possible.

**2** Kick with your legs, forcing your body up and onto the strainer.

**3** Continue to fight to get over the strainer using the arms and legs.

**4** Drop off the downstream side and resume the proper swimming technique.

Courtesy of Steve Treinish.

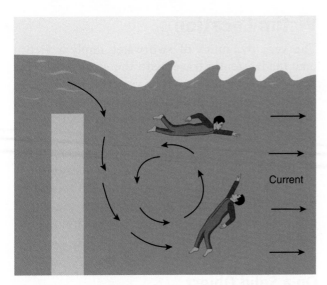

**FIGURE 6-23** Two possible means of escape from hydraulics are to swim across the surface with maximum effort from the downstream edge of the boil line, or to swim downstream underwater at the bottom of the boil.

© Jones & Bartlett Learning.

Once you are forced underwater, trying to swim downstream while underwater may save your life. Another option is to time a heroic surface swim effort while on the edge of the boil, as close to the downstream edge of it as possible (**FIGURE 6-23**). Once the flow is moving you back to the face of the dam, you will likely be fighting a losing battle; you will probably be rolled underwater again.

## Safety Among the Hazards

Generally, any place where you can find refuge in the current is a good option when you are in swiftwater. The bigger the body of water, the more important it is to be able to get somewhere safe. Longer swiftwater sections will keep the rescuer in the hazards for longer times, which then increases the odds of ending up somewhere bad or being injured. Perhaps the swiftwater flow slows downstream, or even becomes a larger, calmer body of water that is easier to move in. Defensive swimming over the course of these shorter runs may be the better bet. Gravel bars or sandbars, eddies, and slower-moving pools may all offer refuge from the current, especially inside the corners of

rivers. In general, you would much rather worry about walking out on the shore than fighting a major flow, especially if you didn't want to get in the water anyway.

Try to position yourself into the downstream Vs, as these flows usually provide deeper water, hopefully preventing injuries. Upstream Vs are a sign that something is in the water, so steer clear. Be aware of your surroundings, and watch for other rescuers who are trying to get a throw bag to you. Once you get into a position to get out or at least stop moving, do not re-enter the water.

# Swiftwater Rescue

Like any other water rescue, swiftwater rescue can be sliced into the categories of "yo, reach, throw, and row," although current forces generally rule out the use of any directly controlled reaching tools such as pike poles and shepherd's crooks. The pulling force of a victim in current who is hanging onto such a solid object can easily pull the rescuer into the water. Aerial ladders may give a rescuer some extended reach to deploy other tools, but for the most part, reaching tools are not the appropriate choice for this kind of rescue.

## Spotters and Safeties

Even if a victim is in sight, two crucial crew positions—spotters and safeties—should be filled before placing rescuers in a hazard zone. Beware of the potential tunnel vision that can occur if rescue crews focus on the victim without seeing the entire scene and performing a full size-up. Spatial awareness is a key part of overall swiftwater safety.

### Spotters

Upstream spotters are given the task of announcing any hazards traveling downstream—usually debris such as trees, docks, or even houses or cars (**FIGURE 6-24**). Debris traveling along the flow can not only knock rescuers in or around the water off their feet, but can also entrap them and transport them downstream. Direct radio contact on a dedicated channel is the best way to ensure that spotters' messages will not be talked over at crucial times.

### Downstream Safeties

Downstream safeties serve as the rapid intervention crews in swiftwater rescues (**FIGURE 6-25**). These crews may range in scale from a few extra operations-level rescuers with throw bags to entire boat crews made up of rescue technicians. While no one expects to slip or fall in the water, these mishaps do occur. Remember the best question in rescue is: "What

**FIGURE 6-24** This typical spotter's viewpoint has good sight lines to the boats approaching and any potential debris upstream. The spotter is standing at a "no further travel" point, which the boat crew can also see.
Courtesy of Steve Treinish.

**FIGURE 6-25** Downstream safeties serve as the rapid intervention crews in swiftwater rescues.
Courtesy of Steve Treinish.

would happen if this evolution catastrophically failed?" It can be rephrased in this context as "How will those rescuers be rescued?"

Safety rescuers positioned downstream need to be aware of two key points. First, they should know the amount of room they have to operate along the bank. Terrain may limit the amount of actual ground that can be used to effect a rescue, and rescuers should space themselves to be able to use all of it if needed. Second, which specific hazards may need to be dealt with: Strainers? Faster water? Worse? Downstream safety personnel must ensure that their strategy and tactics match the scenario at hand, and adjust them if necessary. Perhaps putting the backup rescuers across the river on an opposite bank, or even on watercraft, may prevent injury or death.

**LISTEN UP!**

Many downstream safeties position themselves too close to the actual rescue to that they can see the action. Spreading out resources will let more folks throw lines with less entanglement, and will prevent a wild chase downstream if the ropes miss their target.

## Victim Location

The very dynamics of swiftwater imply a victim may inadvertently travel with the water flow. This kind of moving target can be very tough to manage because it may be difficult to see for long distances in swiftwater areas due to bends and corners. Swiftwater victims frequently present in one of several ways, discussed in the following subsections. Keep in mind, however, that every victim who presents around moving water has the potential to end up moving in the water, so downstream plans and other rescue evolutions should be a part of the initial plan, not a reaction.

## On a Solid Object

People stranded on rocks, boulders, or even islands have one major advantage going for them. While the water around them may be very hard to work around, the object offers a good place to rest, provided the water does not rise enough to wash or force them off it. Rescuers can go to work at that location and have a stationary target (**FIGURE 6-26**).

## Hanging on for Dear Life

Sometimes a victim may be hanging onto a low-hanging branch or other object. This problem will have a definite time factor attached to it, because the river never rests. Eventually, the victim will become fatigued and lose his or her grip, or other objects may wash downstream and dislodge the victim. Even if a rescue tool is presented directly beside the victim, the victim may be using both hands to keep hanging on to the stationary object. Some

**FIGURE 6-26** Generally, victims will strive to find a stable spot higher than the water. This offers respite from the current, and a stationary target for rescuers to approach.
© dani daniar/Shutterstock, Inc.

victims may be so fearful that they refuse to let go and grab the rescue device.

## On a Car or Boat

Some victims may be found on or in a car or watercraft that may be moving or stationary. Any car in moving water should be considered unstable. The flotation found in auto tires, combined with the fact that newer cars are built with very air-tight cabin space, means that these vehicles can float and be moved by the current very easily.

When operating around vehicles, stay away from the area downstream of the car. Even though it may be forming an eddy, the car is not stable and can be grabbed by the moving water. Eddies are usable only if the object creating them will not move in the flow.

Another danger zone with a motor vehicle is where the water is physically striking it. Any person in that area may be swept into or under the vehicle by the force of the flow. Also, a rescuer who is attached to a tether line may have that line swept under the tires or body of the car if the rescuer works his or her way around the vehicle. Whenever you are working with a tether, be very aware of your rope and where it is being pulled.

Picture a car in a fast-flowing creek, with the front bumper facing upstream. The eddy formed behind the car is not safe, as the car may float away. The area in front of the car is not safe, as rescuers may be pushed under it. The far side of the car may mean the rescuer must cross the front and rear danger zones, and may have to manage a tether line while doing so. Thus, the side of the car presented to the shore from which rescuers will approach it is generally the safest area to work around.

If possible, extrication attempts should take place through the vehicle's windows. Opening doors may be impossible, and even if opened, they can be slammed shut on a rescuer's hands, arms, or legs by the force of the water. Opening doors also adds an area where flowing water will push against the vehicle, possibly dislodging it.

**FIGURE 6-27** Vehicles in moving water present several hazards, including instability and entanglements.
Courtesy of Steve Treinish.

# Ropes and Rope Systems

NFPA 1006 addresses the need for swiftwater rescuers to be familiar with rope systems, but only to the point of being able to perform the tasks needed by the AHJ. Ropes are a key part of swiftwater rescue, especially at the operations level, so it is important to review certain rope systems.

Rope systems used on the water can range from a quick, very simple setup to multiple lines, anchors, and control points that may confuse some rescuers (**FIGURE 6-28**). More puzzling to some rescuers is that many ropes or rope setups might be used together during a mission. Just getting the rope systems stretched across the water can frustrate rescuers. Rope slack that is not controlled on the water can often be sucked underwater into snags. The information that follows describes progressively complex rope systems, but does not cover high-angle lowering or raising. Whenever a rope system is in use, a safety check should be performed on the individual components as well as the system as a whole.

**FIGURE 6-28** Rope systems used on the water can range from a quick, simple setup to multiple lines, anchors, and control points.
Courtesy of Steve Treinish.

---

### SAFETY TIP

Rescuers working on tethered lines should be very aware of the rope slack and where it lays when working around cars. If possible, ropes should not cross the car, due to the risk of them being sucked into bumpers, frames, wheel wells, and body panels. If this happens, a rope entanglement may be encountered. Even if the rescuer is wearing a Type V PFD, this situation can be very dangerous (**FIGURE 6-27**).

# Voice of Experience

In my 28 years in the fire service, I have responded to numerous water rescues and recoveries. There are three water rescue incidents that stand out the most. Ironically, two of them occurred on the same day, about an hour apart from each other. They were both during an advanced swiftwater rescue technician course on the Arkansas River in Colorado.

Preparedness was stressed in this class from day one. As soon as any throw bag or other piece of equipment was used, it was immediately repacked to ensure its readiness. We also had numerous safety lines, boats, and personnel strategically placed in upstream and downstream locations. This preparedness contributed to the first successful rescue of the day. We had just finished a few throw bag evolutions and were resting along the bank when one of the students slipped and fell into the river. The section of the river where we were training was Class 5 whitewater and had numerous obstacles in it. We were able to immediately throw him a rope bag and rescue him without further incident.

Within an hour of the first incident, we noticed an overturned whitewater kayak heading downstream. At that time we did not know if there was anyone still trapped in it. Again, we immediately put into play what we had been practicing all weekend. As emergency notifications were being made to the downstream safety personnel, we immediately launched three rescue standby personnel in an inflatable boat and began to intercept the kayak. We also sent personnel upstream searching for any signs of the kayaker. Fortunately, the kayaker was not in the kayak and was found by the search team upstream on the river bank. We were able to provide basic first aid for minor bumps and bruises. The outcomes of these two incidents could have been much more serious for both of the victims and rescuers if we didn't prioritize preparedness.

The third incident happened a few weeks later in a small town in central Kansas. Recent rainstorms had caused the water levels in the rivers and streams to be elevated to historical levels. Many low-lying bridges were flooded and under water. We were dispatched to one of these bridges where a passer-by noticed an overturned tractor in the river. Upon arrival, we found a tractor tire in the river. It appeared that the tractor was trying to cross the flooded bridge and rolled off the side. This river had swiftwater moving about 10–15 mph. We again relied on our recent water rescue training and were able to immediately and effectively establish command, size up the incident, request additional resources, and begin rescue operations.

Within a few minutes, we had crews setting up an upstream boat highline as well as a downstream safety tension diagonal line. Additional safety personnel were deployed with throw bags. During the search for victims, the highline boat began to capsize in the swiftwater. The safety personnel immediately deployed throw bags, and rescue swimmers assisted the crew in the boat. Unfortunately, the victim was found the following day about a half mile downstream. Again, if it wasn't for the amount of effort that was put into preparation, this incident could have been a lot worse.

**Scott VanPatten**
Fire Fighter, Paramedic, Rescue Technician
City of El Dorado Fire Department
Butler County Rescue Squad
Technical Rescue Instructor, Butler Community College
El Dorado, Kansas

## Water Rescue Rope

Water rescue rope is normally made of polypropylene, which does not absorb water. It is much easier to work with than other ropes, whose fibers can trap water and become very heavy. Water rescue rope also floats. Webbing, Prusik rope, and carabiners should all be rated to handle the job. Remember, you may not be hanging directly on this equipment, but it will be keeping you out of a possibly life-threatening hazard. While other types of ropes such as kernmantle and nylon can work, they are not the best choice.

## Throw Rescue

Throwing lines or bags for swiftwater rescue is possible, but as one can imagine, the timing must be good. Also, on many banks or shores, rescuers have only a limited amount of room to maneuver so that they can stay with a moving target. The ability to redeploy a line may be crucial, because a missed throw cannot be easily moved to the victim—the line will, of course, also move on the water. For throw bag and coiled line deployment information, refer to Chapter 4. Multiple bags thrown by multiple rescuers in multiple downstream locations might be a good idea in some scenarios.

Once the bag is ready, the rescuer loudly shouts "ROPE!" to the victim, in the hope that the victim will hear and can help himself or herself. The rescuer times the throw, yells, and tries to put the rope on target. If the victim grabs the rope, the rescuer holds the rope, and the force of the river will swing the victim toward the bank (**FIGURE 6-29**). This maneuver is called **belaying**, or simply holding the rope firm.

If space along the bank permits, a gentler way to swing a victim into shore is to dynamically belay the rope (**FIGURE 6-30**). A **dynamic belay** sounds impressive, but is fairly simple in practice. The rescuer walks downstream along the bank, easing the tension on the line. Perhaps a short walk downstream will let the victim come ashore much closer to a shallow, calmer area. Dynamic belays can make a big difference

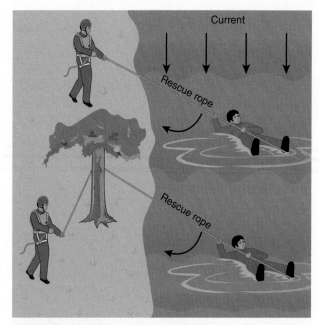

**FIGURE 6-29** A static belay holds the rope steady, whether by the rescuer holding the rope or by anchoring the rope to another object. The victim will swing like a pendulum to shore, but the static belay may give a rather forceful pull to the rescuer.
© Jones & Bartlett Learning.

in the amount of water a tethered rescuer may take over the head when performing a swimming rescue.

Rescue line guns may increase the distance "thrown" and can be equipped with flotation devices, a simple line, or an inflatable harness. Shooters must remember that deploying more line can result in tangles, and eventually more mess.

## Ferry Angles Using Throw Bags

It could be a tall order, but getting the victim to assume a certain position in the water can reduce the effort it takes to rescue someone by a thrown line or rope, and can certainly keep the victim's airway drier during the trip in to the bank (**FIGURE 6-31**). As discussed previously in this chapter, changing the ferry angle of the body while swimming defensively allows for good lateral movement in the water. Once a person has a secure grip on the rope, rolling over on the back to perform defensive swimming will do two things. First and foremost, it will keep water out of the person's face and mouth by letting the water flow past the back of the neck instead. Not many people can or will put up with a face full of rushing water, and some may quickly let go of the rope under that condition. Second, by placing the hand on either the left or right shoulder (depending on the angle direction desired), the created ferry angle will help turn the person and move him or her to shore.

**FIGURE 6-30** Dynamic belaying is done by moving downstream with the victim. The victim will still swing to shore as tension is held, but the rescuer has some control on the force and exit point.

© Jones & Bartlett Learning.

> **LISTEN UP!**
>
> Have the victim hold the thrown rope over the shoulder that is opposite the bank from which the rope is thrown. For example, a rescuer standing on the river right bank should try to direct the victim to hold the rope at the left shoulder while lying on the back. This will automatically set the proper ferry angle.

## Pendulum Rescues

When victims present on a firm, stable structure, it may be possible to rescue these folks without team members even getting their feet wet. The first decision to make in this case would be whether the victim can help himself or herself. Is the victim stable? Is the object holding the victim stable? Is the victim physically capable of helping in his or her own rescue? If so, the

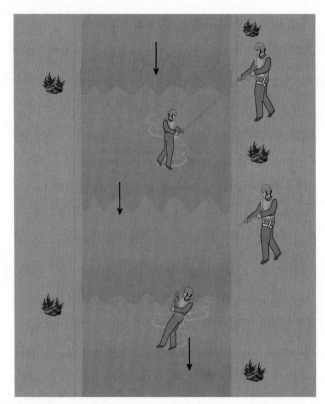

**FIGURE 6-31** Slight changes to the victim's posture and position can make the rescue easier, drier, and safer for the victim by changing the ferry angle.

© Jones & Bartlett Learning.

next step is to try to get a PFD to the victim. Be very aware of the victim's need to stay put should the PFD miss or if the victim cannot get to the PFD, and ensure the victim fully understands what to do in case of a miss. The victim must also don the PFD correctly.

The last step is to toss the victim a throw bag, but only after describing—very specifically—what the person needs to do. Do not throw first, thinking an explanation can follow, because victims may decide to jump in immediately after catching a rope. Ensure the victim knows to place the rope over a shoulder and to lie on their back. Explain the direction in which the river will take the victim, and reassure the victim that rescuers will be downstream to help out.

The case study at the beginning of this chapter was based on an actual rescue, which proceeded much like strategy described here. Crews made verbal contact with a very self-sufficient victim, threw him a PFD, and then threw him a throw bag. They explained what the victim had to do and when, and ultimately performed this swiftwater rescue without getting wet.

## Bridges and Bags

Many rescuers stage on bridges watching for a victim traveling under them. The most natural line of sight is upstream, but lines should *not* be deployed on the upstream side of the bridge. A rope thrown successfully to the victim on that side will then hit the railing, but the victim will keep going. If the rescuer has thrown the bag off the up-current side of the bridge, the line may be pulled so hard under the bridge decking that it cannot be moved along the bridge. The victim is now stuck hanging onto a rope, under the bridge, and often taking the flowing water right in the face and airway.

## Droop Lines

Another option for rescuers, and one that is much better when operating off bridges or spans, is to use a **droop line**. A droop line is a rope held by two rescuers, one at each end of the span, or a distance apart; it is used to drop a line onto the victim. Drooping a rope from two ends allows a wider target to be presented to the victim, rather than one rope. Even if a ring buoy is tied to a single rope and lowered, it does not present a large target, especially to a victim who is not able to move or swim laterally. The droop rope should be dropped or hung off the downstream side of the bridge, which prevents it from being pinned by friction on the upstream side. Spotters on the upstream side of the bridge can guide and provide information to rescuers holding the rope.

Follow the steps in **SKILL DRILL 6-2** to deploy a droop line. In the scenario in the skill drill, the rescuers are on a bike path crossing a shallow river, with a simulated victim washing downstream.

### SAFETY TIP

Never attach yourself to a line when operating around or in swiftwater unless you have the proper training to perform specific evolutions calling for a rope. The force of a victim suddenly tightening up a rope can easily pull rescuers into the water along with the victim they are trying to help.

## Taglines

**Taglines** are very effective means of performing water rescues if rescuers can get personnel on each shore. One or more lines are used to control a flotation device such as a ring buoy, a boat, or even a rescuer in a PFD. Taglines are a very easy way to deploy and control a flotation object from each side of moving water and are helpful tools for water rescuers to have in their toolbox.

### Two-Line Taglines

In a **two-line tagline** evolution, a line is deployed and operated on each side of the current. Each line is connected to the flotation object, which is positioned in the water flow.

A low-head dam entrapment is a prime example of a tagline operation. Operations on the water just upstream of the dam are usually last-ditch efforts to rescue victims, and are very risky to anyone in that area. Working below or downstream of the boil line, though, can provide a place of protection, even to the point of being able to ferry a line across the flowing water. Boats are commonly used to achieve this goal. Rescuers must be staged on either side of the water, and downstream safety personnel, either equipped with proper equipment or staged in watercraft, can serve as a rapid intervention crew (RIC) for the rescuers working upstream.

With the two-line tagline, the first line should cross the water and have a flotation object attached to its end. In shorter, tamer rivers, large buoys can be used for this purpose, as can a cluster of several PFDs. Many teams and departments fill a truck tire's inner tube with expansion foam and add some webbing for handholds. These homemade rings take a lot of abuse, offer great flotation, and cannot deflate. The second line in the tethering system is attached to the flotation (in this case, a Jim-Buoy), and the original line is then hauled from the far bank. The near bank crew feeds their line out as the buoy is pulled into the flow (**FIGURE 6-32**). Moving the buoy upstream places it closer to the dam. If the victim is able to grab it, pulling quickly downstream can usually get the victim and the flotation object out of the recirculation of the boil. The victim can then be laterally hauled to shore, or one line dropped or slackened to let the victim pendulum to shore.

## Boat Taglines

Some missions may take place in very fast water that is too shallow for motorboat operation, or rescuers may not

**FIGURE 6-32** Two-line tethers often utilize some type of flotation controlled with opposite lines from shore. The flotation can be moved into the hazard by pulling or slackening each line.
Courtesy of Steve Treinish.

# SKILL DRILL 6-2
## Deploying a Droop Line NFPA 1006: 17.2.1

**1** Let the line droop in the middle, down to a height just above the water, on the down-current side of the bridge. Just before the victim reaches the line, lower it slightly to the water. This will make it easier for the victim to get a grip on it. (Inflated fire hoses can also be used in this fashion, but they are pulled harder by currents because they have a larger surface area in contact with the water.)

**2** As the victim grabs the rope, a rescuer on the bridge releases one end of the line. In this scenario, the throw bag may provide a better grip if the rope starts to slide through the victim's hands.

**3** It is a much quicker trip to shore when the rope closer to the middle of the river is released first.

**4** The victim swings to shore as the rescuer belaying the rope walks back to the right bank of the river.

Courtesy of Steve Treinish.

**FIGURE 6-33** In a two-line boat tagline, lines are deployed from each side of the shore to the bow of the boat.
Courtesy of Steve Treinish.

**FIGURE 6-34** A four-line tether being used to control a rapid deployment craft at the base of a dam.
Courtesy of Steve Treinish.

have a boat with a motor. Perhaps the victim is an elderly person trapped in a car who will not be able to hang onto a flotation device. Boats can be used in tagline rescues, too, and may be controlled a few different ways.

Two-line tagline operations with boats are the same as described previously, but with rescuers in the vessel to assist the victim (**FIGURE 6-33**). The boat is boarded and equipped on the shore, and a line is used from each side of the flow to direct the boat where needed. The lines can also be used to provide travel upstream and downstream by tightening and slackening them.

## Four-Line Taglines

Four-line systems accomplish the same chore, but two lines are added to the rear of the boat, one off of each corner, giving rescue crews even more control (**FIGURE 6-34**). Any rescuer manning a line attached to a boat should be aware of the added force that will be transmitted through the rope when dealing with a larger object in the water. If the angle of the boat is changed and ferry action occurs, the rope will pull hard away from the boat.

## Using Simple Rope Systems

**Simple rope systems** use only one rope, or a rope and pulley, to change the direction of the pull or to provide a small mechanical advantage system. The largest mechanical advantage that can be obtained using a single pulley is 2:1, which is achieved simply by attaching the pulley to the object being hauled. This creates a moving pulley, which provides the advantage. Simple systems are easy to set up and easy to use. They can often be built with equipment carried on most trucks in one gear bag.

**LISTEN UP!**

Any rope system, no matter how simple, should be examined and tested before it is committed to service in an operation. Anchors, hardware, and the rope itself should be examined for mistakes in the setup, and the entire system slightly tensioned before it is fully placed in the hazard. Also, the members of the team should confirm the roles they have been assigned, and communication types and signals must be known by all team members.

## Natural Anchors

Any anchor used in any system should be secured on something substantial enough to safely handle both the load and any additional expected load during the mission. Trees are commonly used as anchors around the water, but larger trees should be sought, and they need to be solidly rooted in the ground. It does no good to anchor to a tree, only to have the tree snap because it was rotten and dead. In some cases, a tree's root system may be exposed and weaker due to water washing out the surrounding soil. Many times, a tree farther from the water's edge will provide a better anchor. It also lets the rescuers spread out and not congregate at the water's edge, thereby increasing working room and safety.

Multiple or load-equalizing anchors can be used to provide a measure of safety. If one fails, the others may take the load and keep the system intact.

## Human-Made Anchors

Occasionally, anchors must be utilized in clearings or in locations that offer no suitable natural anchors. Anchoring to vehicles is an option, but a few points must be taken into consideration. The size of the vehicle obviously comes into play, as larger vehicles weigh more. Parking a vehicle on concrete or pavement offers more traction than parking it in mud, dirt, or gravel, and air or emergency brakes must be set regardless of the surface. Wheel chocks of some kind must be used, even if commercial chocks are not available. The preferable pull is from the side, to eliminate potential rolling of the wheels.

Another human-made anchor system is a picket system. Pickets consist of multiple steel stakes driven into the ground at an angle to the pull and lashed or bound together. Pickets take some time to drive into the ground, and their security in the ground can be greatly affected by the soil type. Generally, pickets are not the preferred system around water rescue, but they represent another tool in the toolbox if no other options are available.

Some locations that have potential for high-frequency or high-risk incidents may have installed anchors at key locations nearby to the anticipated events (**FIGURE 6-35**). Be aware that these anchors may be subjected to the elements year-round or have little or no maintenance or testing for strength. These anchors must be thoroughly examined before using them in a water rescue operation.

## Pulleys

Pulleys are used to change the direction of the rope pull or to provide a mechanical advantage; they must be

---

**LISTEN UP!**

When using vehicles as anchors, it is imperative that the transmission be in "park"; the gearbox must not be in "neutral." Brakes must be set, and the ignition guarded against starting and moving the vehicle. Consider placing a rescuer at the controls, or placing the keys with the command staff if the vehicle uses them.

---

**FIGURE 6-35** Anchors around water can be human-made objects, such as existing structures or picket stakes, or natural objects, such as trees.

Courtesy of Steve Treinish.

---

rated for the expected load. Changing the direction of the rope with a pulley can be key when lowering a boat downstream. It also can make life easier on rescuers pulling lines, since one simple change of direction may let them walk the entire distance without taking their hands off the rope, or operate on drier pavement offering good traction instead of in wet, slippery mud.

## Boat Attachment

Watercraft used in simple lowers or in other tethered rope systems must be firmly attached to the system. Inflatables have one key weakness in these kinds of scenarios, as the bow rings are often glued onto the bow of the boat itself. The severe pull placed on a single D-ring during a lowering or rescue mission can easily tear it off. To combat this possibility, a load-equalizing bridle of webbing should be laced through the various D-rings and handles along the bow, and even the gunwales if possible. The goal when using such a system is to spread the force of the pull onto several areas of attachment, rather than placing it at a single point that is more likely to fail. If any questions or concerns arise about this process, consult the manufacturer of the watercraft.

## Simple Lowers

Simple rope systems can be used to lower or guide a boat into position, and there are no limitations in terms of how and where these operations take place. Bridges, trees, and even rescue vehicles can all be used as anchors, when properly set up. Rescuers can also be lowered in the same fashion, provided they are equipped with the proper PPE, including a quick-release buckle.

If the boat can be lowered straight downstream, sometimes all that is needed is a line, a change of direction (pulley and carabiner), and a crew (**FIGURE 6-36**). This technique is also known as a boat on a tether. Remember, when the water flows under the boat instead of pushing on the boat's sides, less force is needed to control the craft. Basically, a line is passed through an anchored pulley, the rope is manipulated to maintain

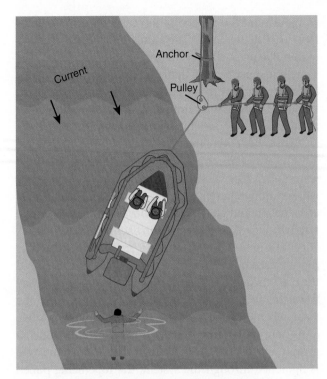

**FIGURE 6-36** A simple lower uses rescuers' muscles or a change of direction to manually control all aspects of the boat. The crew must manually belay the rope at all times.
© Jones & Bartlett Learning.

**FIGURE 6-37** Adding a stopper knot to the rope will stop it from pulling completely through the pulley.
Courtesy of Steve Treinish.

safety if something goes wrong, and a crew belays the boat downstream and then pulls it back upstream. As this is a boat-based rescue operation, an operations-level rescuer may be tasked with being lowered with the boat to rescue someone. Teams must remember that using the minimum number of rescuers will eliminate excess weight, thereby facilitating the lowering and raising steps, and will leave room for the victim. Packing too many rescuers in the boat, in contrast, may hinder the operation.

When the boat is approaching the victim, voice contact should be made as soon as possible, and the victim given instructions on what to do and when. Often victims will leap for the boat before it is close enough to do so safely. Having a plan will enable the rescuers to place the boat in the optimal position *before* the victim is loaded, making the rescue safer.

As with the two- and four-line tethers, a small amount of lateral movement can be accomplished if needed by throwing lines to shore and letting shore crews pull. The pendulum action of the current will likely prevent much movement, but sometimes just a few inches can make a significant difference in a rescue.

A good practice is to add a **stopper knot** (also known as a line-control knot) to the rope, to stop it from pulling completely through the pulley in some other way. Simply putting a figure eight knot in the rope will stop it at

the pulley should a slip occur (**FIGURE 6-37**). Rescuers in the watercraft will then be protected from downstream hazards, and in the event of a failure, this safety will limit the amount of distance the boat travels.

Another option is to tie off and anchor the rope somewhere just past the farthest distance expected in the operation, or to place a knot in the line. Using a Prusik-minding pulley allows the return pull to be held in place in case of a slip, but this pulley will need to be monitored during the lowering operation. Placing a friction belay device into the system can also greatly increase safety and control, by allowing one or two rescuers to handle the weight of the rescuers and victims and the water force to which the boat will be subjected.

## More Complex Rope Systems

**Complex rope systems** are hauling or belaying systems that use multiple lines or multiple pieces of hardware to build the system. The more complicated things become, the higher the possibility of failure and the more vigilant rescuers must be. A handy tool in these scenarios is a small field guide showing the different rope systems and how to build them. Many rescuers make and use their own such field guides by laminating index cards.

Larger or heavier loads may require a hauling system or a lowering device to safely control the load, such as when hauling a victim up a steep, muddy incline after removing the person from the water. Indeed, NFPA 1006 requires that water rescue technicians be able to construct and operate a rope system for such an application. Proper rope techniques should be used when constructing these kinds of complex systems, including safeties.

A 3:1 Z-rig–type system is the easiest and quickest way to set a **static line** across flowing water. Z-rigs are also commonly constructed in rope rescues in general.

Remember, the more hardware added, the more complex the system will be. Keep your rope system simple, if possible.

**LISTEN UP!**

Many rescuers focus on the specific strengths of the equipment used in rope systems, but it is important to remember that the entire system is only as strong as its weakest link. You can build a textbook system, only to have it fail at a weak anchor. No rescue should take place without installing safety backups in rope systems. Placing these safeties on the main line can catch the rope and prevent a catastrophic failure if any piece of rope hardware fails.

## Static Lines

Some boat-based operations are done from a static line. *Static* (meaning "not moving") is a fancy way of saying that the line across the water is in a fixed position and does not move. The static line is the anchor that everything else depends on. Most static lines are constructed with a Z-rig system. The distance across the water also greatly affects the static line: The greater the distance covered, the more the line will droop into the water and be pulled downstream. Often a static line must be constructed with a Z-rig simply to get it tight enough to hang above, but not in, the water it is crossing.

To set up a static line, two anchors must be selected, one on each shore. Safety anchors also could be utilized if they are placed in a good position. One end is firmly attached to one anchor, usually with a tensionless hitch. Tensionless hitches wrap around the anchor a few times, and then the end of the rope is attached back to the main line (**FIGURE 6-38**). The friction of the

wrap anchors the rope, and the end of the rope hangs free with no tension. The other end of the Z-rig is manipulated by the rescuers to tighten the line. The static line must be very tight, as rescuers will be operating perpendicular to the current; if it is too loose, it will be pulled downstream into a *V* shape, making it very difficult, if not impossible, to move laterally.

The other end of the rope is used for the mechanical advantage system, so it should be set up away from the water's edge, if possible. A longer distance from the water will pay off when tightening the line later. Select a suitable anchor, and attach a carabiner and pulley. The rope is fed through the pulley and run back toward the water's edge. The next step is to attach a Prusik line (ascenders may be used by some teams) and pulley there, then feed the rope through it, this time laying the rope back away from the shore. This pulley will travel away from the water when the rope is pulled. The setup should look like a Z—hence the name Z-rig (**FIGURE 6-39**). Such a system establishes a 3:1 advantage; that is, when the rope is tightened through the system, the rope being pulled will travel three times as far as the part stretching across the river.

Give the crew enough room to pull the rope. A change-of-direction pulley can be installed somewhere to make this process easier. Once the system has been

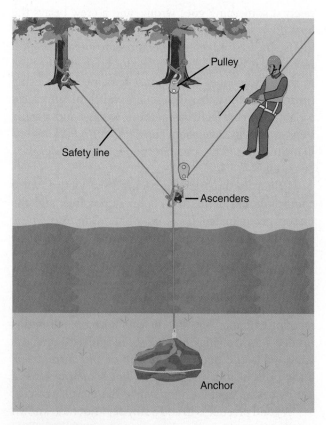

**FIGURE 6-39** A 3:1 advantage system, or Z-rig, used to assemble a static line across the river.

© Jones & Bartlett Learning.

**FIGURE 6-38** Tensionless hitches use friction from a suitable anchor to provide holding power. Trees are excellent anchors for tensionless hitches.

Courtesy of Steve Treinish.

**FIGURE 6-40** The safety anchor can assume the load if anything behind it fails. It also lets the 3:1 system be reset for another pull.

Courtesy of Steve Treinish.

**FIGURE 6-41** Tension diagonals are lines stretched across the flow at an angle. If the victim can grab this line, ferry action will move him or her to the downstream anchor to which the line is attached.

Courtesy of Steve Treinish.

tensioned, the haul end of the rope can be half-hitched around the anchor to hold it.

A safety anchor should be placed in the system, providing a backup for the hardware used (**FIGURE 6-40**). Another Prusik line can be attached ahead of the traveling pulley, and webbing or even another rope attached and run back to another anchor. This way, if anything in the system behind the safety fails, the rope will not fall into the water. This safety anchor should be kept stretched as far as possible toward the water, ready to absorb the pull of the main line if a failure occurs.

Tension is applied with the pull, but operating in tighter areas may mean that the team runs out of room to continue pulling. This is not a problem. The safety placed ahead of all of the hardware is not just a safety, but can also hold the load while the system is reset. Once the haul line is slackened and the safety is holding the load, the traveling pulley will have traveled inshore during the pull. Loosen the Prusik line, slide it back into position closer to the water, and pull again. Keep the safety tight as the system is pulled again. The proper tightness will maintain the line above the water, but be aware of overstressing the system, including the anchors.

In many operations, 3:1 Z-rigs are used as piggy-back haul lines, used in conjunction with a simple belay line. The pulling end of the rope (i.e., the end of the rope not made into a Z) can be attached to the belay line at any point, and the belay line can be hauled by the 3:1 system. This is very convenient, in that the belay line is fed back through the belay device and becomes a safety and a reset point. Aerial backboard lifting systems also often use 3:1 systems as haul lines.

## Tension Diagonal Lines

A **tension diagonal system** is a static line strung across the water, with the descriptor "diagonal" referring to the rope being placed at an angle to the current rather than perpendicular to it (**FIGURE 6-41**). Placing the rope at an angle to the current lets the force of the flowing water push someone both downstream and toward the shore. These angled static lines can enable rescuers to travel across flowing water very quickly and with minimal effort. A pulley or carabiner can be attached to the tension diagonal; the rescuer grasps this device, and the current can then move the rescuer more easily. Rescuers should hold the webbing or carabiner hanging from the line with their hand, rather than clipping in or attaching to it. Adjusting the rescuer's ferry angle while in the water will result in an even faster trip.

Delivery of technician rescuers to a far shore can be accomplished quickly in this fashion, and may be done upstream of a potential rope system setup. A victim suffering from a foot entrapment may have a tension diagonal strung beside him or her; the technician is then swept down, stops at the victim, and offers support and assistance.

Shorter tension diagonals can be rigged quickly using throw bags. Once the line is tossed across the flow, the line is placed in service by selecting and wrapping suitable anchors on each side of the flow. Lines spanning a longer distance may require a Z-rig to tighten them until they are properly tensioned. The rope should never be held perpendicular to the current, as that may cause the victim to become stuck in the swiftest part of the current, with a tight rope on either side.

Teams can improve the victim's ability to hold on by hanging a section of snow fence on the tension diagonal with carabiners and pulleys and letting this snow fence hang into the water a bit. Ropes control it from both sides. In a two-line tether fashion, the victim is ferried to the shore simply by hanging onto the snow fence because the line is angled.

Rescuers should also figure out which bank the tension diagonal will end at. Perhaps the island in the middle of the flow offers calmer water, even if it is farther offshore. Rescuers can always first get the victim to a safe place, and then change to a different mode of rescue, such as using a boat, deploying a static line, or even waiting it out.

## Movable Control Points

The more complex setups are used mostly in **movable control point (MCP)** rescues involving watercraft. Also known as a *boat on a tether*, MCPs can be used to direct and control a boat completely from the shore, thereby allowing the onboard rescuers to concentrate on the rescue. MCP setups also increase safety in that the boat they control does not have to rely on a motor that could fail. With an MCP, the boat can approach the victim from upstream, in line with the current, and stay there.

The first step in assembling an MCP is to stretch a static line across the river, upstream from the victim. It needs to be a tight, strong, static line. With shorter spans, the team may be able to utilize a throw bag or messenger line to bring the static line across. Maintaining the line above water will help keep the current forces from pulling on the line and making deployment harder.

If a static line must be paddled or oared across, tying the rope off at the stern of the boat can be helpful, in that it allows the rescuers to focus on paddling, rather than hanging onto a rope. Be aware that the rope may change the ferry angle as the watercraft is paddled across and the current pulls on the rope. Once the boat reaches the far shore, the boat must be tied off by a separate bow line. Many attempts go awry when rescuers reach the far shore, but then try to hold the boat there while anchoring the static line. With a secure bow line, rescuers will have a good platform that will not float away or be pulled back into the current.

When using a boat to establish the MCP, the current forces can be used to advantage by starting the paddling far enough upstream that the boat can be paddled over while the rope is traveling with the current. The rescuers paddling or oaring should make every effort to reach the desired anchor point before the rope is pulled by the current. Once this happens, the current may pull the boat back into the middle area of the river before the boat is secured (**FIGURE 6-42**).

After the static line is properly established and ready, rigging is built so that movement in all four directions—river left, river right, upstream, and downstream—can be controlled from the shore. The first step is to hang a pulley on the static line, connected to a steel ring or rigging plate. The rigging will have a load applied at four sides, so a carabiner is not a good choice, as the carabiner would be side-loaded. A line is attached on each side of the rigging (sometimes tied on to save hardware, but usually with a carabiner), and below that another pulley is attached (**FIGURE 6-43**).

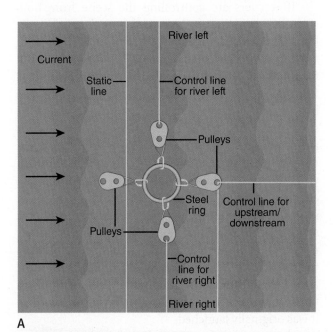

**FIGURE 6-42** A trailing line in the water can add enough force to pull a boat from the shore back to the middle of the flow.

Courtesy of Steve Treinish.

Current

River left

Static line

Control line for river left

Pulleys

Steel ring

Control line for upstream/downstream

Pulleys

Control line for river right

River right

**A**

Three more ropes are needed to move the boat. One runs from the ring directly to the river's right bank, and is used to pull the boat river right. The other runs to the river's left bank, and is used to pull river left. Together, these two ropes give the rescuers lateral control. One option is to hang another Prusik line and pulley on the static line, farther across than the boat needs to be placed. The second rope is then passed through the pulley and back to shore. This gives a change of direction and lets rescuers control the lateral movement of the boat from the same shore. On long runs, Prusiks and carabiners can be hung as guides (also known as festoons), keeping the lines out of the water.

There are two options to attain some upstream or downstream control. The first option is to place a rope in the bow of the boat, run it through the bottom pulley, and anchor it to the boat's bow. The rescuer in the front of the boat can control the line from the bow, lowering or raising the boat as needed. One possible problem with this setup is that one person needs to control this force, and the person who does so may be needed to directly effect the rescue of the victim.

The second option is to anchor the line at the boat's bow and then run it through the bottom pulley attached to the ring and to shore, where it can be controlled by shore crews (**FIGURE 6-44**). This lets the boat personnel concentrate on rescue, and it also enables more than one person to control the rope.

If constructing the MCP system sounds somewhat intense, it is. Several rescuers are needed to set up and man this system, and the rope crews must have good communication during the entire process. A good communication system with the boat must also be in place, including hand signals in case of radio failures.

Stopper knots or attaching a safety to the upstream/downstream control rope, just as in a simple lower

**B**

**FIGURE 6-43** A movable control point. All directions of travel can be controlled from the shores. In this system, river left control could be run through a pulley on the static line back to the river right shore, and all directions controlled from one bank.

**A.** © Jones & Bartlett Learning; **B.** Courtesy of Steve Treinish.

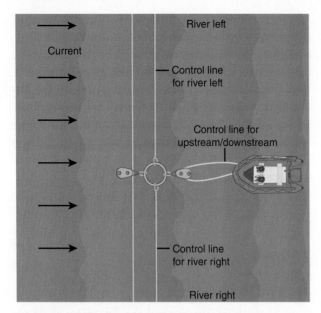

**FIGURE 6-44** This MCP lets the upstream or downstream movement be controlled by the boat crew from the boat itself. Large current forces could present a problem with pulling upstream.

© Jones & Bartlett Learning.

operation, should be mandatory when an MCP system is used. Even a simple figure eight knot in the end of the rope may stop the boat from being swept downstream if something goes wrong (e.g., a slip). Once a wet, slippery rope gets moving, it is hard to stop. Another safety issue is the boat getting away and washing downstream while the system is being set up. Using a carabiner to attach the boat directly to the static line and keeping it in place until the boat is ready to be lowered can alleviate this risk.

**Operating the MCP.** Once the system is in place, attach the boat to the ring using webbing as a load-distributing harness. When rigging the harness, the goal is to spread the force of the pull from one attachment point to many points of attachment. Inflatable boats are basically glued rubber structures, so placing enormous strain on one point can cause a failure. Using various rings and webbing to spread the force to multiple areas is easier on the boat and offers some redundancy.

Another piece of webbing can be used to secure the boat to the MCP rigging directly. This leash keeps the boat attached to the static line until the rescuers are ready to be lowered downstream. Once the boat is placed in the current laterally, the crew can unhook this leash, placing the load on the up/down-current control rope.

Before committing rescuers to the strong current, test the system at the edge of the shore, just as you would in a vertical lifting operation. Find the problems there instead of when the boat is already in the main flow. The

shore rescuers need to coordinate rope movement, as pulling from one shore requires slack from the other side. The upstream/downstream-current control rope also needs slack, as it travels to the point on the static line at which it will be used. Generally, the rescuers in the boat have the best view and can signal by hand, radio, or whistle blasts which way they need to travel.

Once the boat reaches the point where it needs to be lowered, everyone involved should hear the phrase "Boat on rope!" or "Boat on belay!" or something comparable. At this point, rescuers should unhook the webbing or strap between the bow and the static line and transfer control to the MCP.

Operation of the ropes involved can be made much simpler with one person relaying the needed movement from the boat to the shore crews. Orders should be clear and concise, such as "Pull river right!", "All stop!", and "Lower downstream!" A general plan should be in place as to where the boat will be taken once a victim is on board. As always, maintaining control, staying calm, and ensuring effective communication will greatly help things go smoothly.

**Breaking Down the MCP.** Once the victim is rescued, it is imperative that rescuers remain attentive while the incident is terminated. All too many rescuers are injured (or worse) during this stage. Any static line across water can present a hazard to rescuers removing it. Removing the MCP hardware is fairly simple and straightforward, but the static line can be problematic.

If rescuers are controlling the scene from both sides, the static line can be untied from the anchor, the knots removed, and the line pulled to shore. Removing the knots prevents entanglement from trees or debris if the rope catches on them. If the MCP has been operated from one shore, many rescuers choose to paddle the boat across, untie the tensionless hitch, and have the shore crews pull in the rope. They then paddle back across the river. This strategy could be problematic, as the boat crew may be subjected to the same hazard that ensnared the original victim. Consider sending the boat attached to the static line to perform this task. If the static line can be moved upstream from the operational shore, a tension diagonal has now been formed, and the river will do all the work moving the boat across. Use the river! If the tension diagonal can be set up in the right location upstream, it is entirely possible to have the boat return to the exact spot on the river from which it was originally launched.

Once the boat crew reaches the far-side anchor, they can tie the boat bow to the tail end of the rope and simply pendulum back to the operations shore. Care must be used to forecast where they will travel across the water and land on this shore. This strategy gets the rope back to shore, protects the boat and the rescuers, and allows the shore crews to help guide the boat to safety, even if a paddle is dropped or current forces prove too much for the rescuers in the boat.

## Row-Based Rescues

Many watercraft swiftwater missions can be accomplished using river-running tactics, as described in Chapter 2. A well-trained and experienced boat driver can use motors or even oars or paddles to place a boat where he or she wants, and do it safely. Oftentimes getting a boat into the water and to a victim is the best course of action, provided a victim can be approached safely. Using watercraft also limits the amount of water rescuers have to enter and provides protection for rescuers and victims alike. Boats stationed downstream of an active rope-based or swim-based action are one of the best safety nets we rescuers can provide for ourselves, even in training operations. Another option is to station a boat downstream, clear of hazards, and wait for the victim to come to the rescuers, especially if the victim has his or her own flotation and is able to help himself or herself.

A good rule for agencies to follow is this: If there is one boat in the water, there should be two. The second boat backs up the first boat and provides safety should the first boat fail or lose power, or if a rescuer on board has an issue (**FIGURE 6-45**). Complying with this rule may present a logistical challenge if multiple rescues or attempts are occurring, but proper preplanning can help with getting all of these boats in the water. All swiftwater and marine hazards must be accounted for during such operations, and teams should maintain a constant awareness of "What if?" possibilities.

**FIGURE 6-45** If one boat is in the water, a second boat should be in the water to back up the first boat.
Courtesy of Steve Treinish.

## Boat Operations

Not every swiftwater scenario involves the water trying to kill us as soon as we enter it. Some hydraulic boils can be tame enough to let a boat approach them without much trouble. If the proper boat is chosen to attempt this access, problems can often be avoided. Inflatables are good candidates for this task, since they can handle extreme loads of water without sinking. It is not the water inside the boat that sinks many boats in currents, but rather the force of the current acting on the boat itself, or the boat being pushed into debris or debris being pushed into the boat. Of course, if a current is strong enough to capsize or flip a boat, the rescuers who get thrown out of that boat could certainly be in a bad spot. Each AHJ should examine its capabilities, equipment, and training before committing boats into the water. Remember, complacency kills. Even if you have done something dozens of times, the water might be more difficult this time.

## Victim Pick-offs

Victim approaches should be made from downstream if possible. Maneuvering while moving upstream is easier and, in case of failure, the boat will not be swept down onto the victim, possibly striking the victim or the rescuers in the water. Stranded victims should also be brought on board at this side of the boat if possible. Nosing straight into an object can be done, but it exposes the victim to the boat and propeller if he or she falls into the water. Bringing someone on board at the side of the gunwale will keep the person safer. A good operator will travel upstream, under control, to a point that places the boat beside the victim, ferry closer, but stop just short of making the final approach. Rescuers can then verbally instruct the victim on when to come aboard and where the person should plan on entering the boat.

## Two-Boat Tether

The low-head dam is very dangerous to humans, whether rescuer or victim, and the power of the current around these dams also presents a unique hazard

The **two-boat tether** is one of the most hazardous evolutions undertaken in water rescue—or any rescue, for that matter. It is *not* for the minimally trained rescuer, as any small failure or oversight in any part of the system can result in multiple rescuers being pulled into the hydraulic, with potentially fatal results. A two-boat tether can have huge failure factors due to the complexity and the danger that low-head hydraulics inherently offer. It is, however, one of the few viable options on some larger, wider low-head dams.

**FIGURE 6-46** A johnboat can be used as a sea anchor in case of motor failure on the primary boat. Courtesy of Steve Treinish.

to boat crews. Many low-head dams are too wide to effectively use rope evolutions to span them in current. In evolutions where the only response to a low-head dam can be by boat, the two-boat tether may be the choice. The force and physical makeup of the boil lines require a specialized backup in the form of another boat on the water directly attached to the primary boat while the primary boat approaches the boil line for a rescue attempt. The rescue attempt is almost always a throw bag deployment because of the danger of getting sucked into the boil. This strategy lets the rescuers maintain a safer distance from the hazard. Even if the boat does not sink in the face of the boil if it is sucked in, rescuers always face the potential of being thrown into the water.

If the primary boat is pulled into the boil, it will likely happen very fast. The up-current movement of the water being sucked back into the dam hydraulic is a danger in and of itself; adding to this danger is the fact that the boil will create a slope of up to a few feet that will slide the boat downhill into the dam face. Also, because the boil is being churned with air, the boat's propeller may not produce enough thrust in the water to get out. The motor will spin at high speed (i.e., RPM) while the propeller simply cavitates.

The two-boat tether is a tool in the water rescuer's toolbox, but if it is not done properly it could be one of the most fatal evolutions undertaken in a boat. Once a dam has you in its grip, it will not easily let go, and many people have lost their lives in the boils associated with low-head dams. A boat getting close enough to ride up onto a 2- to 3-foot-high boil line should make rescuers somewhat uneasy, as that is essentially the "no man's land" for that dam. Training in high-flow currents is also much different than training when the water is low, and simulating a heavy current does not provide sufficient training for this type of situation. Performing this evolution near a dam that causes standing waves instead of boils can be a good option for preparing rescuers for a real incident.

The two-boat tether always comprises two boats, which are tied together with an adjustable-length rope. The use of johnboats versus inflatables for the backup boat can be debated, but for the purposes of this text, we will assume one of each type is used. Because the johnboat can also be used as a sea anchor in case of motor failure on the primary boat, it offers another "last chance" for the primary boat if it enters the hydraulic (**FIGURE 6-46**). It is possible to use an inflatable or even a larger boat as the secondary boat, but the johnboat is the only boat that can sea anchor effectively.

The primary boat is the one that will approach the dam face. The crew should have whatever tools they deem necessary for the rescue at hand. At a minimum, multiple throw bags, Jim-Buoys, and extra flotation should be carried. Rescuers should be versed in throwing the lines from the kneeling position or with grenade throws, as the turbulent water can get bumpy.

The second boat is actually pulled upstream by the primary boat, and its sole task is to pull the primary boat from the dam if needed. The second boat should be ready with rescue equipment, as it is also the first downstream safety boat.

The boats attach to each other on the water, downstream from the dam, using a rope connected from the stern of the primary boat to the bow of the secondary boat (**FIGURE 6-47**). A webbing harness may be used to attach the line to the primary boat, attaching

**FIGURE 6-47** The boats attach to each other on the water, downstream from the dam, using a rope connected from the stern of the primary boat to the bow of the secondary boat.

Courtesy of Steve Treinish.

**FIGURE 6-48** The two-boat commander should be in a good position to control the boats because on-water visibility may be lacking. Brightly colored gloves or mittens can be worn to make visual signals easier to see.

Courtesy of Steve Treinish.

at the eyebolts or U-bolts found at the stern. Another method is to lace rope or webbing through and around the various handles and D-rings that may be found on the inflatable. This distributes the force of the pull equally.

The length of the tether line depends on the distance needed for the secondary boat to stay in water where it can operate safely and effectively. Water aeration, hazards, and the route of escape all can affect the length of rope needed. Sometimes the incident commander will want to lengthen the tether to match conditions that are being observed from the shore. Friction devices and ascenders can be used to lengthen a line as the rescuer is working. These pieces of hardware can be tied into a floor strut and used to control the length of the rope deployed. The length of the tether will normally be 100 to 300 feet.

Of course, it does no good to have two boats pulled into the boil. The secondary boat must be able to use the water flowing downstream from the boil to assist in pulling the primary boat out if things go wrong.

An interesting problem here is visibility. The primary boat can usually see the dam itself very well, but may have trouble seeing a victim who is now lower than the primary boat, since boil heights can be 1 to 3 feet or more. If there is no one onshore to guide them, the primary boat crew can easily be sucked into the boil because they simply cannot see well. The evolution's commander should be set up on shore and communicate with the crew using hand signals or whistle blasts (**FIGURE 6-48**). The commander can use brightly colored gloves or mittens to better show his or her hands. Radios may work as well, but the noise of the water can be very loud, and a missed verbal communication could cause problems. Each boat keeps watch on the commander, who can see the entire operation, including the length of the tether, the conditions of the boil line, and the victim. The incident commander makes the calls, since he or she can see the entire boil, not just part of it.

The built-in safety of a two-boat tether is based on the secondary boat peeling out. To peel out, follow the steps in **SKILL DRILL 6-3**.

Peeling out does not take a great amount of speed, but requires excellent timing, good throttle control, and a tether that is kept taut between the two boats. A secondary boat using too much throttle will create slack. In such a case, instead of the line providing a good, hard, immediate pull on the primary boat, the primary boat can be sucked into the boil by the time the slack is pulled tight. Once the victim is rescued,

# SKILL DRILL 6-3
## Peeling Out

**1** The operator turns the boat perpendicular to the current.

**2** The operator continues the turn under power.

**3** The attached line on the bow pulls the primary boat backward and away from the boil.

Courtesy of Steve Treinish.

**FIGURE 6-49** In sea anchoring, the secondary boat is swamped perpendicular to the current so that the current grabs it like a parachute and pulls the primary boat from the boil. This is a last-resort procedure in case of motor failure.

Courtesy of Steve Treinish.

**FIGURE 6-50** The snap shackle is designed to release under many thousands of pounds with one quick jerk of the small pin used to keep it together.

Courtesy of Steve Treinish.

the boats can ease off the throttle and return to the downstream flow. After that, the incident becomes a normal river operation, but crews should use caution, as there will still be a tether rope between the boats until they are unhooked.

Most agencies and rescuers believe that an inflatable should be used as the primary boat because it can be filled with water and still remain stable. A metal johnboat, equipped with seat flotation, can be used as a secondary boat because it can serve as a sea anchor. Sea anchoring in two-boat tethers is the act of actually swamping the secondary boat intentionally. It is a last-ditch effort, as a proper peel-out will usually pull the primary boat out from the hazard. Motors can fail, however, and they tend to fail at the worst possible times. For rescuers who face this situation, creating a sea anchor is the only way to pull on the primary boat. The secondary boat is swamped perpendicular to the current so that the current grabs it like a parachute and pulls the primary from the boil (**FIGURE 6-49**). When sea anchoring, the boat will get the most grab in the current when it is held up and on its side as long as possible. Just flipping it will give some grab in the water, but the more floor surface facing the current, the harder it will pull.

Johnboats can be used as secondary boats even if they are not equipped with a motor. If they do not have a motor, however, they should probably be used as secondary boats only as a last resort. Nothing beats combustion engines when you need them, but a sea-anchoring johnboat works as a last resort.

Many teams hook up the tether rope to the floor of a johnboat. When peeling out this way, the rope swings to the side of the boat and usually ends up pushing on the rescuer in the forward area of the boat. Sometimes this rope can even pin the legs of the rescuer to the seat.

An alternative way to hook up johnboats is to use a quick-release snap shackle to keep the tether line connected to the bow, thereby taking full advantage of the peel-out turn. Attaching the rope to the bow also starts the peel-out forces the instant the secondary boat is turned, and the resulting ferry angle and current force on the side of the secondary boat assist with the downstream pull needed. In the event a sea anchor is needed, the rope is still anchored to the floor in the middle of the boat. If the secondary boat gets orders to swamp and sea anchor the primary boat to safety, the rescuer in front merely releases the snap shackle before sea anchoring. The snap shackle is designed to release under many thousands of pounds with one quick jerk of the small pin used to keep it together (**FIGURE 6-50**). Undoing the tether line from the bow then puts the rope attachment in the middle of the johnboat floor, making it easier to hold perpendicular to the current, and maximizing the effects of the sea anchor effort.

## Working Above Low Head Dams

Occasionally, boats and people may become trapped on top of low-head dams. In most cases, if a person or watercraft is stuck on the low-head wall, the currents will be slow or weak enough to allow rescue work in a slightly safer environment than if the boil is very pronounced. There is a very limited set of rescue evolutions to use for this type of rescue. An MCP system would be the best choice, as the watercraft used does not rely on motor power, and the boat can be precisely and slowly maneuvered into place, even inches at a time if needed. A two-line tether, throw bag, line gun, or reaching operation presents the distinct possibility that the victim will wash over the low-head wall when moving toward the device used, and pulling upstream may dislodge a victim's grip on a line.

In a motored operation, it is possible to tether two boats together in the form of a lowering operation. Any motor failure above a low-head dam is a crucial and extremely dangerous failure, as there may not be enough downstream distance available to ensure rescuers can paddle or oar to safety before going over the dam. With the tethering strategy, two watercrafts are tethered in the same fashion as a two-boat tether, but instead of the primary boat approaching the boil, it is slowly lowered downstream to the victim. The secondary, or lowering, boat uses throttle and ferry control to "lower" the primary boat downstream to the target. However, because the rescuers are not able to pull upstream, the secondary boat must have redundant engines, with each engine alone capable of pulling the primary boat, crew, and victims upstream in the current. This powered operation would be considered a true risk to life, and would be attempted only after careful deliberation and a very calculated risk versus benefit analysis.

## Terminating a Swiftwater Incident

In addition to the incident termination activities discussed in Chapter 4, the swiftwater operations rescuer must consider the hazards and equipment specific to swiftwater rescue. As the old saying goes, "The river never rests." Rescuers will be exposed to the hazard as long as current continues to flow.

After a rescue in swiftwater, plan the disassembly before it begins. Since the rescue is over, perhaps equipment and rescuers on an opposite shore could be transported back to a safe place, even if different routes and methods accomplish it. Once equipment breakdown begins, downstream safety should be maintained until the last person is clear of the water. Keeping a couple of boats downstream can cover anyone or anything swept downstream, but these boats must have the ability to move at will. At the minimum, downstream safeties with throw bags should be in place. If ropes are untied and pulled back to the shore, all knots and hardware should be removed. This step keeps the ropes from snagging on bottom features such as rocks, debris, and trees. During breakdown, rescuers are especially prone to creeping to the water's edge unprotected, so vigilance should be high.

Swiftwater rescue gear takes tremendous abuse. Ropes and hardware frequently become caked with mud, sand, and grit, and not cleaning it sufficiently can do tremendous damage to this equipment. After each use, all gear, whether staged or actually used in the operation, must be examined and cleaned before returning it to service. Watercraft (especially inflatables) should be cleaned and rinsed well, as sand and grit can abrade the area where the inflated gunwale meets the flooring. For more information on terminating an incident, refer to Chapter 4.

## After-Action REVIEW

### IN SUMMARY

- Swiftwater is any water moving at a speed faster than 1 knot, or 1.15 miles per hour.
- One important facet of fire service training is the balance of realistic training, availability of training props and environment, and safety.
- Swiftwater rescue is a very dynamic field that presents the rescuer with many hazards. At the same time, moving water is predictable—a fact that can be used to the rescuer's advantage.
- Personal protective equipment for swiftwater rescue is not much different than the PPE for any of the other water rescue disciplines; it includes personal flotation devices, helmets, thermal protection, and cutting tools.
  - Water flows are normally faster on the surface and in the middle of a river or stream running in a straight line, but are altered by objects in the water.
  - All personnel working on or around swiftwater should know how to save themselves if they end up in the water. They should recognize that they will not beat the river, so they must use it.
  - Even if a victim is in sight, two crucial crew positions—spotters and safeties—should be filled before placing rescuers in a hazard zone. Beware of the potential tunnel vision that can occur if rescue crews focus on the victim without performing a full size-up.

- Rope systems used on the water can range from a quick, very simple setup to multiple lines, anchors, and control points.
- Boats stationed downstream of an active rope-based or swim-based action are one of the best safety nets, including during training operations.
- Even after the rescue, risk can remain high in flowing water.
- Termination of a swiftwater rescue should be well planned and deliberate, with alternative methods being considered if they offer greater safety.
- Swiftwater gear is especially susceptible to dirt and debris damage, and should be cleaned and inspected after each use.

## KEY TERMS

*Access Navigate* **for flashcards to test your key term knowledge.**

**Belaying** The act of guiding and controlling rescuers or victims in the water via a tether rope.

**Complex rope systems** Belay or hauling rope systems that use hardware or multiple lines to provide larger mechanical advantage systems or safety systems.

**Droop line** A rope stretched across a span, held at two points, and maintained just above the water. The rope can be dropped to the water level just as the victim touches it, and one end released to let the victim swing to one shore.

**Dynamic belay** Holding a rope steady, but moving with the current to absorb the load more gently.

**Ferry angle** Lateral movement across moving water, caused by angling an object away from being perpendicular to the current. The resulting force, which is applied at an angle, forces the object to move river right or river left.

**Helical flow** Water that tends to circle into the middle of the flow, due to the friction of the flowing water against the bottom surfaces of the stream or river.

**Hydraulic** Water that is forced down in the water column vertically, then recirculated in a rolling motion back into the vertical fall. A hydraulic is very dangerous, as it constantly recirculates objects, including humans.

**Laminar flow** Water flowing faster at the surface than the bottom, due to the same friction.

**Low-head dam** A dam that has no effective means of regulating water flow. Water frequently flows over it and creates a hydraulic, which recirculates objects or people for long periods of time.

**Movable control point (MCP)** A rope system that allows rescuers on shore to control all directions of a watercraft from one or both shores.

**Simple rope systems** Belay or hauling rope systems that include minimal hardware and have no more than a 2:1 mechanical advantage.

**Skull cap** Tight-fitting, beanie-style thermal protection worn under a water rescue helmet.

**Standing waves** Actual waves that form when water is forced up vertically in the flow and then falls back upstream. Unlike in a hydraulic, current passes underneath the waves.

**Static line** A rope system stretched across a span, usually perpendicular to the current. It is used primarily in water rescue for movable control point operations involving watercraft.

**Stopper knot** A knot that keeps a rope from slipping or moving entirely out of the object controlling it; most commonly created by tying a knot in the end of the line, making it impossible to pass through a pulley.

**Swiftwater** Water moving at a speed of greater than 1 knot (1.15 mph [1.85 km/h]) (NFPA1006).

**Swiftwater helmets** Rescue helmets designed for water operations, which usually include a smaller visor to reduce flexion at the neck.

**Tagline** Any rope tied or fastened to an object.

**Tension diagonal system** A rope that is tightly pulled across flowing water, but at an angle to the current (not perpendicular). Objects can pass laterally with the water flow from shore to shore.

**Tri-glide** A quick-release buckle found on most dedicated swiftwater flotation vests, which allows the wearer to release himself or herself from a tensioned rope with one quick and easy movement.

**Turbulent flow** Water flowing in irregular fluctuations, or mixing.

**Two-boat tether** A method of approaching hazards, usually low-head dams, with two boats attached to each other. The downstream boat guards the upstream boat's safety, as it has the ability to peel out and pull the upstream boat from the boil.

**Two-line tagline** Two ropes fastened to two opposite sides of a flotation object to move the object into the current, controlled laterally and upstream or downstream by the crews on each shore.

# On Scene

You and the other members of your crew are howling with laughter at your youngest crew member's stories about his vacation. Everyone is in a good mood, as spring is coming quickly and the air is warming every day. Suddenly, your station is toned out on a run to a local bike path with a reported boat in trouble. This dispatch seems a little odd to you, as this location is not the usual place where these runs occur.

When you arrive at the bike path where it nears the river, you see a boat up against the bridge abutment. There are two people on board, each of whom is wearing a PFD. The river is about 200 feet wide here, and bank access is muddy, but crews are able to work.

A large industrial crane is sitting on the bank, and one of your crew says he can operate it. Your chief quickly says no to that suggestion, and asks you to come up with the incident action plan. You have three operations-level swiftwater rescuers and enough PPE to safely outfit six people. You have a good cache of water rescue rope and hardware, and four boats at your disposal, complete with crews.

A member of the ladder company on scene has actually telephoned one of the victims, after getting his cell number from the mobile data terminal. The victim reports that there are just two people on board; although cold and scared, they are in good shape. You decide to have the victims shelter in place, and plan the first rescue attempt. You tell the chief to launch a primary boat to try to reach the eddy behind the bridge pier, and a secondary boat to back it up. If needed, a movable control point can be set up upstream of the victims' location. You also send a ladder crew to the bridge being constructed, with the idea they may be able to lower a line downstream as a catch line if the victims fall in.

**1.** A rope that spans a river and can be a basis for more complex rope systems is called a:

**A.** span line.

**B.** static line.

**C.** tension diagonal.

**D.** droop line.

**2.** Droop lines should be deployed on the downstream side of a bridge. What should be deployed on the upstream side?

**A.** Safety lines

**B.** Throw bags

**C.** Spotters

**D.** Hanging PFDs

**3.** As a channel narrows, what happens to the speed of the water flowing through it?

**A.** The speed increases.

**B.** The speed decreases.

**C.** The water becomes much more choppy.

**D.** The speed stays the same.

**4.** During a two-boat tether, which rules must be followed?

**A.** Each boat must carry extra throw bags and PFDs.

**B.** Command should have a good view and guide the progress and tasks of each boat from shore.

**C.** Rescuers with throw bags should be stationed around the dam to act as safeties.

**D.** The secondary boat must be in minimally aerated water to be able to peel out properly.

*Access Navigate to find answers to this On Scene, along with other resources such as an audiobook and TestPrep.*

# Swiftwater Rescue Technician

## KNOWLEDGE OBJECTIVES

After studying this chapter, you should be able to:

- Identify additional personal protective equipment used by swiftwater rescue technicians. (NFPA 1006: 17.3.1, pp. 199–200)

- Describe how to protect against hazards presented in swiftwater entry. (NFPA 1006: 17.3.1, 17.3.2, pp. 200–203)

- Identify additional equipment used by swiftwater rescue technicians. (NFPA 1006: 17.3.1, 17.3.2, p. 203)

- Describe swimming rescue techniques used by swiftwater rescue technicians. (NFPA 1006: 17.3.1, pp. 203, 205–211)

- Explain how to perform a swiftwater rescue from a watercraft. (NFPA 1006: 17.3.1, 17.3.3, pp. 211–212)

## SKILLS OBJECTIVES

After studying this chapter, you should be able to:

- Perform a leaping water entry. (NFPA 1006: 17.3.1, p. 206)

- Perform a live bait rescue. (NFPA 1006: 17.3.1, pp. 209–210)

# Technical Rescue Incident

Your department is located in an area populated by rural farms and country estates, and is normally very picturesque. Spring rains have swollen rivers to the point of flash flooding, however. The latest call seems like all the other "stranded motorist" calls your department has handled in the last two days—at least until the first-in truck reports it involves a car with children in it that is actually moving in the current. When the driver was trying to cross a flooded bridge, the force of the water pulled the car into the water, but it has stopped moving approximately 40 yards downstream of the bridge. The local police are using the public address system of the cruiser, but the victims cannot hear the instructions to stay in the car until help arrives. Help in this instance is a boat-equipped swiftwater team. You are instructed to proceed downstream and keep your eyes on the situation. Although you have two Type V personal flotation devices (PFDs) and some rope gear on your engine apparatus, the strategy at this point is to keep the victims in or on the car until the boat crew can set up a movable control point and perform a pick-off rescue.

Suddenly the car lurches a bit and moves another 15 feet. This time, when it stops again, it does so forcefully, and a child loses her balance and falls into the water. Now, it is up to you to rescue this child.

**1.** Which hazards can a rescuer face if he or she is not wearing a PFD with a releasable attachment point?

**2.** Can other evolutions be done to effect a rescue in this scenario?

**3.** Which other resources may be needed if a swimming rescue is attempted in this situation?

*Access Navigate for more practice activities.*

## Introduction

Swiftwater operations at the technician level can be one of the most dangerous undertakings in the water rescue environment. The actual water environment is very dynamic and can change just a few feet downstream. Even a very slight mistake in judging where to swim or how to handle a certain hazard can make the difference between life and death for the rescuer or the victim. Nevertheless, swiftwater swimming can become the needed spoke in the rescue wheel when victims are traveling along the water at quite a pace, or things are progressing too quickly to use other rescue methods. The hazards and water features examined in Chapter 6, *Swiftwater Rescue Operations* may also apply to river swimming, running, or boating. This chapter addresses the issues associated with actually entering swiftwater to rescue a victim, describes intentional swimming and entries, explains ways to manage some commonly found hazards, and provides some helpful tools to help avoid those hazards (**FIGURE 7-1**).

Rescuers should keep in mind that having a swiftwater technician certification does not mean that an entry rescue automatically becomes the first choice in a rescue. In keeping with the principle of "Keep It Simple,

**FIGURE 7-1** Swiftwater rescue technicians are trained to enter swiftwater environments to rescue victims.
Courtesy of Steve Treinish.

Simon," rescuers should mull over all other nonentry, boat-assisted, and boat-based options before entering the water. Even then, proper backup crews should be in place before starting. Often, technicians will be staged as rapid intervention rescue crews, because rescuing the rescuers may entail an even more complicated and dangerous mission than the original operation.

# Additional Personal Protective Equipment for Swiftwater Entry

In addition to personal flotation devices (PFDs) and helmets, some extra gear may be utilized to ensure safety at the technician level, especially when operating great distances away from shore. Similar to a long climb from a mountain base camp, rescuers wading or walking great distances must be able to remove themselves from harm and mitigate risks while on the move. Some safety gear especially suited for in-water use is described in this section.

## Drysuits or Wetsuits

Contaminated water is noted as a key concern in most of this text, and for good reason. Many swiftwater rescues are responses to flood events, which naturally collect more runoff from land areas, thereby creating water that is considered mildly contaminated. As a consequence, flooding in most areas will require drysuits, at a minimum, to protect rescuers. Making matters worse is the lack of information regarding what exactly is in the water.

Even water that is considered cleaner may require special protection. Whitewater rafting rivers, for example, are often much cleaner than flooded rivers; for those areas, a neoprene wetsuit is common personal protective equipment (PPE). In either case, at least gross decontamination of rescuers as they exit the water must be considered. Neoprene should not be used in or around waters known to carry chemicals or hydrocarbons, as it will trap those contaminants against skin. After the rescue, a mild soap solution, copious rinsing, and drying will extend the life of the PPE.

## Swim Masks

Some rescuers prefer to use diving masks or eye protection when swimming in moving water, so as to reduce the entry of water into the eyes. The smaller goggles favored by personal watercraft operators make a good addition to the rescuer's PPE. Any dive mask should use tempered glass and should be treated with anti-fog liquid or gel before using it to reduce the blinding effect of fogging lenses.

## Personal Throw Bags and Tethers

Many swiftwater rescuers carry a personal throw bag for use on the water. **Personal throw bags** are just a smaller version of the typical throw bag; they are small

**FIGURE 7-2** A small throw bag can be kept in a fanny pack, along with a tether leash, and can be valuable when swimming in fast water.
Courtesy of Steve Treinish.

enough to fit into a pocket or a hip belt. Because the bag is smaller, it can be worn comfortably and not present too large of a profile on the wearer. Personal throw bags must be properly secured to the wearer to reduce entanglement risk (**FIGURE 7-2**). They can be used from the shore, but also provide the ability to throw a rope from another place on the river that would be out of reach from the shore. In addition, they may facilitate the rescue of other rescuers. Sometimes a quick, short throw of a personal throw bag to a rescuer in trouble will be all that is possible or needed, especially in wider rivers. For example, a rescue swimmer standing by in an eddy can throw a rope to a rescuer who is caught in current too strong for that person to be reached by swimming alone.

> ### LISTEN UP!
>
> Rescuers swimming in any water, but especially swiftwater, should strive to swim "clean," with minimal gear attached to or hanging on the outside of their PFD or suit. Do not clip on a typical pouch-style throw bag; it can easily become entangled on some part of your body or on features in the current, resulting in injury or death.

## Tether Leashes (Cow Tails)

Short tether leashes (also called cow tails) can be attached to the rear of the PFD and used for support on steeper banks (**FIGURE 7-3**). They should be stowed when not being used to minimize the entanglement risk. Cow tails can also be used for towing other swimmers or even kayaks. Many rescuers stow a longer piece of webbing (around 20 feet) and a couple of carabiners in their PFD pocket that can be used in the same way as a cow tail.

**FIGURE 7-3** Short tether leash (cow tail).
Courtesy of Steve Treinish.

## Cutting Tools

Every rescuer near or in swiftwater environments should carry cutting tools. The risk of getting caught in a line or needing to assist another rescuer caught in a line is very real. A short, blunt-tip, straight-blade knife that can be drawn and used with one hand is considered the best type of tool for water rescuers (**FIGURE 7-4**). Folding knives need to be manipulated to open them, and pointed tips can injure a rescuer. Trauma shears also need a fair amount of dexterity to use, and in the heat of a swiftwater entrapment, a quick-draw blade is needed.

## Walking Poles

Wading in water can be much easier when you use a wading or walking pole. Oars, paddles, pike poles, or even aluminum conduit can be used to sound bottom conditions, makeup, and hazards such as larger rocks or holes, and also offer good support to physically lean on while walking. Although most paddles and oars float, when dropped they may get flushed downstream away from the rescuer. In contrast, pike poles and conduit sections will usually sink, so they may be recovered from shallow water to use again (**FIGURE 7-5**).

## Swiftwater Entry

To complete technician-level training, rescuers must learn about river wading, walking, and swimming and the hazards that may be encountered when a rescuer must enter the water. A rescuer must also perform a "go" rescue. The swiftwater technician performs two main types of "go" rescues: wading rescues and swimming rescues.

**FIGURE 7-4** Smaller, blunt-tipped knives that can be used with one hand work best in swiftwater. Knives should be accessible from any point, and not covered with other gear.
Courtesy of Steve Treinish.

**FIGURE 7-5** Walking poles can be used both to sound bottom conditions and hazards and to support the rescuer while walking.
Courtesy of Steve Treinish.

## Wading or Walking Rescues

Walking rescues are exactly what they sound like—simply wading or walking to a victim and either investigating and waiting for other means of rescue or accompanying the victim to safety in another location. Usually, multiple rescuers work together, with faster currents calling for rescuers to group themselves closer together to handle the stronger water forces. With lesser currents, rescuers may spread out more, but help and backup should always be close enough to assist them if trouble occurs.

## Walking in Current

Although moving water can easily knock down a person, even in very shallow depths, it is often possible to walk or wade through such water, especially if you have something on which to lean. Paddles, pike poles, or aluminum poles can all be used as supports.

When using a walking pole, the rescuer works into the current, leaning slightly on the tool when taking each step. Care should be taken to maintain two points of contact on the bottom, with the legs or one leg and the tool touching at all times. This help maintain balance and solid footing. It also lets the rescuer carefully feel out the bottom before body weight is committed, possibly saving the person from a slip, trip, or fall.

The support tool can also be used as a probe to sound out holes, check depths, or check under logs or rocks for submerged victims. Multiple rescuers need to walk or wade together, leaning on each other for support and to help maintain good footing. Communication of hazards, direction changes, and even steps should be vocalized and confirmed before the movement. In the typical approach, the rescuer most forward utilizes a probe pole to sound for issues for support.

## Group Walking

Group walking can be done in currents as well. It offers the advantage of having more people available to both reduce the amount of force to which everyone is subjected and provide some protection for a victim (**FIGURE 7-6**). Regardless of the type of walking chosen, remember to take any extra PPE needs along to equip any victims who will be rescued.

When group walking, every rescuer must have a firm hold on the others in front or beside them, using straps on the PFD. Ensure the PFD fasteners are cinched fairly tightly, as a firm hold on each individual will strengthen the group as a whole.

**Upstream *V* Formation.** The **upstream *V* formation** lets everyone in the water break some of the current's force, which makes walking easier. The group must travel very close together, in a formation resembling a *V*, with the point of the *V* always pointing upstream, similar to geese flying. Each person grabs the back of the person in front of him or her, and the leader can use a pole (**FIGURE 7-7**). This gives each rescuer some support. The walkers in front can verbalize hazards or conditions to the people in the rear. Forward travel is done in a group, one step at a time, and lateral travel is done the same way. The group must always face upstream and must communicate the next move between the members. Walking a victim to safety can be done very easily by inserting the victim into the shelter provided in the middle of the stack.

**Group Line.** Similar to the upstream *V* formation, a group line simply stacks the rescuers into a single-file

**FIGURE 7-6** Group walking has the group assembled in a tight circle facing upstream, and moving as a whole unit.
Courtesy of Steve Treinish.

**FIGURE 7-7** Multiple rescuers can walk in a *V* formation. Efforts must be made to walk or move in a synchronized manner.
Courtesy of CESTA, Inc.

line, which should face upstream at all times. The line can move upstream, downstream, diagonally, and laterally on command, with any potential victims inserted in the middle of the stack. Victims should not be placed last or next to last. This arrangement lets multiple rescuers tend to a victim and to each other.

**Circle Wading.** Smaller numbers of rescuers, or those without probes, might elect to use circle or tripod wading. In this practice, rescuers firmly grasp each other and wade in a circular motion while moving in the desired direction in the current. The rescuer on the upstream side, who is exposed to the current force, can lean on the downstream rescuers, who will have better stability from the eddy the forward rescuer is creating.

## Foot Entrapment

One particular hazard that may be encountered while walking or wading in a river is a **foot entrapment**. The river is always moving, and one slight slip or twist may result in a rescuer stumbling or tripping, and possibly getting a foot wedged into a crack or hole on a rocky bottom. When this happens, rescuers often instinctively bend at the knees to keep from falling down. The rescuer's body is then immediately forced downstream when the force of the water applies pressure on the larger area of the torso. This places a large amount of pull on the leg and trapped foot. Simply moving back upstream may not be an option if the water is moving with great force (**FIGURE 7-8**). The rescuer is now pinned. Broken or sprained ankles are common in such cases; while fellow rescuers can try to handle these injuries with care, they will be extremely painful.

### LISTEN UP!

Wearing hard-soled shoes with good ankle support may allow better performance in the water than wearing neoprene booties. Even an inexpensive pair of high-top athletic shoes can work well. Using shoes a size or two larger than normal will allow the use of heavier socks, providing insulation from colder water.

**Freeing the Trapped Foot.** A rescuer experiencing a foot entrapment may be able to wiggle loose, especially if he or she has a walking pole. Leaning into the pole and forcing the body up current may allow the rescuer to slip the foot loose. If the rescuer cannot loosen the trapped foot, other options should be used.

The first option is to stretch a rope immediately downstream of the rescuer for support, called a stabilization line (**FIGURE 7-9**). Narrow rivers or channels can be spanned with a personal throw bag, and manned on each bank by other rescuers. A stabilization line takes the initial pull off the trapped foot and may allow the rescuer to free himself or herself. In wider waterways, rescuers staged in safe areas upstream may be able to float a bight of rope down to the rescuer and control it from there.

### LISTEN UP!

This stabilization line can be formed using a tension diagonal in many cases. This arrangement still lets the person with the snagged foot support himself or herself, but also lets another rescuer use the line to quickly reach the victim. Rescuers must be able to stand at the victim's side, however, as deeper waters may sweep them downstream. Grabbing and holding onto a victim with a snagged foot will create an extreme amount of discomfort for the victim.

If needed, another rope used as a snag line can be deployed. This rope is held low in the water and pulled under the trapped rescuer's knees, in hopes of snagging the foot (**FIGURE 7-10**). Once the foot is snagged, it might be possible to pull it loose.

### KNOWLEDGE CHECK

Which group walking technique is best suited for a small number of rescuers?

a. Downstream diagonal line
b. Upstream *V* formation
c. Tripod wading
d. Group line

*Access your Navigate eBook for more Knowledge Check questions and answers.*

**FIGURE 7-8** A foot entrapment occurs when the foot is wedged between two objects, and the current makes it impossible to reposition the body to free the trapped foot.

© Jones & Bartlett Learning.

**FIGURE 7-9** A stabilization line lets the trapped rescuer take the body weight off the foot.

Courtesy of Steve Treinish.

FIGURE 7-10 After supporting the trapped rescuer, a snag line can be lowered and pulled upstream to try to free the foot.
Courtesy of Steve Treinish.

FIGURE 7-11 Swim fins give good propulsion to the wearer but are smaller than standard dive fins so they are easier to handle on shore and in the water.
Courtesy of Steve Treinish.

FIGURE 7-12 Swim boards make swimming easier and provide rescuer and victim flotation.
Courtesy of Steve Treinish.

# Additional Equipment for Swimming Rescue

The swiftwater technician should consider adding several types of equipment to his or her personal gear before getting in the water. While not crucial to the swim, they may make the work somewhat easier.

## Swim Fins

River swim fins (also known as shredder fins) give good propulsion to the wearer but are smaller than standard dive fins so they are easier to handle on shore and in the water (**FIGURE 7-11**). The rescuer should be aware of the risk of foot entrapment while using them, as the fin can become wedged in a place a foot would not normally squeeze into.

## Swim Boards

Swim boards, or "boogie boards," can be used in swiftwater. These small, very buoyant pieces of foam look like half a surfboard (**FIGURE 7-12**). Swim boards provide better flotation, both for the victim and the rescuer. Also, while the rescuer is swimming in swiftwater, they keep the body higher, thereby providing better visibility. Rescuers can also push the swim board to a victim first, keeping valuable real estate between themselves and a panicky person. This equipment can even be used as an impromptu backboard device until the proper c-spine precautions are taken.

# Swiftwater Swimming

Defensive swiftwater swimming is described in Chapter 6, and technician-level training adds forward stroke (crawl) swimming to the rescuer's repertoire. Many swiftwater rescuers consider the forward stroke to be composed of two components: the aggressive swim, which is used for general and deliberate movement

# Voice of Experience

It was our third day on the river doing our swiftwater rescue class. This was my second time taking this class, and we were spending almost the entire day in the water, swimming. I have always been super-comfortable around and in water and consider myself a really good swimmer, having grown up with a pool and working as a lifeguard as a teen. I got involved in swiftwater rescue because of my passion for water.

My day started out really great. We were doing "go" rescues and practicing swimming over strainers. After we got back from lunch, we started the exercise of swimming over a Jim-Buoy that represented a strainer. The Army Corps of Engineers was releasing a good amount of water for us that day, and the water was moving fast. After everyone in our group attempted to swim over the top of the buoy, we realized the water was moving so fast that it was pushing the front of the buoy up in the air and not allowing us to complete the exercise. I volunteered to shimmy out on the rope and attach another tagline to the buoy to help push the front of the buoy down back in the water.

I grabbed a 50-foot throw bag and clipped it to my PFD with one of the carabiners. When I reached the buoy in the middle of the river, I clipped one of the carabiners to the buoy. As I turned to shimmy back on the rope to shore, I slipped off. This is where my day went from great to really bad. It all happened so fast. I moved down river really fast, and the 50 feet of rope from my throw bag quickly reached the end. The carabiner I had clipped to my PFD pulled around to my back. At this point I was at the end of a 50-foot rope that was wrapped incredibly tight around my torso. The force of the water started pushing me underwater.

As scary as the situation was, I did not panic. My training started to kick in. The first thing I did was reach for the quick release on my PFD. I pulled it but nothing happened, because I had clipped the carabiner to the bottom strap, not the emergency release strap.

The second thing I did was reach for my knife. It was not there. Earlier that day we had lost three knives in the "go" swim exercises, so the instructors had all of us remove them so we would not lose any more.

My third step was to reach for my PFD buckles. I tried to unclip them with everything I had in me. The force of the river was so strong that no matter what I did, I could not release the buckles.

I could hear people on shore yelling at the safety boat. At this point I had taken a few gulps of water but was also able to get a couple of breaths in. I remember trying to arch my back, which created a small air pocket with water running over the top of my head. The safety boat came up beside me, and the crew leader told me to do all the things I had already tried. The water was moving so fast, however, that the safety boat was pushed away from me.

A few seconds later, the boat approached me again. A person in the safety boat jumped out and used a knife to cut the line to which I was attached. I popped up to the surface of the water and began floating downstream. With the emergency over, I remember swimming to shore, helped by a few of the people from the class. I felt so relieved.

At this point, the training came to a brief halt so we could get together and talk about what had happened. As scary as this event was, it did not stop me from doing what I love so much. After taking a break and calling my wife and explaining to her what had happened, within an hour I was back in the water again.

I learned some very valuable lessons that day. First, never think an accident cannot happen to you. Second, before you think you can do some task, make sure you take the proper steps to be safe. Eight-dollar knives are replaceable!

I am still a swiftwater technician for my department and look forward to getting out on the water and training each and every year. Remember to train like your life depends on it—because someday it might.

**John Burley**
Lieutenant
Jackson Township Fire Department
Grove City, Ohio

around the river, and the super-aggressive swim, which is swimming at maximum effort to avoid problems. When a person is swimming in a "normal" forward crawl, the use of the word "Hurry" is a command for a swimmer to swim with gusto. When the words "Hurry HARD!" are heard, the swimmer must swim with all his or her might to avoid potential danger such as strainers or washing out. Washing out is simply losing control, changing to a defensive swim position, and letting the river win.

Rescue swimmers should be physically fit and confident in the water, and should have a good working knowledge of currents and how to use them to their advantage. The choice to enter swiftwater should not be taken lightly, due to the dynamic environment in which rescue swimmers will be operating. This kind of swimming is not simply untethered pandemonium on the water, though. River swimming looks uncontrolled, but experienced rescuers use the power of the water to help with the rescue.

## Leaping Water Entry

Getting into swiftwater improperly can be hazardous. Jumping feet first is not recommended, as it may put the rescuer's feet directly on the bottom, resulting in a foot entrapment before the swim even starts. This increases the chances of a broken ankle, because the other foot may not be firmly planted enough to prevent the leg from being bent. Diving headfirst can result in a severe spinal injury, or even death.

The best entry is a low leap, with the head and face protected by crossing the arms in front of the face. The PFD can absorb a lot of the impact to the chest, and the lower the rescuer stays to the water level, the less impact there will be. This technique reduces the amount of vertical drop the rescuer experiences through the water. Follow the steps in **SKILL DRILL 7-1** to make a low-leap water entry. Remember that the water may be moving fairly fast, which will increase the force of the hit. Have a swim plan before making entry, because once the water grabs you, it will take you with it.

Another type of leaping water entry is the compact entry, in which the rescuer mimics a cannonball jump into a pool. With this technique, the arms are held around the knees, which are folded tightly to the chest. This position also keeps the legs as shallow as possible while entering the water, so as to reduce foot entanglements. Once in the water, the swimmer can stretch out and begin the swim. As with any entry, any vertical leap at all increases how deep the rescuer may plunge into the water column.

## Ferry Lines

Not all swiftwater swimming will be performed as part of direct rescue efforts. Sometimes swimming might be used for getting somewhere else to set up another aspect of the operation, such as anchoring a line across a stream, trying to spot victims or investigate another area for victims, or reaching a victim who has made it to an island in the water. For example, a rescue swimmer might swim across a flow, catch a line or throw bag, and start setting up other rope systems, such as **ferry lines**, also called tension diagonals (**FIGURE 7-13**). Tension diagonals are also discussed in Chapter 8, *Flood Rescue*.

## River Swimming

The number one rule when swimming in any moving water is to know which hazards exist in the swim path and how to get to safety. River swimming is not something done as a knee-jerk reaction or by flying by the seat of your pants. Operating in this manner is highly dangerous. A good understanding of river features, both the good and the bad, is paramount to safe operations.

## Eddy Hopping

Eddy hopping can be done in a boat or while swimming. It involves swimming to one eddy, scouting or discussing the next target, and then moving to it. Distance, hazards, travel routes, and locations to go if swimmers wash out should all be considered before

**FIGURE 7-13** A rescuer crossing the river using a tension diagonal line, from river right to river left.

Courtesy of Steve Treinish.

# SKILL DRILL 7-1
## Conducting a Leaping Water Entry NFPA 1006: 17.3.1

**1** To enter the swiftwater environment as safely as possible, get low by bending the knees, and strive to skim across the water before entry.

**2** To maximize protection from shallow objects, leap out—not down—into the water, with the arms crossed in front of the body.

**3** Let the PFD take the brunt of the hit, and keep your head up. Have a swim plan ready.

Courtesy of Steve Treinish.

the next swim section is attempted. Hopping, or stopping in these various locations, offers the ability to slow down, take a "pause for the cause," and confirm options for the next move (**FIGURE 7-14**). During a true rescue, staying put in an eddy and awaiting a better option of rescue is certainly possible, and an eddy provides a good resting spot that may be more stable than many areas downstream.

**Entering an Eddy.** An eddy is an area of water that is backfilling a solid object immediately downstream of it. This backfill water flows up current to the normal water flow and is usually much calmer. Eddies may also pull a boat or swimmer into them through their up-current flow. Swimmers must aim for the water flowing upstream and place themselves in it before the downstream flow pushes them past it. These different flows are often quite pronounced, and some effort may be required to break through into the up-current water, as the down-current flow will usually increase speed beside the object, thereby creating the eddy, as more water passes by.

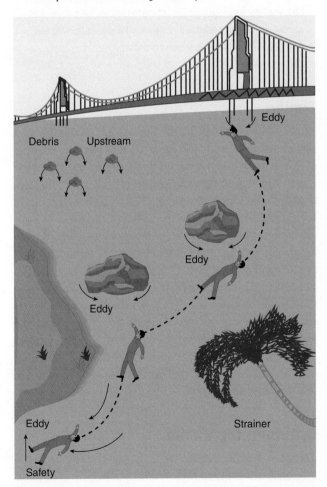

**FIGURE 7-14** Eddy hopping allows rescuers to slow down and take a moment to think about their next move.

© Jones & Bartlett Learning.

**Leaving an Eddy.** If a swimmer leaves an eddy, the route and swim must be planned. Almost immediately after the person leaves the up-current flow, the down-current flow will grab the swimmer. The swimmer enters the eddy swimming upstream, and once the downstream water is encountered, ferry angle can be adjusted to move laterally if needed. At that point, the flow is more than likely to be in control of the swimmer's movement, and the swimmer may choose to change to a defensive swim position to increase his or her visibility.

## KNOWLEDGE CHECK

Which of the following is a preferred form of swiftwater entry?

**a.** Shallow dive
**b.** Low leap
**c.** Giant stride
**d.** Back roll

*Access your Navigate eBook for more Knowledge Check questions and answers.*

# Lowering Rescuers

The releasable attachment on a Type V PFD allows the wearer to be lowered and raised on a rope, similar to a simple lower of a boat. The rescuer dons the PFD, and a water rescue rope is attached to the steel ring, often with a carabiner. Rescuers can then be belayed on a single line, with lateral movement coming from the position on shore directly above the rescuer in the water. A quick and easy two-line tether can be attached and created by utilizing two lines from the opposite banks (**FIGURE 7-15**). This technique requires each bank to be clear enough to let both belay crews travel upstream and downstream as needed. Once the victim is secure, one line may be slackened or released to perform a pendulum rescue, or one rope can be slackened and the other pulled to bring the rescue swimmer to shore.

Another method of rigging a two-line tether is called a "Y" lower. In this setup, one rope is used to control the upstream/downstream movement of the rescue swimmer, and to laterally pull to the bank from which this rope originates. The other rope can

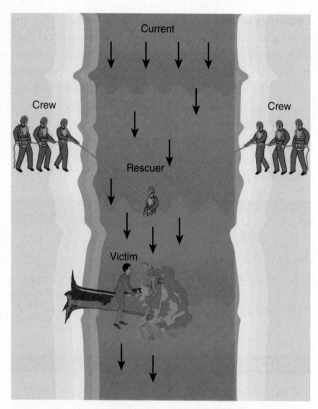

**FIGURE 7-15** Utilizing the steel ring on a Type V PFD, a swimming rescuer can be lowered to a victim to perform a pick-off rescue, then delivered to either bank.
© Jones & Bartlett Learning.

**FIGURE 7-16** In a live bait rescue, the swimmer enters the water tethered to a person tending the swimmer's line on the bank.
Courtesy of Steve Treinish.

be clipped to the line using another carabiner and can be used to manage the lateral river left/river right movement.

## Live Bait Rescues

A tethered swimming rescue in any discipline is often termed a **live bait rescue**. When performing a live bait rescue, the swimmer enters the water tethered to a person tending the swimmer's line on the bank (**FIGURE 7-16**). In this formation, the rescuer resembles a worm on the end of a fishing line—hence the name "live bait." Many swiftwater rescues are performed using this technique, but it has limitations because the tension on the rope with a swimmer and the victim can be quite high and will continue to increase as the water flows faster.

The availability of a good extrication plan affects the decision to undertake a live bait rescue. If rescuers cannot remove the victim and the live bait swimmer easily or quickly, another type of rescue might be considered. Although most shores are somewhat gentler environments, in fast-moving drainage canals in urban areas, the water moves very fast everywhere in the channel. In these settings, live bait rescue may not be possible.

## Timing

Timing plays a huge role in the success of the live bait rescue swim. Leaping too early requires a strong swim to stay in place while the victim floats closer, and going too late means that the rescuer will be playing catch-up with a victim who is already moving. Catching a victim might seem easy, as you can use the speed of the water to your advantage, but the tether length may quickly become an issue. Not getting to the victim before the tether runs out of slack will result in a miss. Even worse, getting there but not getting a good grasp on the victim may suddenly stop the rescuer but not the victim, so that the victim keeps moving with the current. With a live bait rescue, the swimmer is still traveling downstream while trying to secure the victim, as the river never, ever rests.

To perform a live bait rescue, follow the steps in **SKILL DRILL 7-2**.

## Grabbing the Victim

Tethered rescue in swiftwater is, for the most part, a one-shot attempt, so little victim packaging is done in the water. The rescuer has only a very limited amount of time to secure the victim in any way. While the victim is being secured, the rope attached to the rescuer

# SKILL DRILL 7-2
## Performing a Live Bait Rescue NFPA 1006: 17.3.1

**1** After timing the leap to get as close to the victim as possible, approach the victim as he or she is pulled downstream. Timing is crucial, but rescuers can adjust the distance by changing the ferry angle or by down- or up-current swimming.

**2** Close in on the victim. Note how far the victim has traveled—for example, by relating the victim's movement to the log stripped of bark. You must have good timing, but go early enough to get into position while the victim is still traveling.

**3** The swimming rescuer turns his or her body upstream, uses the ferry angle to slow downstream travel, and waits for the victim to arrive at the swimmer's position.

*(continued)*

# SKILL DRILL 7-2 Continued
## Performing a Live Bait Rescue NFPA 1006: 17.3.1

**4** Make contact, and secure the victim from behind by grasping under the victim's armpits.

**5** Leaning back helps keep the victim's airway and face drier. Verbal instructions can also be given from this position.

**6** This rescuer and victim have been belayed to a large tree creating an eddy, and are in much slower water flow. With assistance from the shore crews, the victim can now be rescued.

Courtesy of Steve Treinish.

is deploying farther downstream with the rescuer. Once the entire rope is deployed, it can no longer be used to secure the victim.

The rescuer should make every attempt to get behind the victim and to place his or her arms under the armpits of the victim. This positioning has several advantage. It gives the rescuer a very secure grasp on the victim, even if the victim is in panic or fight mode. It also lets the rescuer lean back, support, and talk to the victim and keep the victim's head out of the water. Finally, being pulled from the rear of the Type V PFD lets the water flow around the rescuer's head and neck, so the airway stays drier.

If the shore tenders cannot dynamically belay the rescuer and the victim, when the rope tensions there will be considerable pressure from the water flow. Be ready for this force.

## Rate of Retrieval

Rescuers retrieving swimmers and victims need to pay attention to the rate of pull when bringing them in. Once the rescuer has a secure grip on the victim, all efforts should be made to move downstream and reduce this pressure with a dynamic belay. Slow down, and let the pair keep their heads above water. If debris or trees block their travel, it may be possible for another rescuer to grab the rope and back away from the water, helping to pull the victim and rescuer into shore.

> ### LISTEN UP!
>
> One aspect to consider is whether the rescue swimmer is planning a quick release from the tether in the event of a miss. If conditions downstream support this option, the rescuer owes it to the victim to detach from the tether, swim with the victim, and get the victim to shore. All parties involved in this operation must know of the potential for it and plan for its completion.

## Multiple Victims

When facing rescues of multiple victims in moving water, rescuers must quickly decide who gets rescued first, and how. Perhaps the first person in the water seems to be floating well, and seems to be able to help himself or herself. This individual would be a good candidate for a throw attempt, with the live bait swimmer then being deployed to help those victims later in the stack. Multiple swimmers must know which victims they are swimming for, and shore tenders must make every

attempt to belay their swimmer and victim out of the "traffic" as soon as possible. Dynamic belaying will do wonders to free up shore space and keep entanglements to a minimum. It may also be advantageous to place swimmers on each side of a current, if possible, to try to reduce entanglement risk and to keep the entire operation running smoothly (**FIGURE 7-17**).

**FIGURE 7-17** Plan the rescue of multiple victims in such a way as to reduce the possibility of entanglements.
Courtesy of Steve Treinish.

# Rescue from Watercraft

While such operations are somewhat uncommon, rescuers occasionally may have to exit and enter a watercraft to perform an in-water swimming rescue. Rescuers may be delivered to an area too shallow for the boat to operate, or someone unconscious could fall overboard. While it may seem as if jumping off the boat and swimming would be straightforward, these actions must be thought out and planned before the rescue swimmer enters the water. The same general communication methods—whistle blasts and hand signals—work well with watercraft, but all parties involved should know all the signals involved for the watercraft and the swiftwater operation.

## Exiting a Boat

When placing swimmers in the water, the watercraft operator must be completely sure which direction the boat will move, which side the rescuers will exit, and where the rescuers will travel. A spinning propeller can quickly kill someone who is pushed under the boat, and a stray line that has fallen overboard and is dragging in the water can entangle swimmers. Swimmers should always leap as far from the boat as possible, off the port or starboard side, and just using one side to place swimmers in the water is preferable for visibility reasons (**FIGURE 7-18**). When possible, the boat operator should turn off the motor but should ensure the motor is ready to be quickly restarted. Motors in neutral may inadvertently be bumped into gear.

## Entering a Boat

Extricating swimmers back into watercraft can be done as in surface rescue, but the same precautions should be used as when placing rescuers in the water. Motors should be shut down, and the boat must still be navigated while in current. The boat crew can maintain some steerage in smaller boats by using paddles or oars if a motor is shut down. Also, do not discount the idea of using throw bags to bring swimming rescuers closer to the boat, thereby reducing their swim effort and increasing safety by reducing the maneuvering (using the motor) needed to get close to the swimmer.

Swimmers can be extricated from either side or the transom of the boat, but never from the bow. A slip or failure of a lifting point may drop the rescuer back into the water, sweeping the rescuer under the boat. Stirrup loops, double bounce, and even parbuckling can all be used to remove swimmers from the water (**FIGURE 7-19**).

Smaller capsized watercraft can often be righted, even in moving water. Chapter 2, *Watercraft Rescue Operations* offers more details on how to flip a capsized boat back over on its keel. In faster currents, it may be better to get to safety by using a swimming technique than to spend too much time focused on entering a boat.

**LISTEN UP!**

While it looks very dramatic and gutsy, the extrication of people into a moving boat, or from current while the boat is held steady, is not generally recommended. A rescue crew could face this possibility in a life-or-death scenario, but the risk of a rescue swimmer being pulled under a boat and into a moving propeller is high.

**FIGURE 7-18** If rescue swimmers must leap from a watercraft, a big leap and immediate ferrying of the boat can increase safety. If the motor is shut down, paddles or oars must be at the ready to assist navigation in current.
Courtesy of Steve Treinish.

**FIGURE 7-19** Extricating swimmers from the water to a boat. The motor should be in neutral to eliminate the propeller strike potential, but still running in order to navigate quickly.
Courtesy of Steve Treinish.

# After-Action REVIEW

## IN SUMMARY

- Swiftwater operations at the technician level can be among the most dangerous undertakings in the water rescue environment.
- In addition to personal flotation devices (PFDs) and helmets, some extra gear may be utilized to ensure safety at the technician level, especially when operating great distances away from shore. This equipment includes personal throw bags and tethers, cutting tools, and walking poles.
- Swiftwater technicians perform two main types of "go" rescues: wading rescues and swimming rescues.
- Wading or walking rescues often use multiple rescuers to reduce current forces on everyone involved, including the victim.
- Additional equipment that can help rescuers perform swimming rescues includes swim fins and swim boards.
- Rescue swimmers should be physically fit and confident in the water, and should have a good working knowledge of currents and how to use them to their advantage.
- Rescue swimming routes should be scouted beforehand, if possible, and can utilize eddies in the current to stage, scout, and decide the next leg of the swim.
- The choice to enter swiftwater should not be taken lightly, due to the dynamic environment in which rescue swimmers will be operating.
- Rescue swimmers can be lowered by water rescue rope attached to the steel ring on a Type V PFD so that the swimmers can pick off victims.
- Rescue swimmers and watercraft crews must be aware and unified as to the dangers and techniques of using watercraft around rescue swimmers.

## KEY TERMS

*Access Navigate* **for flashcards to test your key term knowledge.**

**Ferry lines** Tight ropes stretched diagonally across the current, which let swimmers or boats use the current force to move laterally across the water; also called *tension diagonals*.

**Foot entrapment** The inability to free a foot or leg wedged under crevices and rocks in current, which may potentially be worsened by the force of the current.

**Live bait rescue** A tethered swimmer, controlled from shore, who enters the water to effect the rescue of a victim.

**Personal throw bags** Smaller, belt-style throw bags meant for rescuer use, and short tethers used to belay shore crews. They are often deployed for rescuer assistance.

**Upstream *V* formation** A rescue technique in which multiple rescuers walk through the current, with the lead walker forming the point of a "V", pointing upstream. The rest of the rescuers fall in behind the lead rescuer to create a wedge shape that makes walking easier for everyone.

# On Scene

After a 48-hour double shift, you are happy to get home and find the kids at the neighbors' house, which allows you to take a quick nap. After a couple hours of sleep, you head toward your favorite river for some smallmouth bass fishing. The water in this river is mostly deeper and slower, but one section has a vertical drop that flows water much faster. The water has cleaned the bottom over the years, and offers many holes and cracks that attract fishermen, river waders, kids, and lovers of the outdoors. You are fishing a few dozen yards upstream from a few teenagers who are wading in the river for fun, when suddenly you hear a scream that you know is real. You look downstream and realize two teens are holding onto the body of another teen who has fallen.

You suspect something is wrong based on the noise and move closer. You now see one teen is being pushed downstream, with his body farther downstream than his legs. He is in obvious agony. While his friends are trying to help, it takes all of their effort to stay with him, given the slippery rocks and the current flow. You quickly call 9-1-1 and request a water assignment. It is obvious that this teen will not be able to walk on his own, so the rescuers en route will have to deal with not only managing a foot entrapment, but also extricating the teen as smoothly and safely as possible.

**1.** Which commonly found tools can be used by rescuers to brace themselves while walking or wading in swiftwater?

   **A.** Attic ladders and roof ladders

   **B.** Pike poles, oars, and probe poles

   **C.** Broomsticks and pinchbars

   **D.** Long-handled tools such as shovels and hoes

**2.** When tending a live bait rescue, what can be done by the tender to help reduce the force to which the rescuer and the victim might be subjected?

   **A.** Dynamic belaying of the swimmer, moving downstream along the bank to ease tension

   **B.** Belaying around a nearby tree, leading to more rapid retrieval

   **C.** Using a short leash to combine the pull of two rescuers on shore

   **D.** Digging the heels in, and squatting down to provide anchoring force

**3.** Cutting tools used by swiftwater rescuers should be:

   **A.** short, pointy-tipped, and serrated blades.

   **B.** trauma shears.

   **C.** short, blunt-tipped, and usable with one hand.

   **D.** folded into a pocket of the suit or PFD.

**4.** Foot entrapments are problematic for people in swiftwater because:

   **A.** the weakness of the affected foot hinders movement of the victim.

   **B.** the actual foot is hard to see by rescuers working in the river.

   **C.** communication in rushing water can hinder rescue effectiveness.

   **D.** the force of the flowing water on the body often pins the foot in the rocks and prohibits the victim from moving up current.

*Access Navigate to find answers to this On Scene, along with other resources such as an audiobook and TestPrep.*

Chapter Opener Image credit Courtesy of Public Safety Dive Services, a division of Technical Rescue International; Voice of Experience, maltese cross © awesleyfloyd/Shutterstock, Inc.; On Scene, siren © Bildgigan/Shutterstock, Inc.; Roped rescuer and Clinging rescuer: Courtesy to Public Safety Dive Services, a division of Technical Rescue International; Raft: © Josh Schutz/ShutterStock, Inc.

# CHAPTER 8

## Operations and Technician Level

# Flood Rescue

## KNOWLEDGE OBJECTIVES

After studying this chapter, you should be able to:

- Identify the prerequisites for performing operations-level rescue activities in floodwater environments. (**NFPA 1006: 22.2.1**, p. 217)
- Describe the types of floods and their causes. (**NFPA 1006: 22.2.8**, pp. 217–218)
- Explain how to assess and forecast floodwater conditions. (**NFPA 1006: 22.2.2, 22.2.3**, pp. 218–219)
- Identify the types of personal protective equipment needed in the flood rescue environment. (**NFPA 1006: 22.2.3, 22.2.11, 22.3.1**, pp. 219–220)
- Identify and describe utility hazards in floodwater environments. (**NFPA 1006: 22.2.2, 22.2.4, 22.2.8, 22.2.9, 22.3.1**, pp. 220–221)
- Describe the sources of contamination in floodwater and how to limit exposure and effects on rescuers and victims. (**NFPA 1006: 22.2.2, 22.2.6, 22.3.1**, pp. 221–223)
- Explain the concept of differential pressure and how it affects floodwater rescue operations. (**NFPA 1006: 22.2.2, 22.2.8, 22.3.1**, pp. 223–224)
- Explain which locations have the highest probability of containing victims in different kinds of floods. (**NFPA 1006: 22.2.7, 22.3.1**, pp. 224–225, 227)
- Describe the use of aircraft in floodwater rescue. (**NFPA 1006: 22.2.5**, p. 227)
- Describe the use of watercraft in floodwater rescue. (**NFPA 1006: 22.2.3, 22.2.4, 22.2.9, 22.2.11**, pp. 227–228)

- Describe rescue techniques that require entry into floodwaters. (**NFPA 1006: 22.3.1**, pp. 228–232)
- Explain the search process for structure interiors in the floodwater environment. (**NFPA 1006: 22.2.7, 22.3.1**, pp. 229–232)
- Explain how to terminate a floodwater rescue incident. (**NFPA 1006: 22.2.10**, pp. 232–233)

## SKILLS OBJECTIVES

After studying this chapter, you should be able to:

- Calculate differential pressure. (**NFPA 1006: 22.2.2**, p. 224)
- Search structure interiors in floodwater incidents. (**NFPA 1006: 22.2.2, 22.3.1**, pp. 229–232)
- Perform non-entry victim rescue (operations level). (**NFPA 1006: 22.2.3**, pp. 228–229)
- Perform entry victim rescue (technician level). (**NFPA 1006: 22.3.1**, pp. 229–230)

# Technical Rescue Incident

It has been a typically stormy spring week, but nothing that has stood out as record breaking. Your community is on the western edge of a 3300-acre lake that not only supplies recreational boating and fishing, but supports private housing nearly all the way around it. This lake has a dam that automatically releases water as needed, and an emergency spillway was finished a couple of months ago that is gravity-fed and cannot be controlled or shut off. The state authorities have been applauding themselves for its construction, and it is being hailed as a great asset to the area. At 05:30 on a Sunday morning, your pager going off awakening you. A water response is being toned out to an address just east of the spillway that does not match your knowledge of the local flooding areas, and sounds completely wrong. Turning on your portable radio, you hear the chief asking the dispatchers to check the address. They confirm they have checked the address, and it is correct. The callers advised of water above the front door sill, and filling up their basement. Shortly after, they advise the chief that other calls are starting to come in from the same neighborhood. This neighborhood comprises houses that are built onto a four-mile-long earthen dam that was built in the 1800s and retains water from a large recreational lake. The main street follows the dam, about 12 feet below the water level above it, and there are also houses across the street. Floodwater has never been seen here before.

The first-arriving engine reports water is blocking them from responding from the east. They report water in the road and starting to flood the houses along the road, and they have no idea where the water is coming from. The chief arrives from the west, and reports that the water in the emergency spillway has backed up and is now flowing over the banks and into this neighborhood. Another responder advises that he thinks that the river the spillway empties into was never cleaned out, and that the water coming over the emergency spillway cannot drain into the river fast enough. With nowhere to go, it is now filling up the neighborhood. You are on the second engine, towing your department's boat, and the chief requests that you launch the boat to try to get a scope on the size of the flooding and to address any immediate rescue needs.

**1.** How will you figure out where you can travel and by what method?

**2.** How will the chief determine where the floodwater is going, and how much can be expected?

**3.** How will you handle any evacuations needed?

**4.** What are some of the hazards present?

 *Access Navigate for more practice activities.*

## Introduction

Flood response is truly where the world of water rescue can morph into several types of water and water rescues, all within minutes or yards of each other. Even though the other water rescue types can all result in mission blends, flood responses can entail cold water or weather, ice, moving water, surface water, and all kinds of **contamination**. Floods can also range from a small, localized event that happens every hard rain, to responses and operations large enough to cover several states. Many of the skills and techniques discussed in other chapters can be used in flood rescue; please see these environment-specific chapters for more details on performing rescue in those environments.

Both the operations-level and technician-level content from NFPA 1006, *Standard for Technical Rescue Personnel Professional Qualifications*, are covered in this chapter. The bulk of this chapter focuses on operations-level tasks, with the technician level necessary for entry into the floodwater.

# Prerequisites

NFPA 1006 acknowledges that while the flood response area may be large and present specific hazards, personnel must be trained in surface and swiftwater rescue to perform actual flood rescue (**FIGURE 8-1**). To be considered an operations-level flood rescuer, one must have satisfied both a surface and a swiftwater rescue operations-level class. This provides the flood rescuer with a good basis of understanding of water forces and hydrology. This makes sense, because the rescue techniques, safety concerns, and rescuer personal protective equipment (PPE) are the same as in the surface and swiftwater rescue disciplines. Flood rescues normally occur as part of a larger event, with rescuers needing the skills and knowledge to plan and operate in areas where they may have to travel in the hazard to get to the place where the search and rescue occurs.

Technician-level flood rescuers may need to enter the water to perform a search of a structure or area. There are always instances where verbal self-rescue, throw, and reach rescues may occur, but using watercraft and entering the water to search are technician-based skills. Entering floodwater requires

**FIGURE 8-1** Flood rescue techniques are based on surface and swiftwater rescue, but may require more travel and searches.

Courtesy of Steve Treinish.

surface technician training and swiftwater operations training. Floodwater is typically considered surface water, moving at less than 1.15 miles per hour.

This chapter touches on those requirements, but you should refer back to other environment-specific chapters if more information is desired.

## Supporting Technicians

Supporting technician-level rescuers is mentioned in NFPA 1006 specifically, and this is no different than the support of surface or swiftwater rescuers. The ability to act as a safety or spotter is needed, as is the ability to tend a technician doing a "go" rescue.

The ability to assume support roles requires the ability to use communications and a strong understanding of the floodwater environment. For further discussion see Chapter 1, *Understanding and Managing Water Rescue Incidents.*

# Types of Floods

Generally speaking, flood events can be broken down into several types. Most are rain- or storm-driven events, and unfortunately, rescuers in many parts of the world have now witnessed the power and increasing frequency of flooding. Many people hear the terms **100-year flood**, and **500-year flood** and think a large flood will happen only once in 100 or 500 years. The term comes from efforts to place floods in context and balance public protection and regulation overkill. By monitoring stream gauges over time, a flood that is so high it has a 1 in 100 chance of occurring any given year can be called the "100-year flood." Thus, there can be multiple 100-year floods in any area over the course of time. The 500-year flood has a 1 in 500 chance of occurring any given year. Instead of frequency, consider this a description of the size and effect of the flood.

## Flash Floods

A **flash flood** is a rapid rise in local waters that normally recedes quickly after the event (**FIGURE 8-2**). Because the rainfall associated with flash floods is so quick and heavy, time for warning the public may be minimal. The forces of the flowing waters can be enough to dislodge foundations of buildings, topple utility lines, and easily move or destroy cars and trucks. Flash floods are often associated with slow-moving storms that may sit over one area for hours, or by multiple storms or rains that move along areas without ceasing. Geographic terrain plays a large factor, with mountainous or hilly terrain providing steep chutes for water to flow.

**FIGURE 8-2** Flash floods are a rapid rise in local waters that normally recedes quickly after the event.
Courtesy of Steve Treinish.

## River Floods

More predictably, a **river flood** occurs along most rivers across the world when the amount of water flow simply exceeds the capability of the riverbank (**FIGURE 8-3**). Many areas adjacent to rivers are designated as flood plains, and habitation or development is no longer allowed in these areas. The federal government frequently buys out homeowners who have lost housing, or are at extreme risk, to reduce the toll on human lives and property. One thing that can be somewhat surprising, but is still predictable, is the failure of levees and dam systems along the river. Even with the large amounts of water that can spill, these areas are widely known and can be preplanned using geographical information systems (GIS) and other mapping tools.

## Storm Surge

Caused by extreme winds offshore over large bodies of water (usually oceans), **storm surge** is water that surges higher than normal along coastal areas

(**FIGURE 8-4**). Once it rises over the barriers or levees in place for protection, the water runs inland. Almost every rescuer is familiar with hurricane storm surge, but even the Great Lakes can experience storm surge in the event of powerful winds.

**KNOWLEDGE CHECK**

Which item is a distinguishing characteristic of flood-water rescue responses?

**a.** A relatively low number of technician-level operations

**b.** Dynamic, complex situations blending many skills and specialties

**c.** Mandatory response of the National Emergency Response Team (NERT)

**d.** Frequent use of a unified command structure

*Access your Navigate eBook for more Knowledge Check questions and answers.*

# Condition Assessment and Forecasting

Assessing flood areas could range from local, with just a few houses affected, to huge land areas with floodwater showing up where it has never been seen before. Generally, larger floods will be forecasted, and rescuers will have the expertise of emergency management planners and engineers local to these flood basins. As in any water environment, flood rescuers must have a good idea what the conditions will be during the mission they are undertaking. A mechanical failure of a dam will present a certain amount of flooding and, barring any outside water sources such as rain or river flow, the water will then stabilize. Flood events triggered by rain, however, could be affected by a forecast

**FIGURE 8-3** River flooding occurs when the amount of water flow exceeds the capability of the riverbank.
Courtesy of Petty Officer 1st Class Melissa Leake/U.S. Coast Guard.

**FIGURE 8-4** Storm surge is caused by extreme winds offshore over large bodies of water, usually oceans.
Photo by Lt. Conrad H. Franz/Lake Pontchartrain Causeway Police/FEMA.

**FIGURE 8-5** Rescuers should have a rapid evacuation plan when responding to flood incidents.
Courtesy of Steve Treinish.

of more rain or saturated ground, and may physically cause the mechanical failures mentioned previously, thus expanding the scope and area of flooding.

These failures, even if thought of as possibilities, can still catch rescuers working in their path, and a plan of rapid evacuation may be needed (**FIGURE 8-5**). Rescuers working in areas where the water can rise must consider everything from where to park the truck to how long they may have if an emergency exit becomes necessary, and even more important, they must know *where* to exit.

## Travel Concerns

As swiftwater rescuers understand, the power of flowing water can be incredibly forceful. Enough power is present in severe flow to sweep entire buildings off foundations and downstream. With this in mind, rescuers must be cognizant of the possibility of an undercut or a washout. Undercuts are formed when flowing water erodes the bank or shore beneath the top layer of material. Often, around roadways, this erosion occurs under pavement or concrete and around bridge pilings (**FIGURE 8-6**). Bridges can be damaged to the point of being unusable, and undercuts can give way at any moment, striking those below or sending those above into the water.

---

**LISTEN UP!**

During reconnaissance, it is common for rescuers to make their way to the bank or edge of the water. Be aware that from the "normal" surface, such as blacktop, the undercut may be very large and almost impossible to see from areas above it.

---

## Water and Air Temperatures

After learning about hypothermia, every rescuer should have a good grasp of what water can do to the body's core temperature. Remember that water wicks heat away from wet skin 25 times faster than dry skin. Heavy rain events are often associated with weather

**FIGURE 8-6** A bridge with erosion damage from moving floodwater. This particular bridge is the only means of crossing the river in this location.
Courtesy of Steve Treinish.

fronts moving through, and often air temperatures drop significantly after the rain. Nightfall can also bring a drop in temperature, making thermal protection for rescuers a necessity—not just for the crews, but for the victims and evacuees as well. Flood victims often lose heat in their homes due to the loss of natural gas flow, electric outages, and water damage to furnaces. Time can factor into hypothermia as well. Awaiting rescue for hours or days can have a marked effect on a victim's core temperature. Add this to the likelihood of the clothing they own being wet and the possibility that they are found with just the clothes on their backs. Extra blankets may be needed and welcomed by people who have nothing.

The other extreme can be heat problems. Victims in large-scale floods may be suffering from heat or dehydration issues from a lack of drinking water. Hurricane season often runs from summer to mid-fall, and these storms often combine with the summer heat to make a bad flood problem even worse. To properly protect rescuers from water contamination, drysuits are recommended, but heat stress from their use can quickly sap rescuers of needed hydration.

## Personal Protective Equipment

The PPE used by surface and swiftwater rescuers can be used by flood rescuers in the same manner. A quick list is reviewed here. For more specific information, refer back to the surface rescue and swiftwater rescue chapters.

Flood rescuers should be equipped with, at a minimum, the following items:

- Personal flotation device (PFD). There should be one PFD for each rescuer, plus additional for expected or unexpected victims. Remember, children and even infants may be present and need appropriately sized protection during evacuations.

- Temperature protection. This may entail duty fatigues or thermal wear, depending on the water and air temperature.
- Water rescue helmet. Fire helmets are not recommended.
- Throw bag. In watercraft or with each rescuer, if on foot.
- Drysuit or wader-style hip boots. While exposure to the water may occur, surface or swiftwater drysuits are the suit of choice during flood rescue. Ice rescue suits are sometimes utilized, but they may fill in an accidental immersion and they do not offer good dexterity.
- Signaling device. This could be a portable radio, a whistle, or similar device. A cutting device that is easily accessible.

The use of structural firefighting gear is not recommended.

**LISTEN UP!**

Rescuers may be forced to travel through areas that are overgrown with brush and trees. Wearing a long-sleeve shirt with hip waders may prevent contact with poison ivy, sumac, or other plants that can cause rash and itching. This is also a good point to remember when removing or touching PPE before decontamination.

Other items flood rescuers might need, but may not be required immediately, include the following:

- Hand tools. These can be used for forcible entry, gas shutoffs, etc.
- Insect repellant. Insects can quickly make life miserable on standing water searches, and application of repellant may reduce the risk of disease and infection.
- Flashlight. Chemical light sticks may also be needed for emergency lighting or accountability.
- First aid supplies. These should be kept in a waterproof container.
- Drinking water. This should be available for both the rescuers and victims. Consider that some victims will not leave their premises and may need additional water for survival.

**FIGURE 8-7** Even after gas companies shut off supply, residual gas in the lines or storage tank can cause a gas emergency.
Photo by Jocelyn Augustino/FEMA.

## Utility Hazards

Floods in residential or urban areas can result in massive loss of utilities, including natural gas, propane, sewage, and electrical hazards. Not only can floods move houses off foundations, but they can also fracture transmission lines, float tanks off their mounts, and short out transformers. Natural gas and propane emergencies can occur, and it may not be as easy as shutting off the supply at the meter, as is usually done in dry areas (**FIGURE 8-7**). Imagine an attic furnace fed by a natural gas line with the meter submerged under 8 feet of floodwater. Even if the gas company shuts off the supply remotely, there will still be residual gas in the lines.

Another hazard natural gas or propane presents is damage to appliances inside the structure. While becoming rare, pilot lights can malfunction and let gas flow, and furnaces and water heaters can be filled with mud or debris and cause gas leaks. An unsuspecting homeowner may try to re-light or use furnaces or appliances with horrible results.

**SAFETY TIP**

In larger areas or neighborhoods where multiple gas meters or lines are underwater, it may be easier and safer to have the utility company shut down the entire gas main feeding the area. There may still be residual gas left in supply lines, so rescuers should exercise caution and monitor for gas.

## Electrical Hazards

Flood rescuers may need to walk or float through areas to search for potential victims.

**FIGURE 8-8** Roads that suffer damage from flooding often have associated damage to the utility poles that run alongside them.

Courtesy of Fred Jackson.

Floodwaters, especially from storms or moving waters, can knock over power lines, presenting an electrocution hazard. Roads that suffer damage from flash floods often have associated damage to the utility poles that run alongside them (**FIGURE 8-8**). These lines can injure or kill anyone nearby, including drivers who may be trying to evacuate. In densely wooded or urban areas, power lines can be hard to spot in a mass of tree limbs and other wires. Utility companies should be called in or consulted to try to shut down power, even if an entire grid system must be turned off.

Another danger is from standby or portable generators that many homeowners now use. Even though a transfer switch is required to safely prevent backfeeding voltage into the primary lines that normally feed a structure, many homeowners instead plug the generator feed into a dryer outlet. Failure to switch the main breaker may allow the generator voltage to charge the primary lines, and if they are down and submerged, rescuers can be electrocuted. Any rescuer feeling any tingling or body hair standing up should immediately leave the area. Small electric wands are also available that can be used to help find hazardous voltage.

## Carbon Monoxide Dangers

Homeowners using portable generators may fail to properly vent their exhaust, especially in cold weather. Be aware of the potential for carbon monoxide (CO) poisoning that may be present in enclosed spaces such as garages or basements (**FIGURE 8-9**). This potential is not just for rescuers; homeowners inside a structure may also be subjected to CO. Also consider the fact that neighborhoods may have several families sheltering at one house where a generator is in use, and a multiple-victim incident can easily transpire.

**FIGURE 8-9** Many homeowners and rescuers utilize gasoline-powered equipment; CO should be monitored when in use.

Photograph by Andrea Booher/FEMA.

**FIGURE 8-10** Whatever chemicals or substances are stored by humans can be mixed into floodwaters.

Photo by Jocelyn Augustino/FEMA.

## Contamination

Although there is not a lot of solid data on the amounts or types of contaminants in floodwaters, it is safe to say that whatever chemicals or substances are stored and used by humans will be mixed into waters in which rescuers will work (**FIGURE 8-10**). Picture a typical suburban house. Lawn fertilizers, rodent and insect sprays and powders, cleaning agents, gasoline, and home maintenance chemicals are all typically found in these houses. In a flood, all of these may be present in the water around the neighborhood, with higher concentrations found in the houses where the water is not moving quickly.

A slow-rising flood may not dilute the contamination as fast as one with water flowing around and through the area. Urban and industrial areas can quickly become a toxic brew in which individual chemicals and other substances cannot be easily identified. Biological factors such as sewage or farm wastes may also be present, and rescuers in extended or large-scale events may be exposed to dead animals or humans. Needless to say, the best exposure prevention is to ensure proper protection for anyone working in the water.

## Recommended Vaccinations

The Centers for Disease Control and Prevention (CDC) recommend rescuers working in floods in the United States receive vaccinations for tetanus and hepatitis B. Tetanus vaccination is good for 10 years, and while it is recommended, it is not a disqualifier for employment. Hepatitis B vaccinations are recommended for any rescuer who may be exposed to body fluids. It is an interesting point that this recommendation is only for employment in the United States. Rescuers deploying to or working outside of the United States are urged to investigate additional vaccination needs. Local agencies such as city, county, or state health departments can offer significant information detailing housekeeping or cleanliness needs that can limit the risks of infection.

## Gross Decontamination

Given that floodwater contains an unknown mix of potentially toxic chemicals and substances, it is imperative that proper decontamination be performed on every rescuer and item used in the search and rescue. The simplest of these to perform is **gross decontamination**, using large amounts of clean water. Portable showers, fire suppression lines, and drilled hydrant caps can all offer a quick decontamination to a large number of rescuers in a short time. More specific information on decontamination and cleaning of PPE and gear is covered later in this section.

## Washing Hands

Perhaps the easiest way to limit exposure during floods is simply to wash your hands. During periods of heavy exertion or sweat, rescuers often find themselves wiping their face or eyes. Whatever your hands touched is now on your face. In the case of rubbing your nose, ears, or eyes, this is now a real exposure. Washing hands may be more easily said than done while in the field. Washing hands with clean soap and water for a period of 20 seconds is recommended before eating, using the bathroom, changing clothes, touching animals, and touching others. "Clean" water is water that has been disinfected or boiled. Cleaning your hands and face in water contaminated by the flood does nothing but guarantee exposure. At staging, equipment setup, or rehabilitation areas, hand-washing stations should be made accessible and their use encouraged (**FIGURE 8-11**). Small, self-contained hand and face washing units are now easily deployed and can be filled with clean water.

## Alcohol Gels

Using an alcohol-based hand sanitizer will suffice if you cannot get to clean water and soap. The smaller containers can fit easily into a pack or first aid kit, and it can be used quickly and easily. Having hand

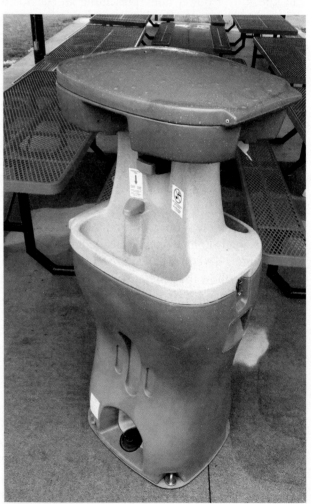

**FIGURE 8-11** Washing hands with clean soap and water for a period of 20 seconds is recommended before eating, using the bathroom, changing clothes, touching animals, and touching others.

Courtesy of Steve Treinish.

sanitizer available in watercraft or rescuer packs can make it available anywhere you go, and having alcohol sanitizer available in latrines or bathrooms and dining areas will encourage its use.

## Gear Contamination

The PPE we wear provides a good base of protection; however, it must be cleaned properly to prevent another contamination when used again. Local hazardous material teams can often provide insight regarding the best methods of decontamination, and many have specific equipment to accomplish this quickly and easily. As with most general prophylactic decontamination, gross rinsing with clean water is encouraged. Manufacturers and local health departments can assist with determining mixtures and ratios of cleaning agents. When using pump sprayers, the cleaning solutions can be sprayed on rescuers while they are still wearing their PPE. (**FIGURE 8-12**). Large-scale rinse tools like fog nozzles or hydrant caps designed to spray water could be a good choice to speed things along. Gear should be kept away from eating or rehabilitation areas for obvious reasons.

**FIGURE 8-12** If approved by manufacturers, cleaners can be sprayed directly on rescuers wearing PPE.

Courtesy of Steve Treinish.

**KNOWLEDGE CHECK**

PPE used in floodwater incidents is most similar to that used in _____ rescue.

a. surf
b. ice
c. shore
d. surface

*Access your Navigate eBook for more Knowledge Check questions and answers.*

# Differential Pressure

One major hazard facing all responders who work around or respond to the water is the phenomenon of **differential pressure**, or Delta P, as commercial divers know it. The simplest form of differential pressure is the typical household bathtub drain. When a body of water is elevated above the drain path and something allows water to flow, gravity goes to work. The suction created at the drain in a bathtub is easily overcome with muscle, but larger drains can exert thousands of pounds of suction force (**FIGURE 8-13**).

Flood, swiftwater, and dive rescue personnel are especially vulnerable to differential pressures because they often deal with water that is normally moving or draining. Delta P hazards in the water rescue world are pressures produced by two bodies of water equalizing, such as a dam gate or storm sewer system. Who gets the call on water runs during flash floods in urban environments? Who must evacuate the occupants of automobiles that end up stranded in low areas that quickly flood? The answer is, mostly

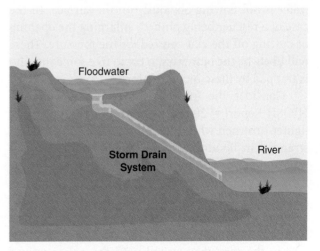

**FIGURE 8-13** Delta P is the same force as a bathtub drain. It can suck a rescuer into the higher opening, such as a storm drain, and that suction force can be thousands of pounds strong.

© Jones & Bartlett Learning.

fire fighters, and many fire fighters have to operate around drains and storm sewers. Thousands of pounds of pressure can pin a rescuer or victim to the entrance of piping or drains.

Delta P can be determined with the equation: the square footage of the pipe opening, times the difference of the depth between the two bodies, times the water column pressure. Fresh water exerts a force of 0.432 pounds per square inch (psi); we will use that value in our sample calculations. Salt water exerts a slightly higher force because it is denser. Here is an example to illuminate the hazards: Let's say we are working in a flooded street, with 18-inch manholes. The manhole covers usually blow off with flooding, so a rescuer finds the open hole. It drains to a point downhill 75 feet vertically into a storm sewer system. Going back to high school physics, the area of the drain is 3.14 × 9 inches × 9 inches (Pi × radius × radius) = 254. Now, multiply 254 × 75, and we get 19,050. Multiply that by 0.432, and the answer is 8,229. There are 8,299 pounds of force holding the rescuer at that manhole, or even worse, pushing him through it. (Area of the drain × the vertical distance the water travels before coming to a rest × 0.432 = force exerted on a person trapped by that water)

## Avoiding Delta P

Perhaps the best way to avoid Delta P is to be aware of it. Many rescuers do not fully understand it. Watch out for potential traps, such as manholes or culverts. Consulting with other agencies that control storm or sanitary sewers and drains can be beneficial; computerized mapping of these systems is common and should be utilized as part of a GIS used in many departments. Be aware of any currents at your feet, especially if the water is not flowing obviously on the surface. In the case of a rescuer being pinned, enlarging the opening or cutting off the exit can reduce the pressure. These will likely be the best ways to try to free someone who is trapped by these incredible forces.

Consider the following story, taken from the NIOSH Report #F2001-02: On August 17, 2000, a fire fighter drowned while attempting to rescue a civilian stranded in floodwaters. The career fire department was notified of several cars that were stranded due to heavy amounts of rain and subsequent flooding. At around 1700 hours, a crew was dispatched to the scene to assist motorists stranded by the floodwaters. After the crew determined that there were no civilians in the cars, they waited until the police arrived to take over control of the scene. While two of the fire fighters from this crew were waiting for the police to arrive, they were verbally summoned by a bystander to help a civilian stranded in the water. The civilian

was observed holding onto a pole in a pool of water that appeared to be about three-feet deep. Due to the flooding conditions, it was not obvious to the fire fighters that she was standing at the top edge of a culvert that was approximately 10 feet deep. Both of the fire fighters responded to the location of the civilian and attempted a rescue.

Fire Fighter #1 was the first to enter the water, and he was quickly pulled into the deeper water by the force of the drainage flow. Then Fire Fighter #2 (the victim) entered the water to aid Fire Fighter #1 to safety, and then re-entered the water to attempt to rescue the civilian. In the process, the victim was pulled under the water, into the culvert, and through a large-diameter pipe. For several hours, Fire Fighter #1 and other crews made numerous attempts to rescue and recover the victim. At approximately 2245 hours, the victim was found several blocks from the original location of the attempted rescue. He was pronounced dead at the scene.

The NIOSH investigators recommended the following procedures in the case:

- Ensure that a proper scene size-up is conducted before performing any rescue operations, and applicable information is relayed to the officer in charge.
- Ensure that all rescue personnel are provided with and wear appropriate personal protective equipment when operating at a water incident.
- Ensure that fire fighters who could potentially perform a water rescue are trained and utilize the "Reach, Throw, Row, and Go" technique.
- Develop site surveys for existing water hazards.
- Municipalities should identify flood-related hazards and take steps to correct them as soon as possible in order to minimize potential for injury.

While not obvious to the fire fighters involved in this case, this was Delta P acting on the pipe. Any rescuer may have done the same thing that they did if he or she was not aware of differential pressure. It is important to look at every flood situation, especially in urban areas or areas with large elevation changes, in terms of this hidden risk.

## High Probability Locations

Generally, people head to higher ground to escape flooding. That being considered, the intensity of the flood and the amount of time people had to prepare or flee can play a factor in where they may go and how long they may have to spend there.

## Flash Floods

People in automobiles or trucks often climb onto the roof, and fast-moving floods in rural areas may have people climbing trees or poles, creating a single rescue event, very specific in location (**FIGURE 8-14**). Time can be a critical factor, as the person may slip, the water may continue to rise, or the vehicle may be pushed farther downstream. With the more powerful water forces often found in flash floods, victims may be pushed along with debris or found in strainers. Many deceased people in flash floods are found in debris piles downstream from where they went in. Even in a single flash flood incident, multiple rescues may occur, because the speed with which these floods strike often catches multiple people off-guard.

The probability of multiple rescues can play into planning and potentially staging resources, as a severe flash flood event may entail several drainages such as creeks or small streams flowing into larger streams or rivers, creating even more flooding in the areas downstream. If a flash flood results in one rescue, there is a good chance more will be needed, especially in the areas into which the water flows.

## Larger Floods

Larger floods, especially in urban areas, may present crowds of victims in higher places. Freeway overpasses were frequently used as dry ground in New Orleans during Hurricane Katrina, as the failure of levees put more water in the neighborhoods than anyone expected or had experienced. Geographical hills in flood areas will naturally be used by victims for shelter, and they may be a good spot to set up temporary shelters. This brings up a good point rescuers must consider—the idea of sheltering in place. Imagine a neighborhood that has never experienced flooding in the residents' lifetimes. Most will not believe it could happen to them. In this situation, you should offer as much guidance as possible to residents, as they might not know the best way to keep themselves safe.

In the event of a flood, single-story houses do not offer much in the way of escaping to upper stories, as most attic space is through ceiling access. Residents of single-story homes will usually go outside and migrate somewhere to higher ground. Home owners with multiple-story houses will climb to stay dry, usually to the second floor. If the water keeps rising, the next place of refuge is the attic space or the roof (**FIGURE 8-15**). Attic spaces in older two-story homes often have livable space or attic stairs, which are now the logical next step for the occupants. This presents problems, though, as many attics are not ventilated and do not have egresses. These folks are now trapped and not visible from the outside. Searching houses is covered later in this chapter.

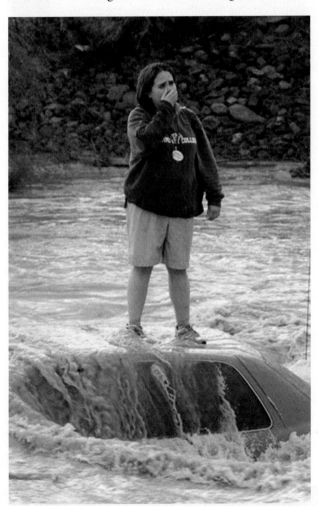

**FIGURE 8-14** People stuck in flash floods often look to get to higher ground as quickly as possible, even if it's just up a tree or on top of a car.
FEMA/Aaron J. Latham/Arizona Daily Star.

**FIGURE 8-15** Residents not accustomed to unprecedented flooding may end up on their roof or similar high ground.
Photo by Jocelyn Augustino/FEMA.

# Voice of Experience

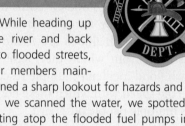

During the height of Hurricane Sandy, my department's dive/water rescue team was dispatched for mutual aid for "multiple people trapped in a building in rising water." Our team was fully staffed in quarters and responded immediately with all our water rescue equipment, including all three boats.

Upon arrival at the scene, our team was already dressed in swiftwater equipment. All members of the team are required to be swiftwater rescue technicians. This flexibility is paramount, as most flood water responses are continuously evolving. Our team quickly decided to launch our smaller boat that is lighter and has a shallower draft. The boat was moved with the motor stowed up out of the water by members in swiftwater dry suits walking alongside, probing as we moved. We were able to quickly reach our objective through the shallow water. Upon reaching the building, all five occupants were placed in PFDs and loaded into the boat. We turned around and made egress in the same fashion we had entered. We unloaded the five occupants on dry land.

At the same time, we received a request to respond further upriver to rescue a person having a medical emergency, who had become trapped by rising water. Our team quickly spun the boat around and headed north, up the flooded-out road, to attempt to access the river. The flooding had reached a high spot where we could no longer push the boat. We now had to carry the craft with members on each side of the boat. The motor was removed and left on the ground. Our members carried the boat up the road to an access point of the river. We then returned for the motor. The boat was put back together and members climbed into the boat as we now could use motor power in the deeper water.

While heading up the river and back into flooded streets, our members maintained a sharp lookout for hazards and victims. As we scanned the water, we spotted a man sitting atop the flooded fuel pumps in a gas station. Our operator navigated the boat into position to rescue the trapped occupant. The man was severely hypothermic and unable to assist our members. Members had to place him into the boat and immediately start administering care for hypothermia. With the occupant onboard the boat, we headed south to drop the patient off for immediate transport to the hospital. After the patient was in EMS care, the boat headed north once more to retrieve the initial patient. Our boat crew retrieved the patient from the police units that had become stuck by rising water. The patient was removed and transported down the flooded roads to awaiting EMS.

While flooding can strike flood-prone areas, often the massive uncommon storms can bring flooding to areas that are not usually affected. All teams should only deploy members cross-trained with swiftwater rescue training, flood training, and medical training. A large supply of medical equipment, spare PFDs of various sizes, helmets, and blankets in dry bags are a must to provide immediate medical care for exposure. Large-scale storms causing vast flooding happen infrequently, but all teams should prepare for the low-frequency, high-risk events. Training should serve to mirror real-life events.

**Timothy Mentrasti**
Assistant Chief-Dive/Water Rescue Unit
Yorktown Heights Fire District
Yorktown Heights, New York

**FIGURE 8-16** Rescuers can enlist the use of a helicopter for searches, supply drops, and assistance in getting boats or rescuers to victims.

FEMA/Aaron J. Latham/Arizona Daily Star.

If large numbers of victims are present in places of safety, the decision may be made to delay evacuation until water recedes or additional resources become available. Of course, ill or injured patients must be treated, but for those simply displaced, keeping them in place with emergency supplies may be advantageous, especially if the duration of stay is expected to be short. In the event rising water turning landmasses into islands, the assessment and forecast conditions plays a large part in this decision. Sometimes it is better to make sure people stay alive and stay put, even at the expense of comfort.

## Aircraft Use

Aircraft use in large area flood operations can range from aerial reconnaissance to lifting rescues, and the potential for aircraft use in a flood rescue should be covered in the incident action plan (IAP). Rescuers may never be a part of a lifting operation, as these tasks are reserved for highly trained rescuers on helicopter crews. A typical news helicopter likely should not and will not be used as a lifting device. Rescuers can, however, enlist the use of a helicopter for searches, supply drops, and assistance in getting boats or rescuers to victims who can be seen only from above (**FIGURE 8-16**). Communication with aircraft is often done through cross-patching or joint radio frequencies, and teams may consider procuring a few hand-held aviation radios for those missions when other aircraft are called into service. Staging of helicopters should include suitable landing zones. Localities should determine the capabilities of the helicopters that could respond to their areas to expedite the ordering process and prevent confusion. More information on helicopters and landing zones can be found in Chapter 16, *Helicopter Rescue Support*.

# Watercraft Selection and Use

Every news outlet has shown dramatic pictures of watercraft being used by rescue crews and civilians in flood situations. Even in areas that are shallow enough for rescuers to walk or wade, boats provide better safety, more equipment hauling capability, and the ability to evacuate or move civilians if needed. The type of watercraft can be tailored to fit the needs of the mission. Do not discount other agencies with specific types of vehicles. Some larger military vehicles offer the ability to load, move, and offload people and equipment in flood situations. Sheltering in place may be a better option if rescuers are unsure about whether to deploy rescue watercraft.

Another aspect that needs consideration is the population that may be riding in the boat. Evacuating a nursing home may require a larger boat with removable seating to accommodate wheelchairs, and the residents may not be able to enter and exit the boat without additional assistance. Caution should be taken not to overload the boat beyond its rated capacity. Furthermore, consideration must be given to ensure there are enough PFDs and peripheral PPE for rescuers and victims. The addition of a warm blanket or two on a person battling hypothermia will make a huge difference in comfort and care. IAPs should be implemented in flood emergencies as they are implemented in other emergencies. These plans should include all aspects of watercraft preparation and use in the floodwater conditions, as well as personnel communication protocols and equipment needs.

## Navigation

Most people can relate to how tough it is to drive an unfamiliar area at night, even with street signs and our smartphones giving turn-by-turn directions. Now, imagine being deployed to search along Main Street in a boat, but not seeing any street signs or house numbers because they are underwater. Navigation can be a huge problem, even if operation is not. Just relating a location can be challenging (**FIGURE 8-17**). Hand-held GPS units can be beneficial to search crews. Today's smartphones and tablets can load and use moving maps and satellite views, and they don't care whether you are looking at the front door or see only a few feet of roofline. Splash protection must be considered due

**FIGURE 8-17** If street signage or other landmarks are submerged, the use of a portable GPS unit can be beneficial to search crews.

Photo by Win Henderson / FEMA photo.

**FIGURE 8-18** Law enforcement officers may be needed to keep rescuers safe.

George Armstrong/FEMA.

to the possibility of dropping these devices overboard. Periodic updates regarding the location and direction of travel must be relayed to command or accountability to ensure they know the last known point. If an address can be obtained, then spray paint it onto a conspicuous location of the building. If it can be accessed easily, also consider spray painting the address on the roof.

In the absence of electronic navigation, objects such as cell towers, larger buildings, or even trees can be used as landmarks. Large events may require marking the tops of buildings with spray paint. Addresses and street names should be large enough to see from a fair distance away. Marking power or telephone poles can be done also, but rescuers must be aware of the risk of working near or approaching electrified equipment. Shallower flooding areas may be marked with sticks or poles.

Poles or markers can also be used to mark features that may damage watercraft, such as fire hydrants and street signs. These objects can slice an inflatable boat or puncture a metal boat, and can be hard to find until it is too late. Speed around these areas should be kept as slow as possible, and if something is felt by boat, rescuers can use probe poles, a paddle, or an oar to help confirm the object. Physically marking the object should be done if possible, as marking the GPS location may not be as precise as needed to prevent damage to other boats.

## Security

Rescuers may need security personnel to travel with search crews. Aside from the criminal activities, such as looting or robberies often found during disasters or disturbances, some civilians will be on edge or near panic if food, water, and shelter seem out of reach to them. Be aware of your surroundings at all times. Rescuers should not travel alone. The use of law enforcement to accompany rescuers may be required, and structures or buildings may need to be cleared by law enforcement officers before a search occurs (**FIGURE 8-18**).

## Rescue Operations Using Watercraft

Removing or rescuing occupants into a boat is a fairly straightforward process, and the same precautions apply, such as PFD use. Watercraft operators must ensure they approach and depart the structure or feature in a manner that keeps the boat safe. The current flow, if present, plays a large part in this. Just as when operating boats in swiftwater, keeping the bow pointed into the current lets the watercraft operator approach with more control and lets the water flow under the boat as it's designed to. Trying to work with a boat that is getting hit by moving water slamming the side of the boat is asking for it to capsize, especially when loading other people that may be nervous or scared.

In the event that a watercraft operator cannot secure the boat to an object or beach to load or unload, the bow should be placed against a firm surface and slight throttle applied to help firmly hold the boat against it (**FIGURE 8-19**). If this is accomplished, all travel in and out of the boat should be done from the bow area. The best example of this would be a boat operator dropping off rescuers that have to create attic access in a house that has only the roof showing. If possible, the safest way to offload crewmembers is to gently nudge the boat against the roof decking, perpendicular to the roof, and apply enough power to hold the boat against the roof. Ferry watercraft use the same procedure when docking to off-load vehicles and occupants.

## Technician-Level Entry Rescue

Technician-level rescue for flood situations is pretty straightforward, as the same procedures used in surface technician rescue apply to flood rescues. The

**FIGURE 8-19** Rescuers unloading search equipment onto a roof. Note the position of the bow of the watercraft.

Jocelyn Augustino/FEMA.

**FIGURE 8-20** Rescuers often need to wade through water to search for and rescue stranded people.

FEMA/Liz Roll.

## Wading or Walking

In shallow floodwater, rescuers will often wade or walk door-to-door, or tow boats that hold and transport victims (**FIGURE 8-20**). A probe pole is a great tool in these situations and can be as simple as a piece of steel or aluminum piping or conduit. Light and low cost, no team is out much if these makeshift probe poles are dropped or lost, and they can help identify hazards from a safer distance. Even in water as shallow as knee deep, rescuers may find holes, drop offs, and trip hazards. Rescuers should be concerned if a drop off or depression in the terrain is found with a probe pole, as it could be an area to collect drainage and could have differential pressure present at the entrance.

When floodwater starts to recede, rescue crews tend to relax because they think the main danger is going away, but this is when Delta P risks can be highest. The storm drains are likely doing their job. If differential pressure is suspected or known, no rescuer should be placed in the area on foot. Deployment in watercraft is a safer option, but remember, if the watercraft capsizes or a rescuer falls out, they will be subject to forces high enough to cause entrapment or submersion.

## Interior Structural Searches

Searching structures during floods is not complicated, but access may vary based on water depths. Entry may be required through second-story windows or by creating access holes in attics. Completely flooded floors or basements may be impossible to search properly until the water has receded. Remember, at a flooding event with active water, a primary search must be done for life. Certain fire service tools can be helpful in this search, including thermal imaging cameras and forcible entry hand tools. Battery-powered lights and hand tools keep the air clear of carbon monoxide,

basics may be expanded depending on the needs of the victims, especially when dealing with home or structure evacuation. Older victims may require medical care or medication, even during evacuation to shelter, and rescuers may have to consider alternate evacuation methods to ensure safety. For example, someone confined to a wheelchair who is safe and stable may be better off in a larger boat for evacuation if one can be obtained. Another option may be to move a victim to a more suitable area of a structure to facilitate packaging and loading the victim into a watercraft.

### SAFETY TIP

Rescuers should not unload or load the watercraft from the sides, as the deeper water and slippery roof surface could cause a fall at the side of the boat and possibly capsize or cause the rescuer to slide under the boat.

**FIGURE 8-21** Using a chainsaw to make entry.
Jocelyn Augustino/FEMA.

which is given off by gasoline-powered saws and some small, generator-mounted lights.

Creating attic access allows rescuers to search from the top down or to free trapped victims in these spaces. Chain saws are the most common tool used, and nothing about cutting open a roof is different from fire service technique. Before the cut, however, consider that there might be victims under the roof decking. Try to make verbal contact before making a cut. Moving the victims as far as possible from the cutting area will increase safety. Make sure the victim knows exactly what you are going to do with the saw (**FIGURE 8-21**).

**LISTEN UP!**

Rescuers should remember they are not creating vertical ventilation, so they may elect to cut a larger hole farther down the roof to make entry and egress onto the attic floor easier for both victims and rescuers.

**SAFETY TIP**

Ensure that the chainsaw guide bar is only just penetrating the roof decking. Not only is it safer for people underneath, it also may help preserve structural elements like trusses and rafters.

## Air Quality and Monitoring

With the hazards of natural gas, propane, sewer gas, electric, and even carbon monoxide all possible, rescuers should use gas and environmental monitoring equipment any time entry into a structure occurs. Small, hand-held meters are extremely easy to use, and many have sample tubes that can be lowered into spaces lower than the entry point to ensure search crews are not walking or climbing into a hazardous environment (**FIGURE 8-22**). Any hostile environments should be noted and relayed to the command post or operations command.

## Oriented Searches

Another aspect of searching are the names of the walls or areas around a structure the fire service has adapted. Naming the side that the building is addressed from (normally the front) can help eliminate confusion from using compass directions or vague terminology. The front is normally the Alpha side, and the Bravo, Charlie, and Delta sides are named rotating clockwise from the Alpha side. Floors are given "Division" designations in some jurisdictions. For example, the Alpha side, division 2 is generally the front of the house on the second floor, but with unfamiliar scenery, and only a roof showing, it may be hard to figure out which side the Alpha side would be if the streets were dry. Consider making the side on which the entry cut takes place the Alpha side. Needing to guide other rescuers or equipment into a specific area of the house may be easier in this fashion.

**LISTEN UP!**

Remember that during serious flooding, utilities can be disrupted, gas mains and lines can be fractured, and chemicals can mix in or react to water. In larger buildings, ensure that a quick retreat can be made if needed.

## Guide Ropes

When searching larger structures, or going down or up floors, a search rope may help ensure a quick path out, or may guide others in. Remember, flood events do not occur just in single-family dwellings, but can occur in high-rise apartments, warehouses, or other large commercial structures. Civilians who can move unassisted may be able to move to safety just by following the rope. If a hostile environment is encountered, the quick retreat the rope offers may come in handy.

## Rapid Intervention Crews

As with any operation where rescuers may need to be removed from harm, the use of rapid intervention crews

A

B

C

**FIGURE 8-22 A**. A small CO meter worn on a PFD.
**B**. A hand-held alternating current (AC) detector.
**C**. A multi-gas meter worn on a PFD.

Courtesy of Steve Treinish.

(RICs) should be considered. Any rescuer going down or encountering problems in a tight attic space with limited entry and exit would certainly tax the rest of the crew. At the very least, any penetration or search activity of an enclosed structure should have a backup team ready to assist. These teams should be equipped with the same PPE needed to perform rescue work and have a plan in place should an emergency occur.

## Building Markings

As structures or buildings are examined for damage or injury potential for rescuers and searched for victims, they must be marked to provide information to other crews and to show crucial information to others who may enter later. NFPA 1006 and the Federal Emergency Management Agency (FEMA) task forces use a standard marking system to identify both the structural assessment and the search results. Spray paint (usually fluorescent orange) can be used, but stickers may be used to reduce blight after the flood incident (**FIGURE 8-23**).

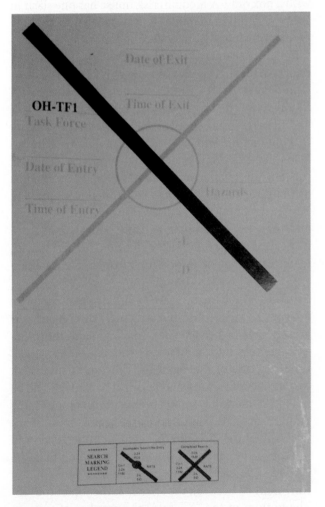

**FIGURE 8-23** Search sticker.

Courtesy of Steve Treinish.

**FIGURE 8-24** A standardized marking system is used by many teams that search structures.

Liz Roll/FEMA.

**Structural Assessment.** Many teams use a marking system comprising a 24-inch square box, possibly marked with a slash or an "X" shape within (**FIGURE 8-24**). A house that appears to be low risk, without further possibility of collapse, is marked with a box only. A medium risk house has one slash in the box, signifying a moderate risk of collapse, or some type of problem that may need monitoring or mitigation before searches take place. Gas leaks, hazardous materials, or shoring and bracing may need to be done before the structure can be searched. A structure with a full "X" in the box poses a significant risk to enter, and flood rescuers should not enter without working in conjunction with a structural collapse technical team. Structural engineers can also provide expertise to teams in structures that have suffered severe damage. A directional arrow can be painted to signify the best entry point if it is away from the box; the date and time, team identifier, and any hazards noted can be painted on the other side of the box. In this manner, teams can get quick information from a safe distance.

**Search Markings.** Search markings can comprise a large "X" painted on the structure, or even in separate rooms or occupancies. In each quadrant of the "X", include information on who is searching, the time and date of the entry, the time and date exited, the number of victims, and personal hazards should be noted. All markings should be made with orange spray paint, if possible. Upon entry, the team should mark one slash of the "X" with the team identifier, date, and military time marked in the left side. Upon exit, the second slash can be made, and the date and military time marked between the slashes at the top of the "X." The bottom of the "X" holds the number of victims, with "L" meaning live, and "D" signaling deceased. A "0" should be marked for no victims found. Personnel

hazards such as rodents or sewage can be marked at the right of the "X." Lastly, below the "X," the number of floors can be marked. For further information on marking structures, consult Annex G in NFPA 1006.

Multiple teams working with others on large-scale incidents should ensure everyone knows the marking systems used. This is one advantage of the work of the NFPA standards, FEMA, and the National Incident Management System (NIMS). Teams in which everyone uses this setup find themselves much further ahead when working out of their jurisdiction.

# Terminating a Floodwater Incident

As with any rescue or fire event, incident termination is a key part of rescuer safety and involves several components.

## Accountability

Because a flood event may be very large, personnel accountability is crucial. Keep in mind on larger incidents that many rescuers and watercraft will look the same, so a verbal or radio confirmation is preferred. When tracking personnel at the end of an incident, remember that crews must be able to travel to the safe point, whether that location is a firehouse or a staging area. Getting a personnel accountability report (PAR) is more than counting the numbers; it is making sure the numbers are okay *at the end location*. If rescuers are moving to different locations, some communication method, such as cell phones or portable radios, should be noted and accountability tracked to each location.

## Gear and PPE

Gear and equipment will most likely be extremely dirty if they are used, as floodwater is often muddy or silty, and likely laden with chemicals. Cleaning and drying equipment will be a large task, but important. If possible, use clean water to rinse gear before attempting to transport it back to quarters. Any rescuer and victim PPE must be decontaminated and ready for reuse in case another call comes in. As the old adage says, "Many hands make light work." Working as a team will make life easier on everyone.

## Rescuer Wellness

Rescuers working in flood conditions should be well hydrated and rested before resuming any additional work loading or disassembling gear or equipment. Taking the time for a hot meal may allow personnel more energy to pack up and end the mission, and during cold weather, just a 10-minute coffee break may be well needed.

One aspect not commonly associated with flood or water rescues in general is time lost after the incident due to illness. Any rescuer showing signs of sickness may need to be seen by a physician. Infections from floodwater are a real possibility, and personnel working in this environment should be encouraged to report sores, rashes, coughing, or any other unusual signs or symptoms. Time off must be documented, and exposure reports completed and forwarded to the proper agencies.

# After-Action REVIEW

## IN SUMMARY

- Flood events can be very localized or large enough to involve many square miles.
- To be considered an operations-level flood rescuer, one must have satisfied both a surface and a swiftwater rescue operations-level class.
- Flash flooding often occurs quickly, can do tremendous damage, and also recedes fairly quickly.
- River flooding is usually over a bigger area, more predictable, and may last for longer periods of time.
- Storm surge consists of water blown into coastal areas by tides, storms, or winds.
- As in any water environment, flood rescuers must have a good idea what the conditions will be during and after the mission they are undertaking.
- Floods in residential or urban areas can result in massive loss of utilities, including natural gas, propane, sewage, and electrical hazards.
- Urban and industrial areas can quickly become a toxic brew in which individual chemicals and other substances cannot be easily identified. Biological contaminants such as sewage or farm wastes may also be found, as may animal or human remains.
- Differential pressure (Delta P) hazards in the water rescue world are pressures produced by two bodies of water equalizing, such as a dam gate or storm sewer system.
- The intensity of the flood and the amount of time people had to prepare or flee can play a factor in where they may go and how long they may have to spend there.
- Aircraft use in large-area flood operations can range from aerial reconnaissance to lifting rescue, a task reserved for highly trained rescuers on helicopter crews.
- Even in areas that are shallow enough for rescuers to walk or wade, boats provide better safety, more equipment hauling capability, and the ability to evacuate or move civilians if needed.
- Interior searches of flooded homes and buildings may be required, and proper search technique such as rope-based or oriented searches should be used.

## KEY TERMS

*Access Navigate for flashcards to test your key term knowledge.*

**100-year flood** A flood event that has a 1 in 100 chance of occurrence.

**500-year flood** A flood event that has a 1 in 500 chance of occurrence.

**Contamination** The process by which protective clothing or equipment has been exposed to hazardous materials or biological agents. (NFPA 1852)

**Differential pressure** The force exerted on a drain intake when water is draining to lower elevation.

**Flash flood** A very fast-rising flood, often localized from heavy precipitation.

**Gross decontamination** The phase of the decontamination process during which the amount of surface contaminants is significantly reduced. (NFPA 472)

**River flood** A flood event based in larger basins that collect water from tributaries.

**Search markings** A separate and distinct marking system used to identify information related to the location of a victim(s). (NFPA 1670)

**Storm surge** Water pushed ahead of a storm area. The energy from these waves does not dissipate readily.

# On Scene

You are in charge of your department's rescue company today, and you are dispatched to a neighboring district that is experiencing severe flooding. You are directed to a local mobile home park that is home to 60 trailers. None have been searched, and most residents are trying to shelter in place. The population is generally older, and many do not get around well. The park abuts a large commercial truck repair facility, and the water flowing into the area smells of sewage. You are asked to take charge of the operation to search this park, and have four boat crews with watercraft available. Two inflatable boats and two large johnboats are on scene. Some rescuers are equipped with swiftwater drysuits, some have ice rescue suits, and all have a PFD. The water is warm, as is the air temperature.

1. What are some additional concerns you must consider before committing your crews to the search?

   A. How much more water is expected, how deep will it get in the park, and where does it drain to?

   B. How many boats can fit down the streets in the park?

   C. What will they need for lunch or dinner if the incident goes that long?

   D. Who will pay for damages and fuel for the watercraft?

2. One of the boat teams advises you that several residents are on home oxygen and do not get around without assistance. What may this mean for your assessment?

   A. They may have to be sheltered in place with supplemental oxygen.

   B. Additional watercraft that can carry these people may need to be called in.

   C. Additional EMS resources may need to be dispatched to support their needs.

   D. Both B and C are correct.

3. During search operations, what hazards will the boat crews face?

   A. Hostile homeowners who do not want to leave

   B. A lack of knowledge of the local street names

   C. Utilities such as natural gas, electric, and sewage

   D. Mud and debris blocking travel routes

4. What can be done to ensure that rescuers wading in the floodwaters can be protected from contamination?

   A. Performing gross decontamination with an engine company staged in a safe location

   B. Ensuring proper hand-washing technique and alcohol gels are used as needed

   C. Preventing them from touching the floodwater

   D. Both A and B are correct.

5. What hazard may suck the rescuers into storm sewers or drainage pipes?

   A. Pumping stations

   B. Septic tanks

   C. Differential pressures, or Delta P

   D. A lack of manhole covers in the area

*Access Navigate to find answers to this On Scene, along with other resources such as an audiobook and TestPrep.*

# CHAPTER 9

## Operations Level

# Ice Rescue Operations

## KNOWLEDGE OBJECTIVES

After studying this chapter, you should be able to:

- Identify and describe types and formations of ice. (**NFPA 1006: 19.2.2, 19.2.3**, pp. 236–239)
- Discuss environmental effects on ice. (**NFPA 1006: 19.2.2, 19.2.3**, pp. 239–240)
- Discuss environmental factors in ice rescue. (**NFPA 1006: 19.2.2, 19.2.3**, p. 240)
- Identify and describe victim considerations common to ice rescue. (**NFPA 1006: 19.2.2, 19.2.3**, pp. 240–242, 244)
- Identify and describe tools and equipment used in ice rescue. (**NFPA 1006: 19.2.1, 19.2.3**, pp. 244–246)
- Describe how to self-rescue at an ice incident. (pp. 246–247)
- Identify and describe additional equipment and techniques used by operations-level ice rescuers. (**NFPA 1006: 19.2.1, 19.2.3**, pp. 246–250)
- Discuss the use of row rescue equipment and techniques during ice rescue. (**NFPA 1006: 19.2.1, 19.2.3**, pp. 250–251)
- Describe how operations-level rescuers support technician-level rescuers at ice incidents. (**NFPA 1006: 19.2.1**, pp. 251–252)
- Describe basic care of the hypothermic victim. (**NFPA 1006: 19.2.2**, pp. 252–253)
- Describe how to terminate an ice rescue incident. (**NFPA 1006: 19.2.4**, p. 253)

## SKILLS OBJECTIVES

After studying this chapter, you should be able to:

- Self-rescue on an ice shelf. (p. 247)
- Perform a non-entry rescue of an icebound victim, recognizing hazards to rescuers. (**NFPA 1006: 19.2.3**, pp. 246, 248–251)
- Support a technician-level rescuer by tending attached tether lines. (**NFPA 1006: 19.2.1**, pp. 251–252)
- Terminate an ice rescue incident. (**NFPA 1006: 19.2.4**, p. 253)

# Technical Rescue Incident

After a long night of junk runs and a couple actual fires, you want nothing else than to get home and crawl into bed. On the way home, you always pass a busy park that many people use, with a sledding hill that has a pond at the end of the run. You see a commotion at the water's edge, and then realize some people are waving their hands at traffic like something is wrong. You pull in, and now see 8 or 10 people out on the partially frozen pond trying to reach a young woman whose snow sled has fallen through the ice. You don't have much in the way of rescue gear in your truck, but there is a ring buoy on a post nearby, and one civilian is throwing what looks to be an old clothesline to the woman. Behind you, you hear one person tell someone else that he is going to swim out to her. At that time, another would-be rescuer falls in after approaching from the other shore.

**1.** Even though the dangers of hypothermia are well known, should a "go" attempt be done, even with the rope and ring buoy available?

**2.** What resources will you need?

**3.** How will you control the number of people who are starting to attempt their own rescue of the young woman?

*Access Navigate for more practice activities.*

## Introduction

Obviously, ice rescue is specific to those areas of the world that get cold enough to form ice on a water surface. Some colder climates have areas where activity on the ice is the norm, and other areas may have ice only once every few years. Similarly, ice rescue to some departments consists of a quick crawl onto a farm pond and back, and ice rescue to others is a five-mile trip to an ice floe that can be acres large. For the sake of this text, ice rescue will be assumed to be an immersion event. Larger ice floe responses often turn into a large area search-type mission with evacuation challenges, and the victims found on these floes are frequently in a shelter-in-place situation. Regardless of the conditions or environment, one of the highest priorities (after rescuer safety) is getting the victim some sort of flotation, or not letting the victim submerge before you can get there. In certain situations, Operations-level rescuers may operate from the shore in such a fashion as to buy the victim time on the surface until a more direct rescue can take place.

Ice rescue is also unique in that rescuers are guaranteed to be dealing with cold water. Sure, the other types of water rescue may deal with near-freezing water, but it is a certainty in this one. Aside from hypothermia concerns and protection, there are other facets that rescuers must deal with to successfully perform a non-entry rescue of an icebound victim in a safe and well-planned manner.

## Types and Formations of Ice

The cardinal rule in dealing with any ice surface is to remember that it may fail at any time. This way, the rescuer should be prepared if it does indeed give way. The fact that a victim is in the water is already proof that something went wrong on the ice surface. Ice rescue incidents are actually rare on strong, clear ice. Rescues often occur on newer, thinner ice; older, rotten ice; or where conditions of the ice change and leave no evidence of it. Ice sheets several inches thick may not offer the same thickness just a few yards away, and weather changes, underwater currents, animals, and shore run-off can all change ice conditions. A good understanding of how ice is formed, the different types of ice, and environmental effects on ice is beneficial to the rescuer. Just like learning to read conditions in a house fire, the initial observations of trained rescuers will likely have an effect on the resulting operation. **FIGURE 9-1** shows a chart of recommended ice thicknesses for different vehicles and activities.

### Frazil Ice

**Frazil ice** is the initial freezing action of water (**FIGURE 9-2**). It is clear, but has a jumbled look to it. This look is from the way the initial ice crystals form, and it breaks very easily. Frazil ice is the first step in forming clear ice, and it often moves and flexes on the surface with wave or ripple action from wind or waves. Moving water

**FIGURE 9-1** Different ice thicknesses can support various activities, but all ice should be viewed as potentially failing.

Minnesota Department of Natural Resources.

**FIGURE 9-3** Clear ice is transparent and very strong. It may be colored similarly to the water where it is found.

Courtesy of Steve Treinish.

**FIGURE 9-4** Rotten ice, shown here on top of clear ice, is usually whitish and opaque.

Courtesy of Steve Treinish.

**FIGURE 9-2** Frazil ice is clear, very new ice.

Courtesy of Steve Treinish.

remains liquid for longer, but it will eventually freeze in cold enough conditions.

## Clear Ice

**Clear ice** is the strongest type of ice that can form (**FIGURE 9-3**). Good clear ice can be transparent to several inches and supports the largest amount of weight relative to ice thickness. Clear ice with a thickness of greater than four inches is recommended if people wish to walk on it, but only if they are spaced out and do not concentrate their weight together on the ice. One key point to remember is that ice, and anything placed on it, is supported by floating on the water surface. Thus, if the water level is below the ice sheet, it may break. Clear ice can pose problems to folks who venture onto it too soon after it forms, because although it may look strong, it may not be thick enough to hold weight.

## Rotten Ice

**Rotten ice** is what rescuers frequently have to deal with (**FIGURE 9-4**). Rotten ice has been thawed slowly and is very weak. It has a milky or opaque look. It can fail without warning and can be a very difficult surface on which to operate. The jumbled and rotten surface may slow ice travel and make other gear harder to drag over the ice surface. Rotten ice also may break into larger pieces, which may hin-

FIGURE 9-5 River ice often forms on the banks of flowing water, where the current is lighter. The middle channels often are open water due to the increased flow.

Courtesy of Steve Treinish.

FIGURE 9-6 When ice forms and the water levels drop, the unsupported ice can fracture easily.

Courtesy of Steve Treinish.

FIGURE 9-7 Drift ice is chunks of ice of any size from just a few square feet to acres large.

Courtesy of Steve Treinish.

der forward movement when trying to travel on it. Some pieces can be so large the rescuer must push them behind his or her body to get farther along.

## River Ice

**River ice** is very dangerous, especially if it is exposed to strong currents. Currents erode ice from underneath, so rescuers may not be easily able to see how thick the ice is. River levels rise and fall with temperatures, melting, and runoff. As the river rises, the ice is stressed as it is forced upwards; as the water recedes, it may sag, leaving dead air space in between the river water and the remaining ice. This is extremely hazardous to rescuers because if someone goes through the ice, the current may sweep the victim under the ice downstream. River ice formed along the edges of the river may also be called *border ice* or *edge ice* (**FIGURE 9-5**). This ice forms in the water that is not moving as fast as the center current. It may be thicker at the bank, but thinner at the center of the channel, as the faster currents found in the center erode the ice.

## Unsupported Ice

The fluctuation of water levels in lakes, flood reservoirs, rivers, and creeks can cause unsupported ice. Ice is much stronger when it is floating on the water below it. When ice forms and the water levels drop, the unsupported ice can fracture easily (**FIGURE 9-6**). Generally, larger sections cannot stay in one piece when they droop, but in areas with objects that help brace the ice when it forms, or smaller sections close to shore, unsupported ice can look like normal ice.

## Drift Ice

**Drift ice**, also known as floe ice, is a large piece of ice not directly attached to a shore. Drift ice is sometimes acres large, and sometimes just a few square feet (**FIGURE 9-7**). A prime example is the ice found in the Great Lakes region.

Ice fishermen may become trapped on drift ice and not even know it until they retreat to shore and find open water where they once drove or walked. While an immediate, life-hazard–type rescue may not be warranted, the floe, along with the people stranded on it, may travel great distances while they are marooned on it, and they will need to be safely evacuated. Helicopters, hovercraft, or airboats are often used for this type of rescue. Drift ice can also be blown onto other pieces of ice, or crushed together so that the ice buckles and fractures and ridges are formed from the pressure. Smaller pieces of ice with victims on board

**FIGURE 9-8** Snow ice is ice with a layer of frozen snow on top, and it can weaken through trapped air or melting snow.
Courtesy of Steve Treinish.

may drift with river currents or wind, and may fracture if they strike other pieces or objects. Once fractured, the victim may not be able to stay on top, may face being struck or trapped between ice pieces, or may fall into the water. Once this happens, victims can be trapped underwater or pulled under by other pieces of moving ice.

## Snow Ice

**Snow ice** is ice with a layer of frozen snow on top of it. Melting snow can degrade the ice layer and may insulate the ice enough to cause melting (**FIGURE 9-8**). It can also trap air in the mix, weakening the ice.

## Environmental Effects on Ice

If nature were like our kitchen freezers it would be simple. There would be no wind or water, constant subfreezing temperatures, and clean, clear ice. Unfortunately, Mother Nature is not nearly so kind or consistent. From day to day, ice can be affected by forces of nature or by human-made influences. Most people think of warmer temperatures melting the ice as the largest factor that causes problems on ice, and to an extent they are right.

## Melting Ice

Melting changes the ice back into liquid form, which adds weight to the ice surface, stressing it further and making more breakage or sagging possible. Imagine an aerial ladder pumping thousands of gallons of water into an elevated floor in a fire. The floor may be stable when empty, but the added weight of the water may cause the floor to sag and collapse.

The warmer air temperatures of spring often tempt people into venturing onto the ice when they are most at risk. Ice can melt because of warmer runoff water, the sun, or animal feces. Geese and ducks can change the ice

**FIGURE 9-9** Conduction melting occurs when other objects warm the ice past the melting point. Here, metal steps have warmed from the sun and melted the ice near them.
Courtesy of Steve Treinish.

drastically by doing their thing in a concentrated area, thereby causing a mess for the rescuers. The effect is **conduction melting**, which is caused by the heating of other surfaces warming the small, immediate area around an object where it enters the ice (**FIGURE 9-9**). Conduction melting can cause breakthroughs right where most people congregate—at the docks and boat ramps that are natural staging areas.

## Wind and Wave Action

Wind blowing across the ice does not melt it per se, but it does churn up any existing open water, which causes wave action, which erodes the edges of the ice. The edges of the ice sheet get hammered even more as the ice chunks that are broken by the wave action beat on the ice edge from the waves and winds. The wind can force water under the ice shelf, eroding the edge. The wind also can move huge chunks or sheets across open water, and will offer rescue teams a challenge in that the victim may be a moving target. Wind and wave action can erode or change conditions much more quickly than most rescuers realize. It also can exert more force than you might think. A sheet of ice being blown by the wind or pushed by waves can certainly destroy wooden dock posts in its path, and rescuers must be aware of staging on these structures.

### LISTEN UP!

Ice weighs a lot for its size and can easily injure rescuers pinned between it and other objects. Larger pieces of ice may be so big that they cannot be guided or maneuvered due to the force of the wind or current on them. Being aware of your surroundings is a must.

## Other Factors

Other factors affecting ice include the human touch, for example, dock bubblers. **Bubblers** blow air or bubbles to churn the water, keeping it from freezing in order to protect docks or boats. When ice freezes, it expands, doing a lot of damage to structures such as dock pilings and boathouses.

Other things affecting ice conditions are underwater springs, recreational vehicle travel, and human-made holes used for training or recreation. Underwater springs can present a special hazard in that they erode ice from below, leaving no visible signs of erosion on the surface. Warmer runoff waters from large industrial complexes or warmer areas can greatly affect ice conditions where that flow enters the water area (**FIGURE 9-10**). Examine a large commercial district with a storm water retention pond the next time the sun hastens the melting action that occurs on the pavement and roofs of the complex. There will usually be an area of thinner ice, or even open water near the runoff drain. The warmer parking lots and rooftops warm the water that has run off, thus melting the ice in the retention pond in the outflow area.

**FIGURE 9-10** Warmer runoff waters from large industrial complexes or warmer areas can greatly affect ice conditions where that flow enters the water area.
Courtesy of Steve Treinish.

rescuers cannot pull them across the ice, or they can flash freeze to the ice. Because the protection needed to operate on the ice normally includes a thermal hood, which may affect a rescuer's hearing, communication becomes problematic and hand signals must be used. And if a rescuer falls in, even protected, he or she must get out. This sounds easy enough, but on a super-slick wet ice surface that is breaking up underneath someone, it can be much harder than one might think.

## Victim Considerations

As you may assume, victims trapped in the water through the ice are in a very bad spot. They usually do not have the training or education needed to save their own life, and often by the time the call comes in and the rescuers are deployed, the cold or fatigue will have removed some of their ability to help themselves. Also, do not forget to look for reported victims who are now floating in or around the ice or water. These bodies will often appear as a small patch of color, depending on the clothes worn. Cold weather gear can often trap air, and sometimes this air keeps an unconscious person on the surface, but they can be hard to spot, especially with other ice pieces present in the hole.

## Hypothermia

The biggest problem any victim faces in any cold-water incident is **hypothermia**, a condition in which the body temperature falls below 95 degrees Fahrenheit (35°C). Hypothermia immediately starts setting in once the victim enters the water, which conducts body heat away 25 times faster than air. **TABLE 9-1** lists the signs and symptoms of hypothermia as it progresses.

Many victims also panic and thrash violently during an escape attempt, causing them to lose precious body

---

**LISTEN UP!**

After any ice training, it is a good idea to secure the ice hole if possible. Not only can the piece that is cut out be replaced, but scene tape and signs may also be used to signal danger. People have entered the water while investigating holes that were used for training.

---

**KNOWLEDGE CHECK**

Large ice floes are the most common sites of immersion rescues.
a. True
b. False

*Access your Navigate eBook for more Knowledge Check questions and answers.*

## Environmental Factors in Ice Rescue

A wet, melting ice surface becomes so slick that rescuers can easily slip and fall while trying to maneuver on it. Not only can this result in bumps and bruises, but more severe injuries are possible, such as a broken wrist or tailbone, or even a head injury. Rescuer injuries complicate the mission, because now not only the original victim(s), but also the injured rescuer needs to be rescued and removed from the ice. In addition, tender ropes can soak up water and become so heavy the

**TABLE 9-1** Signs and Symptoms of Progressing Hypothermia

| Stage | Core Temperature | Signs and Symptoms |
|---|---|---|
| **Mild Hypothermia** | 95°–93°F (35.0°–33.9°C) | Shivering<br>Rapid breathing<br>Numbness in limbs |
| **Moderate Hypothermia** | 92°–89°F (33.3°–31.7°C) | Intense shivering<br>Loss of coordination<br>Muscle stiffness<br>Mild confusion<br>Difficulty speaking<br>Unable to use hands |
| **Severe Hypothermia** | 88°–80°F (31.1°–26.7°C) | Shivering stops<br>Exposed skin blue or puffy<br>Muscle rigidity<br>Loss of awareness of others<br>Pulse and respiration rate decrease<br>Death possible |

heat even faster. The victims who actually do get themselves up on the edge of the ice may also try to stand up and walk or run to shore. This concentrated weight on the ice may cause another breakthrough, starting the process all over again.

Another hypothermia-related issue that rescuers need to consider is that the victim in this environment will lose motor skills. Fine motor skills are lost in a matter of minutes, so even a victim who has the perfectly placed throw bag tossed over his shoulder may not be able to grip the rope. The person in the water may not even be fully aware of rescue efforts, let alone able to assist in them. This factors into the decision regarding whether technician-level help is needed. A good idea on any ice-related call is to get technician-level rescuers on the way in case this scenario develops.

## Multiple Victims

It is human nature to want to help others in distress. This instinct kills many people because they want to help so badly that they place themselves in the environment from which they are trying to rescue someone without proper training, protection, or even second thoughts. Who hasn't heard of or even responded to a fire where bystanders

**LISTEN UP!**

On January 13, 1982, Air Florida flight 90 crashed into the Potomac River in Washington, D.C. One passenger struggled in the ice water, and even though a helicopter was able to drop a ring buoy beside her, she could not hang on due to the cold. A man jumped in and rescued her. At one point, she is seen on video moving her arms to swim, but not moving, and in fact, barely staying above the surface. This is a prime example of hypothermia causing the reduction of gross motor skills, and how an ice rescue can require different methods.

or family have gone in after hearing people screaming, only to succumb to the smoke themselves? Ice rescue is no different, and many well-meaning people complicate missions by providing rescuers multiple opportunities to perform rescues.

It is entirely possible to find that another victim is either in or under the water between the shore access and the original victim. This can be very hard to spot, especially if the ice is degraded enough to allow open water holes in the area. Another hole in the ice, especially between the victim and the shore, needs to be cleared, either by physical means, or by ruling it out based on witness statements. A witness who watched the entire episode unfold may be a good means of doing this, but if the witness arrived later, he or she cannot rule out the possibility that there was a first-arriving person trying to perform a rescue who fell in the water as well. Shore personnel should make an effort to find witnesses and to document a timeline of events.

## Trapped Under Ice

Some victims may be underwater at the ice hole and trapped underwater by the ice. In this instance, the ice may not shatter, but instead it may fail in a large sheet. A section of ice may act as a trap door, creating a "slide" where the victim essentially slips down and into the water. The sheet can float back into place, likely sealing a victim's fate. Even if it doesn't happen this way, people can and do get trapped underwater and become disoriented, missing the open water from which they entered.

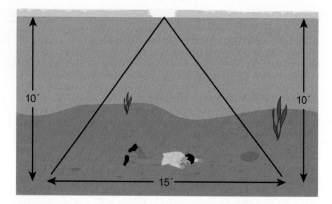

**FIGURE 9-11** Most victims will be found in an area 1.5 times the depth of the water at the entry point.
© Jones & Bartlett Learning.

---

> **LISTEN UP!**
>
> For ice pieces that are easily handled, especially in training, consider moving the pieces of ice out of the way by moving them under the ice shelf. This opens up the hole for more effective movements, allows easier extrication, and results in less danger of being bumped by or under the ice. Pushing one end of the ice sheet down and just under the ice shelf usually just requires a good push to send it a good distance away and out of the mix. For training, many smaller pieces may need to be cut out to create a hole rather than one large piece. Thick, clear ice weighs more than many people realize.

Entanglements or moving objects can snag a person while they are sinking or on the bottom, and even if the person shakes free underwater, they easily can become disoriented, especially in poor visibility. Imagine what it would be like to escape from a sinking car and try to make a blind ascent, trying to hit a hole in the ice ceiling above you. Even if you hit the ice a few feet away from the open water, one or two wrong kicks in the wrong direction and you are likely going to die. Generally, a person submerged in calm water with no current will be found in an area equal to about 1.5 times the depth of the water (**FIGURE 9-11**). For example, in water 10 feet deep, with no current, the victim will most likely be found in a 15-foot-wide circle directly underneath the entry point. Remember, this is an area formed when the victim sinks, not when he enters the water. If evidence is found that suggests these conditions, surface-based searches with tools like water jaws and pike poles may be possible.

One huge consideration is marking the original point last seen (PLS) with a marker buoy, or doing a good triangulation with several points on shore. A quick photo or two with a cell phone can also go far in documenting the victim's potential location later. With the arrival of other rescuers, the ice around the original entry hole can quickly become broken up, and this point of reference must be in place for later operations.

## Moving Water

Victims who fall through the ice with water currents moving below the ice have low chances of survival. This is another example of the need for an immediate risk/benefit analysis upon arrival, and conditions such as this must be looked at realistically. The hazards of operating around conditions like this are incredible; not only can people be swept under the existing ice by the current, but there are numerous cases of rescuers getting trapped under drift ice as it travels, or crushed between ice pieces as the flowing water applies force on the ice sheets.

Downstream hazards also need to be considered. Multiple dangerous conditions might be found downstream, such as low head dams, strainers, or even more ice.

---

> **LISTEN UP!**
>
> Rescuers who deal with moving water and ice floes must use extreme caution. The force of moving water and the mass of large ice floes are no match for any rescuer, and while it may look inviting to ride a chunk of ice downstream, it has resulted in rescuer fatalities. When faced with a rescue of a victim on moving ice, helicopter rescue may be the safest option.

# Voice of Experience

A call came in one afternoon reporting an ATV through the ice and a possible victim. A first-alarm assignment was dispatched along with the dive team. I was the officer of Rescue 1, an EMS/Rescue Company, and began donning the ice rescue suit en route to the scene. Upon arrival, two victims could be seen in a large hole in the ice approximately 300 feet from shore. A second ATV was close by the hole and was being operated by an intoxicated individual trying to pull the two victims out with a rope tied to the rear of the ATV.

After a second rescuer donned an ice rescue suit and became my backup rescuer, I made my approach tethered with a rope. I noted that the ice around the hole was beginning to collapse under the weight of the second ATV, and water was coming up onto the ice. This caused the wheels of the ATV to slip and lose traction. I ordered the ATV off the ice, to which the operator responded with a few choice vulgarities. I then cut the rope attached to the ATV, knowing that the victims were clutching onto the ice shelf and were safe for the moment. The ATV exited the area with a very unhappy, but safe operator.

I entered the hole to the side of the victims and was able to secure both victims to the rope, then called for a haul. Multiple rescuers on shore pulled the 300 feet of rope extracting the two victims and myself. At this point, two more rescuers wearing ice rescue suits approached and assisted getting the victims back to shore. They were immediately stripped of their wet clothing and placed into warm EMS transport vehicles. Both were transported to a local hospital, treated for mild hypothermia and alcohol intoxication, and released later that night.

This incident shows that we should always plan for the unexpected. Finding two victims and the potential of a third was a bit of a surprise, but our training helped make this a successful rescue. Training plays an important role in any situation, as does having the proper equipment. Currently, all members of our department are certified ice rescue technicians and we conduct department-wide ice rescue drills at least once during the winter months.

**Robert Shields**
Captain, Water Rescue Team Coordinator
Cumberland Rescue Service
Cumberland, Rhode Island

## Drifting Ice

One unique aspect to drift ice may be that the victim has trouble relating his or her exact location to rescuers. This is especially common on larger bodies of water, such as the Great Lakes. Drift ice can be blown miles through open water, and a change of wind direction can move the ice in any direction. If there are no landmarks visible from the ice, people trapped when ice floes start to move can have a very hard time relating an exact location. If drift ice takes them far enough offshore to lose landmarks, things get even worse. There are simply no landmarks to describe. In extreme cases, people must report where they traveled onto the ice sheet, then rescuers must figure out wind direction or current action and try to determine where the ice is traveling. With drift ice ranging in size from a few square feet to square miles, in extreme cases overhead search and rescue aircraft must be used to locate trapped victims. The U.S. Coast Guard often flies sorties of this nature.

Given the smaller size and lower price of handheld locator beacons and GPS units, dispatchers would be wise to ask victims if they have these devices. Getting a quick GPS coordinate or a cell phone "ping" can help enormously in the early stages of an incident response. Advising the victims to stay put on the ice rather than try to move may also help them in the long run.

## On-Scene Evidence

Once again, common sense can play into operations. A well-intentioned caller stating they saw children playing on the ice does not mean they are in the water, no matter how concerned the caller sounds. Is there even a credible hole? Fishermen, kids playing, etc. may have created and used an existing hole, but does the timeline fit the scene? Holes that are skimmed over with one to two inches of ice do not fit a call that says someone just fell in minutes ago, but it might fit a call of an ice fisherman missing overnight, especially if there is evidence to support it. Footprints in the area on the bank or on the surrounding ice offer many clues, and more than once rescuers have found animal tracks beside a hole in the ice, with no obvious human evidence to support anything but an animal entering the water.

# Tools and Equipment of Ice Rescue

Much of the same thoughts and equipment used in other water rescue can be used on ice. The biggest concerns are with submersion and cold temperature. Operations and rescuer personal protective equipment (PPE) reflect this and do a great job protecting the rescuers from this hazard.

## Personal Flotation

First and foremost, flotation must be provided for each rescuer working in the hot zone, usually the area within 15–25 feet of the water's edge. Personal flotation devices (PFDs) are required. In addition, flotation coats, or **float coats**, combine great flotation with good topside thermal insulation (**FIGURE 9-12**). Ice rescue suits may be worn by rescuers, and are examined more closely in the next chapter. With the PFD should be an accessible whistle or other signaling device.

The other main personal protection needed for operations-level rescuers is thermal protection. Many rescuers underestimate the wind conditions when

**FIGURE 9-12** Float coats combine great flotation with good topside thermal insulation.

Courtesy of Lance Anderson/Peterborough This Week.

working in open spaces, the temperature, and the amount of time they may be deployed. Quite simply, overdress for the environment and wear layers.

## Ice Rescue Suits

Technically, operations-level rescuers are not supposed to be entering or operating in the environment where a victim is trapped. But for operations-level ice rescuers, especially those in boats, hypothermia is still a risk, even if entry into the ice or water is accidental. **Ice rescue suits**, generically (but mistakenly) called Gumby suits, offer great thermal protection, and some suits offer some flotation as well. It is highly recommended that operations-level rescuers operating in boats wear them. (PFDs are still required if the suit is not certified for buoyancy.) The farther the distance from the shore, the longer it will take to get back, even if the craft is tethered. This increasing time and distance allows hypothermia to creep up on the rescuers unless they are properly protected. Most ice suits are one-piece coverall-type suits, with vertical dry zippers running from the crotch up to the neck (**FIGURE 9-13**). They have attached boots and hoods, insulated gloves, and some type of face shield that helps keep the face and neck drier and can be fastened by hook and loop material. They are the descendants of equipment used in the maritime industry, and a similar version is used on the oceans and waterways to protect workers bailing out sinking ships.

Neoprene construction suits offer good thermal protection and inherent flotation; however, they absorb water into the material and can be damaged by petroleum and chemicals. Nylon-shelled suits are easier to decontaminate if the water requires it, and they are easier to dry. They utilize an insulating liner to give more warmth. Nylon suits may allow some rescuers to sink unless the rescuer wears a PFD. Even though these suits offer some inherent flotation, it may not be enough to float all rescuers, and adding a victim's weight must be factored into operations. Wearing a PFD with either type of suit offers the ultimate in protection by providing warmth, flotation, and the ability to stay dry in case of an accidental immersion. Using these suits takes some practice and training, as dexterity suffers somewhat from the heavy gloves and hoods, but it is worth it for the warmth.

---

**LISTEN UP!**

Inverting an ice rescue suit in the water can trap air on one or both legs, trapping the rescuer upside down in the water. This is a very dangerous position. Operations-level rescuers using ice rescue suits should be trained and able to right themselves in case of an inversion in the water. This situation is examined further in Chapter 10, *Ice Rescue Technician*.

---

## Ice Awls or Picks

The next most important piece of gear with which rescuers can equip themselves is a set of **ice awls**. Ice awls, or picks, are simple tools that mostly consist of two ice picks attached to the rescuer or to each other with a leash (**FIGURE 9-14**). When victims fall through the ice, one of their main problems is getting a grip on something to pull them up and back onto the ice surface enough to roll out. Ice awls are used to stab the ice and provide that grip. For years, rescuers and outdoorsmen have been using wooden dowels and nails, or sharpened screwdrivers connected by cord, but most factory-manufactured ice awls offer a way to sheath the points, protecting the user. When the awl is jammed down onto the ice, the sheath automatically

**FIGURE 9-13** Most ice suits are one-piece coverall-type suits constructed of neoprene or coated nylon.

Courtesy of David Gillespie.

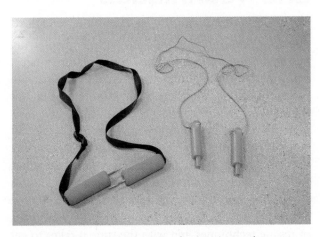

**FIGURE 9-14** Driven into the ice, ice awls provide a good method of pulling rescuers along on their stomachs.

Courtesy of Steve Treinish.

retracts. Several styles of commercial awls are available. It is not only a good idea to equip yourself with a set, but to recommend to anyone venturing onto the ice to have some, too.

# Emergency Self-Rescue

No rescuer should be working around the ice, even to support an operations-level rescue, without the knowledge of how to get out of the water should an accidental fall through the ice take place. It is possible to get out of the water and back on the ice surface without any equipment, which is where self-rescue comes into play. Getting out quickly is important, but staying out is the crucial ingredient to surviving an immersion in the cold water. To self-rescue on an ice shelf, follow the steps in **SKILL DRILL 9-1**.

Your body weight may break the ice several times before you find ice strong enough to support you. The worst part of this is pulling your body forward, instead of slipping backward into the water again. Using ice awls makes this process easier. One common mistake users make is to try to set the picks as far away as possible. This doesn't let them use the chest muscles effectively. The picks should be planted into the ice closer to the hole, in order to use the larger pectoral muscles to pull the load. Once a few inches have been gained, one pick is removed and reset a bit farther away, and once rescuers get moving out of the hole, they need to keep going quickly to get to better conditions.

# Other Equipment and Techniques

Other ice rescue equipment that may be used by the operations-level rescuer includes throw bags, ring buoys, reach objects, line guns, hose inflator systems, and watercraft. Operations-level rescuers generally do not enter the actual environment the victim is in, but there are many ways to effect rescue by indirectly dealing with ice safely and efficiently. Rescuers may not have every tool explained here, but preplanning a district's water areas and the ice-related activities found on them can allow for good use of certain tools and techniques. The standard adage of "Reach, Throw, Row, and Go" can be used to describe different types of rescue techniques, but with the time it takes to get to an incident scene, rescuers should mentally prepare to have to get out to the victim to perform rescue. The cold will have a head start on most victims by the time

rescuers get there, and the victim may not be able to help himself or herself.

# Reach Tools

The imagination can be used in figuring out which tools that typical rescuers have access to can be used as reach tools, and combining other equipment into the mix may be possible (**FIGURE 9-15**). Ladders, pike poles, oars or paddles, and even aerial ladders can be used to reach a victim. Aerial ladders or platforms may allow crews to get closer to the victim before using something else (**FIGURE 9-16**). Ground ladders can be extended out with a ring buoy or multiple PFDs attached to the end to float the tip, and more than one good rescue crew has figured out a way to hinge a roof ladder so that a victim might be able to actually climb onto the rungs, and be pulled out via a tagline.

Stiff cable systems are commercially available that allow a rescuer to thread a small cable out to a victim. Hose inflators can be used in a similar fashion. Hose inflators have an advantage in that they offer inherent buoyancy when air is added to the hose.

---

**LISTEN UP!**

Remember that aluminum or fiberglass ladders will quickly sink if the ice gives way underneath them. If a victim or rescuer is using a ladder to spread weight, the rescuer will end up in the water with it, or even worse, it will entangle the victim and pull him or her under water.

---

# Hose Inflation Systems

The use of hose inflator systems is becoming more and more common as fire companies get a good use from a tool that works with equipment every fire

**FIGURE 9-15** Here, a rescuer is extending a PFD with a long reach pole. Pike poles could be used, as could ring buoys or similar equipment.

Courtesy of Steve Treinish.

# SKILL DRILL 9-1
## Performing Emergency Self-Rescue

**1** When you start to feel the ice break underneath you, concentrate on your fall. Cover your face with your hands. This may somewhat stifle the gasp reflex that can happen when the cold water hits your face. Instead of thrashing around and flailing your arms, float on your stomach while resting your arms on the ice shelf. Kick your feet and try to "beach" yourself on the ice, lying prone.

**2** When you get onto the ice surface, do *not* immediately stand up. This is the same ice that just failed, and it will likely do so again. The key is to spread your weight over the ice surface and minimize any concentrated loading, which is what happens when you stand on your feet.

**3** Once you get enough forward movement, about to your waist or hips, try to roll 90 degrees, and roll yourself away from the hole, or even to the shore if that is what it takes.

Courtesy of Steve Treinish.

**FIGURE 9-16** Aerial ladders or platforms may allow crews to get closer to the victim before using something else.
Courtesy of Steve Treinish.

**FIGURE 9-17** A typical hose inflator system being used to push a buoy onto the ice.
Courtesy of Steve Treinish.

fighter should know and be comfortable with. A 50- or 100-foot (15.2- or 30.5-m) section of 2½- or 3-inch (64- or 76-mm) hose and a self-contained breathing apparatus (SCBA) bottle are the minimum tools used with an inflator system. The system consists of a threaded cap for the male end of the hose, most likely with an eye ring bolted through it, and a plug with an inflation valve or quick connect hose installed for the female end of the hose (**FIGURE 9-17**). A quick connect hose or valve system that uses air from an SCBA cylinder to inflate the hose completes the setup. Most systems have a pressure relief valve installed that limits working pressures to around 110 PSI. Taglines may be used on either side of the inflated hose as it is pushed out on the ice, or a ring buoy may be lashed onto the hose end. Inflated fire hose is very buoyant and pretty easy to push long distances over the ice.

## Deploying an Inflated Fire Hose

There are a few key points to remember when using hose inflator systems, the first of which is that someone needs to carry and control the SCBA bottle. A hose inflated to the maximum pressure will be very stiff, making a straight push into the water much easier. The rescuer controlling the SCBA bottle may have to open the valve occasionally to keep the pressure up. The ring buoy is a great addition to the inflated hose. Ring buoys can be attached in such a fashion that the hose end is held up, keeping it from digging into loose snow or ice and jamming up. Webbing can then be used to lash the buoy to the hose with a clove hitch and a few half hitches. The key is getting the buoy attached tightly. Lash the buoy onto the hose before it is fully inflated. A tight attachment to the hose with the ring buoy lets the rescuers twist the hose and flip

the buoy to either side on the ice, moving it laterally as needed (**FIGURE 9-18**). If tag lines are attached and used as guides, remember that the harder the pull on the tagline to direct the hose end, the harder it will be to push the hose out. This will cause a tendency to put "S" bends in the hose, instead of it moving in a straight line.

## Two-Line Tethers

Another method, called **two-line tethering**, combines reaching rescue with using ropes. Connecting two lengths of rope on either side of a ring buoy, PFD, or other flotation device lets two groups of rescuers walk that flotation into place, normally from opposite banks (**FIGURE 9-19**). The flotation object can be controlled very effectively over long distances and can be placed right up to a victim. If the victim is able to hold on, one of the rescuer groups can then pull the line, helping the victim up onto the ice shelf, or even to shore if the water is not totally frozen. The opposite crew must release the rope or give slack.

## Throw Equipment

Although throwing from the shore is done most often, do not rule out throwing or dropping lines or flotation from bridges, docks, or even boats. Even if a boat is being pushed or poled out to where the victim is, victim flotation is still the number-one concern, and throwing a well-placed bag or buoy before the boat reaches the victim may prevent a submersion. There are multiple types of **throw tools** that can be thrown or launched at a victim to provide a means of extraction or flotation.

**A**

**B**

**FIGURE 9-18** A ring buoy can be lashed to the end of the fire hose, and by twisting the hose on shore, rescuers can flip the buoy, moving it laterally across the ice.

Courtesy of Steve Treinish.

**FIGURE 9-19** Two-line tethers can be used to help guide flotation to victims. In this case, a rope is tied to the inflated hose and guided from the far shore.

Courtesy of Steve Treinish.

**FIGURE 9-20** Thrown rope may be used to apply a slight pull to a severely hypothermic victim to help pin them to the ice edges until rescuers arrive. Be cautious about pulling too hard, as the victim might lose grip on the rope.

Courtesy of Steve Treinish.

## Throw Bags/Coiled Rope

Throw bags or coiled ropes can be deployed like at any other water incident. (See Chapter 4, *Surface Water Rescue Operations*.) However, there is one twist—due to hypothermia, the victim may not be able to hang on to it, and rescuers may end up using the throw bag to provide time while other rescuers respond to the victim. Deploy the bag as usual, but remember that the victim will have to be pulled up and onto the ice shelf in order to be brought in. With heavy clothing or lack of flotation on the victim, this may be hard to do. Most of the victim is below the plane of the pull, and it takes a great amount of leverage to "flip" or "pop" the victim up and onto the ice. Making matters worse is the cold water that makes it hard for the victim to grip the rope. If the victim cannot grip the rope, there is virtually no chance of getting him or her in on a line, so try to get the victim to wrap an arm

around the rope, and apply gentle pulling pressure (**FIGURE 9-20**). This technique will pin the victim to the ice shelf, and buy some time while a "go" rescue is performed.

## Rescue Discs

Also used for surface water rescue, a Frisbee-style disc with a small, nylon cord is available to ice rescuers (**FIGURE 9-21**). The disc is thrown like a Frisbee, and as it travels through the air the cord is deployed behind it while the rescuer holds the end. Although rescue discs can be blown off course in the wind, they can travel far. There may also be a possibility of bouncing it or sliding it on the ice surface to increase the distance of the throw. They are directional, so a user throwing right-handed must pay attention to which way the wrapped line will spool off. Using the Frisbee-style disc upside down lets it slide and bounce along slushy or rotten ice, because in

**FIGURE 9-21** A Frisbee-style rescue disc.
Courtesy of Steve Treinish.

this position the edges are curved upward from the ice. Picture the snow discs children use for sledding and you have an idea how it can travel across the ice.

## Line Guns

Line guns, especially guns with an attached flotation device, can be beneficial to getting something out to a victim if he or she is beyond the reach of a throw bag or when lines cannot be stretched across the water or ice surface (**FIGURE 9-22**). Most line guns are air powered and resemble a rifle. They can shoot a sling, harness, rescue rope, or messenger line great distances. Again, a victim being able to don flotation or a harness is unlikely, but if crews can talk the victim into wrapping the rope or sling around his or her arms or body, the crews may be able to hold the victim against the ice shelf until further help arrives.

# Row Rescues

Yes, we are dealing with ice, but a row rescue simply means using a boat to get to the victim. Many people would call this a "go" rescue, but "go" rescues involve entering the actual environment in which the victim is trapped, and this is not an operations-level evolution. The difference is, if a rescue boat breaks the ice, the rescuers are protected by the boat's flotation.

## Boat Types

Basic types of watercraft are explored here briefly in light of ice rescue. They are also discussed in Chapter 2, *Watercraft Rescue Operations*. Not only does flotation

**FIGURE 9-22** An air-powered line gun can shoot a messenger line or an automatically inflated harness to a victim.
Courtesy of Steve Treinish.

need to be considered, in case the ice breaks, but also the fact that a deadweight, large victim may need to be wrestled into the boat by the rescuers. Quite simply, the less stable the boat, the more likely it is to capsize.

**LISTEN UP!**

As rescuers, we must deal with risk versus benefit decisions on every mission. Is a poor first-choice watercraft better than nothing? That is an on-scene call. If someone were drowning in front of rescuers, and all they had to work with was a small canoe, most would enter the hazard zone in that canoe. On the other hand, if the mission timeline or lack of evidence suggests no victim or one that is surely dead, rescuers should wait for a better watercraft choice. Preplanning your response and knowing what tools you will get when you ask are invaluable.

For many operations-level ice rescues, sliding the boat to the victim can offer speed and protection to the rescuers and provide a good base from which to work. A boat can be loaded and equipped with the tools rescuers need to handle different scenarios they may

**FIGURE 9-23** A johnboat equipped and ready for ice rescue.

Courtesy of Steve Treinish.

**FIGURE 9-24** Rescuers using poles to push the boat across the ice surface.

Courtesy of Steve Treinish.

encounter on the ice, and used to eliminate trips shuttling equipment to and from the shore. Flat-bottomed johnboats work the best, as the metal runners on most bottoms slide very easily (**FIGURE 9-23**). Johnboats offer adequate flotation, and the hull can take some abuse. Inflatables are an option, but are harder to slide across ice surfaces, especially slush. Compare a tennis shoe and an ice skate—which is easier to slide across slush? Canoes are very unstable, and if the ice surface breaks with the weight of the boat and crew (as it most likely will, at least closer to the victim), the sudden jarring of the canoe can flip it easily.

Propelling a boat on the ice eliminates most motor use; so plain old rescuer muscle will be needed to get it moving. Ropes can span shorter distances so that shore crews pull the rescuers into position, very similar to a two-line tether operation. This can be assisted by the rescuers using pike poles to push themselves and the boat across the ice.

> ### LISTEN UP!
>
> Rescuers should operate with a tagline attached from the shore to the boat at all times. Poles and oars or paddles can break or fall overboard, and if this happens without a rope attached, rescuers will be stuck. Rescuers should also be tethered to any watercraft with at least a short leash of webbing. Falling overboard is fixable, but falling overboard under an ice shelf with no direct link to safety might not be.

## Poling a Boat

**Poling a boat** is propelling it by using long poles, pike poles, or paddles to push it across the water or ice surface (**FIGURE 9-24**). The poles are pushed down into the bottom of the ice surface, and then pushed by the rescuer to effect motion. Hint: When poling across

ice, work in unison. Two rescuers working with different timing will usually end up moving inefficiently or not at all. A very small jump—just enough to take the weight off the feet—can start the boat moving, and once it starts moving, it moves much more easily than from a dead start.

Since the boat went onto the ice with a tagline, the trip out might be long and hard, but the trip in should be much faster. Consider traveling across the ice stern first (as in Figure 9-24). It is not impossible to travel bow first, but the curved shape of the bow keel may make it harder to push the boat up and onto an ice shelf without sliding back down when approaching from open water areas. When approaching the ice shelf stern first, the rescuer in the stern may shift his or her weight toward the bow to allow the stern to rise slightly above the ice shelf, and then the push or pull is made. Another advantage to this setup is with the tagline attached to the bow: the boat is in a position to go straight back to shore, without a 180-degree turn. If the ice was so bad it needed to be broken up to get to the victim, the path back will now be mostly water, so it should be quicker and smoother. This allows the rescuers to immediately focus attention on patient treatment.

## Operations-Level Support

Operations-level rescuers can do a lot to attempt a rescue from the shore or with a boat. However, because victims in icy waters may not be able to self-help, many ice rescue attempts will proceed directly to a technician-level response, with ice rescue technicians going out on the ice itself or into the water to get to a victim. A multipoint rescue may even be attempted, with operations-level efforts used in conjunction with an ongoing technician-level effort. Remember, it doesn't matter who gets there first; it is the rescue that counts. Operations-level personnel should be able to support the technician-level evolutions, usually by tending tether lines.

## Tending Tether Lines

Every ice rescuer must be tethered to some type of safety line, whether to shore, to a boat, or to another rescuer (**FIGURE 9-25**). Long-distance rescues, in which rescuers walk hundreds of feet, do occur, but many rescues will be tended from shore. Shore tenders must have a working knowledge of the hand signals used by the authority having jurisdiction (AHJ) and the limitations and abilities of the technician-level rescuer.

Some teams utilize an ice rescue technician staged between the shore and the victim to monitor the packaging and progress of the rescue. This rescuer can relay signals to shore if needed. However the signal reaches the shore crews, a slight delay in pulling will give the rescuers time to position the victim and themselves before the force of the pull is applied.

## Night Operations

Nighttime missions can present a challenge, especially if rescuers are operating in remote areas and not using radios. Consider calling for helicopter support for lighting, using aerial ladders for lighting, or using chemical sticks for hand signals. Chemical sticks or personal marker lights are also used to act as markers for each

**FIGURE 9-25** All rescuers must be tethered by rope to the shore, a boat, or another rescuer.
Courtesy of Lance Anderson/Peterborough This Week.

---

### LISTEN UP!

Be aware that some teams and agencies use the OK signal for "Go!" If a crew on the bank pulls on a rope on a mistaken signal, the rescuer may not be in a position for it, and there is a chance of injury. Three people on a bank can exert a lot of force—force that the rescuer may not be ready for. Many hand and body signals are in use. Different teams may use different motions for different signals, but here is a generic example that seems to work well.

- Stop: Raise one arm up with a clenched fist, and hold it until the stop occurs.
- Give slack: Raise arm and with a chopping motion, point to the direction of needed slack.
- Go: The typical lasso motion, rotating the arm in a circle above the head.
- OK? or OK!: The same signal many SCUBA teams use, a large circle formed with the arm, with the hand on the head.

---

rescuer in the area. They are normally clipped to a PFD, ice rescue suit, or helmet, and different colors can differentiate between crews or training levels. Flashlights also work for this purpose, but in the course of being bumped around they can be broken, and night vision is very easily disturbed with an accidental blinding by a high-powered light. This presents a problem for the on-ice rescuers who have well-meaning crewmembers shining lights at them trying to illuminate the area. When the rescuer looks back at the bank, those lights are right in his or her eyes.

### LISTEN UP!

Radios might not be used, but consider using handheld radios in drybags attached to an arm. If there is hesitation to put expensive radios on the water, then cheap, lightweight Talk-about–type radios in a 10-dollar drybag work very well. If they fail or get wet, the cost is minimal, but voice communication is still maintained on the mission.

## Hypothermic Victim Care

Hypothermia is the primary problem for the victim. The victim is already in a precarious state and must be evacuated not only from the cold water, but also from the cold air. The airway, breathing, and circulation (ABC) priorities apply, as they do with any victim. Nonbreathing patients must have airway interventions to allow rescue breathing, and cardiopulmonary resuscitation (CPR) must be started immediately if there is no pulse. If the victim has a pulse and respirations, rescuers must act quickly but gently. Treating a hypothermic patient too roughly can cause cardiac arrhythmias.

The next task on the list is to stop the patient from getting any colder. Providing blankets and removing wet clothes is OK, but what most truly hypothermic victims need is hospital intervention. Hospital personnel can rewarm patients much better than emergency medical services (EMS) can in the field, so a rapid transport is the best call. Consider helicopter evacuations in remote areas or in areas where transport times will be long.

**SAFETY TIP**

The following information is general care and should not supersede local EMS protocol or care. For good preplanning, consult the local medical director or protocol used by the AHJ.

Wet clothing should be removed from the victim, and the victim should be wrapped in blankets. Warm, humidified oxygen can be given; if intravenous fluids are given, they also should be warmed. Commercial oxygen and liquid warmers are available, but EMS crews can do a lot just by keeping the EMS supplies somewhere warm. It makes no sense to pump fluids that are chilled to 50 degrees into a patient who is already too cold. Rescuers may be colder than they thought, especially after the rescue excitement has ended, or if they ended up wet. They also need to be warmed, but will generally not be as cold as the victim. Warm trucks or buildings offer a place to get out of the cold, and hot liquids can be sipped to bring the body temperature back up. Most rescuers just need the proper environment to rehabilitate, and will be fine with that.

# Terminating an Ice Rescue Incident

As with any type of rescue, terminating an ice rescue incident can cause harm to rescuers, but generally, any equipment directly used in the rescue—such as ropes, ice awls, or even boats—make it to shore with the victim. This can reduce the risk if a simple, one-faceted rescue attempt is successfully used, but if multiple methods are used, or other rescuers are exposed to the water or temperature, they must be returned to safety. An accountability system can help ensure this takes place.

Ice rescue suits should be checked after use, particularly for tears and abrasions, as the ice can lacerate a suit when it is slid across sharp ice cracks or edges. Equipment and PPE should be washed with clean water with a mild detergent and dried completely inside and out before returning to storage. Nylon-coated ice rescue suits have removable liners that facilitate this task.

One issue that rescuers frequently discuss after a training or rescue evolution is the hole in the ice itself. Some are of the mind that marking the hole may incite curiosity from the public, possibly to the point where people venture onto the ice to get a closer look. Not marking the hole may result in a nuisance call at the best, or, at worst, a well-meaning civilian attempting a rescue, resulting in injury or death. Perhaps the best solution is to post signage at the water's edge advising of the activity and warning of the potential hazards. These signs are an inexpensive investment in public safety and education. Ensure all incident information is documented and reported appropriately. See Chapter 4, *Surface Water Rescue Operations*, for more information on terminating incidents.

# *After-Action* REVIEW

## IN SUMMARY

- Ice rescue to some departments consists of a quick crawl onto a farm pond and back, and ice rescue to others is a five-mile trip to an ice floe that can be acres large.
- The cardinal rule in dealing with any ice surface is to remember that it may fail at any time.
- From day to day, ice can be affected by forces of nature or by human-made influences, including warming temperatures, wind and wave action, underwater springs, recreation, and human-made holes in the ice.
- Ice incident victims usually do not have the training or education needed to save their own life, and often by the time the call comes in and the rescuers are deployed, the cold or fatigue will have removed some of their ability to help themselves.
- The biggest concerns in ice rescue are submersion and cold temperature, making PPE such as PFDs and ice rescue suits especially important.
- No rescuer should be working around the ice, even to support an operations-level rescue, without the knowledge of how to get out of the water should an accidental fall through the ice take place.
- Operations-level rescuers generally do not enter the actual environment the victim is in, but there are many ways to effect rescue by indirectly dealing with ice safely and efficiently, using throw bags, ring buoys, reach objects, line guns, hose inflator systems, and watercraft.
- Due to their stability and construction, johnboats are one of the best watercraft for reaching victims in an ice rescue.
- Operations-level rescuers can do a lot to attempt a rescue from the shore or with a boat. However, because victims in icy waters may not be able to self-help, many ice rescue attempts will proceed directly to a technician-level response.
- Hypothermia is the primary concern with ice rescue victims. Once the ABCs are managed, the next task is to keep the victim from getting any colder.
- Ice rescue suits should be checked after use, particularly for tears and abrasions, as the ice can lacerate a suit when it is slid across sharp ice cracks or edges.

## KEY TERMS

*Access Navigate for flashcards to test your key term knowledge.*

**Bubblers** Pumps that blow air or circulate water around structures or boats to keep ice from forming.

**Clear ice** Strong, transparent ice, with no rotting or melting. Strongest of all ice, it offers good working conditions if the thickness of the ice is matched to the function being performed on it.

**Conduction melting** Melting of ice around objects that carry heat into the water. A prime example is metal stairs being warmed by sunlight and melting the ice immediately around the metal; it may weaken surrounding ice, even in below-freezing temperatures.

**Drift ice** A large piece of ice not directly attached to a shore. Also called *floe ice.*

**Float coats** Coats or jackets that offer surface warmth and flotation in the same garment.

**Frazil ice** Ice that is in the formation stages and looks very ragged. Weak ice, but continued cold temperatures change it to clear ice.

**Hose inflator systems** Small kits comprising a fire hose cap, plug, and valve to allow inflation of a section of hose. This hose is then snaked or pushed out to a victim on the ice.

**Hypothermia** Condition in which one's body temperature is lower than 95 degrees Fahrenheit (35°C).

**Ice awls** Small, handheld tools used to dig into the ice to provide a grip that helps a rescuer pull himself or herself out of the water and onto the ice surface. Also called *picks.*

**Ice rescue suits** Coverall-type suits used by rescuers or workers around ice that provide warmth, flotation, and the ability to stay dry.

**Poling a boat** Using pike poles or paddles to push a boat across the hard ice surface.

**Reach tools** Any device for water rescue that can be extended to a person in the water so that he or she can grasp it and be pulled to safety without physically contacting the rescuer. (NFPA 1006)

**River ice** Ice formed on moving waters in creeks, streams, or rivers. It can be very hazardous to rescuers in that the water levels under the ice may vary, weakening it, and may subject rescuers to moving water under the ice surface.

**Rotten ice** The typical ice that rescuers have to deal with on missions. It can be several inches thick, yet still weak enough to fail without warning. It usually has been subjected to melting, sunlight, warmer temperatures, or other environmental factors to destabilize it.

**Self-rescue** Escaping or exiting a hazardous area under one's own power. (NFPA 1006)

**Snow ice** Ice with a layer of snow on top.

**Throw tools** Anything that can be thrown or launched to a victim to provide a means of flotation.

**Two-line tethering** An evolution with one piece of flotation equipment being tied between two lines. The lines are then used on opposite banks, with one bank pulling the flotation out to the victim.

# On Scene

As you arrive at the city park, you see a very young boy being whisked into the police cruiser. He is soaking wet, but crying hard. The officer states that she pulled up to see him stumbling on the shore, and her first reaction was to simply get him in the warm cruiser. The dispatch call was a reported unknown emergency, and the caller cannot be reached upon call-back. Upon questioning, dispatch advises the caller was a male who was very excited and convinced the child was drowning. The other civilians in the area did not call 911, and hikers, joggers, and walkers frequently use the area. The pond is about ½ acre, with rotten ice showing around the perimeter, except for some open water areas by the fountains the town installed a couple seasons ago. You see a dog struggling at the edge of the ice out by the fountain, but no one else. Your department has a small johnboat, but not much water rescue equipment with it.

**1.** As the first-arriving officer, what is your first priority?

  **A.** Ascertain whether anyone else might be in the water.

  **B.** Call for EMS and start warming the child.

  **C.** Attempt to get a helicopter flyover to check for signs of other people.

  **D.** Start mutual aid, as they have a well-trained technician-level team.

**2.** The child seems OK, and is about 8 years old. What information can he give you to help your decision-making process?

  **A.** Whether the dog on the ice is his

  **B.** Whether anyone else was with him

  **C.** Whether he went through the ice trying to get the dog

  **D.** Whether he heard or saw anyone else around before he fell through

**3.** At this point, what can be logically assumed?

  **A.** The dog appeared trapped, and the child probably went onto the ice to try to bring it back.

  **B.** A passerby probably called this in, but didn't want to be identified.

  **C.** There is a good chance of a second victim in the water or through the ice.

  **D.** If the law enforcement officer did not see any other evidence of anyone in the water, the incident can be contained with your on-scene resources.

**4.** Can the dog be rescued?

  **A.** Only by the Humane Society representatives that need to be dispatched.

  **B.** Only by the incoming team of technician-level rescuers.

  **C.** By your department, but only from the boat wearing acceptable PPE and maintaining safety.

  **D.** It depends on the conditions, your department's SOPs, and rescuer safety.

*Access Navigate to find answers to this On Scene, along with other resources such as an audiobook and TestPrep.*

# CHAPTER 10

## Technician Level

# Ice Rescue Technician

## KNOWLEDGE OBJECTIVES

After studying this chapter, you should be able to:

- Operate on or around ice for a rescue. (**NFPA 1006: 19.3.1, 19.3.2**, pp. 257–259, 261–262)
- Explain how to reach a victim on or around the ice. (**NFPA 1006: 19.3.1, 19.3.2**, pp. 262–265)
- Explain how to use rescue sleds. (**NFPA 1006: 19.3.1, 19.3.2**, pp. 264, 266)
- Explain how to travel long distance on ice. (**NFPA 1006: 19.3.1, 19.3.2**, pp. 266–268)
- Explain how to handle multiple victims at an ice rescue. (pp. 266–268)
- Describe how to perform an auto extrication on or under the ice. (p. 268)
- Explain the importance of realistic training. (pp. 268–270)

## SKILLS OBJECTIVES

After studying this chapter, you should be able to:

- Safely move along the ice to a victim or victim entry point. (**NFPA 1006: 19.3.1, 19.3.2**, pp. 258–259, 261–262)
- Attach a victim to a tether line. (**NFPA 1006: 19.3.2**, 263–265)
- Construct and use a 2:1 removal system for victim rescue. (**NFPA 1006: 19.3.2**, pp. 266–267)

# Technical Rescue Incident

After completing classes to become an ice rescue technician, you are working a shift at a quiet station when you receive a call to a local beach. This seems a bit strange, since it is mid-March, but you respond anyway. When you arrive, a news crew tells you that the news station's helicopter has spotted a dog wandering on the ice about ½ mile offshore, and it appears it cannot get back to shore. The news media seem to be very excited that they have stumbled on such a story, and are reporting live on Facebook.

You have trained and traveled around drift ice, and even though it is mid-March, the ice pack looks jumbled but still passable. A crowd has gathered in response to the news station also tweeting the story on social media, and some people are starting to murmur about starting their own rescue attempt. Your department does not have a policy on animal rescue for any technical rescue response.

**1.** Where do animals fit into the typical risk/benefit decision?

**2.** Will you attempt this rescue? If so, what tools will you take?

**3.** What means of communication can be used?

**4.** What are some rescue alternatives in this situation?

 *Access Navigate for more practice activities.*

## Introduction

Technician-level ice rescuers do not have many major tasks to learn. As a matter of fact, if operations-level rescuers learn enough about self-rescue and entry-level personal protective equipment (PPE), ice rescue is likely the easiest transition to a technician-level certification of all the water rescue subdisciplines. However, evolutions at this level can be fairly dangerous to the untrained or unprotected rescuer. After all, technician-level training prepares the rescuer for operating in or on the same hazard that has put the victim in danger. One prerequisite for technician-level certification at any discipline is being able to perform the operations-level tasks. Choosing the proper PPE is one of those tasks, so that should not be an issue. This chapter will assume the reader is already certified as an operations-level ice rescuer.

## Operations on or Around Ice

Although the ice rescue world can offer somewhat strange runs—anything from ATV crashes on ice to simple falls suffered by ice fishermen—the typical emergency that most ice rescuers will face is people who have fallen through thin or rotten ice. Just like the popular quote about garbage collectors never being surprised when they turn a corner and see garbage, good rescuers should never be surprised when they break through the ice themselves. After all, people don't call 911 because things went right. It's important to make this environment safer for rescuers.

## Rapid Intervention Crews

Before anyone ventures onto the ice, the rescue of rescuers must be planned and in place. Because the ice is already considered weak, frequent breakages will cost time if shore-based safety needs to assist the crews on ice. Therefore, it makes sense to cover ice technicians with additional ice technicians who can travel fast and do not rely as much on other rescuers, boats, or peripheral equipment. One thing to keep in mind is that even the rapid intervention crews (RICs) need to be tethered, so more rescuers on shore will be needed to control and support them.

## Personal Flotation Devices

The first ice rescue suits were modeled directly from cold-water immersion suits used by mariners and workers exposed to cold-water immersion threats. These suits are still made today, and offer a lot of inherent flotation due to the neoprene construction. Other models on the market today are made from a coated nylon fabric and use an inner liner for warmth. It is recommended that these vinyl-constructed suits be used with a Type III or Type V personal flotation device (PFD) to bolster buoyancy. Although most rescuers can float in these vinyl suits, some will sink. Add the extra dead weight of a victim with no flotation, and this could prove dangerous. Even the neoprene suits can be used with a PFD if desired, but rescuers will have to deal with the potential discomfort of material bunching up around the chest. This problem can be somewhat alleviated by using a plain Type III PFD with a clean front and buckles, not zippers. This allows the front harness connection on the suit to be used properly. Using Type V PFDs can result in more problems when moving or being pulled across the ice due to the tri-glide buckle and/or the gear compartments frequently found on them.

**FIGURE 10-1** Ice rescuer being tended with rope.
Courtesy of CESTA, Inc.

> ## LISTEN UP!
>
> Ice rescuers should be aware of the dangers of using ice rescue suits in moving water conditions. Ice rescue suits can be used to swim short distances of open water, but these swims should be conducted using a backstroke and/or swim fins. Forward crawl swimming in moving water may be required, but can easily fill the suit with water.

## Helmet Use

Some water rescue helmets may be frustrating to wear with an ice rescue suit, as the thermal head protection adds some thickness to the head. Wearing the hood off or undone can result in a suit full of water and lead to hypothermia. A good choice is to wear a helmet with a quick-adjusting headband, just like those found in fire suppression helmets. This lets the helmet fit different rescuers, and it keeps the hood in place. The ice is an unforgiving surface to strike, and slips and falls are inherent to the work. Rescuers should strive to reduce falls on the ice, and they should operate low to the ice to reduce injury risk from falls, but helmets remain key PPE.

## Tether Lines

One thing that is considered crucial in most "go" ice rescue evolutions is the use of a **tether line**. Most tether lines in ice rescue are ⁷⁄₁₆" or ½" in diameter and connected either to the front of the rescuer or to the equipment the rescuer is using to assist in the rescue (**FIGURE 10-1**). Ease of rescue and speed of return of the rescuer (and hopefully a rescued victim) are the two main reasons to use a tether. Rescuers on shore tend to this line, and every rescuer on the ice or

in the water should use or be attached to one. It is possible to tag into an existing rope for safety, or use one rope between two rescuers.

For example, when traversing large distances, rescuers can tag into each end of a throw bag and act as a safety to each other. Specifically constructed water rescue rope is the best line to use because it does not absorb a lot of water and it floats. Heavier kernmantle ropes may be used if nothing else is available, but they can become heavy when wet, and can flash freeze to the ice surface.

Just as when operating around water with any type of rope, another piece of highly recommended PPE is at least one cutting tool that can be reached, drawn, and used with one hand. Tethers can and do become tangled around ice pieces, trees, or other objects in the water. In some circumstances, they can drop under large items, and when the line is retrieved, the rescuer is pulled under where the rope is tangled. Even worse, a pumped-up shore crew pulling a rescuer in may inadvertently pull a rescuer underwater or against the ice shelf. You must be able to cut yourself free.

## Moving on the Ice Surface

Even though we humans like to walk upright, in many public safety scenarios that may not be the best way to get around. Hot fires make us crawl on our hands and knees; steep, slippery slopes make us use our hands and heels; and breaking or crumbling ice puts us on our bellies, or at least it should. It is possible to walk on thicker ice, and many rescues begin with rescuers walking to the general area of the victim. **Crampons**, which are metal spikes or cleats that can be attached to the soles of shoes or fins with straps or rubber harnesses, can help with traction and stability (**FIGURE 10-2**). However, standing or walking upright concentrates all the body weight of the rescuer in one small area on the ice, which often results in breakage. Rescuers should spread

**FIGURE 10-2** Crampons can help with traction and stability on the ice.
Courtesy of Steve Treinish.

out their body weight, either by adjusting their body position or by using some mechanical means to help them do so. The victim has already punched through the ice once, and from that point on the ice will only get weaker.

## Spreading Body Weight by Positioning

While it may not be required for the entire distance to be traveled, at some point rescuers will likely have to slide across the ice in order to stay on top of it instead of falling through it. Simply lying down on your stomach does two things. First, it spreads more body weight across more of the ice surface, which increases the weight-bearing area of the ice (**FIGURE 10-3**). Second, if the ice does fail, the rescuer is much closer to the water in this position, making the entry gentler and safer. A sudden fall from a standing position can really rattle your head if your chin hits the edge of the ice on the way down.

Another problem with falling through the ice is that the ice can act like a trap door, giving way and then floating back into place after the rescuer is submerged.

**FIGURE 10-3** Simply lying down on your stomach spreads more body weight across more of the ice surface.
Courtesy of CESTA, Inc.

If the ice starts to break while belly crawling, there are two options. First, the rescuer may be able to lunge forward and find better ice, and second, relax! With flotation from a PFD, you will not sink. Fighting it will only increase the odds of getting your face wet or submerged. Ice supports moving objects better than stationary ones, so the rescuer should keep moving if possible. Letting the feet and legs drop first reduces the possibility of inverting in the suit underwater. (The problems of suit inversion are discussed later in this chapter.) Often a rescue will begin with rescuers walking until they see or hear cracking ice, then they drop to the ice and slide the remainder of the distance.

## Spreading Weight Using Objects

The bigger the footprint employed on the ice, the better. There are a few ways to increase your footprint on the ice. Ring buoys, backboards, boogie boards, and even Stokes-style litters may be used to spread the weight of the victim and the rescuer across more surface area of the ice. Rapid-deploying inflatable crafts or paddleboards are easily carried in apparatus, and can deploy in under two minutes. These, as well as motorized watercraft, can also be used to traverse open water and ice, which is especially important where ice is fractured and rotten. Even larger boats or inflatable boats may be a faster way to the victim, and make travel easier by spreading the weight of the rescuers. A boat can also be used as a work platform while an in-water rescue is made. Ladders used to be a common method of spreading weight, but ladders sink in open water. If the entire ladder falls through the ice, one must be cautious of entangling a tether line in it.

Specialized **ice-rescue-pontoon–style sleds** are also available and work well (**FIGURE 10-4**). They employ two large floating pontoons to minimize the pressure

# Voice of Experience

Our department is part of the York County Advanced Technical Rescue Team and responds on all technical rescue incidents in our region. We responded to an incident during the winter months for a reported person who fell through the ice. En route to the incident, several updates were given confirming a subject fell through the ice, eventually went underwater, and did not resurface. On arrival, a location was determined for the breakthrough, and a quick recon was done to possibly find the subject just under water or the ice shelf and recover him for possible hypothermic resuscitation.

The "quick recon" turned into a six-hour event. Once we realized the subject was not in reach of our search tools, additional resources were needed. A dive team was dispatched, arrived over an hour later, and began their operation. Again, another component was added to the command structure and people who had never worked together seamlessly accomplished the task of a safe recovery. This was accomplished through training and following established safety guidelines. The subject was recovered and delivered to EMS and law enforcement officials once brought to shore. After a "hot wash," personnel gathered their equipment and returned to service.

This rescue scenario was unique in that it occurred along a state border. Two technical rescue components from different states were on site and had to integrate into one. Additional issues arose that could complicate the situation. A family member was on site and had to be kept informed while the incident continued. Also, the site of the rescue was a secure area near a power plant, so law enforcement and security were involved in all movements. As with any incident, media officials wanted to be close and obtain interviews.

You must make sure a strong command presence is established and objectives are clearly defined. Updates must occur to all personnel on site to keep them focused and aware of what is to occur. Make sure during a lengthy incident that you can rehab your personnel. During ice rescue incidents, you must be able to keep your personnel warm while staging. Rotate your teams on an established time frame. Make sure you have adequate personnel on site to do the job, and call for additional resources early. Communication is imperative. If calling for additional resources, be very clear what resources are needed. Do you need equipment or just trained personnel? You can congest a rescue area with apparatus that wasn't needed. Establish a staging area and have someone control it. Freelancing and wandering people are a recipe for confusion and possible injury.

There is more to ice rescue than sliding across an ice shelf, slinging a victim, and bringing him or her to shore. Your personnel must be trained, have proper PPE, know how to use specialized equipment, and understand the ICS. Safety is number one, and all precautions need to be taken to allow an incident to have a successful outcome.

**Anthony J. Myers**
Fire Chief
Shrewsbury Volunteer Fire Company
Shrewsbury, Pennsylvania

**FIGURE 10-4** A pontoon device that spreads the rescuer's weight and provides flotation is a great method of getting across the ice quickly.
Courtesy of Steve Treinish.

**FIGURE 10-5** With practice, ice rescuers can move quickly on the ice.
Courtesy of Steve Treinish.

on the ice, and they float if the ice breaks. The rescuer walks out holding onto two rails that rise up off the pontoons. For ice conditions that are so bad that the rescuer is unable to do this, a special paddle equipped with a pick is used to pull the sled and the rescuer across the ice. The paddle can also be used for propulsion across open water. Rescuers must be sure to use watercraft and other tools only as approved. These rescue crafts are quick to deploy, provide a minimal footprint on the ice, and offer a sling-type victim harness that, once the victim is extricated to the sled, helps keep the victim's airway above the water and ice, and therefore drier.

# How to Get from Point A to Point B

When rescuers are down on their stomachs, crawling on the ice, traction can become an issue, especially when the ice surface is wet. Ice awls are the best way to provide the traction needed to propel rescuers across the ice when on their bellies. (Ice awls are discussed in more detail in Chapter 9, *Ice Rescue Operations*.) It is possible to crawl and snake yourself across the ice but, with practice, a rescuer can make much better time with the "sticks." Again, motion is the key. Once forward motion is started, don't stop (**FIGURE 10-5**)! Use that momentum to keep going. Remember, it is easier to keep an object in motion than it is to start it, and moving objects tend to stay on top of the ice. Once you stop, the concentrated, static weight can cause the ice to sag and break.

With practice, ice rescuers can move quickly on the ice, and even move backward and sideways, too. Utilizing good technique with ice awls and belly sliding is much faster than walking on smooth ice. However, it may be slower and harder to move across slushy ice due to the increased drag of the suit across the

slush. The awls also make it easier to move large pieces of fractured ice out of the way when the ice will simply not hold body weight. Stabbing the ice and sweeping it back behind the rescuer clears a forward path.

## Swimming in Ice-Laden Water

A rescuer may have to do more than simply travel across ice to get to a victim. Perhaps a better path for victim access would be to cross a section of open water or to approach from a different direction while other rescue efforts are ongoing. Either way, it is possible to swim in ice rescue suits. The best swimming method is to lie in the water like you were going to do the backstroke, sweep your arms from up around your shoulders down to your hips, and glide. A forward crawl swim is not recommended due to the open neck area of the suits. Even when using the face guard, the suit will fill with water while doing a forward swim stroke. Also, hypothermia, a reduction in dexterity, and an inability to self-rescue onto the ice surface can occur. A rapid, thrashing motion will tire you out quickly, and every time you extend your arms to swim, drag is created, slowing any forward motion. Hold the tucked position even if just for a few seconds, and take advantage of the gliding action through the water.

**Drysuit Fins.** Drysuit fins, commonly used in SCUBA diving, can fit onto ice rescue suit boots and make a huge difference in both the speed of the swimmer and the amount of work needed to get to the victim (**FIGURE 10-6**). A forward walk when wearing fins can be hazardous due to the possibility of tripping, so on surfaces like shore and thicker ice, one should use a backward shuffling walk or else remove the fins altogether. (They are easy and quick to don.) A typical belly-down slide or crawl will have the fins splayed out behind the foot and they will not be in the way. A

**FIGURE 10-6** Drysuit fins.
Courtesy of Steve Treinish.

sharp crew will keep the swimmer concentrating on moving through the water by having the shore tender use his or her arms to direct the swimmer to swim left, right, or straight. This way, the swimmer does not have to constantly turn to look at his or her target. Another advantage to this position is that the face is elevated. This will help keep the swimmer drier.

### Ice Rescue Suit Inversion

One of the worst things that can happen when wearing an immersion suit is getting air trapped in the legs and flipping upside down underwater. Many rescuers fail to realize how much air these suits can hold in the legs when flipped upside down. Inverting a suit can happen during a fall into the water or when trying to control a victim. If the rescuer does not get flipped right side up, humans' limited air capacity will cause problems. If you find yourself upside down, you must roll into a tight ball and force your knees and legs back under you. Use your arms to roll yourself in the water and "burp" the suit when you get back up. A pull of the neck flap will help bleed the air off quickly.

**LISTEN UP!**

Righting an inverted suit is much harder than rescuers think, so it is a good idea to practice in a pool or warmer water with an instructor or fellow rescuer present.

**Burping a Suit. Burping an immersion suit** purges most of the air up and out through the neck. Some ice rescue suits are available with a one-way purge valve installed in the neck to allow this. If your suit does not have a purge valve, lean back and tilt your head back. This allows the space in the neck area to open up a bit, and the air will exit the suit. There will be a feeling of pressure on you now, but this is only the water pressure squeezing the suit around your body. Burping a suit on shore before and while entering the water can

**KNOWLEDGE CHECK**

What is the term for the metal spikes or cleats that can be attached to the soles of shoes for use in ice rescue?
a. Runners
b. Piolets
c. Crampons
d. Cleats

*Access your Navigate eBook for more Knowledge Check questions and answers.*

reduce the chances of inverting it and improve motion and control in the water.

# Reaching the Victim

All the work getting through the water and over the ice should result in reaching the victim. There are a few points to abide by. First, count on anything from the initial contact on as the victim needing flotation to stay above the water. Second, many hypothermic victims are unable to help themselves by the time they are reached, and rescuers must be ready for this.

## Initial Contact

Some victims may be found with their arms on the ice surface, sometimes with arms frozen to the ice shelf. This could be the result of a last-ditch effort by the victim to stay above water, and many agencies teach this. Laying the arms on the ice shelf and letting them freeze there may hold the victim in place even if the victim's strength is depleted so much that he or she cannot hold on any longer. Rescuers must be careful not to break this bond without having a good grip on the victim, or the victim can easily sink before rescuers are able to get a hand on him or her. For this reason, strive to make the initial approach from the side or the rear of the victim. Two rescuers can each take a side if they are on the ice together (**FIGURE 10-7**). Making approaches from two different sides also doubles the rescue effort, and if one rescuer falls through the ice, the other may not. Either way, the victim is, hopefully, reached faster.

**FIGURE 10-7** Two rescuers can each take a side if they are on the ice together.
Courtesy of Steve Treinish.

**FIGURE 10-8** Use your flotation to your benefit. Grab the victim under the arms, lie back, and simply float.

Courtesy of Steve Treinish.

When a straight-on approach is made, the ice frequently breaks just before the rescuer gets there. But ice floats, and even if an ice fracture occurs, the piece the victim is hanging onto may be large enough to buy the victim some time before he or she sinks. Consider yourself lucky if this happens and you are able to grasp the victim. Once the rescuer grabs onto or releases a victim from a hold, they must keep the victim positively buoyant!

Protecting the airway also becomes a prime concern. It does no good to reach a person who is still breathing only to allow the person's airway to submerge while untangling and attaching a harness. These deadweight victims feel heavy in the water, and it is hard to manipulate them. Remember, you are wearing the suit, and you have the flotation, so use it to your benefit. Grab the victim under the arms, lie back, and simply float (**FIGURE 10-8**). This technique gives the rescuer some control, protects the victim's airway, and allows the rescuer to use his or her hands for whatever task is next. If the victim is held vertically in front of the rescuer, this tends to pull the rescuer down and forward in the water, which will let water into the neck of the suit. And, the rescuer might find his or her mouth bobbing underwater, too.

## Attaching the Victim to a Tether Line

If ice rescue technicians are lucky enough to have water only three to four feet deep and a lightweight victim, they may be able to physically shove the victim up and onto the ice. Most water entries at these depths become self-rescues for the victim, and teams either don't get called, or they arrive to find a victim standing at the shore or somewhere warm. In deeper water, it is much harder to get the victim up onto the ice shelf, because any push up on the victim will push the rescuer down into the water.

If no sled or ice rescue equipment is available at the hole, the rescuer can simply attach the victim to the tether line and let the shore crews do the extrication. There are many commercial **rescue slings** on the market, but they all have one general trait in common: They wrap under the victim's armpits, attach back into the tether line with a carabiner, and are used to attach the victim to the tether rope while he or she is being moved. A sling can also be fashioned from 1-inch (25-mm) tubular webbing, although it will not float.

The typical tether line is set up with a figure eight knot on the rescuer's end, with a carabiner used to clip into the rescuer's suit or body harness. A quick and inexpensive way to attach a victim to the tether line is to tie a figure eight knot about seven feet (2 m) from the tail end and use a carabiner to clip in the end of a piece of webbing four to six feet (1.2–1.8 m) long. During the rescue, the webbing is wrapped around the victim, and another carabiner is used to clip the tail end back into the figure eight tied in the main line. This completes a loop around the victim.

**LISTEN UP!**

The bigger the carabiner used to complete the webbing loop, the easier it will be for the rescuer to clip it back into the figure eight knot. Between the suit gloves and a lack of visibility (the rescuer's view may be blocked if the victim is lying on the rescuer's chest in the water), the larger carabiners work best.

The reason the webbing is attached a distance from the end of the line is to space the victim from the rescuer. Spacing the victim from the rescuer just a foot or two can help spread out their combined weight. If the victim is attached directly to the rescuer, the total weight applied to the ice could be enough to break it not only at the edge of the hole, but also in other places on the way to shore that a single body could cross without breaking. There is not a lot the rescuer can do while being pulled by the shore crew, other than making sure that the victim does not come out of the loop. Trying to pull a pair of bodies across the ice when it constantly breaks is tough on both the rescuer and the victim, and it wastes a lot of time. It also subjects the airways of both people to unnecessary submersions. One inhalation of water, and the gasp reflex could be triggered, or aspiration could occur.

The rescuer, once the webbing loop is completed, gives the signal for the shore crew to pull the tether line, then grabs the victim down low around the crotch area or legs and tries to get the victim horizontal in the water, along or close to the same horizontal plane as the ice surface. The pull on the tether from the shore crews should allow the rescuer to maneuver the victim up and onto the ice. As the tether is pulled to shore, the rescuer brings his or her legs up parallel to the ice, the pull is transferred to the suit harness, and he or she is pulled onto the ice. Once this has occurred, the rescuer can let the shore crew pull him to shore. At this point the rescuer is in a position not only to spread the total weight on the ice, but also to grab the victim if the victim somehow comes out of the webbing. Generally, once the pair is up on the ice and moving, the shore crews take them all the way in with no trouble. To attach a victim to a tether line, follow the steps in **SKILL DRILL 10-1**.

This evolution takes a lot of practice by rescuers, but can be done quickly and efficiently on the ice.

## Specialized Floating Slings

Taking the webbing evolution a bit farther, adding flotation to the sling is entirely possible. Cinch slings are used in surface and swiftwater rescues, and also can be used in ice rescue. They are essentially floating versions of the webbing slings described previously. Cinch slings can be placed over a victim's head, and the victim can then be pulled to shore.

---

**LISTEN UP!**

Any time a victim is asked to move arms or hands that are grasping ice, it is possible for the victim to slide into the water or lose whatever grip they have. Rescuers must ensure that positive flotation is established before rigging any sling, webbing, or sled. Your victim may not be strong enough to keep himself or herself afloat.

---

## Ring Buoy Use

Rescuers may be faced with a "go" rescue with only a rope and standard apparatus equipment available. A ring buoy can be used in much the same fashion as detailed in Chapter 9, *Ice Rescue Operations*, where it is lashed to the end of a hose inflator. In this case, the rescuer takes the ring buoy directly to the victim. The ring buoy is placed between the victim and the ice edge, the victim grasps the ring or the ropes attached to it, and the ring is hauled to shore with the victim riding on top. The rescuer can enter the water to assist in bringing the victim horizontal with the ice, making it easier for both the victim to get on top of it and for crews to pull the ring buoy up onto the ice edge. If the ice gives way, good flotation is already under the victim, keeping the airway higher and hopefully drier.

One disadvantage of using the ring buoy is that the victim is not securely fastened to the ring, making it possible for the victim to fall off before reaching the shore. Unconscious or weak victims are especially risky to transport in this manner.

## Backboard Use

A typical backboard used in emergency medical services (EMS) can help spread body weight on the ice, and on clear ice a backboard can slide fairly easily across the surface. Plastic boards can be propelled with ice awls. Once the victim is reached, it is possible to load the victim onto the board and extricate, but it can be tricky to keep the victim on the board while working to get the victim out of the water and onto the ice. Most boards can be tethered to shore and retrieved by crews manning the tether rope. Victims can also be strapped to the board during transit to shore, but this takes time and keeps the victim immersed in cold water and clothing, increasing hypothermia risk.

---

**KNOWLEDGE CHECK**

Which is a disadvantage of using a ring buoy to remove a victim who has fallen through thin ice?

**a.** Additional flotation will be needed.
**b.** Two rescuers must enter the water.
**c.** A rope tether is required.
**d.** The victim is not securely fastened to the ring.

*Access your Navigate eBook* **for more Knowledge** *Check questions and answers.*

---

# Using Rescue Sleds

The use of a rescue-type sled can make life much easier for the rescuers. Similar to a backboard in ice rescue use, rescue sleds offer superb flotation, and a flat, smooth bottom or hard runners allow sleds to slide across the ice easily (**FIGURE 10-9**). Using an ice rescue sled helps keep the victim drier, keeps the mouth away from standing water, and allows speedier return to shore. Rescue sleds are generally used by laying the victim on the sled and attaching a harness or webbing system built into the sled. The rescuer lies behind the victim. Once the rescue crew has the victim secured on the sled, the shore crew can pull the entire group into shore. Stokes baskets can be used similarly—as a litter the victim lies on or is packaged in and returned to shore. The sleds generally have the advantage of having a way to run the tether line through the board, which makes pulling the victim up and out of the hole much easier. With Stokes baskets, the victim must be secured in the Stokes before it is pulled.

# SKILL DRILL 10-1
## Attaching a Victim to a Tether Line NFPA 1006: 19.3.2

**1** Lean back, supporting the victim and protecting the airway. Bring the harness around and clip it in.

**2** The rescuer is hauled up onto the shelf behind the victim, ready to be pulled to shore.

**3** As the shore crew hauls the lines, push/pull the victim onto the ice shelf.

Courtesy of Steve Treinish.

**FIGURE 10-9** Rescue sleds offer superb flotation, and a flat, smooth bottom or hard runners allow them to slide across the ice easily.
Courtesy of CESTA, Inc.

**FIGURE 10-10** Rescuers should be tethered to each other and spaced out so that no one puts too much pressure on the ice at any one spot.
Courtesy of Lance Anderson/Peterborough This Week.

# Long-Distance Travel on Ice

There is no magic number for how long a tether line can be. Ice conditions, rope size, traction on the ice, and strength of the rescuers all play into this. If the rescue or search calls for a crew to walk beyond the reach of a shore-based tether line, rescuers should be tethered to each other and spaced out in a line so that no one puts too much pressure on the ice at any one spot (**FIGURE 10-10**). This also provides a margin of safety, because if one rescuer breaks through, the others have a physical rope already attached to work with.

Each rescuer must have the proper PPE, including ice awls, so getting a rescuer out of the water should not be a problem. A typical load out of gear for a long-distance ice rescue would be PPE (including crampons for each rescuer), a sled or Stokes basket, a Halligan bar or ice anchor, a few throw bags, a PFD for the victim, an emergency warming blanket, and some basic rope rescue equipment to allow the building of a mechanical system if needed, perhaps just a pulley and a couple of spare carabiners. Pike poles or water jaws might also be added to the mix in case the victim submerges while rescuers are en route, and attempts must be made to rescue a victim from the bottom. Lights, radios, and signaling equipment also might be needed as conditions dictate.

## 2:1 Removal System

During long-distance rescues, groups of three to four rescuers will need to get to where the victim is, remove the victim, and transport him or her back to safety. Using three to four rescuers allows a two-rescuer approach to the victim, with the other two rescuers

**FIGURE 10-11** Ice screws can be used as anchors.
Courtesy of Steve Treinish.

providing pull strength and backup to the rescuers in operation. One way to remove someone with no direct tether line or to remove another rescuer with no footholds or grips is to build a quick mechanical advantage system on the ice. To use a 2:1 removal system, follow the steps in **SKILL DRILL 10-2**.

Rapid intervention crews (RICs) often use a similar system in large area rescues of downed fire fighters.

Ice screws can be used as anchors, and rescuers should consider having one or two in the hardware bag or pack (**FIGURE 10-11**). An **ice screw** is a piece of hardened steel that can be deployed in seconds by screwing it into the ice like an auger. For long-distance work, the use of an ice screw as a safety anchor or extrication device is a good addition to any cache.

# Multiple Victims

Dealing with multiple victims in ice rescues presents many challenges. Time is on no one's side with hypothermia, and rescuers on scene will have to do

# SKILL DRILL 10-2
## Using 2:1 Removal Systems NFPA 1006: 19.3.2

**1** Drive the pointed end of a Halligan bar into the ice surface 20 to 40 feet (6–12 m) from the hole. This becomes an anchor and/or a foothold for the hauling rescuer. Because the pull is coming from the same plane on which the ice is lying, even weaker ice may provide a decent anchor.

**2** A rope, usually from a throw bag brought out with the crew, is tied or attached to the Halligan bar at the forked end. The rope is then looped through a carabiner attached to the victim's harness, which then becomes a traveling pulley.

**3** The hauling rescuer sits behind the halligan, using it as a foot stop, and pulls the remainder of the line in, creating a 2:1 haul system. Once the victim is out, packaging can take place and evacuation can proceed.

Courtesy of Steve Treinish.

some quick triage. A victim who is still talking to rescuers and seems stable may need to wait until other victims who are in worse condition are extricated. Verbal coaching can be used to help victims assume better heat-retaining and flotation positions. Multiple rescue attempts must be coordinated, especially if originating from the same area on shore. It is easy to cross tether lines during complex rescues. Consider rescuer approach with a slightly longer crawl, but from a better angle, to reduce congestion at the ice hole.

It may be advantageous for the first rescuer at the edge to provide flotation in the water and wait for other ice rescue technicians to arrive and extricate. If multiple victims are observed, it may be beneficial to send the first rescuer with extra flotation devices such as PFDs or rescue buoys.

## Auto Extrication

Auto extrication on the ice or just under the ice surface can be a very difficult task. While not something many rescuers have faced, cars and trucks can end up on the ice after rollover accidents, sliding on slippery roads, or even suicide attempts by their drivers. Rescuers should train for the worst cases possible—once-in-a-career events that some folks think will never happen.

### SAFETY TIP

Using fire service extrication tools, including hydraulic rescue tools, can be very hazardous around or under the water surface. Just losing sight of purchase points and cutting locations can cause injury. Removing the vehicle from the water should be the first option, if possible. Consider requesting wrecker trucks or heavy rescues with winch capability early in the incident.

Many cars will be unstable on the ice shelf, and the ice may block some or all access to doors or windows (**FIGURE 10-12**). Winches, struts, comealongs, or rope systems might all be used to lend some stability to the vehicle. Once the vehicle has been stabilized, a chainsaw is useful in this situation, especially for thick ice. Ice will cut just like wood, but users must be aware of not only the saw, but the position of the tether line they are on. If hydraulic rescue tool (HRT) extrication is to be performed in the water, get a chainsaw in operation to make room to open or remove doors with the HRTs. When cutting the ice, though, remember

**FIGURE 10-12** Auto extrication on the ice or just under the ice surface can be very difficult.
Courtesy of Steve Treinish.

that if the car is not sitting firmly on the bottom, it stands a good chance of moving when the ice is cut. It is imperative to get the car stabilized.

Also consider cutting the ice away to allow room for a winch to pull the car out of the water, or at least to pull it closer to the shore to gain access. It is often difficult for a winch to make this sort of pull, because the car is often being pulled directly against the horizontal plane of the ice. It can be done by a well-trained, experienced operator. It also might be prudent to set an ice screw somewhere close, set a pulley with it, and run the rope back to shore. This creates a "clothesline" that can be used by the shore crew to shuttle equipment to the car. This saves a lot of travel by the technicians. Always tether the rescuers working on the ice in some fashion.

### LISTEN UP!

If the rescuer is working on the offshore side of the car, tether lines may drift under the car, and the signal to pull the line may drag the rescuer underwater. Slacking the rope at this point may not loosen the rope enough for the rescuer to come back up. Consider the angled gap created between the door and front fender area under the bottom hinge when a typical car door is opened. A rope can be pulled up and into this angle so tightly that it cannot easily be freed. If the rescuer submerges and this happens, the rescuer is in deep trouble. Keep in mind that this is the case not only for ice, but for any tethered water operation. Account for the rope!

## Realistic Training
### Weather-Related Factors

One of the challenges with ice rescue training is the lack of realistic ice and realistic victims. How many books or classes have shown a rescuer in an ice

**FIGURE 10-13** Consider using a water rescue mannequin for ice rescue training.

Courtesy of Steve Treinish.

rescue suit playing the role of a victim? Of course we must protect our people when they have to act as victims, but consider using a **water rescue mannequin** (**FIGURE 10-13**). Many are made of vinyl and sewn into the shape of a human, and they can be stuffed with old blankets or clothes and weighted a few pounds negatively buoyant. This not only makes them feel like dead weight on the surface, but in the water they sink like real victims. Once a rescuer lets a mannequin slip through his or her hands and watches it disappear underwater, it indelibly illustrates the point of maintaining victim contact.

The other problem is environmental. In today's time-constricted world, it is hard for some regions of the country to schedule ice training. Departments feeling manpower crunches or committed to other training requirements struggle to get good ice training due to scheduling conflicts. Some departments or companies are fortunate enough to be able to take advantage of the ice when it begins rotting and becomes more realistic. However, for many, some years it stays warmer than normal and nothing freezes, and some years it gets so cold the ice can be *too* thick.

For the days with super-thick ice, bring a chainsaw along on the training truck. Use it not only to open up the typical training hole, but also to score the ice in a few places on which rescuers will have to travel during the training. Scoring the ice is simply cutting lines in the ice to weaken it. The scored ice, while thick, will eventually break or sag, forcing the rescuer to get low and feel what it is like to go through the ice and into the water.

Another trick when the weather will be subfreezing for days at a time is to remove large sections of ice and then let the open water refreeze. Timing may be an issue, but generally, a new thickness of up to one to two inches will be found in 24 hours. This newer ice may fracture much faster than the thicker ice surrounding it, resulting in more realistic training conditions. Some locations may be affected by

heated runoff, or natural drainage, and these locations can potentially offer more realistic conditions by keeping the ice thinner. Marinas using bubblers often have open water areas, and do not have the current hazard that may be problematic with other locations. Retention ponds that drain large roofs or parking areas may also have better areas around piping locations.

**SAFETY TIP**

Moving water under or around ice can be fatal! All potential ice training sites must be investigated for hazards. If there are any doubts, err on the side of caution, and find a better, safer location. Power plants that have warm water discharges will likely have a water intake also, and conditions this creates must be observed for safety before practice occurs.

## Animal Rescue

Every winter, pets or other animals end up stranded on the ice, or are found struggling to get back to shore after falling through (**FIGURE 10-14**). These animals can range from small dogs to horses or livestock. And every winter, the fire service takes calls for these incidents. It is a commonly asked question: "Do we do anything?" The situation can be extremely emotional, especially for the public, who may not be educated about the dangers of ice. Some departments do not allow any response onto the ice unless there is a human life at risk. Another view of these incidents might be to treat them as a good public relations opportunity, and since we have the gear and training, we are much safer in attempting rescue than the civilians who are likely to venture

**FIGURE 10-14** Response to an animal stuck on the ice varies based on local regulations.

Courtesy of Andy Spitler.

out immediately after we leave. Frightened animals can injure rescuers, but many technicians view these runs as exceptionally realistic training. A dog that is instinctively paddling will certainly be struggling, require a side approach just like a human, and the "victim" must be securely held while the shore crews pull the line for retrieval. Often these situations should be considered on a case-by-case basis.

## Snare Devices

Rescuers can be better protected from animals by utilizing a snare. A snare is a loop of cable on a pole that can be looped over an animal's head from a short distance, cinched, and then used to pull the animal in while maintaining space between the rescuer and the animal. Using snares can also keep the rescuer farther away from the edge of the ice.

# *After-Action* REVIEW

## IN SUMMARY

- Technician-level training prepares the rescuer for operating in or on the same hazard that has put the victim in danger.
- The typical emergency most ice rescuers will face is people who have fallen through thin or rotten ice.
- Strive to make the initial approach from the side or the rear of the victim. Two rescuers can each take a side if they are on the ice together.
- Commercial harness systems can be used to assist in victim extrication. They wrap under the victim's armpits, attach back into the tether line with a carabiner, and are used to attach the victim to the tether rope while he or she is being moved.
- Using an ice rescue sled helps keep the victim drier, keeps the mouth away from any standing water, and allows speedier returns to shore.
- Ice conditions, rope size, traction on the ice, and strength of the rescuers all play into how long a tether line can be.

- Cars and trucks can end up on the ice after rollover accidents, sliding on slippery roads, or even suicide attempts by their drivers. Many cars will be unstable on the ice shelf, and the ice may block some or all access to doors or windows.
- One of the challenges with ice rescue training is the lack of realistic ice and realistic victims.

## KEY TERMS

*Access Navigate* **for flashcards to test your key term knowledge.**

**Burping an immersion suit**  Squeezing the air from the lower to the higher areas of the suit, releasing trapped air from the neck area. This reduces the chance of a suit inversion in the water.

**Crampons**  Small spikes that can be attached to the sole of a boot or fin to provide traction to walking rescuers on ice.

**Drysuit fins**  Standard SCUBA fins, but with a larger foot pocket to allow larger boots to be used. Very effective with an immersion suit when swimming in open water.

**Ice-rescue-pontoon–style sleds**  Small, ice-specific craft pushed onto the ice by rescuers. It spreads the weight effectively and lets rescuers operate with flotation, which is built into the pontoon.

**Ice screw**  A small device twisted into the ice surface to be used an as anchor. Very quick to deploy and offers good strength in fairly thin ice.

**Rescue sling**  An assembly used to haul ice rescue victims up and across the ice, with some commercial models offering built-in flotation.

**Tether line** Rope used to physically connect the rescuer to the shore, with control of the rope by the shore crew. Used to pull the ice rescuer back to shore when signaled to, and to help remove the victim and rescuer from the ice.

**Water rescue mannequin** A training prop with the flotation characteristics adjusted to provide more realistic training. It will sink if contact is lost by the rescuer.

# On Scene

Your rescue company has been dispatched to a local park for a report of a person in the water. En route, the dispatcher informs you there is at least one person in the water. You know from preplanning this park that it can be heavily used, especially in late spring when the air temperatures start to warm up a bit. This is one of those days, and folks simply have cabin fever. You arrive to find one child 30 yards offshore, clinging to the ice shelf. The ice is fractured from the shoreline to about 15 yards out, and then looks unbroken. The police officer on scene says no one made themselves known when she arrived, and when asked, the dispatcher says the caller is not answering.

The water is within easy eyesight of the walking trail. You have two fellow fire fighters on board, one of whom is trained to the ice rescue technician level, and one to the operations level. There are multiple sets of ice rescue gear on the truck.

**1.** A quick and logical glance at this scene may indicate which of the following?

**A.** The possibility of multiple victims, including one underwater

**B.** A quick and fairly simple rescue

**C.** A need for multiple rescue teams, including dive capable teams

**D.** More resources will be required, including an aerial view.

**2.** You are stretched thin, but your first priority is:

**A.** establishing command and denying entry to the hot zone.

**B.** starting a dive team response.

**C.** the life visible at the ice edge.

**D.** trying to find witnesses to confirm or dispel a second victim.

**3.** An operations-level rescuer could be expected to function, at least temporarily, as an incident commander.

**A.** True

**B.** False

**4.** Based on conditions presented, you decide the best move is:

**A.** to set up a fixed command on the shore closest to the trapped child.

**B.** to try to use a hose inflator system to push a hose to the child.

**C.** to place your operations rescuer in service as a tender and perform a "go" rescue with your other technician.

**D.** to send one technician on a tether with the operations rescuer tending and to perform Incident Commander duties yourself.

*Access Navigate to find answers to this On Scene, along with other resources such as an audiobook and TestPrep.*

# CHAPTER 11

## Operations Level

# Surf Rescue Operations

## KNOWLEDGE OBJECTIVES

After studying this chapter, you should be able to:

- Describe the surf environment and how to prepare for it. (**NFPA 1006: 20.2.1, 20.2.2, 20.2.3**, pp. 273–275)

- Identify and describe the components of open water. (**NFPA 1006: 20.2.2, 20.2.4**, pp. 275–278)

- Identify and describe surf features. (**NFPA 1006: 20.2.2, 20.2.4**, pp. 278–281)

- Describe watercraft operations in the surf environment. (**NFPA 1006: 20.2.4**, pp. 281–282)

- Explain the use of motorized rescue watercraft at surf incidents. (**NFPA 1006: 20.2.4**, pp. 282–285)

- Explain the use of nonmotorized rescue watercraft at surf incidents. (**NFPA 1006: 20.2.3**, pp. 285–286)

- Describe the management of conscious and unconscious victims. (pp. 286, 288–290)

- Describe how to support technician-level surf rescuers. (p. 288)

- Terminate a surf rescue incident. (**NFPA 1006: 20.2.5**, p. 288)

## SKILLS OBJECTIVES

After studying this chapter, you should be able to:

- Communicate with other rescuers via radio, hand signals, and any other means used in your jurisdiction. (**NFPA 1006: 20.2.3; 20.2.4**, p. 281)

- Extricate a victim using a tow sled. (**NFPA 1006: 20.2.4**, pp. 283–284)

- Operate motorized watercraft in the surf environment. (**NFPA 1006: 20.2.4**, pp. 282–285)

- Operate nonmotorized watercraft in the surf environment. (**NFPA 1006: 20.2.3**, pp. 285–286)

# Technical Rescue Incident

Your department, based on the shores of Lake Huron, receives a call of a person in distress at the local marina. You are thinking it might be a medical issue inside the harbor, because the call details are pretty sketchy and dispatch has no other information. The weather today is certainly "nautical," with gusty winds to 30 mph and air temperatures in the 50s. The manager, who meets you at the gate, points to the break wall where several marina customers are watching something offshore. He states that someone is in the water a quarter mile offshore, in waves that are three to four feet tall. No boat traffic has left the harbor, but he did see several personal watercraft (PWC) launch from the public boat ramp a few blocks away. Wave conditions like this frequently attract hard-core PWC enthusiasts.

You park your apparatus close to the break wall and climb onto your hose bed to get a good view. You see one PWC drifting with no operator, one PWC barely afloat, and a person in the water waving to the shore. You cannot hear him directly, but think he is yelling for help. Using the binoculars from the truck, you see he is holding another person afloat, but unconscious. They are both wearing PFDs and wetsuits. There is normally a watercraft patrol boat somewhere in the vicinity, but you do not see one at this time. The marina manager offers to take a boat out to them with your crew.

**1.** At this point, what is the best next move?

**2.** What must be considered about the manager's offer to assist with a boat?

**3.** What type of information can be given to the responding emergency medical services (EMS) crews?

Access Navigate for more practice activities.

# Introduction

Surf rescue has evolved from a public safety effort that once saw many civilian casualties each year to a highly specialized water rescue field; this has resulted in thousands of saved lives each year. Surf rescuers have been wise enough to adopt and utilize new technology and equipment, much of which has been developed from various water and surf sports found right on the beach. These surf rescue chapters are not meant to encompass a true ocean lifeguard course. Instead, these chapters examine some of the techniques, equipment, and hazards found in the surf rescue field. The United States Lifesaving Association listed 82,000 documented saves in 2016. Of course, lifeguards on duty at beaches performed most of these, but that number alone should suggest the fire and rescue service could be pressed into duty at any time, especially at areas that do not have lifeguard protection.

# The Surf Environment

Humans have been drawn to water for thousands of years. The body depends on it, transportation and trade use it, and for many of us, water areas are a fun and relaxing place in which to play. One of the requirements in NFPA 1006, *Standard for Technical Rescue Personnel Professional Qualifications*, is the ability to evaluate and produce a site survey, including different water levels, hazards, and the users of that water. While that sounds fairly straightforward, consider how many surfers congregate in an area showing higher than average surf when a storm is somewhere nearby. In some areas, they flock to it by the hundreds (**FIGURE 11-1**). In California, a location called Mavericks draws hundreds of people just to watch the surfers, who catch waves that are 20 to 30 feet (6–9 m) high. These giant waves are caused by a specific response to storms and swells offshore and only occur a dozen or so times each year. This does two things to the site survey: It identifies a known population increase in and along water showing tremendous power and therefore hazards, and it provides some of the toughest, largest surf in which to operate. These surfers tend to be at the professional level and bring in their own rescue boats, PWCs, and even helicopters to shoot video. But what if those support folks, or even the bystanders, call for help? What is the plan then?

The range of the customers around the surf is broad, from body boarders to babies, to local citizens, to tourists from across the world who may not speak

**FIGURE 11-1** Open water is any large body of water, and it can hold large numbers of people and a host of different activities.
Courtesy of Steve Treinish.

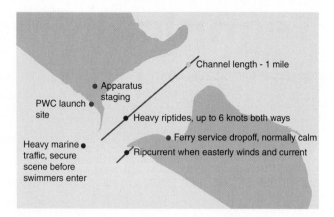

**FIGURE 11-2** A site survey for known hazards or heavily used locations should be performed.
© Jones & Bartlett Learning.

the same language you do. It all adds up to quite a challenge for many rescue organizations.

## Site Surveys

Coastlines attract not only recreational users, such as surfers, body boarders, fishermen, boaters, and beachcombers, but also transients or homeless people. Given the challenges of surf rescue in general, a preplan of the rescue is an excellent undertaking (**FIGURE 11-2**). Not only should low and high surf and tide conditions be noted, but also things like boat ramps, access to the beach or shoreline areas, optimal entry and exit areas of the water, and any specific hazards in the location. For example, during summer months, a beach with a rock jetty a few hundred yards away may be docile most of the time. However, in winter, the prevailing surge directions may change and be affected by storms, causing rip currents or longshore currents along the jetty. The hazard for both rescuer and victim is now increased, and perhaps watercraft should be the first line of rescue, rather than swimming or rescue boards.

### SAFETY TIP

When planning responses around areas with development, consider that docks and piers often have electric power installed on or under the structure. These power lines, especially if the structure is damaged or not maintained properly, could potentially cause the water in the area to become charged, resulting in electrocution. High water periods may also bring the water up to the power line level. Lock out/tag out procedures may be called into service before a rescue commences.

### LISTEN UP!

The general rule in surf rescue is getting flotation to victims, whether a rescue flotation device, a rescue board, watercraft, or even a rescuer. Once that is accomplished, especially with conscious and uninjured victims, the extraction phase may be done at a different location, with a significant reduction in urgency.

## Rescuer Personal Protective Equipment

Certain personal protective equipment (PPE) common to surf rescuers can go a long way toward protecting operations rescuers if they fall off or lose control of their rescue board. Remember, the rescuer must be able to save himself or herself if they enter the water. One might think a PFD would be in this mix, but in heavy surf, a PFD can actually increase the risk of injury by eliminating the ability of the swimmer to dive under a plunging wave, reducing swim effectiveness, and exposing the swimmer to the power and weight of the wave. This is another reason rescuers must have extremely good swimming and survival skills.

### Swim Fins

**Swim fins** used in surf rescue are shorter, fit either foot, and should float in case of loss (**FIGURE 11-3**). They can greatly increase swimming effectiveness in case of accidental entry. A nonadjustable heel strap works better, because there is no chance of a buckle breaking or loosening, causing the fin to be lost. Swim fins can also offer some protection around rocks or cliffs.

**FIGURE 11-3** Swim fins, a rescue tube, and a rescue buoy.
Courtesy of Steve Treinish.

**FIGURE 11-4** Physical fitness and swimming ability are paramount in the water, and many AHJs test for skills and ability in the surf environment.
Courtesy of Steve Treinish.

## Rescue Flotation Devices

Two main types of **rescue flotation devices (RFDs)** are common to surf rescue. A rescue buoy is made of hard plastic and resembles a torpedo, and a rescue tube is a longer piece of foam covered in vinyl (see Figure 11-3). Both comprise a harness, leash or lanyard, and the float. The RFD is the throw bag of surf rescue. RFDs are valuable for both self and victim rescue, and ensuring one is in your possession may save your life in the event of an accident. In heavy surf or plunging waves, the tube will result in less injury if it strikes a person, given the soft foam compared to the hard plastic of the rescue buoy. An RFD can easily be worn and trailed behind the operator, or attached to a kayak or rescue board, providing safety and protection in case of an accident or losing the watercraft.

## Rescuer Fitness

Rescuers undertaking surf rescue training or missions must be powerful swimmers and absolutely comfortable in the water. A 500-yard swim in less than 10 minutes is required by many agencies, and other skills such as treading water and underwater swimming can be added by the authority having jurisdiction (AHJ) as deemed necessary. Even operations-level rescuers committing to an operation in low surf with flotation aids or

paddling a kayak or rescue board may end up in the water and need to save themselves in extreme conditions. Many AHJs now require operations-level rescuers to demonstrate survival skills in expected surf conditions in case of a capsized watercraft or fall off of any nonmotorized watercraft they may utilize. Similar to structural fire fighters, physical training sessions to maintain fitness are encouraged and even required by some AHJs (**FIGURE 11-4**).

## Open Water Makeup

Most surf agencies define **open water** as any large natural water environment. This makes sense, as waves and surf are created primarily when the wind blows across large expanses of open water. The higher the wind velocity, the larger the waves that are kicked up, and these waves can travel hundreds of miles before they reach shore. There is not a lot offshore that can absorb the energy created by this action, so wave travel from distant storms is frequently seen. This is one reason many surfers are great at using weather formations hundreds or thousands of miles away to predict conditions at their favorite spot.

This open water environment also attracts huge numbers of people to the shores or the water in the form of swimmers, fishermen, vacationers, boaters, and divers. Many people underestimate the power and the danger of the water, and because they may be enjoying the water only knee or waist deep, use of lifejackets or other flotation is not as common as it might be offshore, increasing the chances that rescuers may be needed.

## Surf Line

The **surf line** is the line where swells start to build and break (**FIGURE 11-5**). Surf lines are important

**FIGURE 11-5** The surf line is where swells come into contact with an obstruction or shallower bottoms and change their makeup. Swells are seen toward the top of the picture, with no white areas of agitated water.

Courtesy of Steve Treinish.

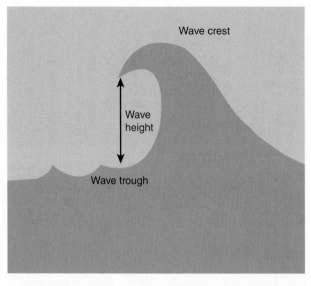

**FIGURE 11-6** Wave height is the distance between the trough of the wave and the crest of the wave just before it breaks.

© Jones & Bartlett Learning.

to rescuers because swells, while their heights change vertically, are not as hard to handle as the breaking surf. Beyond the surf line, the waves are not breaking and crashing around as much, so conditions are usually easier to manage. This means the victim might be better off being transported by rescuers to an area farther offshore to await help, or to escape much more hostile conditions in the surf line, especially when working in conjunction with other rescuers in boats or helicopters.

## Wave Makeup

The National Fire Protection Association (NFPA) defines various surf rescue terms in its NFPA 1006 standard. **Low surf** is considered to be normal for the area considered, and **high surf** is when waves are larger than normal for the same area. Wave heights are measured from the bottom point, or **trough**, of the wave up to the top, or **crest**, of the wave just before it breaks (**FIGURE 11-6**).

Judging wave heights can be difficult, but it can be helpful to use references nearby, such as a boat in the area, or swimmers. One effective way to estimate wave height is to preplan the area and compare known wave heights at the time of preplanning to what you are seeing now.

Some other terms that might come in handy are:

- **Wave period**, or length: The time between two passing wave crests.
- **Wave train**: Waves from the same energy source, showing similar tendencies and timings.
- **Set waves**: A group of waves that are larger than the others, formed when waves mingle and fuse

offshore. Water and wind currents, different meteorological areas, and shore formations can all affect wave sets.
- **Lull**: The time between wave trains, presenting the best time for swimmers or paddlers to get through the impact zone.
- **Rogue waves**: Waves that are formed when conditions and forces in the water intermix and create a wave much larger than the others in the train. Rogue waves can be extremely dangerous, and rescuers must be aware and on the lookout for them.

**LISTEN UP!**

Wave sets are an important feature for boat-based rescuers in that by working between large wave sets, a well-timed and rapid rescue attempt may take place.

## Wave Action

Waves break when the traveling energy, in the form of a swell, finally comes across something to change the shape of it. A **swell** is a large, hump-shaped pillar of water that resembles more of a sand dune than the image of the typical wave many people imagine. Hitting shallower bottom surfaces can cause the wave to rise, and when the wave cannot hold its form, it collapses on itself while still traveling forward (**FIGURE 11-7**).

The rising bottom can make a big difference in the type of wave and its characteristics. The place where waves are breaking is considered the **surf zone**. This is

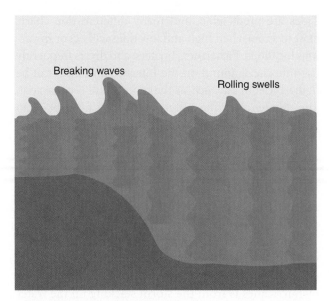

**FIGURE 11-7** Breaking waves are formed when swells move into shallower water, and the wave builds and falls onto itself.
© Jones & Bartlett Learning.

**FIGURE 11-8** Spilling waves simply wash up on the shoreline, making a lot of foam from the agitated water, but they do not crash into the water or shore from an elevated position.
Courtesy of Steve Treinish.

**FIGURE 11-9** Plunging waves present a lot of energy from above, when the wave actually breaks ahead and falls down on itself.
Courtesy of Steve Treinish.

the area of greatest hazard as the surf can hide or injure a victim, force them underwater, or the returning water can sweep them out to an open area. Swells can also have smaller wave features on them, such as smaller chop or breaking waves created by wind, or current differentials. Tides running at different directions than the wind also can greatly change wave makeup.

## Wave Types

Working in the surf exposes the rescuer to some of the most powerful forms water can take. Different wave makeup will make life easier or harder on us, and if not handled properly, could prove fatal. Let's take a look at the main types of waves we may deal with, from the fairly benign ones to the more hazardous ones.

## Spilling Waves

**Spilling waves** are waves that flow up and onto a gently sloping, mostly unobstructed surface, such as a beach that gradually gets shallower as the waves travel inshore (**FIGURE 11-8**). As the wave rises, the crest of the wave rises slowly enough that the wave simply collapses gently on itself, and it spills into the shore where

it fizzles out. A small amount of air is mixed into this collapse, creating a line of foamy water. The picture of a gentle, tropical beach usually shows spilling waves.

## Plunging Waves

**Plunging waves** are much more dramatic and are what most people associate with the sport of surfing (**FIGURE 11-9**). The wave passes over the bottom ground much faster, forcing the wave up and over onto itself so fast that the crest collapses on the wave. These waves mix a lot of air and water to form lots of foam, and can slam a person down into the water or onto the bottom surface with incredible force. This also compounds rescue problems because these forces can easily injure rescuers and victims.

**Impact Zone.** The area where these waves actually plunge is referred to as the **impact zone**. It is common

**FIGURE 11-10** Surging waves do not have any obstructions or bottom changes to dissipate their energy, and can strike cliffs or seawalls with enormous power, making water-based rescues very hazardous.
Courtesy of Steve Treinish.

to have swells developing into plunging waves, then changing into spilling waves after some of the wave's energy is dispelled, but operations in the impact zone can be harmful and dangerous to rescuers and victims alike. A surf rescuer must strive to remove the victim either closer to shore or farther offshore, but remaining in the impact zone increases the likelihood of injury or death.

## Surging Waves

**Surging waves** are waves that do not break, but instead are pushed into rocks, cliffs, seawalls, or other large objects (**FIGURE 11-10**). The energy does not fizzle out like on a beach, but rather gets bounced into other waves and objects, making the water choppy and rough. The huge, spectacular scenes of ocean water meeting rocky shores and spraying high into the air are surging waves.

Surging waves are also where much of the exciting video of life-and-death rescues comes from. Suicides are common in these spots because of the cliff heights, and intentional or accidental falls require a longer response time to get to the victim. In fact, many of these waves are so dangerous that rescue crews often attempt to rappel to the victim from the cliff, rather than try to get close to the rocks or cliffs from the water. The water smashing the cliff will pound anything that gets in the way.

## Tides

The gravitational pull of the moon creates tidal flow. This means the areas around oceans and seas have both high and low water levels that change twice per day. These changes take place because of the moon's pull on the water, which moves it to different areas of the ocean.

Tides are predictable, and many different tide charts exist to show when high and low tides will occur in various locations. Fishermen, boaters, and divers frequently use tide charts found on the Internet, and tides can be figured well in advance, as the moon's pull is consistent. The difference in the water level varies from place to place and time to time as the moon orbits the earth.

## Storm Surge

Storms cannot be predicted as easily as tides. Large storm areas cause **storm surge**, which is when water is pushed into a landmass as a result of a storm's energy. This surge can be several feet deep. Couple a big storm surge with a high tide, and the surf may increase considerably in size and force. Much of the damage done by hurricanes is from the storm stacking up the water blown into shore areas in the form of storm surge, rather than from wind.

---

**KNOWLEDGE CHECK**

The _____ line is where swells start to build and break.

a. boil
b. surge
c. swell
d. surf

*Access your Navigate eBook for more Knowledge Check questions and answers.*

---

# Other Surf Features

All this wave energy dissipating on the shore also leads to other problems. Some people, such as surfers and body boarders, like to harness this energy for recreation and entertainment. Many feel that the bigger the waves, the better. Others may flirt with the ocean at the water's edge and walk the beach while staying clear of the water. The people unfamiliar with waves and shores may get swept off their feet and then swept to sea.

To assist victims, operations-level rescuers must be proficient in emergency swimming, able to complete a distance swim with any stroke, using no surf rescue aids. Operations-level rescuers must also have a good working knowledge of the surf environment in order to be able to save themselves, particularly from features described in the following sections.

## Upwash and Downwash

**Upwash** is the rush of water coming up the shore from a wave after it has broken. This can be a pleasant stream of water that toddlers can handle, or a rush of water so forceful that adults cannot stay upright even if braced and ready for it. Stronger upwash presents the possibility that people could lose their footing and possibly be pulled back into deeper water or injured when the subsequent waves crash down on them. Knee and leg injuries are also possible, affecting swimming ability if deeper water is encountered.

Of course, if there is an upwash, there also must be a **downwash**, which is the water draining back off the shore. Downwash can be somewhat gentler because the sand can absorb some of it, but even then it can knock people off balance. Downwashes can be worsened by the angle of the beach, as steeper slopes create a faster flowing and more forceful downwash. The term *undertow* implies that current action will suck a person underwater, but there are no known current formations capable of this. A person may be swept offshore by downwash or a rip current, and submerge not from downward current action, but from a lack of swimming ability.

## Rip Currents

The United States Lifesaving Association reports 80% of all lifeguard responses are related to rip currents. **Rip currents**, or "rips," are the result of the waves or currents pushing a great deal of water up to the shore. The water must go somewhere, and a rip current is this water draining back to the open water from which it came on the surface of the water (**FIGURE 11-11**). Rips can be ferocious and much stronger than even a skilled swimmer. They can be identified by the eye as a foamy or stained area on the water, with reduced wave-breaking action, but this is not always the case.

**FIGURE 11-11** Rip current problems make up the majority of surf rescues, and occur when water piled up onshore drains back out to the open water.
Courtesy of NOAA.

**FIGURE 11-12** Even though signs are commonly posted along high-hazard rip current areas, they are frequently ignored, and many rip currents occur in places with no direct supervision.
Courtesy of Steve Treinish.

Rip currents are composed of feeders (the water that is supplying the rip), the neck (which is the water physically traveling back out to the ocean), and the head (where the neck current dissipates). Rip currents can form when the incoming water hits the shore, travels laterally with it, and flows back offshore along a pier or shore break. Another key place for rip current formation are sandbars found just off beach areas. Enough water can be pushed past the sandbars that, eventually, the water will find a drainage point in a low area of the sandbar. Once it does, the rip current forms.

Rip currents are also somewhat predictable, and many agencies are familiar with locations that present them on a regular basis. A large problem is that much of the general public, especially tourists, are not aware of where the rip currents may be found, so they venture into rip current areas without knowing or recognizing the power or the danger. In addition, there are always the "doesn't-apply-to-me" types who simply disregard warning signs (**FIGURE 11-12**).

## Dealing with Rip Currents

Many surf rescue problems arise when people do not know how to handle getting caught in a rip current. Most people will immediately turn and try to swim to shore. Because this is where all the water is draining back offshore, swimming toward shore is akin to running on a treadmill. Hence, they make little forward progress and also get very tired, very quickly. Keep in mind that many rips will be large enough not to be obvious when swimmers are in them, so a point of view must be maintained. If you are struggling to swim in to shore and it does not seem like progress is being made, or worse, the shore is getting smaller, you are likely in a rip current. To escape a rip current, swimming parallel to the shore until out of the rip current is the best option.

Most rip currents subside as the water deepens, often just past the plunging waves. Knowledge provides options in the water. A better option to preserve strength, especially for weaker or fatigued swimmers, may be to simply ride the rip until it dissipates. The focus can be on maintaining flotation, situational awareness, and signaling for help. Just like in swiftwater, you rarely beat the current, so saving energy and maintaining awareness will go a long way toward self-rescue. Rescue swimmers or boaters can use a rip current to their advantage by using it to get out to the victim faster. This is discussed further in Chapter 12, *Surf Rescue Technician*.

## Rip Tides

When water is pushed through channels such as harbor entrances, rivers, or between islands, it can travel faster due to tidal action. **Rip tides** are simply pulling water through a narrower space, and rescuers can use this to their advantage due to the fact that tides are predictable. Rescuers still must be able to forecast victim and rescuer movement, but normally this movement is well-known.

## Longshore Currents

Another hazard is **longshore currents**. Thinking *along the shore* will give a good idea of what these currents entail. Wave action dumps the water along the beach, and if the waves are approaching at an angle, this water may travel laterally along the shore (**FIGURE 11-13**). At the right angle, they can move along fairly quickly, but for most swimmers, getting back to land from a longshore current is not as difficult as returning from a rip would be, because they are now swimming across a current, rather than straight against it. This is similar to ferry swimming in swiftwater. You'll get back to shore, but it will likely be in a different place than you planned.

The larger problem that longshore currents may present is that they may be feeding a rip current, especially if the longshore current is approaching a jetty or land that juts offshore.

## Inshore Holes

**Inshore holes** are holes in the bottom surface, normally sand or fine gravel that are hollowed out by wave action or currents. Holes are dangerous because they may be present in shallow areas, even in water only two to three feet deep, and children playing in the area

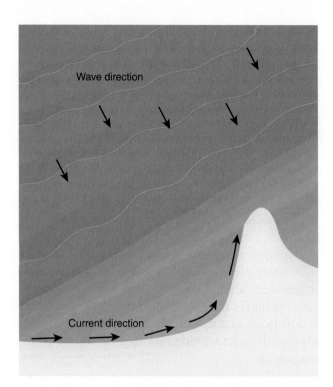

**FIGURE 11-13** Longshore currents run at an angle to the shore and are caused when water is pushed diagonally along the shore before it drains back to open water.
© Jones & Bartlett Learning.

may suddenly, with one step, be in over their heads. Also, rescuers running into the water may be injured when they are not able to plant their feet as expected. Sprained ankles and knees are possible.

## Rocks and Vegetation

Also on the list of surf problems are rocky areas, coral, and vegetation. Getting slammed into a rock cliff or coral outcropping can require stitches or worse, and vegetation can entangle lines, harnesses, boat propellers, or even swimmers. Becoming entangled can lead to not being able to move properly while swimming, or losing propulsion on a watercraft, and thus losing the ability to keep the watercraft pointed into the waves.

## Weather and Lighting Issues

Surf areas are not just attractive during daylight hours. Nighttime often finds beach walkers, parties, and sportsmen fishing and crabbing, all using the surf areas. Fog is often present in coastal or water areas, and both fog and darkness make surf rescue more difficult. Night rescues are extremely hazardous when undertaken on nonmotorized rescue aids, or by rescue swimmers. Lighting becomes paramount, if possible, but the open areas of the beach or ocean may render lights on shore useless. Searches that provide more rescuer safety, such as boats or helicopters, should be considered.

Lightning is a distinct possibility, and the National Weather Service supplies a good general rule. If you can hear thunder, you are close enough to get struck by lightning. Rescuers should avoid tall objects such as trees or lifeguard towers when around beaches and lightning, and should consider seeking shelter.

# Surf Rescue and Watercraft

Surf rescue in NFPA 1006 is a bit different, in that operations tasks at the low surf level allow a rescuer to provide support measures in a motorized boat in surf rescue, including performing crew duties, communications, and assisting victims into the watercraft.

Operations-level surf rescuers may also deploy a nonmotorized watercraft, most often a rescue board. Although deploying a rescue board into the surf zone places the rescuer in the physical environment (which is a level of risk typically reserved for the technician level), the environment, culture, and hazards of surf rescue require a slightly different thought and skill set than the other disciplines. A technician-level certification is still needed to enter high surf as a swimmer or watercraft operator. Many AHJs require the watercraft operator to be able to demonstrate swimming skills in the surf in which they expect to operate the watercraft in. This way, the rescuer can save himself or herself in case of failure or capsizing the watercraft.

## Communication

In this day of modern electronic radio, it is rare to use hand signals, but they still have a place. Communications with the shore, command post, or other boats are required, and may be as simple as using a small hand-held radio, or a complete radio stack in the wheelhouse of the boat. In heavy surf or extreme conditions, the pilot of the boat may choose to concentrate solely on operating the boat and let someone else talk. Whistle signals may also be used, but with the typical noise from crashing surf and longer distances found in surf rescue, hand signals are preferable.

Hand signals can be used and adapted from whatever signals the AHJ has adapted for use in other areas. Most commonly they would reflect the signals used by the lifeguards who enter the water. Why make it harder by having two sets? Flags, flares, and other methods can be used, and even though we rarely need these things because of the radio, it is important to have backup methods.

## Boat Deployment

One good example of operations-level rescuers working with technician-level rescuers would be getting to a victim via a boat. Operations-level rescuers may be tasked with various jobs as crewmembers on a boat, with technicians possibly entering the water to assist if needed. All operations-level rescuers should have a good working knowledge of watercraft operations. (Refer to Chapter 2, *Watercraft Rescue Operations*, for more information.) Motorized boats, of course, can be inflatable boats, rigid hull runabouts, or even PWCs. One word of advice: Be well-versed in the watercraft on which you will be crewing. Different boats handle differently in different surf conditions.

## Keep It Pointed Offshore

All boat operators have one general rule while operating in surf, regardless of the type of boat or propulsion method: Keep it pointed into the waves! Keeping the bow into the waves lets the boat cut through the water and ride over the waves or swells. If the boat turns parallel to a large wave, and the wave breaks, the boat can easily roll over or swamp, potentially flipping upside down with the wave break. An effective and controlled way to move toward the shore is to ease off the throttle slightly and allow the waves to help bring the boat in. Some thrust must be used to maintain steering control. Many boat operators prefer to stage facing offshore, and back the boat in, making for an easy route back out past the surf zone. The less maneuvering that needs to be done in the surf, the better and safer it will be for the boat.

---

**LISTEN UP!**

The operation of watercraft inside a surf line is a highly refined skill and must not be attempted without training. Many boats have been lost after rolling or losing power and being swept to shore.

Consider also using a safety boat outside the surf zone. It may never be deployed, but at least it is there, watching, and may have the ability to come in between waves if something goes wrong. In this environment, it can be a good thing to have someone watching.

---

Coming inshore is usually easier. The boat can be kept at a speed that lets the boat "surf" somewhat between the waves, but caution needs to be kept to the rear, as a wave that breaks will need to be outrun. Running with swells can even be done at a speed where the boat slows to climb the swell, then speeds up on the downside.

Running parallel, or at an angle, to the swells or waves is somewhat difficult, and can cause a flipping action when the wave passes under the boat. The boat will launch over the wave, but is not balanced in a side-to-side manner. Therefore, it can be tipped over in the direction of the back of the wave. Good operators, with a capable motor and boat, can time operations in the waves to allow some operations between waves.

# Motorized Rescue Watercraft

The sheer distance possible in surf rescue makes a motorized watercraft an attractive option for rescue, if available. As with other water rescue environments, motorized watercraft allow long distances to be traveled quickly, with the advantage of eliminating swimmer fatigue. Watercraft may also be staged farther offshore in swells, with rescuers transporting victims away from the plunging waves and out to the boat.

---

**SAFETY TIP**

The use of prop guards to protect people in the water, or the use of jet drive motors, is urged. Propulsion may be needed to maintain directional control in the surf, creating a higher degree of danger for those working nearby.

---

One of the most hazardous places to operate watercraft is in the surf zone. The breaking waves can not only cause a wild ride, but may also cripple or destroy a boat that might even lose power as a result of the force of the waves.

It is recommended that prop guards or jet drive motors be used to protect people in the water. Propulsion may be needed to maintain directional control in the surf, creating a higher degree of danger for those working nearby.

---

**LISTEN UP!**

Operating a boat in surf is not something that the weekend skipper who doubles as a rescuer wants to learn by fire. If you want to operate a boat, find a qualified boat instructor or take a class specifically for this type of operation.

---

## Larger Boats

Specialized surf watercraft exist and are used by agencies such as the U.S. Coast Guard, fire departments, and specialized surf rescue agencies (**FIGURE 11-14**).

**FIGURE 11-14** Some larger, runabout-type rescue boats can handle large surf, even to the point of rolling over and self-righting.
Courtesy of Gary Robertshaw.

Larger watercraft offer good protection from the waves, and victim care can often begin en route to shore after the rescue. One problem with using larger, rigid hull boats is getting them to the area of response quickly. Watercraft on patrol can travel somewhat quickly, but otherwise, responding to a launch area, getting underway, and responding on the water all take up time.

Another problem is visibility. Moving toward a victim in the water in high surf can be dicey if the victim is not in sight. This leads to the need for another operations-level crew position: the lookout. Someone, or sometimes everyone, must be looking for the victim, and must also keep their eyes on the victim even after the initial sighting. Many times, just keeping an arm pointed at the person in the water is enough to provide steering direction for the operator.

Other positions on larger motorized runabout-type boats might include spotlight operator, tender for a swimmer, and EMS care for the patient.

## Throttle Use

The throttle plays a key part in surf operations. Even with the boat pointed into the surf, how you handle the gas can play a huge part in surf navigation. If the boat runs at even a seemingly slow speed, the angle and speed of the wave coming the other way can easily launch a boat into the air. While larger watercraft can handle this, launching a PWC off a 10-foot wave is not especially helpful on a rescue mission.

"Chopping" the throttle, or rapidly closing it immediately after the bow drops over the wave lets the throttle drop and slows this launching action. There will still likely be a strong hit, but not as much as if the throttle stayed open. This works for any boat, large or small.

## Smaller Boats

Smaller boats, such as tiller-steered inflatables or PWCs, are more common, especially in jurisdictions with a

**FIGURE 11-15** PWCs make excellent watercraft to utilize in surf rescue.
Courtesy of Shawn Alladio, K38 Rescue.

**FIGURE 11-16** Many PWCs can be moved around on shore using a dolly equipped with large wheels.
Courtesy of Steve Treinish.

sparser population. With the drop-in users, everything else, including fire and rescue agencies and their equipment lists, becomes smaller. Most beach areas are not suitable to have docks or ramps installed, so getting a larger runabout to a surf rescue may entail launching it miles away and then running out to open ocean to the rescue. Inflatables and PWCs are easily moved on beaches using flotation tire–equipped dollies.

## Personal Watercraft Use

The PWC is one of the best tools for agencies that respond to water emergencies (**FIGURE 11-15**). They are not new to the arena; they have been used by agencies for almost 40 years. The more popular style of PWC for water rescue is the sit-on motorcycle-style, but stand up-style PWCs are used, too. Most PWCs can handle a multitude of water conditions, including surf, swiftwater, flood, and surface rescue, and they are a true multitasking tool for many agencies. They can be flipped upside down in the surf, but will not normally sink. They can be righted in a similar manner to a johnboat, with the rescuers reaching up and rolling it right side up with their body weight.

> **LISTEN UP!**
>
> The willingness of many manufacturers to place their PWCs with a rescue team at a reduced cost is a big plus. Programs that provide loaner watercraft to an agency are available, and can be a win-win for all involved.

For surf rescue, PWCs are quick to launch, and many can be moved around on shore using a dolly equipped with large wheels (**FIGURE 11-16**). PWCs are also quick with throttle response, making a dash between wave sets to rescue someone a viable option. One person can operate them, although using them solo to perform a rescue can be tricky in rougher water. Alone, the operator must control the boat, keep an eye out for changing conditions or hazards, and load or secure the victim up and onto the PWC. A victim who is unable to fully help himself or herself can really make it tough on a single operator. The boat frequently circles victims who can help themselves, with the victim being grabbed arm-to-arm and pulled onto the rear of the PWC. The addition of tow sleds changes this.

## Tow Sleds

A tow sled is exactly what it sounds like: a towable litter fastened behind a PWC (see Figure 2-26). This allows two things. First, a second rescuer can go along to assist with the rescue, and second, victim loading and transport, particularly of unconscious or injured victims, are much quicker and easier. Picture getting to a victim with a PWC, but you are alone. How do you even get the victim, who is incapacitated, on the PWC, and keep him or her there while you return to shore?

Most tow sleds attach at three points at the rear of the PWC, letting it turn freely but not letting it whip around too far when turning. The driver of the PWC operates the watercraft, and can also assist somewhat, and keep an eye on the conditions, which are likely to change quickly. The second rescuer can ride on the sled behind the operator, and deploy into the water if needed or, in the case of victims capable of helping themselves, pull them up and onto the sled. The rescuer then lies on the sled, over the victim, and hangs on. This keeps the victim in place and protected for the ride to the safe zone. This is shown and detailed in **SKILL DRILL 11-1**.

Keep in mind that this is a very quick rescue evolution that can be used for evacuating victims in hostile water. There may be some bumping and bouncing, but that is unavoidable. In case of a suspected cervical spine injury, field treatment will have to do until further care can be provided safely. One positive feature

# SKILL DRILL 11-1
## Extricating a Victim with a Tow Sled NFPA 1006: 20.2.4

**1** The PWC operator approaches the victim, keeping the victim on the port side. This lets the operator use the left hand to help, but keeps the right hand on the throttle control.

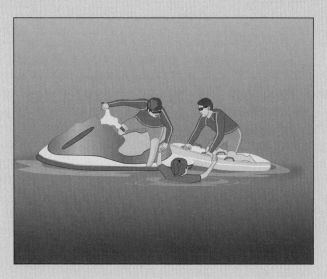

**2** The PWC operator grabs the victim's arm and lets the victim be pulled close to the sled. Having the operator do this instead of the rescuer on the sled keeps the victim closer to the front of the sled.

© Jones & Bartlett Learning.

in our population is that most of the people found in the big water and surf probably have experience with it. They will hopefully be conscious enough to help themselves, be familiar with watercraft and surf rescue, and have an idea of what is happening.

## Inflatable Boats

Smaller inflatable boats can be a good option for a surf rescue crew. (A photo of an inflatable boat can be found in Chapter 2.) Easy to launch with only a few people, they can be carried to the proper area to be launched to provide the quickest route of response. They do not draft much water, offer good flotation even when full of water, and can provide buoyancy assistance for many people in the water when necessary. They can operate in much shallower water than the larger craft, making operations closer to shore a good possibility. On the downside, they do get somewhat bouncy due to the air in the tubes, and they can be blown around by the wind, especially if the keel gets bounced or lifted off the water.

One person must operate the boat, and the other crewmember (usually one) sits ahead of the operator, ready for rescue. The victim is pulled by their arms, clothing, or hopefully, PFD into the boat, and the boat transports everyone either back to shore or to a predetermined safety area, possibly out of the surf zone. Unconscious victims can be wrestled into the boat, and because the inflatable sits lower to the water, this is usually easier. In nasty wave conditions, it may be a true load-and-go situation. There generally isn't a lot of room in an inflatable, but lifesaving treatment can be given, even if space is tight.

# Nonmotorized Rescue Watercraft

Nonmotorized watercraft can include rafts, surfboards, kayaks, and similar equipment. One drawback is that victim care cannot be performed effectively on these watercraft. For this reason, transporting or towing the victim to a place where care or resuscitation may start is a high priority in the incident. This may be a larger boat or helicopter offshore in swells, or back on shore.

Rescue boards, discussed next, are generally the tool of choice, due to the high flotation, stability, and the ability to propel it without paddles. Kayaks are sometimes used, but the higher center of gravity makes them susceptible to rolling over, especially if the paddles must be stowed when assisting a victim. As with any watercraft, it should be quickly checked for damage or obvious issues before deploying into a hazard zone. Ensure that the needed rescue gear and rescuer PPE are present and a plan is in place should the watercraft or the rescuer encounter a problem. Rescuers should try to avoid launching or recovering in locations unknown to the AHJ as approved launch locations, but in a true rescue, launching from these locations may be necessary. The launch and recovery phase can be dangerous with the urgency on scene, and rescue crews should be aware of potential injury or damage that may occur in the event of improper launch and recovery. For more information on launching and recovery of watercraft, refer to Chapter 2.

**FIGURE 11-17** A rescue board can be a primary tool of rescue.
Courtesy of Steve Treinish.

## Rescue Boards

While it may have crossed over from the sport of surfing, and the agencies that carry them on the rig hear all the jokes, the use of a specialized surfboard for rescue is absolutely serious and effective. NFPA 1006's treatment of operations-level low surf rescue is unique because while a **rescue board** technically places the rescuer in the water, it remains the primary tool of rescue (**FIGURE 11-17**). It takes balance and practice, but motorized boats and PWCs notwithstanding, it can be the quickest way to get to and remove a victim from trouble. Paddling a rescue board is much quicker in the water because it is light and fairly small, and is also faster than swimming. Rescue boards also give the rescuer the ability to transport the victim back to shore.

## Paddling

Professional lifeguards are taught to watch how the customers play in the water. A recreational user kicking a surfboard like a swim aid is a sign of inexperience, and a red flag for that lifeguard to keep an extra eye on that person. Rescue boards have handles built into the board to assist in rescue operations. Like their true sport cousins, they are designed to be paddled by the arms from a kneeling or prone position. (**FIGURE 11-18**). Launching them from shallow water is a combination of jumping out and onto the board, trying to skim the water for maximum starting speed. Calmer water is easier for surf rescue, in that breaking, plunging waves can knock the user from the board, and also make it harder for the rescuer to keep the victim on the board. Experienced board users can submerge just before a wave breaks on them to spare the full effect of the break, and then surface after the wave passes overhead. Smaller, spilling waves can be tackled with a board, but even this should be practiced.

**FIGURE 11-18** Although they do have built-in handles, rescue boards are designed to be paddled by the arms.
Courtesy of Steve Treinish.

**FIGURE 11-19** Paddling a rescue board from the knees offers better visibility and uses the back and abdominal muscles as well as the arms and shoulders.
Courtesy of Steve Treinish.

Weight distribution is important as well. Lie too far forward, and the board will nose into the water and possibly even slow so much in the waves that the rescuer will slide forward off of it. Lie too far back, and the board will plow through the water more than glide. A good balance is to keep the nose slightly high, and the back in the water. If the skeg (the fin on the bottom in the back) doesn't have water flowing around it, the board will not track as straight and will zigzag in the water.

Another difficulty with rescue boards is staying on them. Keeping low on your belly creates a lower center of gravity, but this also limits visibility, and because you have to paddle primarily with the shoulders, it causes fatigue. Rising up to a kneeling position lets you paddle with the entire upper body and provides a better line of sight, but the center of gravity is much higher, and it becomes easier to be thrown off the board, especially in bigger surf (**FIGURE 11-19**). If you are having issues staying on the board by yourself, imagine what it will be like with a victim.

# Victim Management

As with any victim approach, the rescuer must get a feel for the victim and his or her state of mental and physical ability. While most people think of surf rescue as a warm water event, do not discount the possibility of hypothermia from water temperatures or the elapsed time of the immersion. Alcohol use is common among beach users and may affect the victim's ability to assist in his or her own rescue. Rescue crews should also question all available witnesses to try to determine the number of victims. Well-meaning civilians may have attempted a rescue before the arrival of rescue crews.

## Conscious Victims

Using a rescue board to approach a victim is like any other evolution: the swimmer makes a judgment call regarding the victim's well-being and state of mind, then goes in for the actual rescue. The front of the board will likely be the first thing they grab for, and this is OK. This gets your victim what he or she needs, and there is no fight. The victim can then be instructed to lay off either side, grab the handle, float up or be assisted to a horizontal position in the water, and either scoot or be helped onto the board. The rescuer then positions the victim in a way that would produce the best ride and gets on behind or on top of the victim's legs. This position holds the victim down, but still lets the rescuer paddle to a safe zone. Remember, the safe zone may be back to shore or farther offshore.

A victim who is conscious but injured, or who may not be able to help in a productive manner, can be helped to the board by the rescuer approaching and grabbing the victim's wrist. The wrist is held high, and the board is turned toward the direction of travel. Next, the rescuer reaches down and grabs the leg closest to the rescuer, pulls the victim horizontal, and places the victim on the board. This is similar to positioning for an able-bodied victim, except the rescuer is helping to place the victim horizontally (**FIGURE 11-20**).

## Multiple Victims

Multiple victims can also hold onto the rescue board while backup is on the way. Once a victim is loaded onto or hangs onto the board, its lack of mobility in the water becomes much more of an issue. In the event the victims are scattered in the water, a good tactic might be to utilize an RFD towed behind the board, commonly a rescue tube. The tube can be left with the first victim as a flotation aid, and the rescuer can then head for another victim. Keep in mind, in-water triage may become necessary. Deciding which victims are in the most jeopardy will guide a rescuer

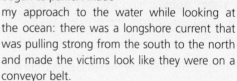

# Voice of Experience

As professional rescuers, we should never underestimate the power of natural phenomena. Furthermore, we should use the environment to our advantage whenever possible.

Before I became a Police Officer/Fire Fighter assigned to the Ocean Rescue division in my city, I worked as a lifeguard for 5 years. During that time, I gained incredible experience in the rescue of victims in open water. I found it invaluable to get to know the waters in the area in which I served, as they changed constantly, and being able to "read" the water was a skill that came with time. Rip currents can appear quickly and then dissipate just as fast, which many beach-goers are not aware of. In fact, it is the general absence of knowledge about rip currents that I find to be at the root of rescues, second only to lack of swimming skills. I should mention that just as fire fighters must be physically prepared to climb stairs in full bunker gear and on air, if an emergency responder is employed by an agency that responds to calls for service in open water, they must be fit to fight in the surf as well.

An instance that comes to mind in which I had to read the surf and have knowledge of rip currents was a rescue that I performed a few years back. This had taken place before I became a professional emergency responder, so I had the advantage of being on scene already instead of being informed of the incident via dispatch. Nevertheless, my actions once I recognized the victim were the same.

There were actually two victims: a small boy and his mother who had swum out to help him. Both were struggling in the surf, as the same rip current that had pulled the boy out where he could not stand also pulled his mother away from shore. Many people do not consider that rip currents are often not very long. This one

was just long enough to bring them to water where the mother could not stand, and thus, she began to panic. I made my approach to the water while looking at the ocean: there was a longshore current that was pulling strong from the south to the north and made the victims look like they were on a conveyor belt.

Taking their lateral movement into consideration, I bounded down the beach and entered the water about twenty yards north of their location. By the time I swam out to them, we almost collided. The two victims were clawing at the water and barely staying above the surface. I immediately pushed my rescue buoy toward them, keeping it between myself and the victims. They grabbed hold, and I knew that the most urgent part of the rescue was done. I offered a word or two of comfort, stating that I was there to help them and to hold on. Sometimes just a few words can make the difference between a hysterical victim and a calm and cooperative one.

I side-stroked to shore so that I could keep an eye on both victims while making progress. Once on land, the victims were quickly evaluated for any respiratory issues, but thankfully there weren't any. These kinds of situations happen every day, and often it is unfortunately in early mornings or late evenings, long before or after lifeguards are on duty.

Keep your victims safe, keep yourself safe, and know your waters.

**May Lauzon**
Public Safety Officer
North Myrtle Beach Ocean Rescue
North Myrtle Beach, South Carolina

**A**

**B**

**FIGURE 11-20** Approaching a victim with the side of the rescue board provides a natural position of flotation and keeps the victim away from the rescuer.

© Jones & Bartlett Learning.

in the effort. Many factors will play into this, including age, size, swimming ability, and status of injury.

## Unconscious Victims

Unconscious victims in the water can be flipped onto the board. This makes it much easier to keep the victim's head and airway out of the water. The rescuer pulls the board up along a victim and turns it upside down. The swimmer enters the water and places both arms of the unconscious victim across the board, from side to side. Then, the swimmer goes to the other side and flips the board while holding the arms of the victim. This will roll the victim over onto the topside of the board, thus doing the lifting for the rescuer. The rescuer will have to swing the feet and arms into position and take a better hold of the board to keep the victim on it, but once this is done, traveling with the unconscious victim is just as with a conscious victim. This situation is a perfect place to consider removing the victim to the swell zone, because trying to travel inshore through rougher surf may throw the unconscious victim from the board. To place an unconscious victim on a rescue board, follow the steps in **SKILL DRILL 11-2**.

Because a rescue board is a long board, rescue boards can be used to assist with cervical spine management. There are times when nothing is available or on scene yet, and situations involving mass casualties are when we have to make do with what we have. Tape, some small pieces of rope, and blankets or pillows can all be used to do the best c-spine management possible in those conditions if needed.

## Technician Support

One task that may be performed by operations-level rescuers is tending lines used in swimming surf rescue.

The use of a line is somewhat controversial, as the distance covered in a typical rescue can put a very long length into play in water that is churning and rolling, potentially causing entanglement. However, some agencies use lines during night or low-visibility rescue attempts by swimmers. Rescuer effort may be reduced somewhat in low surf, wherein once the rescuer gets to the victim, the shore crews may retrieve the pair by pulling the line in. The rate of retrieval is crucial to the safe recovery of everyone involved. Pulling too quickly, or with mechanical winches, can easily pull multiple people underwater and trap them there until the pressure is released. As mentioned before, once a rescuer provides a victim with positive flotation, much of the initial emergency might be over, and time can be taken to ensure a safe extraction from the water.

## Terminating a Surf Rescue Incident

Termination of a surf rescue incident is really no different than terminating other water rescues, but a few specific things to watch for include a high risk of damaged equipment or injured rescuers from operating in plunging waves or heavy surf. The forces placed on rescue tubes can tear or gouge the foam, or crack plastic buoys or kayaks, as well as crack or break the fiberglass most rescue boards are constructed from. Rescuers are also subjected to an increase in sprains and strains, with knees and legs getting pushed on laterally by surf. The physical exertion of surf rescue might also require ample hydration for rescuers; drinking a sports drink or water is recommended. For more information on terminating water rescue incidents, refer to Chapter 4, *Surface Water Rescue Operations*.

# SKILL DRILL 11-2
## Placing an Unconscious Victim on a Rescue Board NFPA 1006: 20.2.3

**1** The rescuer approaches the victim, placing the side of the rescue board between the rescuer and the victim. The rescuer will be on the opposite side of the board as the victim.

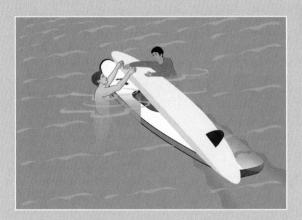

**2** The rescuer flips the board upside down and places it against the victim's chest, with the rescuer grasping the victim's hands.

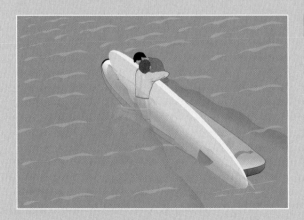

**3** Pushing down on the near side of the board, the rescuer flips the board right side up, continuing to hold the victim's hands and arms.

*(continued)*

# SKILL DRILL 11-2 Continued

## Placing an Unconscious Victim on a Rescue Board NFPA 1006: 20.2.3

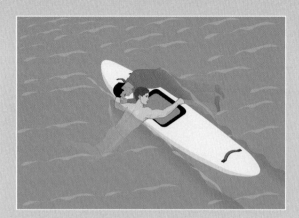

**4** As the board rolls over, the victim will now be rolled onto the face of the rescue board, face down, perpendicular to the board.

**5** The victim can be pushed and turned ninety degrees so that he or she is lying on the board, ready to be secured and moved by the rescuer.

© Jones & Bartlett Learning.

# After-Action REVIEW

## IN SUMMARY

- Surf rescue has evolved from a public safety effort that once saw many civilian casualties each year to a highly specialized water rescue field; this has resulted in thousands of saved lives each year.
- The range of the customers around the surf is broad and can lead to a variety of challenges.
- When preplanning a surf rescue, note not only low and high surf and tide conditions, but also things like boat ramps, access to the beach or shoreline areas, optimal entry and exit areas of the water, and any specific hazards in the location.
- A PFD can increase the risk of injury by eliminating the ability for the swimmer to dive under a plunging wave, reducing swim effectiveness and exposing the swimmer to the power and weight of the wave.
- Rescuers undertaking surf rescue training or missions must be powerful swimmers and absolutely comfortable in the water.
- The higher the wind velocity, the larger the waves that are kicked up, and these waves can travel hundreds of miles before they reach shore.
- Operations-level rescuers must be proficient in emergency swimming, able to complete a distance swim with any stroke without using surf rescue aids, and have a good working knowledge of the surf environment in order to be able to save themselves.
- Although deploying a rescue board into the surf zone places the rescuer in the physical environment (which is a level of risk typically reserved for the technician level), the environment, culture, and hazards of surf rescue require a slightly different thought and skill set than the other disciplines.
- Operations-level rescuers may provide support measures in a motorized boat in surf rescue, including performing crew duties, communications, and assisting victims into the watercraft. They may also deploy a nonmotorized watercraft, most often a rescue board.
- As with any victim approach, the rescuer must get a feel for the victim and his or her state of mental and physical ability.
- Termination of a surf rescue incident is really no different than terminating other water rescues, but a few specific things to watch for include a high risk of damaged equipment or injured rescuers from operating in plunging waves or heavy surf.

# KEY TERMS

*Access Navigate for flashcards to test your key term knowledge.*

**Crest** The top, or highest point, of a wave, just before the wave breaks.

**Downwash** Water returning to the open area, but along the path the wave came to shore.

**High surf** Surf higher than its normal condition.

**Impact zone** The area where plunging waves crash down onto the water or land surface.

**Inshore holes** Depressions dug out by wave or current action in shallow water.

**Longshore currents** Water flowing along a shore, pushed by the waves incoming behind it. May feed rip currents.

**Low surf** Surf at its normal condition.

**Lull** The period of time between set waves, creating an opportunity for safer and more effective movement by rescuers

**Open water** Any large, natural water environment.

**Plunging waves** Waves that flow onto a steep bottom or object, resulting in faster and harder breaks.

**Rescue board** A long, floating rescue device, designed to be paddled by hand, resembling a sport surfboard.

**Rescue flotation device (RFD)** A tethered floating tube or buoy used by rescuers for self and victim rescue.

**Rip currents** The water flowing back out to open water past a surf line, after being pushed onshore. Very strong, and can easily overcome swimmers. Also called *rips*.

**Rip tides** The water flowing back out to open water past a surf line, after being pushed onshore; also *called rip currents*.

**Rogue wave** A higher wave than others found in set waves.

**Set waves** Waves showing different traits than normal. Often formed from merging energies from storms, earthquakes, and other factors.

**Spilling waves** Waves that flow onto a gradual slope bottom, resulting in a slow, controlled break.

**Storm surge** Water pushed ahead of a storm area. The energy from these waves does not dissipate readily.

**Surf line** The line at the offshore edge of the surf zone. Swells are outside this line; waves are inside it.

**Surf zone** The area in which waves are actively building or breaking.

**Surging waves** Waves that do not break, but pile into obstructions in the water, such as cliffs, seawalls, etc.

**Swells** Waves that rise up, but do not crest or break; found offshore in deeper water.

**Swim fins** Rubber or plastic blade-like propulsion, worn and kicked by the feet.

**Trough** The bottom of a wave.

**Upwash** The water rushing into and onto a shore area from wave action.

**Wave period** The amount of time it takes for two crests to pass a set point.

**Wave train** Waves from the same source of energy, showing similar speeds and traits.

# On Scene

You are asked by your chief to preplan a response at a beach area that is known for its local popularity. Swimming, beach parties, and surfing often go well into the night, and its secluded nature makes keeping an eye on it difficult. The local county lifeguards are familiar with the area, and have made several rescues there in the past.

**1.** Preplanning with the county lifeguards should include which of the following components?

   **A.** Communications, both radio and hand signals, known and practiced between the teams

   **B.** Who will be in charge when a run is dispatched

   **C.** Which resources each team brings to the table, and the best use of them

   **D.** Both A and C

**2.** Which of the following is a great source of information about the surf, beach areas, and offshore hazards?

   **A.** The local marine industry

   **B.** The county lifeguard team

   **C.** The swimmers, surfers, and users of the beach area

   **D.** Geographic information system mapping from the county or state emergency management agency

**3.** Surf conditions should be examined:

   **A.** at normal conditions, usually in summertime.

   **B.** only when storm fronts are close by.

   **C.** often, as weather and energy offshore, wind directions, and tides can all change conditions.

   **D.** before the summer season starts and the user population swells.

**4.** Your department has a PWC housed at the station, and is versed in its operation. Based on the area being secluded:

   **A.** you should ensure that the PWC launch capability is examined, an alternate launch point is known, and the estimated time to this area noted.

   **B.** it should be dispatched immediately on any water run in this area.

   **C.** the county lifeguards should be trained in tow sled rescue.

   **D.** it may be better to keep it at the local marina in order to reduce launch times.

*Access Navigate to find answers to this On Scene, along with other resources such as an audiobook and TestPrep.*

# Surf Rescue Technician

## KNOWLEDGE OBJECTIVES

After studying this chapter, you should be able to:

- Describe advanced swimming techniques for navigating the surf zone. (**NFPA 1006: 20.3.1, 20.3.2, 20.3.3**, pp. 295–296)

- Explain the role of flotation aids in surf rescue. (**NFPA 1006: 20.3.2**, pp. 296–297)

- List considerations to keep in mind when choosing a water entry point. (**NFPA 1006: 20.3.1, 20.3.2, 20.3.3**, pp. 297–298)

- Explain how to adjust tactics based on conditions during a surf rescue. (**NFPA 1006: 20.3.1, 20.3.2, 20.3.3**, pp. 298–300)

- Describe how to rescue submerged victims. (**NFPA 1006: 20.3.1, 20.3.2, 20.3.3**, pp. 303–304)

- Explain the requirements for offshore communications. (pp. 305–306)

- Describe the three methods for extricating victims at a surf incident. (**NFPA 1006: 20.3.2, 20.3.3**, pp 305–306)

- Identify the risks and benefits of conducting a tethered rescue. (**NFPA 1006: 20.3.1, 20.3.2**, pp. 306–307)

## SKILLS OBJECTIVES

After studying this chapter, you should be able to:

- Choose a water entry point based on surf conditions and any additional hazards present. (**NFPA 1006: 20.3.1, 20.3.2, 20.3.3**, pp. 297–298)

- Conduct a surf rescue using advanced swimming tactics. (**NFPA 1006: 20.3.1, 20.3.2**, pp. 298–300)

- Perform a free dive. (**NFPA 1006: 20.3.1, 20.3.2, 20.3.3**, p. 304)

- Communicate with your team during a surf rescue. (pp. 305–306)

- Extricate a victim using inshore removal. (**NFPA 1006: 20.3.1, 20.3.2, 20.3.3**, p. 307)

- Conduct a tethered rescue. (**NFPA 1006: 20.3.1, 20.3.2**, pp. 307–308)

# Technical Rescue Incident

You are riding "up" today, and instead of the officer's seat on the engine, you are riding in the Battalion Chief's car. You've been in the department for 18 years and love the warm weather, sand, and recreation that the ocean areas you cover offer. Today, an engine and medic are dispatched to a drowning at a local beach that has a large swimming area and a pier. The pier has a small store, a restaurant, and a fishing area on it, and extends out about one-eighth of a mile from the sand. You request to be added to the run and arrive first. You find a crowd on the pier, leaning over the railing, watching below. There are no lifeguards on duty at this beach because it is late in the evening.

The witness you approach says a teenager was playing around, trying to balance and walk on the rail. He then fell about 20 yards to the water, where he struggled a few seconds, then went under. No one claims to have noticed any other people enter the water. You do not see him by looking over the rail, and you have every reason to believe the witnesses are right. Your vehicle has a PFD; two throw bags; and a mask, snorkel, and fins specific for water rescue on it. Your incoming engine has a rescue board under the hose cover, and at least two crewmembers are operations trained in surf rescue.

**1.** Can you safely attempt a quick free dive rescue before the engine and medic arrive?

**2.** What hazards might be present in or under the water in this situation?

**3.** What other resources or actions may be needed to handle this run?

 JONES & BARTLETT LEARNING NAVIGATE 2 *Access Navigate for more practice activities.*

## Introduction

Although other types of surf rescue equipment are used in conjunction with rescuers—such as watercraft, personal watercraft, and helicopters—frequently it comes down to one person working in possibly extreme conditions to save another. Surf rescue technicians must have a supreme knowledge of their environment in order to decide the best practice for a rescue, where to evacuate a victim, and how to supply flotation to the entire team (**FIGURE 12-1**). Of course, technician-level rescuers must also meet the operations-level job performance requirements (JPRs), so technician rescuers are also tasked with being able to operate nonmotorized and motorized watercraft in the surf zone. See Chapter 11, *Surf Rescue Operations*, for more information.

## Swimming Ability

The U.S. Lifesaving Association (USLA) recommends that a lifeguard candidate be able to swim 500 meters in 10 minutes at a minimum. The real world can be much tougher, and technician surf rescuers must show and possess better-than-average swimming skills because not only do they have to get to the victim, but they often need to tow the victim to a surf-free area. The surf-free

**FIGURE 12-1** Often surf rescue becomes one person braving the waves to rescue another.
Courtesy of Steve Treinish.

area may be at the shore, or farther offshore where the surf is not breaking. Either way, it is paramount that the free swimming surf technician possesses exceptional skills and stamina, and the swim test should reflect this.

Not only should swimming rescuers be able to swim well, but they should also be totally comfortable on and below the water. Some authorities having jurisdiction (AHJs) allow or require the use of personal flotation devices (PFDs). It may sound odd to water rescuers versed in other disciplines, but using a PFD in heavy surf can cause injury by preventing the rescuer from going under a plunging wave. Rescuers using PFDs should be able to doff them in water if needed, and continue the rescue by getting away from the forces of the waves. A lack of flotation can also come in handy if a panicked victim attempts to use a rescuer as flotation, and a lack of flotation is crucial to properly executing a subsurface retrieval.

# Technician Tools

Because surf rescue often deals with distances too far for direct reach or throw rescues, such as a shepherd's crook or throw bags, swimming rescuers can load their tool box with a few items that will do double duty as personal protective equipment (PPE) for the rescuer and as a flotation device for the victim once reached.

## Rescue Flotation Devices

There are two main types of rescue flotation devices (RFDs) in use today by professional lifeguards and rescue personnel—the rescue tube and the rescue buoy (**FIGURE 12-2**). Both can be towed hands free when swimming using a harness that is placed over the shoulder. The harness attaches to a tether, which allows some

**FIGURE 12-2** A rescue buoy (left), and a rescue tube. Both are appropriate ocean rescue flotation devices (RFDs).

Courtesy of Steve Treinish.

space to be maintained between the rescuer and the RFD—or, in case of a victim using the flotation, between the rescuer and the victim. The area where the tether attaches to the shoulder harness should be on the rescuer's back, just to one side of the spine. This will keep the float of the RFD behind the rescuer, reducing risk of entanglement.

The tether is a huge advantage when approaching someone in distress in the water. The length of the tether can play into how comfortable the rescuer is using it. If the leash is too long, it will be harder to control the slack on shore and could be an entanglement risk in the water. If the leash is too short, it may hit the swimmer's feet when swimming. Either way, tethers are made and used with different lengths of line.

## Rescue Buoys

The first of the two types of RFDs is a rescue buoy. Made of a hard plastic, it has molded handles and resembles an elongated football. Almost all rescue buoys are red in color for visibility and offer enough flotation to support multiple victims. However, the buoy cannot be fastened to a victim securely. It is a good choice when facing multiple victims who are capable of helping themselves. One downside is that when it is tossed around in heavy surf, it can strike the rescuer, possibly causing minor injury.

## Rescue Tubes

Another RFD device more commonly used in surf rescue is a rescue tube, also known as a Peterson tube. Rescue tubes are a long piece of foam rubber coated with vinyl that has one huge advantage over the buoy. The tube can be wrapped around the victim's chest and secured with the snap and ring found on the ends of the tube. This allows a swimming rescuer to physically attach the tube to a victim, increasing the ability to tow a victim in the water. The victim is more secured to the flotation, and the rescue swimmer is free to swim a forward crawl, using both arms.

The tube is similar to the buoy in that it also has a quick-release shoulder harness and a tether line and deploys in the same way as the buoy. It offers less drag through the water, due to its slimmer profile, but also offers one big advantage in the water with the victim, especially with unconscious and calm victims. It can be wrapped and fastened around a victim like a belt, creating flotation that does not require maintaining a grip. This means the rescuers can swim more effectively while transporting a victim.

One disadvantage of the tubes is that they are not as buoyant as the buoys, and if your victim is presenting with panic, the tube can easily be pushed aside, giving the victim access to you. Tubes also cannot take as much wear and tear as buoys, as the vinyl can rip or split.

When swimming while towing an RFD, remember that the float will not submerge but, because it is behind you, it will be tossed or pulled by the surf after you pass through or over the wave. It may feel strange, but this setup allows you to submerge and dive through waves, rather than fight the full force of them. A PFD will not let a rescuer submerge and dive through or under a wave, potentially keeping the rescuer in the downward path of the plunge of a wave.

---

**KNOWLEDGE CHECK**

Which statement about the use of a PFD in surf rescue is correct?

**a.** As with other forms of water rescue, a PFD should be worn at all times.

**b.** The PFD should provide sufficient buoyancy for both the rescuer and the victim.

**c.** Use of PFDs in heavy surf can cause injury.

**d.** Only Type I PFDs should be worn.

*Access your Navigate eBook for more Knowledge Check questions and answers.*

---

**FIGURE 12-3** Surf fins are quick to don, have no adjustable strapping to fail, and can also be used as hand paddles. These fins are staged with an RFD to maximize swim speed.
Courtesy of Steve Treinish.

## Surf Fins

It only makes sense to believe that the more effective the swimming technique is, the safer the rescuer will be. Equipping rescuers with a set of properly fitted surf fins, even if they do not wear them in the rescue initially, will provide much greater propulsion both laterally and vertically in the water. Swimming ability, water treading, and subsurface diving will thus be more effective. Typical dive or snorkeling fins should not be used due to their physical size. Fins can be donned in deeper water, once swimming ability is established. Surf fins are slipped on, and most have no adjustable straps to loosen (**FIGURE 12-3**). Some AHJs equip their rescuers in dive booties, giving some protection to the feet in rocky areas or on pavement. Fin size should consider booty use. The fins can also be used as hand paddles, if needed, and they normally float on the water, making recovery possible if lost or discarded.

## Choosing an Entry Point

Getting to the victim comprises not only navigating the water, but also getting *to* the water. For most open shore areas, a run to a point closest to the victim will save time spent in the water and will let rescuers enter the water with less fatigue. However, if the victim is caught in a longshore current, it may be necessary to anticipate the victim's movement by waiting to enter until estimating where the victim will be *after* the swim. Another advantage of entering with a direct line is being able to keep the victim in your line of sight. The longer you can keep visual contact with the person you are swimming for, the better. Most people cannot swim a naturally straight line without reference points, so lining up something on shore can give a quick, triangulated route to the point last seen (PLS). Another way to get there quicker might be to use signals from someone on shore to guide your swim, instead of constantly looking in the water.

Running beats swimming, even in shallow water. In water that is roughly knee deep and shallower, running with high strides will make better time than swimming, and can also possibly help you keep the victim in sight. Once you commit to the swim, the height of the waves, along with your lowered point of view, may block any further sight of the victim for awhile.

---

**SAFETY TIP**

Remember that upwash, downwash, and holes are all commonly found in the surf area. Use caution when high stepping, as injuries, especially sprains and strains, can occur.

## Entry Using RFDs

The tether on the RFD allows the rescuer to drop the float when entering the water. It can therefore be carried while running on shore and dropped just as the rescuer enters the water, allowing it to drag along behind the swimming rescuer. This frees up the rescuer's hands almost immediately to start swimming, high-stepping, or porpoising.

---

### SAFETY TIP

If rescue buoys are not controlled, the hard plastic can strike the head or face when it gets tossed in the surf. Beware of facial lacerations or injuries.

---

## Porpoising

The entry into the surf zone can be a difficult chore. The waves are crashing into the shoreline, and the upwash is streaming up the beach. You want to get out to where the problem is, and going under the waves as they plunge is the most desirable way to get the best forward progress through these waves (**FIGURE 12-4**). A breaking wave has a great deal of force and speed on the surface and can push you back toward the shore. The pounding action of a plunging wave can injure a rescuer. Trying to get through the top of plunging waves is asking for injury, and a PFD would only keep a swimmer on top of these waves.

**Porpoising** is the act of diving into the oncoming surf in fairly shallow water, instead of running through it. The wave action and upwash may prevent effective forward movement when a rescuer is on his or her

**FIGURE 12-4** Porpoising, or diving under or through a wave rather than up and over it is easier and faster for rescue swimmers.

© Jones & Bartlett Learning.

feet, so the object here is to get *under* the wave, where the force of it is greatly reduced. Once the water starts to get waist deep or so, the swimmer leaps horizontally and aims for the water in the trough of the wave. Since the rescuer is still in shallow water, the arms should be extended in such a way to protect the face and head. After the leap, the bottom can be grabbed with the hands and pulled to get more distance. As the wave passes over the rescuer, the legs are brought back under the body, and another horizontal leap, timed to the next wave, is made. Once the rescuer is in water that is deep enough that he or she cannot push off the bottom with force, the swim can commence. The goal is to get swimming. Swimming offers better speed and effective travel compared to porpoising.

Remember, technician-level swimmers are expected to have better than average, if not expert, swimming skills.

## Rescue Swimming

Although there are agencies that require surf rescuers to wear flotation at all times, surf rescue is often done with no PFD worn by the rescuer, even if the rescuer is directly entering the surf. Even though this contradicts the normal water rule regarding PFD use, there are some good reasons for it. First and foremost, flotation should be available, but it is usually not in the form of the standard PFD. The rescue tube or buoy will provide flotation for the swimmer if things go wrong. Of course, unconscious or injured rescuers can result in problems, but your fellow rescuers should be watching or actively backing up the operation by whatever means possible.

Another aspect of using no flotation is defensive tactics—it may allow a rescuer to escape an attacking victim. Victims most certainly do not want to go under, so a rescuer who does so will likely be released. If the victim reaches or grabs for you before you are ready, or before you can place other flotation on him, simply submerge, get away, and resurface a few feet away to try another approach. Rescuers wearing flotation may not be able to get underwater to duck a rushing victim, and they may be pushed underwater and literally climbed upon.

An additional reason is the amount of drag a typical PFD presents to a swimmer. The distance covered while physically swimming in other water disciplines is typically short, usually on a tether line, and frequently assisted by current. Surf rescuers may swim hundreds of yards to a victim, and they need all the speed and effectiveness they can get during that swim.

# Using Rip Currents to Gain Speed

Because rip currents are so prevalent in surf rescue, let's examine one up close as it relates to rescue swimmers. Rescuers who need to rescue people from rip currents may be able to use the water characteristics to their advantage. The typical victim will be trying desperately to swim straight to the shore, unaware of the current they are trapped in or how to escape it. It would make sense, then, to enter the rip current and use its carrying ability to get closer to the victim, thereby using less effort and attaining greater speed. Not a problem, but the rescuer must also respect the power of the current and not be swept past a victim without getting the victim under control.

Once a victim is reached, flotation can be established, the victim calmed or secured, and a path to safety selected. A victim who is able to hold his or her head above water might be moved laterally out of the rip into the surf action and brought to shore, whereas an exhausted or incapacitated victim might be stabilized to keep the airway protected, and the rescuer and the victim can both ride the current to the head of the rip, to be picked up by watercraft or helicopter.

# Approaching Victims

On any mission, we have to take care of two people: our victim and ourselves. The first priority on any run is rescuer safety. The first glance will often determine whether a victim is in the active or passive stage of distress. A victim in active distress is physically trying to save themselves. Active victims have the potential to be so panicky that they fight the rescuer. Panic is a very strong reaction, and the victim can grab and hang onto anything viewed as providing safety, including any part of a rescuer. Getting out of a victim's grasp can be hard, and

may require a good deal of force from the rescuer. A proper approach to the swimmer is better; therefore, quick review of defensive swimming is in order here. Refer to Chapter 5 for a more in-depth explanation.

## Active Victims

The best victim approach keeps the rescuer at a distance from the victim until he or she is able to decide whether the victim is calm enough to approach safely. About two arm lengths will maintain good verbal and visual contact, but not let the victim grab at the rescuer. It is highly recommended for rescuers to face the victim at all times, until total control and confidence allow the rescuer to turn away. Sometimes the simplest prevention is to not get into the situation in the first place. If a rescuer does get closer than two arm lengths and finds himself with a victim holding on, a block can be done. Simply raise one arm across your body, placing the arm directly at the chest of the victim. The arm should be presented parallel with the victim's body to maximize the amount of arm blocking the victim. While blocking with one arm, reach back, and sweep with the other arm and scissor kick your legs. This will create an initial space for protection.

As mentioned earlier in this chapter, not using a PFD allows a rescuer to submerge at will. Often, a victim will get an arm or hand around the rescuer's head or neck. If this occurs, the rescuer can bend the head down into the chest and push off the victim's body, or snap both arms up into the victim's arms, breaking the hold. Often, just submerging will cause a victim to release.

Feet are also an important tool for pushing a victim back, and using your legs to push a victim backwards also will provide a quick window to kick a few times, rapidly putting distance between yourself and the victim. Defensive tactics can be practiced in a pool, and rescuers should be comfortable being "attacked" in training.

When approaching the victim, use the tether line to your advantage and maintain a distance from the victim. The float can be pushed into or handed to a person in trouble, and keeps the rescuer from the hazard. This is a great way to stay out of a fight on the water, and also allows the victim to grab for the flotation and keep their hands on it. Once they hang on to it, normally they will not let go, and the less they have to release that hold, the better they will cope. Most people will calm

**SAFETY TIP**

Do not discount alcohol or drug use in your population. Not only can alcohol or drugs impair judgment, but they can cause a lot of "fight," too. Physical force may be necessary, even punching or kicking a victim.

down enough to follow further instructions. Even if they come after you, you will have the device between you and them, allowing you to block with it.

Once the victim is controlled, the rescue tube-type RFD can be pulled around the victim below the armpits, and the buckles fastened. This places the victim in a secure ring of buoyancy and puts some space between the victim and the rescuer. The swimmer can then tow the victim to the safe zone with a forward swim stroke, whether that is back to shore, parallel to shore, or farther offshore.

## Passive Victims

Passive victims present in water as unconscious or unable to help themselves. Injury, hypothermia, and even emotional shock can all prevent a victim from attempting any self-help. What if victims can't help themselves? No problem. If using a rescue buoy, simply slide your arms under theirs, from behind, and grab the float, holding it to their chest by the handle. (**FIGURE 12-5**). This places you in a great position, on your back and able to use your legs to swim or tread. It also pins the victim between the two flotation devices available: the buoy and you. This is one drawback to the buoy; if the victim cannot hold onto it, some energy from the swimmer will need to go toward keeping the victim afloat all the way to the safe zone.

If the rescue tube is used, the victim can be placed face up with the rescue tube between the victim and the rescuer. This lets the rescuer control the head (and therefore the airway), keeping it drier. The rescue tube advantage is clear here: it can be fastened with the snap and ring, making the provided flotation more secure.

## No Tool Rescue

Although the USLA recommends an assigned RFD for every lifeguard on duty (and extras in the cache), there could be a possibility of a rescue attempt with no supplemental flotation. Of course, it goes without

**FIGURE 12-5** Unconscious victims can be held against the rescuer's chest, using a rescue buoy to "pin" them.
Courtesy of Steve Treinish.

saying that this is a high-risk scenario. With neither the victim nor the rescuer having flotation, it will not be easy to transport the victim any distance. Efforts should be directed at calming a victim down, maintaining a dry airway, and perhaps awaiting more help. Assistance in the form of other rescuers or teams must be underway to the incident before direct assistance is rendered.

Getting to the back of the victim is important, as the tows and stabilizing positions all result from the victim lying back on the rescuer and thereby getting the airway and head up and out of the water. A rescuer can throw an arm over the victim's shoulder, across the chest, and grab the other side of the victim, lie back, and kick with the legs and remaining arm. The hip corresponding to the arm doing the reaching should support the victim, with the rescuer lying on their side.

Another method has the rescuer reaching under the armpit in the same fashion, but this places the rescuer's head much closer to the water. Yet another method places both hands under the victim's armpits from behind, leaning back, and kicking the legs. Any of these will be nearly impossible to do in surf conditions. Stabilizing the victim and awaiting rescue rather than traveling may be the better option.

## Submerged Victims

Even with fast-moving watercraft and highly skilled rescue swimmers, rescuers may arrive to find a victim submerged. It is entirely possible to witness a submersion while a rescue attempt is ongoing. If this occurs, witness interviews should be done as soon as possible by supporting rescuers, and triangulation used to mark a potential PLS. Victim timelines also must be weighed against the incident. A common elapsed time to stay in rescue mode is 60 minutes. Some AHJs will stay in rescue mode for 90 minutes. Whichever time is selected, it should be based from the time of the submersion, not the time the incident started. A witnessed submersion can often be estimated by using the time of the call. These skills are discussed in much greater detail in Chapter 1, *Understanding and Managing Water Rescue Incidents*.

## Human Chains

The use of a human chain, especially in cases where the submerged victim may be close to shore or suddenly went missing, is extremely effective. Rescuers simply place themselves in a line and either join hands or lock elbows, then proceed to shuffle along the shore. It is not unusual to see human chains of 25 to 30 rescuers working a beach that is shallow. Thirty rescuers walking along the bottom will cover a *lot* of ground very quickly. See Chapter 5 for a detailed discussion of human chains.

## Free Diving

**Free diving**, also known as submerged victim rescue, is the act of descending below the water surface to retrieve a submerged victim, without using SCUBA gear. The many physiological problems that can arise from using compressed air are not a factor in free diving. Because no compressed air is filling up the lungs while they are under pressure, they will expand to only their normal size at the surface. With some practice, free dives down to 30 to 40 feet are possible, and because there may be some visibility in the water, especially offshore, it may be possible to free dive to a victim and bring that victim up to the surface.

## Free Dive Gear

Free diving is normally done with a mask, fins, and a snorkel. The dive mask should be a standard dive mask, made with rubber or silicone and equipped with glass lenses to reduce scratching. Dive goggles can be used, but only in depths of around 10 feet or less, as the airspace between them and the eyes cannot be equalized. Mask squeeze, caused by increasing water pressure on the face, can be alleviated by exhaling a small amount of air into the mask through the nose.

Snorkels attach to the mask and are used to breathe while swimming facedown in the water. They will fill up when submerged, and must be cleared, either by a one-way valve built into them, or by the

**LISTEN UP!**

When on the water, dive masks should either be kept in place or pulled down around the neck when not in use. Placing the dive mask on the forehead may be faster, but the mask could easily pop off the wearer's head and be lost in the water.

user blowing forcefully through them on the surface. When surfacing, the first thing that should be done is to clear the snorkel with air from the lungs. **Clearing the snorkel** is done by forcefully exhaling air through the snorkel. Failure to do so may result in the swimmer inhaling a large dose of water. When taking a breath or breathing through a snorkel, holding the tongue behind the teeth causes any residual water to hit the tongue, and it can be spit back out through the mouthpiece. Dry valves may be on the end of the snorkel, and are supposed to shut when submerged to keep water out, but in rough surf these valves may be damaged. Even snorkels with dry valves should be cleared. This prevents inhalation of water if the valve fails.

While swimmers responding to a live victim may not bring a mask, fins, and snorkel in the water, if they have advance notice that the victim has submerged they can bring these things along when they respond. While these pieces of equipment are not absolutely crucial to the dive, no one likes salt water in their eyes, and they do make the dive much easier to handle. For the sake of this text, free diving will be explained with the assumption that mask, fins, and snorkel are in place.

## Surface Searches

With good visibility, it is possible to place several free dive–equipped rescuers on the water together to search bottom areas. This is commonly known as *line searching*, and uses a length of rope to control a search pattern. It is similar to tethered dive searches. A circular search comprises a rescuer who is stationary over the PLS, with other rescuers swimming in a circle around the "anchor" rescuer. This keeps all divers the same distance apart over the entire circle. The width between divers can be adjusted based on bottom visibility at the time. As the searching rescuers complete the circle, they can widen the circle pattern and search again.

Similarly, groups of snorkelers can search in a straight-line fashion; much in the manner of mowing grass. A group of rescuers swims side by side at a set distance, then the same line of rescuers is moved laterally

# Voice of Experience

I started a new job as a lifeguard-paramedic along Florida's Panhandle after spending over a decade as an ocean guard in San Diego County. California guards learn early that surf rescue fins are irreplaceable and carried while on duty at all times. The use of surf fins, however, was not yet a standard among my new Florida lifeguard colleagues, and I was often the butt of many snarky jokes poking fun at my affinity for my duck feet. Sometimes I found myself avoiding the use of my fins on routine rescues to mitigate the hazing.

At the end of a busy August weekend filled with 6-ft ground swells and numerous rescues, my partner and I were dispatched to a report of multiple victims caught in a rip current just down the street from our station. I was assigned to the "wet seat" position, which put me first in the water. Upon arrival on scene we spotted only one victim caught in the head of a big recirculating rip current. After a quick size-up I entered the feeder current leading to the neck of the rip, deploying my Peterson RFD tube and donning my trusty surf fins. My partner remained beachside to handle comms and to back me up. Although this was setting up like a routine rescue, I could not have been more mistaken.

Using the rip current, I was able to navigate to an intercept position quickly to effect rescue. As I approached the victim, he appeared exhausted as he suddenly lifted the arm of his 240-lb friend who was submerged just underneath him. Now I had two victims, one of whom was a legitimate resuscitation job. Quickly I grabbed the submerged victim's arm and pulled the body to the surface, where I clipped him into the RFD. Once I had control of the situation, I signaled my partner for assistance.

Next, I grabbed the fatigued conscious victim by the waistband using the back of his swim trunks as a fulcrum to keep his upper torso above the water while using my other arm to control my unconscious victim securely clipped into my RFD.

Caught in a monster rip 150 yards from shore with two victims, all I could do was kick with my fins to tread water and wait for my partner to arrive.

Seeing my arm signal, my partner communicated to dispatch the need for additional personnel and EMS resources just before entering the water to back me up. Within 1 minute, also using surf fins, he reached my location and took control of the conscious victim. As we began our swim diagonally out of the rip toward shore, we got caught in the impact zone for what felt like 5 minutes. After taking numerous plunging waves on the head, I made the final swim to shore using the side stroke and kicking as hard as I could with 240 pounds of dead weight. I was tapped, exhausted, and on the verge of throwing up. All I could see was a mass of emergency lights gathered on the beach as the sun deeply set under the west horizon. My victim was taken from me at water's edge and moved quickly to dry sand where he was resuscitated. My partner's victim was taken to the BLS staging area for evaluation.

The memories of that day remind me of the importance of surf rescue fins as primary safety equipment imperative to safely controlling multiple victims in a hazard. Never forget your fins as a surf rescue tech. Treat them as you would treat your turnout gear, helmet, and air pack on a fire call: never leave them behind.

**Mike Hudson, NREMT-P**
Captain, Surf Rescue Team 43-88
Sea Bright Ocean Rescue
Borough of Sea Bright, New Jersey

and they return back to the original line. Preserving or marking the PLS on shore documentation charts is paramount in subsequent surf or subsurface searching later in the incident.

## Free Diving Concerns

Free diving does pose some inherent dangers. Entanglement is a significant risk, especially in areas with thick vegetation in or below the water surface (**FIGURE 12-6**). Boat crashes can result from ropes or lines in the area, and if a rescuer becomes entangled underwater with only a snorkel, there are literally only seconds to mitigate the problem. Cutting tools, if attached to the body cleanly, can be used to cut free of entanglements, but surf rescuers do not commonly carry them. A dive knife and sheath could be kept with the free dive gear and used, and may save a rescuer later. Some surf rescuers carry a knife or pair of trauma shears so that they will have the ability to quickly cut a line in the water if needed.

**FIGURE 12-6** A host of hazards can lurk beneath the water, including seaweed, fishing lines, rocks and coral, and cable and rope.
© Jones & Bartlett Learning.

### SAFETY TIP

Water clarity can affect safety underwater, as objects that may entangle a diver may not be seen until the diver tries to ascend and is trapped. This limited visibility can be from natural sources such as silt or algae, or may be from rescuers kicking up bottom material as they ascend.

## Shallow Water Blackout

Free diving can last only as long as the rescuer can hold his or her breath. After swimming to the location and fighting surf and currents, that might not be too long. Many people envision the act of hyperventilating before submersion, but according to Lifesaving Resources, a maximum of two deep breaths should be taken before submerging. This limits the amount of oxygenation that occurs. If too much oxygenation occurs, the urge to breathe will be delayed, and the extended time underwater can lead to shallow water blackout.

**Shallow water blackout** occurs when the brain is deprived of the oxygen it needs to function—it is nothing more than lapsing consciousness. The problem for rescuers, though, is being in the water while unconscious.

When the body is submerged, all the gas pressures in it increase because of the surrounding water pressure. Assuming we are at sea level, the pressure of the oxygen in the air we breathe at the surface of the water is 2.94 pounds per square inch (psi). When we submerge, that pressure increases. The pressure drops as

we burn the oxygen in our system, but the problem comes when we surface. As we surface, the pressure drops some more, and when humans are exposed to an oxygen partial pressure of lower than about 2.4 or 2.5 psi, we lose consciousness (black out).

Another problem is holding the breath for long periods of time. The urge to breathe isn't caused by a lack of oxygen; it is caused by a buildup of carbon dioxide. That urge can be fought somewhat by exhaling small amounts of air throughout the dive. So, to free dive, we need to get down, and the quicker we get down, the longer we can hunt for the victim.

## Getting to the Bottom

Another issue when free diving is getting down to the bottom. A rescuer struggling to dive underwater using improper technique will not really get anywhere, and will be quickly frustrated. Rescuers can descend feet first or head first. Feet first descents can be done in shallower water and present an advantage in that the rescuer does not need to change his or her body position when contact with a victim is made and the victim is brought up to the surface. The disadvantage is that the swim fins can only be used on the ascent. For deeper water, diving and descending headfirst offers good visibility of the bottom area, allows the rescuer to use surf fins to kick himself or herself down, and is faster. **SKILL DRILL 12-1** shows the easiest and most efficient way to get underwater.

# SKILL DRILL 12-1
## Performing a Free Dive NFPA 1006: 20.3.2, 20.3.3

**1** The swimmer lies prone on the water, looking at where he or she would like to submerge.

**2** The swimmer bends at the waist 90 degrees and raises his or her legs up vertically.

**3** The legs are brought up vertically and held motionless, and the arms are pointed down to the bottom to streamline the swimmer's profile in the water.

**4** The swimmer's torso and arms form a dart shape, and the weight of the legs will drive the swimmer into the water. When the legs and ankles submerge, the swimmer can start swimming with the fins.

## Ear Squeeze and Other Squeezes

The pressure from the water also squeezes other parts of the body, and it is mainly felt in the ears. Discomfort from pressure felt in the water is known as **squeeze**. This pressure must be equalized to relieve the discomfort. There are different things that divers can do to achieve this. Some people can simply swallow and move the jaw around a bit, while others need to pinch their nostrils shut and gently blow against them. Still others may have to tilt or lean their head while doing both. Doing whatever works to clear your ears for you on an airplane is a good place to start underwater. When a diver ascends, the pressure usually equalizes itself easily and automatically.

Although ear squeeze is the most common squeeze on the body when underwater, other areas of the body are also susceptible. Sinus, stomach, and even tooth squeeze can all occur. They cause discomfort, but are usually alleviated by surfacing, or even just ascending a few feet.

## Underwater Recovery

If the victim is found on the bottom, the main task is grabbing and getting the victim back to the surface. There is no perfect hold or capture for these victims, as both the rescuer and the victim may have minimal clothing. Grab from the back, placing your arm under one of the victim's armpits, around the front of the chest, grasp the other side of the victim, and kick for the surface.

This is where the fins really help, and are why many consider them a necessity in surf rescue. Kicking up with bare feet is very ineffective with a dead weight victim, even underwater, but getting up, staying up, and getting in are much easier with the fins on. Bodies underwater generally weigh 8 to 15 pounds, but when the victim is on the surface, the true body weight becomes apparent. It is often much harder to keep an unconscious victim on the surface than to bring them up, and this is the one area where not having a PFD on is a disadvantage. Equipping the rescuers on the surface with PFDs may be the best of both worlds.

Unconscious people cannot hold their breath, so expansion injuries are not a factor for them. Besides, they are not breathing, and even in the water, the ABCs of emergency care still apply. Bringing a victim up may be a down and dirty evolution, and it is important to have other rescuers and abundant flotation at the surface to prevent a resubmersion. Immediately securing the victim into a rescue tube

will at least keep the victim from slipping from a rescuer's grasp and resubmerging. Additional rescue tubes can be used by rescue swimmers in the same fashion, providing everyone with flotation while the victim is packaged and extricated from the water. Extrication via a Stokes-type litter, parbuckling, or even plain old muscle into a watercraft is preferred, if possible, so that emergency medical treatment may begin based on AHJ protocols. At a minimum, chest compressions should be performed if possible. Returning an unconscious victim to shore through an impact zone is nearly impossible. However, if it must be done, the focus should be on the speedy return to shore. Emergency care (in the form of cardiopulmonary resuscitation [CPR]) may need to be delayed to expedite the extrication from the surf.

**LISTEN UP!**

At least one rescuer must maintain flotation for the victim at the surface. Time will be of the essence with a patient not breathing, but rescuer roles must be known. Preventing all the rescuers from reaching for a litter basket at once is crucial. Know your role!

## Offshore Communications

Each agency should employ a set of hand and whistle signals used to communicate the status of the swimmer to a person who is somewhere else, be it on shore or a boat. Because it might be possible that lifeguards are responding along with technical rescuers, a mention of the signals they have accepted as standard would be pertinent here, as they are used in many places around the world.

- OK! (or under control): This is signaled by forming a large circle with the arm, pointing the hand down onto the head, or by connecting both hands in a large circle above the head. This is the "OK" signal used in dive operations (**FIGURE 12-7**).
- Need assistance: One arm is lifted and held up steady (**FIGURE 12-8**).
- Resuscitation needed: One arm is lifted up and waved from side to side (**FIGURE 12-9**).
- Missing swimmer: Both arms are lifted up and crossed, forming an X (**FIGURE 12-10**).

Keep in mind that there are some differences in the signals that fire or rescue agencies might use. This

**FIGURE 12-7** The "OK" signal is given by raising both arms into a circle over the head.

Courtesy of Steve Treinish.

**FIGURE 12-9** Waving one arm above the body from side to side means resuscitation is needed.

Courtesy of Steve Treinish.

**FIGURE 12-8** An arm raised straight above the head motionless means assistance is needed.

Courtesy of Steve Treinish.

**FIGURE 12-10** An X shape by crossing both arms means a swimmer or victim is missing.

Courtesy of Steve Treinish.

is a classic example of the need for preplanning with the responding agencies at a scene. Can you imagine a lifeguard on a tether signaling "help needed," and the fire crews think it means "stop!" That might end up being a major problem, and it would obviously be a safety issue.

# Surf Extrication

Getting to a victim and providing positive flotation is only two-thirds of the battle. There are only a few options to rescue swimmers: taking them inshore, taking them offshore, taking them to a watercraft (which may be staged in better conditions), or staying put.

## Inshore Removal

The good news is that a victim who is conscious and able to help himself or herself can be taken to shore. The bad news is that getting through a pounding, plunging surf zone could injure one or both of you and you already have enough on your plate trying to handle everything. An injured rescuer is certainly not what the victim wants, and a victim being knocked silly may change everything for a swimmer who was counting on the victim helping with his or her own rescue. What is usually done is to head outside the surf zone; however, there are some things that might make an inshore swim easier.

First, dealing with rip currents, especially those known to occur with some frequency, can be tricky. Remember that they can dig out channels when flowing offshore. Swimming parallel to the beach may offer shallow water sooner, allowing you and the victim to stand or wade sooner. The sooner you get your feet underneath you, the quicker the transport to calmer water.

Second, your victim will not be in any mood to have his or her head forced underwater, even by the waves, but it is likely that this will happen. Explaining what is going to happen and advising the victim that he or she should stay with you and the provided flotation might make a difference between a victim making it through the surf conditions with you or the victim panicking and thrashing instead. If trying to get into shore using a rescue board, remember you have to balance a victim on it now, and it is easy to be thrown off the board. It is extremely difficult to paddle any water-borne device, such as an ocean rescue board, in surf with an unconscious or incapacitated victim aboard. An RFD should be attached to the victim from first contact, even if the board is used to attempt transport, and rescuers should consider waiting for help to arrive before moving victims who cannot help themselves.

Either way, once your feet are back on terra firma, the worst is over, especially if you are a member in a crew. Be careful that the upwash and downwash do not knock you or the victim down and make sure that you stop at a good distance inshore. Stopping immediately after reaching land might not drown someone, but another mouthful of water is annoying. Even worse would be CPR being interrupted by a wave hitting everyone.

## Offshore Removal

Often it will be easier and safer to get the victim to a calmer place, even if that area is farther from shore. Beyond the surf lines, where the water is only showing swells, lies water that is normally easier to deal with. There may be some vertical rising and falling with the swell on the surface, but the crashing and churning action of the breaking waves can be avoided.

Helicopters and watercraft both can be deployed for victim removal, and after they arrive, it should be a routine egress from the water to safety. (These means of victim removal are covered in more detail in Chapter 2, *Watercraft Rescue Operations*, Chapter 3, *Watercraft Rescue Technician*, and Chapter 16, *Helicopter Rescue Support*.)

## Surging Wave Areas

One of the highest risks in surf rescue is working around surging waves. These waves do not fizzle out on shore; instead, they crash into it. Found around cliffs, seawalls, or even anchored large ships, they can easily tear apart boats and people. Making matters worse is the fact that many of these areas contain sharp and jagged rock or coral. Staging watercraft far offshore, away from these hazards, is a good way to go, and coming in from above is frequently the best bet. The San Diego Lifeguards utilize a specialized high-angle rope rescue rig, because operation around surging waves is so difficult. See Chapter 11, *Surf Rescue Operations*, for further discussion of rescue in surging waves.

If this is impossible, think about combining rescue methods. Maybe take a throw bag along on the PWC, and throw it from a distance. If the victim hangs on, a slow tow to safer water may let the PWC sweep in between wave sets and make a rescue. Remember, if that line snags on a rock, the wave action and the boat moving out will pull very suddenly and very hard. Be ready to let it go. Another possible problem is the boat or motor failure. Be on the lookout for lines, plants, or anything else that may cause the boat to lose propulsion, thus losing the ability to stay out of the crushing action at the wall.

## Land Line Surf Rescue

Tethered swimmer rescue in surf can be done and is another variation of "live bait" rescue. Used mainly in smaller, calmer surf, it may help with safety during night or limited visibility rescues by physically attaching the swimmer to the shore, but it is not generally used as a front-line evolution. Tethered surf rescue can be dangerous because you are now pulling a line that can become entangled anywhere from the shore to the rescuer. The distance from the shore is often much longer than the typical surface rescue, with line lengths of 200 to 300 yards possible. The issue is towing the weight of the line through the surf and where the current or wind may take it. Hundreds of feet of line, even the small-diameter variety of water rescue line, creates a lot of drag, and it can be tossed around in surf and create an entanglement. However, it can be done. Rescuers undertaking this evolution should be sure to use surf fins to add propulsion and increase swim effectiveness.

## LISTEN UP!

Small-diameter (quarter-inch [6–7 mm]) polypropylene water rescue rope is a good choice for tethered line rescues in surf. Polypropylene floats, of course, which reduces entanglements on the bottom, but the smaller diameter makes swimming it out easier.

These lines are often stored on reels, kept in rope bags, or in some cases, in plastic storage type totes.

When deploying a line rescue, one rescuer swims, one or two rescuers or crewmembers on the shore deploy the line at the water's edge, and another rescuer stays at the reel, bag, or basket, and pulls line off as needed. The swimmer dons an RFD, in this case a rescue buoy. The tether line is attached to the handle opposite the buoy's tether, and the swimmer enters the water. Shore crews should try to keep the line as high as possible to keep as much of the line out of the water as possible and reduce drag. Once the victim is reached, the packaging and stabilization occurs as without the tether. When the swimmer gives the signal to retrieve, the line is pulled in like any other live-bait rescue.

The rate of the return pull is crucial to success and safety. Overeager crews that get the signal to pull to shore can easily cause both the rescuer and the victim to submerge, similar to swiftwater rescue. This submersion will not cease until the tension on the line is reduced or eliminated. In the event of entanglement or a submersing force, the rescuer should doff the strap of the RFD and secure the victim.

## SAFETY TIP

Some agencies tow a tethered swimmer out to a victim, and shore crews pull them back in to shore. This does save some rescuer exertion, but can be extremely dangerous. The watercraft towing the swimmer will likely have to maneuver a great deal in the surf, and with a line behind the watercraft, entanglement risk is high.

# *After-Action* REVIEW

## IN SUMMARY

- Although other types of surf rescue equipment are used in conjunction with rescuers—such as watercraft, personal watercraft, and helicopters—frequently it comes down to one person working in possibly extreme conditions to save another.
- Technician-level surf rescuers must show and possess better-than-average swimming skills because not only do they have to get to the victim, but they will often need to tow the victim to a surf-free area.
- There are two main types of rescue flotation devices (RFDs) in use today by professional lifeguards and rescue personnel—the rescue tube and the rescue buoy. Both can be towed hands free when swimming using a harness placed over the rescuer's shoulder.
- For most open shore areas, a run to a point closest to the victim will save time spent in the water and will let rescuers enter the water with less fatigue.
- Surf rescue is often done with no personal flotation device (PFD) to let the rescuer submerge in the surf to escape potential injury from wave action.
- If a victim submerges while a rescue attempt is ongoing, witness interviews should be done as soon as possible by supporting rescuers, and triangulation should be used to mark a potential point last seen (PLS).

- Each agency should employ a set of hand and whistle signals used to communicate the status of the swimmer to a person who is somewhere else, be it on shore or on a boat.
- There are only a few options in extricating victims from surf: taking them inshore, taking them offshore, taking them to a watercraft, or staying put.
- Tethered swimmer rescue in surf can be done and is used mainly in smaller, calmer surf. It may help with safety during night or limited visibility rescues by physically attaching the swimmer to the shore, but it is not generally used as a front-line evolution.

## KEY TERMS

*Access Navigate* **for flashcards to test your key term knowledge.**

**Clearing the snorkel** Blowing forcefully through a snorkel full of water with exhaled breath to expel water, so no water is inhaled on the next breath.

**Free diving** Submerging and swimming down into the water column, without any breathing air other than what is in the lungs.

**Porpoising** A faster way to enter the surf zone, by jumping in headfirst, using shallow dives over shallow waves, instead of running.

**Shallow water blackout** Passing out from reduced partial oxygen pressures in the brain while free diving.

**Squeeze** Pressure felt in or on the body from an increase in the surrounding water pressure. Unless equalized, pressure differentials can cause serious injury.

# On Scene

Your chief officer has committed your crew as the primary rescuers to an ocean rescue. Your PPE cache includes a rescue tube, surf fins, mask, and snorkel. Your engine carries an ocean rescue board, and your entire crew is trained to the technician level. When you arrive, you see a person in the water, waving for help. The water area he is struggling in has less wave action and is foamy compared to the surf around it. He is about 125 feet (38 m) from shore, and is in the area of a corner formed by the beach and a jetty, with a longshore current pushing water toward the jetty. His wife meets your crew on the beach and says her husband is a good swimmer, but he cannot make progress to shore. You have a boat capable of launch three miles (4.8 km) away at an inland harbor, but they have not been dispatched at this time.

**1.** In what type of hazard does the swimmer appear to be struggling?

**A.** The longshore current pushing laterally along the beach

**B.** The waves that likely washed him offshore

**C.** A rip current created by the longshore current pushing against the jetty

**D.** Surging waves bouncing off the jetty to open water

**2.** Your best action at this time may be to:

**A.** make verbal contact if possible and coach the victim to better conditions.

**B.** place rescuers on the jetty with throw bags.

**C.** use any available civilian surfers to effect rescue using their surfboards.

**D.** perform a swimming rescue using your PPE.

**3.** Thinking ahead, what is a good way to plan for possible rescue issues?

**A.** Have the boat dispatched in case the victim cannot break from the rip current, and use at least one of your rescue crew to back-up a rescue swimmer, using the rescue board.

**B.** Cover the jetty with your crew, spread out with their PPE, and let them decide when to commit.

**C.** Call in the U.S. Coast Guard for helicopter support.

**D.** Consider staying offshore, out of the surf zone, and await pick-up from the boat.

**4.** You decide to enter the water yourself and perform a swimming rescue. The best practice would be to:

**A.** stage your backup rescuer in close proximity.

**B.** paddle the rescue board, wearing the mask, snorkel, and fins in case you need to free dive for the victim.

**C.** put yourself as close as possible to the victim's estimated location considering current, perform a swim rescue using the free dive gear, and have a rescuer deploy the rescue board as backup and assistance.

**D.** wait for the ocean watercraft to transmit they are en route in case the rip current pulls you both farther out to sea.

*Access Navigate to find answers to this On Scene, along with other resources such as an audiobook and TestPrep.*

CHAPTER **13**

## Operations Level

# Dive Equipment and Use

## KNOWLEDGE OBJECTIVES

After studying this chapter, you should be able to:

- Describe the relationship between water contamination and personal protective equipment. (**NFPA 1006: 18.2.6**, pp. 313–316)

- Identify the components of the SCUBA rig. (**NFPA 1006: 18.2.6**, pp. 316, 318–320)

- Describe the role of SCUBA cylinders. (**NFPA 1006: 18.2.6**, pp. 316, 318)

- Identify types and purpose of regulators. (**NFPA 1006: 18.2.6**, pp. 318–320)

- Describe the components of a full face mask. (**NFPA 1006: 18.2.6**, pp. 320–321)

- Describe the purpose of manifold blocks. (**NFPA 1006: 18.2.6**, pp. 321–322)

- Explain the purpose and use of submersible pressure gauges. (**NFPA 1006: 18.2.6**, p. 322)

- Explain the purpose and use of dive computers. (**NFPA 1006: 18.2.6**, p. 322)

- Explain the purpose and use of body harnesses. (**NFPA 1006: 18.2.6**, pp. 322–325)

- Explain the purpose and use of masks, fins, and snorkels. (**NFPA 1006: 18.2.6**, p. 325)

- Identify types and purpose of cutting tools and punches. (**NFPA 1006: 18.2.6**, pp. 325–326)

## SKILLS OBJECTIVES

There are no skills objectives for this chapter.

# Technical Rescue Incident

You are new to the station that houses the dive team for your department. Although you are well versed in most ladder work you will be assigned while with Truck 2, you are not well educated on the dive team or its gear. You have just a few days here, and between moving in, runs, and division training on the computer, you have not had time to ask the other crewmembers about dive callouts. Of course, right after dinner, the team deploys. You help secure the scene, and the lieutenant of the engine asks you to help dress a safety diver. This diver is a good friend and tells you he can walk you through the process.

**1.** Should an untrained rescuer assist assembling the life safety equipment worn by dive rescuers?

**2.** What are the limitations of being used as shore help as a dive rescue team member?

**3.** What if the diver tells you he can help you help him and insists you can do it?

Access Navigate for more practice activities.

## Introduction

Public safety diving is a large and expensive part of water rescue, and because it is not nearly as common as some of the other types of water rescue, it can be somewhat of a spooky mystery to many untrained rescuers. For many dive rescue teams, especially those inland, incidents are not common, and they can suffer from "invisibility" until they are needed, making it hard to keep members trained and the team supplied. Dive rescue is training and equipment intensive, and while the crew members on the surface can be protected by the usual flotation equipment and thermal insulation, diving equipment is a whole new game.

This chapter examines the different equipment and its uses. There is a huge range of equipment available to rescue divers, and this text acknowledges the differences in dive philosophy worldwide. Whatever method a dive rescuer deploys, exposure to temperature, chemical factors, and biological factors must be considered, along with a means of communication and the idea that a diver who becomes trapped underwater is on a finite time plane. Air management during a dive is crucial and must be planned for and adhered to. Divers must also learn how to handle physical and mental stressors they may encounter underwater, prior to needing these skills in real life.

In the next two chapters, we examine how to plan a dive and how to safely control a diver, as well as the actual dive. Keep in mind that not every dive requires every piece of gear, and many agencies around the world operate differently in this discipline. Equipment and skills may work for one team that works in hazard-specific water, but may never be used by a different team

in the water in their jurisdiction. The standards have changed to reflect encapsulation diving, and this can be somewhat controversial. A public safety diver retrieving a weapon on a beach in Florida may not need a drysuit, whereas northern divers may need to wear drysuits year-round. This is why the authority having jurisdiction (AHJ) has the ability to tailor dive personal protective equipment (PPE). For many areas around the world, chemical and biological contamination in the water are factors enough to consider diver protection.

The next two chapters also focus on dive rescue and include information on what can affect dive operations later in the mission. Many, if not most teams are tasked with the other two facets of public safety diving, which are body recovery and evidence recovery. While true dive rescue technicians may not be called on to perform these duties, procedures during rescue can greatly affect later operations.

NFPA 1006, *Standard for Technical Rescue Personnel Professional Qualifications*, is rather specific when addressing diver needs. Factors such as safety, redundant air systems, communication between the diver and tender, and chemical and thermal protection are listed as things the AHJ must consider when undertaking dive operations. Many agencies outfit divers with various types of gear, and very different thoughts on water contamination are found around the country. The job at hand also plays into the gear that could possibly be used. A weapon recovery at a beach with clear, tested, clean water might be undertaken with just SCUBA equipment and a set of trunks. Alternatively, diving and swimming through the bottom sediment in an inland retention pond or pulling a victim from a car under the ice warrant much more protection, both chemical and thermal.

# Shore PPE

As always, protecting the shore personnel needs to be automatic, not an afterthought. Any rescuer working in peripheral roles should be using proper thermal and flotation protective equipment, namely warm clothing and a personal flotation device. Gloves can help prevent rope burns and scratches from working shorelines with brush or trees, and long-sleeve clothing in these areas may prevent poisonous plant exposure such as poison ivy, oak, or sumac. The use of structural firefighting equipment around the water is not recommended. For more information, refer to Chapter 4, *Surface Water Rescue Operations*.

# Water Contamination and PPE

The U.S. Navy is considered to be one of the world's foremost diving units. They dive incredible missions in every type of water known to man. When a freeway bridge collapsed in Minneapolis, Minnesota, in 2007, U.S. Navy divers entered the water at 2:00 AM, immediately after their arrival on the scene. This was not ego; it was absolute confidence in skills, ability, and equipment, and a fierce desire to get the job done. In 2004, the U.S. Navy released a report that basically says that all waters, especially inland waters, are considered contaminated to some degree, and makes recommendations on diver protection for varying degrees of contamination. This speaks volumes, especially in a world where information and hard data on contaminated waters are not always clear. A risk-versus-benefit decision is made for each dive location, and this decision should take into account the perceived benefit weighed against the protection the diver has available.

This concern is magnified when one considers that public safety divers operate on the bottom (remember negatively buoyant victims?), searching in or on top of the sediment. The Navy report states that many chemicals settle into the bottom sediment even though they may not be present in the water above it.

If nothing else, consider that one type of mission that rescue teams may encounter frequently is removing someone from a car that has entered the water. Vehicle submersions are often fairly close to shore and present a large target that is often easy to find quickly. Now think of the petroleum that may be burping up from the cylinders, along with the engine oil, transmission fluid, and whatever else the car is carrying. Also, think of all the sediment that is stirred up when the car hits the bottom. Finally, think one more time about the unexpected things that might be carried in vehicles. Protection from all of these is a must.

## Exposure Protection

Divers may wear a range of clothing while operating underwater. Swimsuits, wetsuits, and drysuits are all possibilities. Clean waters are considered to be safe to dive in using recreational SCUBA gear, but if the act of recovering humans with unknown injuries is considered, you may be dealing with at least some possible biological hazards present in the water. Chemicals are ever-present in most water, and a census of safety data sheets (SDS) now declares that there are over 5,000,000 chemicals in existence. One chemical, tributylin, is considered by many across the world to be one of the most toxic chemicals ever introduced into the marine environment. What is it used for? Antifouling paint commonly used on oceangoing ships. A good understanding of possible water contamination or pollution in your area will help you make an educated decision on the type of protection warranted for a given rescue.

## Wetsuits

Wetsuits are constructed with open-cell and/or closed-cell neoprene, and are worn tight against the skin. They offer protection from cold water but do little to protect from chemical contamination (**FIGURE 13-1**). In fact, wetsuits work by trapping a layer of water between your skin and the suit, and allowing body heat to warm the water to a comfortable level. If there is a chemical in that water, such as gasoline, it is trapped against the wearer's skin, and in all probability, a chemical burn of some kind will be sustained. Wetsuits also are hard to clean and decontaminate after a dive, and they can fall apart if subjected to certain chemicals.

**FIGURE 13-1** Wetsuits offer protection from cold water but do little to protect from chemical contamination.
Courtesy of Steve Treinish.

In many waters, however, especially oceans and clean reservoirs, rescuers do use wetsuits. Wetsuit thicknesses range from one-half millimeter for very warm water, to seven millimeters for very cold water. A two-piece suit provides a double layer of neoprene on the body's trunk area, and a two-piece suit with thick gloves, booties, and a hood can protect a diver under ice for extended periods of time, though that time might vary between divers. Some divers wear a larger set of denim jeans over the wetsuit to protect the neoprene. Another point to mull over in wetsuit use in near freezing water is the possibility of dive time being extended, purposely or not. If an entanglement adds 15 minutes to a dive, and a rescuer started the dive with a below-normal core temperature, hypothermia may set in and reduce the diver's dexterity or worse.

## Drysuits

A **diving drysuit** can protect a diver thermally better than neoprene wetsuits, because the diver wearing it does not normally have water held against the skin. Even though they are commonly referred to using the single term "drysuit," diving drysuits differ from surface or swiftwater drysuits due to the addition of an inflator valve and a deflator valve. These valves are controlled by the diver and help manage buoyancy and pressure squeeze underwater. Drysuits are mainly made from three types of fabric: neoprene, nylon tri-laminate, and butyl rubber (**FIGURE 13-2**). **Neoprene drysuits** keep the wearer dry, but do absorb water into the fabric and can be hard to decontaminate. **Trilaminate drysuits (tri-lams)** and **butyl rubber drysuits** are more protective, and they are easier to clean after use. Another big advantage to tri-lams is that because of their light weight, many teams are starting to use them for other disciplines, such as ice or swiftwater rescue, when matched with other PPE. Some manufacturers have published test results documenting the effects of various chemicals on their suits. Keep in mind that the zipper, the seams, and the latex all will have different permeation rates for different chemicals.

All drysuits are inflated underwater by the user with a small amount of air during the course of the

**FIGURE 13-2** Drysuits are mainly made from three types of fabric: neoprene, nylon trilaminate, and butyl rubber.
Courtesy of Steve Treinish.

dive. This further insulates the diver from the cold water and increases diver comfort by relieving some of the water pressure that squeezes the suit while the diver is wearing it. **Dry hoods** and **dry gloves** that are just as chemical resistant as the suit must be worn to provide **diver encapsulation**. Diver encapsulation means the diver is operating in a totally dry state, provided no leaks in the system occur (**FIGURE 13-3**). Divers sometimes wear neoprene hoods and gloves with drysuits; however, this allows an exposure through the ears, even with a full-face mask (FFM). Neoprene gloves allow whatever chemical may be in the water to be trapped against the skin of the hands, exposing the diver to the risk of contamination and infection of the nicks and cuts that are common to the hands.

In hotter sun or warmer temperatures, divers may risk *hyper*thermia, especially the backup diver, who may be staged in shallow water. Even with no thermal protection worn, the body can quickly overheat due to a lack of sweat evaporation. Consider shade for the diver in the form of an umbrella or staging location, or letting the diver sink his or her body underwater. Many divers will splash cool water on their head when

**FIGURE 13-3** Dry hoods and dry gloves that are just as chemical resistant as the suit must be worn to provide diver encapsulation.
Courtesy of Steve Treinish.

topside and dressed, as the latex hood is black and heats skin quickly.

## Dry Hoods

Dry hoods are latex or rubber hoods that attach to the suit around the neck area. They protect the diver's ears and head by covering them as part of the main suit. They also provide a good seal to an FFM. Thermal insulation should be worn underneath the hood in colder water. Divers must also be aware of issues that may arise from the ear canal being sealed off by the squeeze of the rubber, blocking the ability of the diver to equalize his or her ears.

## Dry Gloves

Dry gloves are just what the name implies: gloves worn to keep the hands totally dry and, in theory, protected from the chemical and biological contaminants that teams are trying to keep off the diver in the first place. Most gloves used are a gauntlet style that attach to the suit by being forced up and over a plastic ring installed in the wrist seal. This style of glove is fairly inexpensive, easily replaceable, and can offer great protection. Other styles attach "bayonet" style, using locking rings or sliding O-ring joints. Some insulated dry gloves are sealed or glued onto the mating system bayonet; however, once they are badly cut or torn, they become useless. Gloves can be installed so that air in the suit travels all the way into the fingers, but this leaves the wrist seal compromised. If a glove is ripped underwater, the whole arm or even the suit may flood, resulting in water reaching the wearer's skin. If a glove is torn (it is a pretty safe bet this will happen at some time), suits with intact wrist seals prevent the suit from taking on water through the wrist area and the amount of leaking water is limited to the hand area.

## Drysuit Insulation

**Drysuit insulation** packages are commonly manufactured as coverall-type garments, commonly available in two thicknesses: one for 70°–50°F (21.1°–10°C) water, and thicker insulation for 50°–32°F (10°–0°C) water. Above 70°F, (21.1°C) divers should consider skipping the thermals, and above 80°F, (26.7°C) they are not recommended. At water temperatures that high, using minimal insulation minimizes the risk of hyperthermia, but remember that the water temperature may drop as the diver descends deeper through the thermocline. This could result in an uncomfortable dive, and summertime dives placing divers in cold water present unique challenges with both hyper- and hypothermic risks. On the flip side, the use of cotton undergarments in colder conditions can be uncomfortable, as sweat will quickly soak the cotton and chill the diver quickly. Fabrics that wick moisture away from the diver, such as polypropylene, are highly recommended. Another comfort issue comes when the insulation rides up the leg, exposing the diver's skin to the suit. The suit will conduct cold, and any place it touches bare skin it will feel cold. Take the time when dressing to make sure the insulation is tucked into the leg fully, and consider using thermals with foot strapping, which keeps the insulation in place on the legs.

Diving in a drysuit equipped with a dry hood, dry gloves, and FFM provides much better protection than using recreational equipment, but even a drysuit/FFM setup is not a guarantee against exposure. In a true hazardous materials dive, such as raw, untreated sewerage or harmful chemicals concentrated at high-risk levels in the water, the diver must be outfitted in an encapsulating drysuit with dry gloves and a **surface-supplied air (SSA) system**, complete with a diving helmet (**FIGURE 13-4**). Diving helmets are the ultimate in diver protection and are designed to handle these types of dives. However, because surface-supplied systems generally have a longer setup time and a large umbilical cord to deal with, they are not preferred for true rescue work. If the water conditions or environments are truly contaminated, it may be necessary to come to a "No Dive" decision and use or call in other agencies that have the protection and capability of an SSA with helmet system.

**FIGURE 13-4** Surface-supplied air (SSA) system.
Courtesy of Mark Bender, Brecksville Fire Department.

# SCUBA Equipment

Fully assembled, typical public safety SCUBA equipment can easily weigh upwards of 75 to 100 pounds, depending on the equipment needed for a particular dive. Next, we will break it down and take a look at it, piece by piece. For the sake of this text, a typical public safety dive setup will be examined (**FIGURE 13-5**). Quick deployment packages are also available and are used by some agencies. They will be examined in the next chapter.

## Checklists

Checklists are becoming more and more common to the fire service, especially in high-risk, low-frequency runs like dive rescue. The checklist is used to check off diver PPE, equipment status and setup, emergency equipment, and a final check. Many divers learn the BWRAF acronym in beginner dive classes. This stands for Buoyancy, Weight, Release, Air, and Final check. While meant to be used verbally, many times parts of, or even the entire verbal check can be overlooked. Applying this type of checklist into a written document can help prevent seemingly obvious mistakes in dive preparation or equipment setup, especially when under the stress of an actual rescue.

**FIGURE 13-5** Assembled public safety dive equipment.
Courtesy of Steve Treinish.

## SCUBA Cylinders

The most common SCUBA cylinders hold 80 cubic feet of breathing gas. Ambient air is by far the most common gas compressed into the cylinders, but advanced divers may use different gas mixes on specific dives. Although they are not commonly used on public safety dives, some teams are using mixed gases as a way to increase dive safety at moderate depths.

The most common of these different gas mixes is **Nitrox**, which is a breathing gas that contains an increased percentage of oxygen. Nitrox is used to displace excessive nitrogen intake when breathing, and it is usually 32% or 36% oxygen, rather than the normal 20% to 21%. At deeper depths, nitrox can cause problems in divers, so be sure of what breathing gas is in the cylinder. Most nitrox cylinders have bright green markings or striping, but this is not always the case. SCUBA cylinders are usually made of steel or aluminum, with the aluminum 80-cubic-foot cylinder being the dive industry standard (**FIGURE 13-6**). Some divers use steel cylinders, which range in size from 60 to 120 cubic feet. SCUBA cylinders can also be used in a manifold system with double cylinders on a diver, with the diver controlling which cylinder and regulator he or she breathes from. The use of two cylinders is often referred to as diving "doubles" and is used by some teams. The smaller set of double cylinders supplies more gas underwater while offering a smaller physical profile.

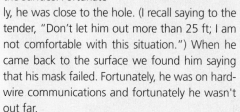

# Voice of Experience

It was a cold early spring morning, the last ice dive of the year. The ice was thin, and we would be boat-based, as the ice would break if we stepped on it. Divers were being checked, a checklist was being used, we had consistent equipment among team members, and then there was two. Two individuals arrived from another department in a hurry; they had to get back to shift because of a manpower shortage. They needed to get the ice dive for their requirements and needed to get out of there as soon as possible.

They had some equipment that was standard issue for their department but was different than what everybody else was using. We told them to go ahead and check off each other as they were both diving the same gear and should be familiar with it.

Then the domino effects started. It was one thing after another: forgotten weights, forgotten chest harness, discomfort. I had a bad feeling, but we pressed on. We had to get the dive in and make it count. The divers' shift coverage was being paid overtime so they could go to this dive. How could they go back and tell the chief, "I didn't get my dive in," when I was sent to do the training. Press on, get it done, we can handle it, right?

I wasn't running this dive; the person who was is an outstanding team leader. The two late rescuers were not from his department either. We had a training safety officer on site too. We had some very experienced rescuers on site, and the safety diver was a member of my team. In fact, I was the 90% diver in this case, but we (the team) missed critical steps and we almost had an issue that would have been devastating. We had a near-miss simply because we were in a hurry. Checks weren't done as we normally do, standardized gear wasn't being used by all, and ultimately this would come down to the lack of training.

What ended up happening is that the diver dove his pony bottle under the ice. When he ran out of air, instead of switching the switch block he panicked and pulled himself back to the surface. Fortunately, he was close to the hole. (I recall saying to the tender, "Don't let him out more than 25 ft; I am not comfortable with this situation.") When he came back to the surface we found him saying that his mask failed. Fortunately, he was on hard-wire communications and fortunately he wasn't out far.

As it turns out, the checks weren't done correctly; if they had been, the responders would have figured out that the switch block was in the wrong position. As it turns out, training was never done, this was the first time this person had ever used that equipment, and he had never had a pool session with it. The gear was issued, but the department hadn't trained in a controlled environment first.

After this incident, we took a hard look at the way we were doing things. First and foremost was to stop hurrying! We realize that, yeah, there are time constraints, but this would have been the ultimate time constraint if it had led to a fatality.

We realize that doing standard checks and using standard equipment is vital. We realize that using supplied air is highly beneficial, especially when diving under the ice! But it really all comes back to training: the basics, a good tender, and most importantly the recognition that divers push the limits and somebody needs to stop them.

**Todd Rishling**
Lieutenant/Paramedic, Water Rescue Team Leader
MABAS Division 1
Elk Grove Village Fire Department
Elk Grove Village, Illinois

**FIGURE 13-6** SCUBA cylinders are usually made of steel or aluminum. Here is an 80-cubic-foot cylinder on the left, and a smaller, redundant cylinder on the right.
Courtesy of Steve Treinish.

Deep (deeper than 30 to 40 feet) no-visibility public safety dives can be a much more complex mission, and teams may elect to use larger twin cylinders for these dives. A cylinder that is more than 10% depleted should be filled or replaced. For example, a 3000-psi cylinder that is under 2700 psi should not be used in a public safety dive until it is refilled. Divers simply do not know when they may need every last breath of air available in the bottle. A fire fighter probably would not keep an SCBA on the truck after a morning check if it were half full, as that air means life in dangerous atmospheres. It is no different for the diver; a person can live days without water, weeks without food, but only minutes without air.

**SAFETY TIP**

Most aluminum SCUBA cylinders require 3000 psi to fill to the rated volume completely. Steel cylinders have a greater range of pressure capacity, ranging from 2450 psi to 3450 psi. There are also carbon fiber SCUBA cylinders on the market that operate around 4200 psi. The fill pressure is stamped into the cylinder somewhere on the top. The important fact is that most high-pressure firefighting SCBA systems fill to 5500 psi. If you fill a SCUBA bottle off an SCBA cascade, and you do not reset the fill pressure, you will likely rupture the burst disc on the SCUBA cylinder valve. The burst disc is there as a pressure relief for safety, and it can easily damage a person's hearing if it ruptures.

## Redundant Supplies

Many public safety divers use smaller SCUBA cylinders as a redundant air supply. Frequently called **pony bottles,** or "ponies," they are usually attached to the diver's SCUBA rig as a completely independent gas supply and delivery system, and can be plumbed to an FFM or used with another regulator the diver can breathe from if needed. Ponies range from 6 cubic feet to 40 cubic feet, and are usually made from aluminum. There are steel ponies on the market, but they are not as common as the aluminum ponies. Figure 13-6 shows a pony bottle attached to SCUBA equipment. They are commonly equipped with their own regulator systems, and if rigged properly, divers can hand them off underwater to other divers needing air. Donating a redundant air supply can be as simple as exchanging second-stage regulators underwater and continuing the dive. For users wearing FFMs, plumbing the redundant air supply into a manifold or gas switching block offers the ability to switch air supplies while keeping the mask on, preserving communication and reducing the gasp reflex that may occur when the face is subjected to cold water. Manifold blocks are discussed further later in this chapter.

Pony bottle systems can be set up to allow a diver to bail out from entangled gear and make a safe ascent to the surface. One concern that arises with this releasable setup is that if a diver is so entangled he or she must leave the other gear on the bottom, that diver might well find that fins, harnesses, or body parts become tangled while on the way up, and then may have a very limited air supply.

**SAFETY TIP**

Bailing out of an entanglement is considered by many agencies to be a last chance, extreme evolution, and divers and teams should be well trained before attempting this evolution.

## Regulators

All that gas has to be delivered to the diver, however no one can breathe from a high-pressure cylinder without some help. Regulators take the high cylinder pressure and make it workable for the next stage regulator. Only then is it ready to be inhaled, just like a fire fighter's SCBA. Some regulators supply air for various parts of the SCUBA rig, too. Next, we will examine different types of regulators.

### First-Stage Regulators

A **first-stage regulator** takes the high-pressure gas from the cylinder and reduces it to around 125 to 140 psi or

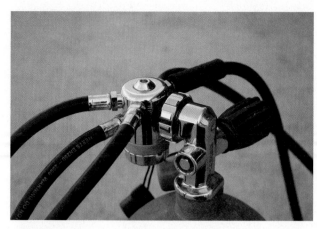

**FIGURE 13-7** A first-stage regulator takes the high-pressure gas from the cylinder and reduces it to around 140 psi.
Courtesy of Steve Treinish.

so (**FIGURE 13-7**). First-stage "regs" are the first step in supplying the gas in the supply to the rest of the SCUBA equipment. First stages operate with a piston or a diaphragm-type pressure reducer, and both types can be equally effective. First-stage regulators for contaminated water should be environmentally sealed. This simply means the regulator is built in such a way as to not have any contaminated water touching the internals of the regulator. It is also resistant to internal chilling, which can allow a free-flow condition. Free flows are discussed later in this section. Many first-stage regulators attach to a SCUBA cylinder with a **yoke-and-screw system** and seal against the cylinder valve with a partially exposed O-ring. A common mistake some divers make is to tighten the yoke screw too much, which damages the O-ring. A damaged O-ring can result in a gas leak at the regulator where it seals to the cylinder.

**DIN regulators** (DIN stands for Deutsches Institut fur Normung, a German standardization group) screw into the cylinder valve directly, and cover the O-ring completely. They will not interchange with a yoke-style cylinder or regulator, but there are adapters on the market.

**Free Flows.** Free flowing is a phenomenon in which the regulator cools internally, which allows the internal pieces of the regulator to become coated in frost, causing it to stick open. This lets gas blow uncontrolled from the mouthpiece. As air is decompressed, it cools. Think of an SCBA bottle being bled down quickly for maintenance. As the air is pushed out, it cools, and will often chill the valve to the point that it becomes coated with frost. Put a first stage in very cold water, flow gas (whether breathing or supplying valves) through it and it may get so chilled that it starts to flow air continuously, thereby depleting cylinders in a very

short time. If this happens, modern regulators are designed to fail in a free-flow position, so that the diver does not suddenly find himself without air to breathe. Environmentally sealed regulators reduce the chances of free flow, but any regulator can free-flow in cold enough conditions. A well-trained diver can handle a free flow easily and simply abort the dive. Be advised that a well-built first-stage regulator can easily drain an air supply in less than a minute.

Keeping the regulator as warm as possible may help reduce free flows. Often, staging a backup diver in very cold air temperatures will pre-chill the regulator enough to cause a free flow within a few breaths underwater. Heat packs can be held on regulators if conditions are extreme. Once a regulator free-flows, it cannot be stopped until it is warmed up enough to melt the internal frost, allowing the regulator valve to re-seat and stop the flow.

### SAFETY TIP

Regulators are designed to flow air when a failure occurs, but this may not always happen. Occluded or plugged filters on first-stage regulators can block the air supply.

## Second-Stage Regulators

**Second-stage regulators** are the business end of the system—the end that delivers the gas to the diver's mouth (**FIGURE 13-8**). They attach to the first-stage regulator with a low-pressure hose (140 psi or so) and provide the diver with the proper amount of air he or she needs to stay healthy underwater (physiology will be discussed in more detail later). Most divers carry two second stages. The first is the primary air source, which the diver uses as his or her main air source. The

**FIGURE 13-8** Second-stage regulators deliver the gas to the diver's mouth.
Courtesy of Steve Treinish.

other second stage is a backup regulator, also known as the alternate air source, or an **octopus**. The diver's backup second stage is primarily used for other divers in trouble. By donating the backup air source, two divers can breathe from one diver's cylinder. This lets them make a proper, slow, controlled ascent to the surface. One problem that the public safety diver faces when using these types of second-stage regulators is that they are held in the mouth by the teeth, thus making them susceptible to being pulled out, and creating an exposure point by possibly letting water into the mouth. This type of second-stage regulator is also known as a recreational-type regulator.

### SAFETY TIP

Even divers diving in recreational gear who make a focused effort to clench the regulator in their mouths could suffer from an exposure. Water that has seeped in as the diver moves his or her head around and/or water leakage occurring through the exhaust ports can contain enough bacteria or chemicals to sicken that diver.

Another issue with using the octopus in public safety diving is the fact that it has an open mouthpiece. Most second-stage backup regulators are clipped onto the diver somewhere in the golden triangle, which is the area between the neck and each hip. The problem is that the front of the diver is usually dragging or operating in mud, debris, algae, and who knows what else from the bottom, thus dragging the mouthpiece of the second-stage regulator through it. This will likely fill it up with the same gunk. Divers must be aware of this if they need to use the second-stage regulator underwater, and should take the necessary steps to clear it before breathing from it.

Another type of second-stage backup regulator setup is a regulator mounted on the inflator valve that hangs off the diver's left shoulder. The idea behind this setup is that it eliminates one extra hose, because the diver donates the second stage he or she is breathing from to the diver who needs it, and then switches to his or her backup on the inflator tube.

### SAFETY TIP

Many divers use lift bags to bring up objects from the bottom. Larger lift bags can be plumbed into an air supply on shore, but smaller bags are frequently filled underwater. Separate air cylinders are recommended for this. Recreational divers often use the backup second-stage regulator for this, but cold water free flow can occur from the air moving past the first stage.

### KNOWLEDGE CHECK

What usually causes a regulator to free-flow?
a. Frost forming inside the regulator
b. Poor maintenance
c. Sticking of the secondary regulator exhaust valve
d. Excessively dry cylinder air

*Access your Navigate eBook for more Knowledge Check questions and answers.*

## Full-Face Masks

Positive-pressure **full-face diving masks** protect the wearer's eyes, nose, and mouth from water entry. Rather than letting water in, air should bleed from leaking spaces along the sides of a mask, just like the modern firefighting SCBA face piece protects from smoke and gases. The other protection an FFM can offer is the ability to maintain a dry airway in an unconscious diver, allowing a rescue diver to bring up the unconscious diver without having to keep a second-stage regulator in place in the unconscious diver's mouth (**FIGURE 13-9**). Thermal protection is also much greater with the airspace created between the facial skin and the water. Most FFMs have a silicone or rubber skirt material that seals well to a rubber drysuit hood, and some have a double silicone seal to help properly fit different-sized users.

FFMs are a specialty diving certification within many agencies, and divers should be trained in their

**FIGURE 13-9** A full-face mask can help maintain a dry airway in an unconscious diver, allowing a rescue diver to bring up the unconscious diver without having to keep a second-stage regulator in place in the unconscious diver's mouth.

Courtesy of Steve Treinish.

**FIGURE 13-10** Once a mask has been removed, it becomes contaminated by whatever is in the water.
Courtesy of Steve Treinish.

use before using them in the field on an actual mission. In the unlikely event of doffing an FFM underwater, the diver may have to fight a gasp reflex underwater. Cold water hitting the face of a diver may cause an involuntary gasp due to the body's reflex of airway spasms sealing and protecting the airway. Even water as warm as around 60°F (15.6°C) can trigger gasp reflex. Splashing cold, clean water on the face before the dive may help prevent a gasp reflex.

Another problem that divers commonly cause for themselves is taking the mask off at the surface, especially if the mask does not have hard-wire communications. It is easy for a diver to pull the mask off at the surface and talk to the tender or another diver, especially in training evolutions. But a mask that has been removed and left to hang has now been contaminated with whatever is in the water, the very contaminants the diver is trying to stay away from (**FIGURE 13-10**). Newer divers often put the mask back on after shaking the water out of it, not realizing what they have just done. This water can be atomized and pushed deep into the lungs. A good team rule is that each diver wears their mask into the water and leaves it on until he or she is clear of the water or, if needed, the diver is washed or decontaminated.

## Ambient Breathing Valves

**Ambient breathing valves** are mentioned here as they are used only with FFMs. If a diver dons an FFM, he or she has to breathe, even if staying above the water surface as a backup diver. This would drain the gas supply from the cylinder and would not allow the diver

**FIGURE 13-11** Ambient breathing valve on a full-face mask.
Courtesy of Steve Treinish.

**FIGURE 13-12** This manifold block is mounted on the diver's full-face mask and can be used to control gas supply from two sources.
Courtesy of Steve Treinish.

to enter the water with a full gas supply. To let a diver breathe outside atmospheric air while masked up, the ambient breathing valve can be installed in the side or the nose pocket of most masks, and the valve can be closed when the diver goes underwater and starts breathing from the gas supply (**FIGURE 13-11**).

## Manifold Blocks

**Manifold blocks** are used to switch between two gas sources carried by a diver. Some blocks are mounted or hung on the diver's side, and some blocks are attached to the FFM (**FIGURE 13-12**). There are two main types of gas switching blocks. One can be mounted anywhere on the SCUBA rig within reach of the diver and is used by turning a knob one way or the other. The other type is a block that attaches directly to an FFM and is used by pushing a small cylinder up or down. Once again, personal preference comes into play. Caution should be used with either system, though, to make sure the gas supply is set to the desired cylinder before the dive begins. You should also make sure that modifying any other equipment does not cause liability or legal issues.

**KNOWLEDGE CHECK**

Which is the function of a manifold block?
a. To add air to the buoyancy control device or vest
b. To provide air supply access for buddy breathing
c. To clear fog or water from a face mask
d. To switch between two gas sources carried on a diver

*Access your Navigate eBook for more Knowledge Check questions and answers.*

# Submersible Pressure Gauges

The only way to know how much air is in the cylinder is to check the gauge. Each first-stage regulator should have some way of showing how much air is in the cylinder to which it is attached. The most common method is a hose coming from the regulator attached to a **submersible pressure gauge (SPG)** (**FIGURE 13-13**). This hose is plumbed into a high-pressure port on

**FIGURE 13-13** A submersible pressure gauge is used to determine how much air is in the cylinder.
Courtesy of Steve Treinish.

the regulator, and it will give a psi (or a pressure bar reading) normally in the hundreds of psi. Most SPGs glow in the dark, and many have digital readouts. Depth gauges are usually attached in a console with the SPG, but are not connected into the air supply or regulator system directly. Depth gauges are normally found above the SPG to allow for the SPG hose. A diver with visibility in the water can tell at a glance how much air remains, and how deep he or she is. Many depth gauges have a maximum depth indicator, or "tattletale." This is a small wire that travels around the depth gauge with the depth indicator needle but stays at the deepest point reached. Tattletales are handy, but must be reset between dives.

More and more analog SPGs are being replaced with wireless units that transmit data from the first stage to a small computer on the diver's console or arm. As technology progresses, heads-up displays are coming onto the market, just as they are in some SCBAs found in the fire service now. This display should prove invaluable for the public safety diver, because information such as depth and gas remaining can be seen inside the mask, regardless of the visibility in the water. This will greatly increase dive safety.

Of course, diving with no visibility means the diver may not be able to see the SPG. This is where shore crews will need to plan the diver's estimated air consumption, and adhere to a strict timeline to ensure the diver surfaces with enough air, and early enough to keep some air as a backup. This topic is examined more closely in the next chapter.

# Dive Computers

Most recreational divers learn to make dive profiles, or plans for each dive. Computer technology has developed, and now **dive computers** put depth, time underwater, time remaining for the dive, and a dive log all in the same amount of space in which a depth gauge once fit. Dive computers also can be strapped to the diver's wrist. Many teams are utilizing wrist-mount computers and feel there is less entanglement risk than with a console unit. Any dive computer can recall many aspects of the dive that can be recorded in the logbook later. A good tender, though, will be able to tell the diver most of these things without the computer.

# Body Harnesses

Most every dive in the public safety realm is done with the diver attached to one end of a rope, with a surface tender on shore holding the other end. This rope can be an electronic communication rope or a line quickly grabbed off a truck. Whatever rope is used, it is only as good as how you attach it to the diver. Some divers tie a loop or knot in it and simply hang on. Not only does this potentially take away one hand the diver may need to work underwater, but the million-dollar question now becomes, "*How does the backup diver find the primary diver if the primary loses his or her rope in blackwater?*" In a typical ice dive, the diver does not have a direct route back to the entry hole if the rope is dropped or inadvertently detached from the diver. Some divers attach the line to a buoyancy compensator buckle or D-ring with a small carabiner. Once again, if the D-ring is made of plastic, and it breaks, or if the diver doffs his or her rig to untangle himself or herself, is there a direct path for the backup diver to find the primary? Negative! The goal here is to be firmly attached to the rope. This is done with a locking carabiner and a **body harness**.

Harnesses can be a special-made diving harness or can be a piece of rescue webbing tied in a water knot and worn over the shoulders, figure-eight style (**FIGURE 13-14**). They can attach to the diver's tether line with a carabiner or a quick-release snap shackle, which gives the diver the ability to quickly bail off the rope with one hand. A snap shackle gives the diver the ability to free himself or herself of the rope and make an ascent with one quick tug. This might be especially useful in current diving when dealing with the added forces caused by the moving water, but can also present problems if accidentally tripped while trying to find the attachment point to help orient a diver underwater. Each team will have to investigate these options and make that decision. Regardless, it is strongly recommended that the body harness be worn under the **buoyancy control device (BCD)**, attached to the tether line. Another concern is making sure the diver's harness is not adjusted too tightly.

**FIGURE 13-14** Harnesses can be a special-made diving harness or can be a piece of rescue webbing tied in a water knot and worn over the shoulders, figure-eight style.

Courtesy of Steve Treinish.

To check this, the diver can take a large breath while wearing full insulation to ensure respirations are not impaired.

### SAFETY TIP

With hundreds of BCD styles and models on the market, teams should consider standardizing the gear they wear and how it is set up. Imagine trying to know all the different buckles, straps, weight pockets, and gear pockets of divers under duress when those divers are all using different setups. If everyone dresses in the same equipment, and all tools are in the same location, muscle memory can be used to your advantage!

## Buoyancy Control Device or Vests

The BCD, or buoyancy control vest (BCV), is the vest or harness-type device the diver wears to offer positive buoyancy at the surface, buoyancy control in the water column, and also hold the gas cylinders. Divers need to control buoyancy underwater to hover at any depth. Divers use buoyancy control to place themselves above objects or debris such as aquatic growth and fallen trees, allowing them to swim and search without getting tangled. This is especially crucial when the water offers some visibility for the searching diver. The major parts of the BCD or BCV include the strapping system to hold it on the diver; an air bladder, which the diver inflates and deflates to control his or her buoyancy; a cylinder strap or two that hold the SCUBA cylinder firmly in place; and an inflator hose, which is used to inflate or deflate the bladder (**FIGURE 13-15**).

Some BCDs also have an integrated weight system built into them, so the diver may elect not to place lead weight on a belt. Most BCDs also have several pockets, and public safety divers generally place cutting tools, window punches, flashlights, and other gear in these pockets. The air bladders can be placed in one of two areas on the BCD: either a jacket style, which places the air around the diver at rib level and also behind him, or a back bladder, which places all the air behind the diver, normally in a horseshoe shape around the top of the cylinder. Either BCD can be worn, but generally the jacket-style BCD is easier to manage with a victim in the water, as the air around the ribcage will provide a better body attitude in the water. Back bladder BCDs, if not set up properly, have a tendency to float the wearer a bit facedown in the water, making the diver work to stay vertical and keep the victim's airway dry.

**FIGURE 13-15** Buoyancy control device.
Courtesy of Steve Treinish.

Some BCDs have chest straps and crotch straps, and at least one BCD on the market has a full harness system built in for aerial operations. The tenders should know how the BCD "dresses" on the diver and should be comfortable helping the diver in and out of it.

### SAFETY TIP

Diving under an ice shelf or in a confined space can become a technical and dangerous dive if things go wrong. A locking carabiner attached directly to the harness should always be used on these dives, with no snap shackles. *Never* simply hold onto a rope under ice. Coming off rope under an ice shelf in blackwater is a grave danger.

## Weight Belts or Harness Systems

Most every diver will have to use some kind of **weight system** (also called a weight harness, a weight belt, or a weight pocket) to overcome the positive flotation caused by the exposure suit he or she chooses to wear. The amount of weight required by the diver depends on the individual's body composition, cylinder and gear setup, and other factors. Some divers place this lead on a nylon belt, with the belt laced through the lead and worn just like a duty belt (**FIGURE 13-16**).

A

B

C

**FIGURE 13-16 A**. Weight belt with weights held by lacing the belt through the lead. **B**. Weight harness with weights held in pockets that are releasable. **C**. Integrated weight pocket used as part of the BCD. The entire pocket containing the weight is released if needed.
Courtesy of Steve Treinish.

Some divers use the BCD's integrated weight system, and some use a suspender-style system that holds a weight belt or pouch under the BCD. The harness systems should have a quick release pocket or quick release buckles on the suspenders to dump the weight. Weight-integrated BCDs often use a clip or buckle, Velcro tabs, or a combination of the two to hold the weight in the BCD, but allow divers to drop it if needed. The idea is to be able to jettison the weight if an emergency arises. It is beyond the scope of this text to recommend a specific weight system; however, regardless of what weight system they ultimately choose, divers must be able to get rid of it, as their lives could literally hang in the balance.

---

### SAFETY TIP

Some divers use a combination of weight-integrated BCDs and a harness or belt. It spreads the weight around and makes doffing and donning underwater easier, as neither the diver nor the SCUBA rig will have such a tendency to rise to the surface. Many times a weighted diver will sink, and the air in the BCD will cause it to rise above the diver. This can be dangerous if the diver loses his or her grip on the BCD, as the regulator could be pulled from the mouth or head. Divers, surface swimmers, and backup divers all should be able to dump whatever weight setup is used in the blind, without fumbling.

---

## Recreational Masks, Fins, and Snorkels

At least one thing is common to divers, whether they dive for public safety or recreationally. They all need fins to propel themselves in the water. The fin straps can be rubber straps with a pull-type adjuster or a set of stainless steel springs. Springs seem to work better for public safety diving for two reasons. First, they are easy to don and doff, always hold tension on the foot, and are more durable than the rubber heel strapping found on many fins, which can break if pulled too hard or if the rubber degrades. (Fins are discussed further in Chapter 4.)

Snorkels are generally frowned upon in most public safety dives. That is not to say a surface swimmer or surf rescuer should not have them—quite the contrary. However, a diver wearing a snorkel in no-visibility water, or blackwater, is asking for major entanglement trouble by wearing one. The snorkel is attached to the dive mask, so an entanglement with it may pull the mask off. This would expose the diver's nose and eyes to the water and take away good vision. A snorkel could be used in clear waters while being used with a recreational mask, and that is a team decision. One item that may be worth noting is a pocket snorkel. Pocket snorkels are small, thin snorkels that can be rolled and stored in a pocket then pulled out and used if needed.

Recreational dive masks should be defogged before every use using a commercial defogger paste or, in a pinch, mucus from the nose. The defogger is applied to the glass inside the mask and rinsed lightly. Many divers will point out that blackwater conditions may not let them see anything, but even 4 to 6 inches (10 to 15 cm) of sight may come in handy at times, especially if you are tangled.

## Cutting Tools and Punches

The blackwater diver's best friend is a good cutting tool, or even two or three cutting tools. In public safety diving, it is not *if* you get tangled, but *when*. Most water contains trash, and even oddball, everyday things can tangle a diver. Some teams and departments recommend carrying a minimum of two cutting tools; some three; and some leave it up to the diver to carry even more if they feel the need (**FIGURE 13-17**). Blunt-tip knives should be one-hand drawable, and the blunt tip can reduce accidental punctures of suits or skin. Trauma shears can be found in many firehouses, and they are inexpensive to replace if lost. Cutting tools can be kept with the SCUBA equipment in pockets, attached to the body harness, or attached to the BCD. If using gear retractors, remember that the retractor may not reach down to the legs or feet to cut some-

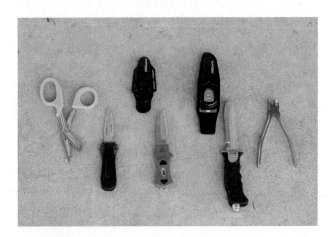

**FIGURE 13-17** Cutting tools.
Courtesy of Steve Treinish.

thing. Many dive rigs have one set on a retractor for security, and one set free for longer reaches. Whatever is carried, at least two tools should be in the golden triangle. To minimize entanglement risks, especially in no-visibility conditions, cutting tools should not be strapped to the outside of the leg. Many divers also carry a seat belt cutter in a pocket, and/or an automotive window punch. There is not a lot of leverage underwater, and the punch makes shattering automotive glass much quicker.

## After-Action REVIEW

### IN SUMMARY

- For many dive rescue teams, especially those inland, incidents are not common, and diving skills can suffer from "invisibility" until they are needed, making it hard to keep members trained and the team supplied.
- All waters, especially inland waters, are considered contaminated to some degree.
- Ambient air is by far the most common gas compressed into SCUBA cylinders, but advanced divers might use different gas mixes on specific dives.
- Regulators take the high cylinder pressure and make it workable for the next stage regulator, or ready to be inhaled, just like a fire fighter's SCBA.
- Positive-pressure full-face diving masks protect the wearer's eyes, nose, and mouth from water entry.
- Manifold blocks are used to switch between two gas sources carried on a diver. Some blocks are mounted or hung on the diver's side, and some blocks are attached to the FFM.
- More and more analog submersible pressure gauges are being replaced with wireless units that transmit data from the first stage to a small computer on the diver's console or arm.
- Computer technology has developed, and dive computers now put depth, time underwater, time remaining for the dive, and a dive log all in the same amount of space in which a depth gauge once fit.
- Most every dive in the public safety realm is done with the diver attached to one end of a rope, with a surface tender on shore holding the other end, but the rope is only as good as how you attach it to the diver.

- At least one thing is common to divers, whether they dive for public safety or recreationally: they all need fins to propel themselves in the water.
- Some teams and departments recommend carrying a minimum of two cutting tools; some three; and some leave it up to the diver to carry even more if they feel the need.

## KEY TERMS

*Access Navigate for flashcards to test your key term knowledge.*

**Ambient breathing valves** Small valves on a full-face mask used to breathe outside air so a suited up backup diver does not exhaust his or her gas supply.

**Body harness** Webbing or strap harness worn by the diver and attached to a tether line.

**Buoyancy control device (BCD)** Jacket or vest that contains an inflatable bladder for the purposes of controlling buoyancy.

**Butyl rubber drysuits** Drysuits constructed of butyl rubber and glued together. Offer excellent all-around chemical protection, but must be worn with insulation for warmth. Usually durable, easy to clean, and field repairable.

**DIN regulators** A type of first-stage regulator in which the sealing O-ring is completely covered in the cylinder when attached. It is common in highly technical dives.

**Dive computers** Devices that track a diver's depth, maximum depth, time of dive, and time allowed to dive. Can be mounted on the SCUBA rig console or worn on the body.

**Diver encapsulation** Personal protective equipment that offers a high degree of facial, hand, and body protection from thermal, biohazard, and chemical contamination by enveloping the user in nonpermeable suits, gloves, and masks.

**Diving drysuit** A drysuit adapted for SCUBA use by installation of an inflator valve and an exhaust valve.

**Dry gloves** Gloves worn by divers to keep the hands dry and free of water exposure.

**Dry hoods** An attached hood worn with a drysuit, offering similar protection as the suit.

**Drysuit insulation** Thermal garments the diver may use under the drysuit to offer warmth.

**First-stage regulator** Apparatus that attaches to the tank valve and delivers air at 125 to 150 psi to the other lines on the SCUBA rig.

**Free flowing** A regulator condition that allows breathing gas to escape the cylinder quickly and uncontrolled; it is a common occurrence in cold water.

**Full-face diving masks** A regulator that delivers air to a diver while covering the face and attaching to the head with strapping, similar to a firefighting SCBA mask.

**Manifold blocks** Units that permit a diver to control which tank the gas is supplied from; it may be mask mounted or hung on the body.

**Neoprene drysuits** Drysuits constructed of crushed neoprene; they can be damaged by chemicals and are hard to clean after exposures, but they do offer more inherent warmth than other kinds of drysuits.

**Nitrox** A special blend of breathing gas with higher than normal oxygen content, used to change dive decompression requirements.

**Octopus** Nickname for the alternate air source carried by divers as a backup source or emergency source for other divers.

**Pony bottles** Smaller SCUBA bottles used as a backup air supply, usually carried with the diver. Also called *ponies*.

**Second-stage regulators** The mouthpiece from which a diver breathes. They deliver air at a usable pressure and volume as needed.

**Submersible pressure gauge (SPG)** A unit attached directly to the first-stage regulator, showing how much gas remains in a tank.

**Surface supplied air (SSA) system** A SCUBA system delivering air to the diver via a hose controlled from the surface.

**Trilaminate drysuits (tri-lams)** Drysuits constructed of nylon layers bonded with cement that offer protection against some chemicals, but require insulation to provide warmth.

**Weight system** Releasable system that holds lead weights the diver uses to counteract positive flotation.

**Yoke-and-screw system** Device that attaches a first-stage regulator to the SCUBA tank.

# On Scene

You have been stationed at the dive team's firehouse for almost 10 years. Although you have no desire to dive in the waters of your local reservoir, you do enjoy helping your co-workers that do. You have a good grasp of the dive gear used by your team and frequently assist during team missions. Almost every diver admits to the manpower concerns that occur on high-stress incidents involving underwater rescue, and while not a diver in the direct sense, you realize that by helping the divers and tenders you are helping to keep them safe while diving operations are underway. A good understanding of the equipment used has the team using you as a "go to" fire fighter when training or working rescues.

**1.** What hot weather problem can occur in direct sunlight in summer?

   **A.** The diver may suffer from thirst, and should stay hydrated.

   **B.** Hyperthermia may set in, and the diver should be kept cool by shading or placing in the water.

   **C.** The diver can get lethargic with heat exhaustion.

   **D.** Equipment can get physically hot enough in sunlight to cause injury.

**2.** You see a cylinder with green striping and the word "Nitrox" on it. What does this mean?

   **A.** It is a brand name of a cylinder manufacturer.

   **B.** It is used for lift bag inflation only.

   **C.** It is a mixed gas and has specific parameters of use.

   **D.** It is a nickname of a diver on the team and is used to label personal gear.

**3.** Divers should enter the water with how many cutting tools?

   **A.** One

   **B.** Two

   **C.** Three

   **D.** At least two, but is dependent on the AHJ

**4.** Encapsulation diving is diving with basic protection from the water and chemicals, and consists of:

   **A.** wetsuit, full-face mask, gloves, and hood.

   **B.** drysuit, wet gloves, dry hood, full-face mask, and boots.

   **C.** drysuit, kevlar rescue gloves, full-face mask, and BCD.

   **D.** drysuit, dry gloves, dry hood, and full-face mask.

*Access Navigate to find answers to this On Scene, along with other resources such as an audiobook and TestPrep.*

# Dive Rescue Operations

## KNOWLEDGE OBJECTIVES

After studying this chapter, you should be able to:

- Determine where to begin dive operations.
  (NFPA 1006: 18.2.1, 18.2.4, 18.2.5, pp. 330–332)

- Describe how to conduct scene investigations at a dive rescue. (NFPA 1006: 18.2.1, 18.2.5, pp. 332–336, 339–341)

- Explain diver physiology and how to plan a dive accordingly. (NFPA 1006: 18.2.2, pp. 341–343)

- Describe how to dress a diver. (NFPA 1006: 18.2.4, 18.2.6, pp. 343–347)

- Identify the components of a pre-deployment diver checklist. (NFPA 1006: 18.2.4, 18.2.6, pp. 347–348)

- Explain the process of deploying divers. (NFPA 1006: 18.2.4, pp. 348–349)

- Identify signaling methods used at dive incidents. (NFPA 1006: 18.2.4, 18.2.6, pp. 349–352)

- Identify and describe search patterns used by divers. (NFPA 1006: 18.2.4, pp. 352–355)

- Describe how to monitor a diver during an operation. (NFPA 1006: 18.2.4, 18.2.7, pp. 355–359)

- Explain decompression illness and why it's a danger to divers. (pp. 359–361)

- Describe how to retrieve a diver from the water. (NFPA 1006: 18.2.3, 18.2.4, pp. 361–364)

- Identify and describe operations-level duties at a dive operation, other than diving. (NFPA 1006: 18.2.4, pp. 364–367)

- Describe how to terminate a dive rescue incident. (NFPA 1006: 18.2.8, pp. 367–368)

## SKILLS OBJECTIVES

After studying this chapter, you should be able to:

- Ready and document dive equipment for diver use. (NFPA 1006: 18.2.4, 18.2.6, pp. 343–348)

- Assemble and perform a pre-dive safety check of the dive rescue personal protective equipment utilized by the authority having jurisdiction. (NFPA 1006: 18.2.4, 18.2.6, pp. 347–348)

- Communicate with a diver or dive tender at an incident scene. (NFPA 1006: 18.2.4, 18.2.6, pp. 349–352)

- Act as tender to a diver. (NFPA 1006: 18.2.3, 18.2.4, 18.2.7, pp. 352–358)

# Technical Rescue Incident

You and your friend Dave have been active on the fire department dive team for several years. You love the activity and the work, and you and Dave have handled many dive missions before this one. The team is well equipped, and training is not an issue. A few months ago, your team was called out for a possible stolen car in a lake at a local apartment complex. There had been several stolen cars dumped there in the past few years. Although the weather during that mission was downright hostile, a few months have passed and your team is pining for something different. You decide to take them to that pond and explore a new training site.

From the shore, it looks like visibility is good for the divers, and it's a beautiful spring day. When you set up, you notice a lack of fish, frogs, or any other life at the water's edge. During the dive, the scribe tells you it seems odd that no Canada geese are around. As a matter of fact, there is no goose waste around either. This is unusual for this terrain, as the slope between the manicured apartment grass and the water is tall and has not been mown in weeks. You commit two of the typical dive crews into the water to train on patterns. You are supervising two primary divers running patterns, two backup divers fully suited and in the water in case of trouble, as well as two tenders, supervisors, and two scribes. Dave is now reporting he is seeing many dead fish lying on the bottom of the lake. After questioning him, the supervisor reports that neither diver has seen anything living at all during their dives. No diver is reporting symptoms, but one tender is saying his hands now itch.

**1.** What personal protective equipment (PPE) will keep the crews safest in this situation? What PPE will the shore crews require?

**2.** What are your actions to protect the team?

# Introduction

Across NFPA 1006, *Standard for Technical Rescue Personnel Professional Qualifications*, operations-level rescuers often function as support to technician-level rescuers. In dive rescue, the operations-level rescuers truly support the divers, who are the technician-level rescuers. Although the dive technicians do the actual diving, they often experience lack of visibility and must therefore rely on the crews working topside to ensure a smooth and safe dive mission. Operations-level dive rescuers must be able to select proper personal protective equipment (PPE), assist the diver in dressing for the mission, select the proper search pattern and where to use it, and have a good understanding of drowning physiology and the human factors that will determine where to start a search. These members must also have an expert understanding of how to handle dive emergencies. An old fire service adage says "the fire goes as the first line goes." Similarly, the tone and feel of a dive emergency will result directly from how the first few seconds are handled.

# Victim Location and Starting Points

Operations-level rescuers in dive rescue must be able to perform witness interviews and develop a location to begin dive operations. Even with witnessed events, some homework must be done to decide on the **point last seen (PLS)**, and as with any homework, knowledge of what may happen before and while the victim is submerging is paramount.

## Drowning Process

Although a person struggling above water may be noisy and dramatic, it is generally thought that the actual act of drowning is a quiet process and, according to some researchers, a fairly peaceful one.

The mental processes of each victim may vary before the true state of self-preservation takes over, but they have to be considered in a mission. This, along with the physiology of drowning, may have a pronounced effect on where dive operations are started.

## Victim Posture

Being a good swimmer does not guarantee survival in the water. The best swimmer may fall victim to cold water, injury, fatigue, or any combination of these or other factors. The classic drowning posture may have the victim instinctively fighting to keep the airway above the water surface, and human instinct does not have the victim wasting energy on screaming. Often the head is tilted back and the arms are out to the side to try to keep the airway above water (**FIGURE 14-1**). It is likely that they may not see or understand instructions or attempts at rescue. Victims are truly in self-preservation mode, and may not even think about actions that can save them. Some drowning victims do not even have the presence of mind to drop an object they were holding, such as a fishing pole, beverage, or glasses.

## Types of Drowning

According to the World Health Organization, the modern definition of drowning is "the process of experiencing respiratory impairment from submersion/immersion in liquid." In any submersions there are physiological reactions that occur, and from these reactions the body may inhale some of the liquid the victim is in. There are entire texts devoted to the physiological aspects of drowning, but for the sake of simplicity, this text will address and discuss each episode as either a wet or a dry drowning, as it relates to the amount of liquid the victim has inhaled.

## Drowning

In 80% to 90% of recorded drownings, the victim's lungs contain some liquid. Commonly termed **wet drownings**, these drownings start out with the airway

**FIGURE 14-1** Instinct will force a victim to enter self-preservation mode, with the head back and arms out. They are unable to yell, because all efforts are now focused on keeping the airway above water.

Courtesy of Steve Treinish.

being protected automatically by the body, through a process called **laryngospasm**. This is basically the airway clamping shut in an effort to keep fluids out. Asphyxia will eventually override this, and the airway will relax. Once the victim inhales, even in fluid, two major things happen.

First, the panic of not breathing subsides. The general assumption is that once the urge to breathe is satisfied, the process of drowning is somewhat peaceful, as the victim becomes confused and relaxes. The blood chemistry, having changed due to the lack of oxygen, changes even more with the inspiration of water. This decreases a victim's movement and thrashing, letting the victim sink somewhat as all efforts to keep the airway above water now go by the wayside.

Second, and more important for rescuers, the victim will likely become **negatively buoyant**. Negative buoyancy is any condition where an object sinks. Upon drawing in fluid instead of air, buoyancy is lost from the person's lungs. This fact plays a major role in determining the PLS, as submerged victims will almost always be found within a certain distance of where they go underwater. The idea of victims traveling distances downstream in current can be perpetuated by the location of where they are found later in time. While it is possible for a person to be transported downstream while drowning, this distance will often be much less than expected. Search times can greatly increase when rescue teams overestimate the distance traveled. Perpetuating this myth is the fact that many deceased victims float to the surface and are found farther downstream; however, this does not happen until after decomposition forms gases in the body, causing it to float and only then be carried by water current.

Clothing can also play a part in drownings in moving water, as heavy clothing can trap air and delay submersion. On the bottom, victims will *usually* be found face down, with their arms and legs on the bottom, because those parts of the body are heavier in the water (**FIGURE 14-2**). In other situations this may not be the case. SCUBA divers will likely end up face up, as the heavier tank affects body position. Infants may float due to baby fat and disposable diapers, and people wearing several layers of clothing may float due to air being trapped in the clothes.

## Dry Drowning

Although the medical community disparages the term **dry drowning**, it is a term used to describe the other 10% to 20% of drownings in which the victim asphyxiates but the lungs stay drier than in a wet drowning.

**FIGURE 14-2** Many drowning victims end up face down on the bottom. Because the legs and arms are the denser parts of the body, they will settle on the bottom before the trunk, which may hold gases.
© Jones & Bartlett Learning.

According to research from the American Journal of Forensic Medicine and Pathology, this percentage could be even lower, around 2%. Remember, the laryngospasm is the body's attempt to seal off the airway when water is sensed in the throat, and can be so forceful that the victim suffocates. The urge to breathe cannot overpower the laryngospasm, and death occurs by suffocation. Eventually the spasm may subside, and some fluid may be drawn into the lungs, but the amount can vary. This may keep the victim **positively buoyant**, or in simple terms, floating, especially if the victim is wearing clothing that provides some buoyancy. These victims can be an easy recovery in still waters, and dive teams may not often witness this, because watercraft or surface swimmers can easily recover the victim.

It can be a challenging, long-distance search if the water is flowing and there is little to snag the body. Victims who are dead before entering the water may float, because breathing has already stopped before submersion. This keeps the lungs drier, resulting in better flotation, often enough to keep the victim on the surface. This is especially true for victims who are wearing clothing that is water or wind resistant. The same technology used to keep us dry and warm can trap air and provide some flotation. This can also be a flag for criminal activity, as people placed in the water to cover a homicide can present the same way. Possible legal issues will be addressed later in the chapter.

## Scene Investigations

Underwater rescue crews are generally faced with two types of scenes: witnessed and unwitnessed submersions.

It can be much harder to arrive at an event when there is no one to provide location information, but using sound logic, reasoning, and good investigational tactics, it is possible to narrow things down enough to enable the go/no-go decision. A **go/no-go decision** is the initial decision to either deploy personnel underwater or to search by other means, after considering all aspects of the impending mission and its parameters.

## Crime Scene Concerns

Every drowning should be considered a homicide until proven otherwise. Although family members shouldn't automatically be handcuffed and seated in the back of a police cruiser, drowning scenes should be treated with special consideration, much like every working fire is considered arson until proven accidental. There are many ways scenes can be manipulated into making a homicide appear to be an accident, and not many rescuers are fully versed in those aspects of investigations.

It is not the job of rescuers to arrive at a decision of guilt or innocence. However, proper scene preservation can easily protect what could be crucial evidence in a crime scene, or save any chance law enforcement has to prosecute or investigate a criminal case. Make note of any potential evidence, including the victim, and document the scene as thoroughly as possible. This might include taking photographs of the scene, interviewing witnesses with the potential of criminal activity in mind, and marking locations where potential evidence was found.

Also remember, the mission may not be criminal, but a proper investigation may be the difference between a death being declared a suicide or an accident. That can factor into life insurance payouts, and it can make a huge difference in the quality of life for children who may now be without a father or a mother.

## Unwitnessed Drowning and Clues

An unwitnessed submersion can still get rescue crews a callout. Information at this time may be a bit overwhelming if there is a lot of it, or it may be scarce enough to cause frustration. The following information will help form an initial plan of action. First, find out who called for help, and determine where they are now. Why did they call? Is there a person who was supposed to have been home by now, or did someone make a good-intent call based on seeing a shoe lying on an ice-covered lake?

**FIGURE 14-3** Scene clues, such as the fishing pole, cooler, and radio shown here, can confirm reports of people in the water. A lack of such evidence may postpone operations until further investigation takes place.
Courtesy of Steve Treinish.

Take a good look around for starters. What is your initial impression of the scene? Are there any signs that anyone was around? An old fishing pole covered in mud may not throw up any flags, but a new pole, sitting in a holder with a fresh bait container would be a good clue that someone was in the area (**FIGURE 14-3**). While the victim might not be in the water, it would be a safe bet that someone was recently at that location. Footprints, especially if they fit the recent environmental conditions, give a good indication of someone at least being in the area, as do beverage cans or food packages. The liquid in them still being hot or cold could support your assumption. Other things to look for are:

- Water surface or bottom sediment that has been disturbed or a muddy cloud in otherwise clearer water. These would indicate a very recent submersion, but coupled with supporting evidence, it would probably warrant a rescue dive if conditions and safety allow.
- Anything unusual floating on the surface or on the shore. Of course, hats, coats, and papers, but also shoes and flip flops. Flip flops naturally float and may be kicked off easily by struggling victims

early in the fight to stay afloat. A pair of shoes with a t-shirt found on the bank could be evidence suggesting that a victim went for a swim, or the items could have been placed there by someone covering up a criminal action. Rescuers generally have no way of knowing this immediately, but evidence such as this *must* be protected and documented. Cars and trucks may have oil or gas leaking from them after they submerge, and oil in most water can be spotted easily from the air. Consider using a helicopter to take a look for this.

- Algae beds that have been broken or disturbed on pond surfaces. High reeds could also be damaged. Many smaller and shallower ponds will have an algae ring wrapping all around the edges of the water. Searches that find this mat undisturbed might lean toward no victim being in the water.
- Look for bank marks. Slips and falls account for a large number of water entries. Some of these will leave definite marks on the bank surface, especially in mud or loose grass.
- A lack of wet bank could mean a person went in the water but didn't make it out. An area of wet bank could also be a sign that the victim got out, but do not rule out the possibility of falling in again. Remember, they are now wet and could be exhausted.
- Blood marks left on walls or docks are another good sign of someone slipping and submersing. Unconscious people don't hold their breath, so as soon as they inhale, they will become negatively buoyant and may be lying on the bottom directly below the entry point.

Any of these items should be kept in place for later investigations, if possible. What may look like trivial pieces of the puzzle may fit perfectly when more is uncovered later, and these pieces also may be used later by investigators to corroborate or counter a person's story. Consider using an adjacent entry point for rescuers, if possible, and taping off the scene and denying entry for anyone except whoever completes the investigation. If any objects must be moved, keep them as intact as possible, record who moved them and to where, and if possible, take a few pictures of them before touching them, preferably from different vantage points (**FIGURE 14-4**).

## Active Searches

A decision will have to be made regarding whether to commit divers based on the evidence and the scene presented at the time. When the evidence on a scene points

**FIGURE 14-4** Try to preserve the shore scene as much as possible by restricting entry. This also lets the dive crew focus on the mission at hand by keeping distractions away.
Courtesy of Steve Treinish.

to an active rescue possibility and the operation can be done safely, a "go" decision is ordered, and the operation becomes an active search. Generally, this means that the PLS is located, the peripheral clues and evidence support a submersion, the timeline is less than the agency's accepted maximum time underwater (usually 60 to 90 minutes), and the personnel and equipment on scene meet the requirements to perform the dive safely.

## Passive Searches

When the evidence points to another possible location of a victim (such as home, their car, or a nearby restaurant), the scene is too unsafe to dive, or there are not enough qualified people to fill the roles needed to dive safely, it may be better to perform passive searches and check the potential locations where the victim might be, search the surrounding land or water areas, and try to make contact with family and friends, if possible. This will require the use of other rescuers, including spotters who may be posted around the water areas watching for signs of a victim. In a typical river search, spotters will often be posted downstream, or at heavy use locations in hopes of finding or seeing the victim. In some cases, a rescue that was thought to be underwater in nature may suddenly change to a surface or swiftwater rescue when a victim who was thought to be submerged appears in another area topside. In these cases, typical surface rescue can be done. For more information on surface and swiftwater rescue, please refer to Chapters 4 through 7.

When the timeline deems an incident a recovery operation rather than a rescue, the passive searches will be expanded to searching for a victim in one of two ways. Using electronic equipment such as sonar or remotely operated vehicles to locate the victim on the bottom will limit the amount of time a diver is underwater,

subjected to risk. Once a body is located, recovery dives can occur at a slower pace, allowing for a higher degree of safety, because the urgency of a rescue timeline is removed. Once a longer period of time has passed, rescuers may be tasked with searching for a body that has floated to the surface.

## Float Time

Many bodies float to the surface as they decompose. As decomposition gases form, a body becomes positively buoyant. There is no magic timeframe for this, because the float times can vary with water temperature, body composition, food and beverage intake by the victim before death, and underwater debris. Floating bodies are often moved by currents, wind, or wave action, and can end up under or snagged by brush or trees alongside the water, or held in current formations such as hydraulics. Bodies are frequently found some distance downstream from where the submersion occurred, because the current carries the floating body downstream before it is noticed. When searching for floating deceased victims, look for legs, arms, and head hanging below the water surface, and minimal torso area above the water. Victims can be hard to spot, so searches must be thorough and detailed. In deeper and/or colder waters, it might take weeks or months before a victim's body floats, if at all. Similarly, in oceans, marine life may consume a body before flotation occurs or the body is found.

**LISTEN UP!**

Remember, every time a diver goes below the surface there is a chance of injury or worse. Looking at the other side of the coin, though, our job is to clear underwater areas, and on small waters, it may be beneficial to go ahead and clear the area. These types of callouts can often be used as an opportunity to provide training and experience. It is unfortunate to have a body float a few days after a team did not see enough evidence to dive.

## Most Probable Point

If evidence is found that would suggest a victim is in the water, but no witnesses are found, logic will have to enter the process. The **most probable point (MPP)** is the point an investigator thinks a victim would try to get to while under stress in the water, or the place where a missing person would have entered the water. A person in the water who is truly in trouble may not necessarily try to find the easiest way out; more than likely they are headed for the closest shore regardless of how rough the shore condition is.

When a victim is in this type of situation, he or she simply sees any shore as safety. No other thought will be given to actually getting up a steep wall or through the fallen trees. This MPP can also be considered if no evidence is in place to suggest a submersion, but there is a missing person. For example, a missing child last seen in the backyard playing is reported missing. The neighbors are outside, but no one has seen the child since the mother last glanced out the back window. You pull up as a rescuer and find a pool in the backyard that has not been filtered or cleaned, and you cannot see the bottom. A logical MPP for that child *at that time* is in the pool, and divers would most certainly be committed. The child may very well walk back across the street 20 minutes later, but as a rescuer, an educated guess will provide this MPP.

An MPP cannot always be reached, however, or may change drastically with later intelligence reports. Many times, that educated guess is all that is possible. It can be maddeningly frustrating to work for hours or even days, only to get a new piece of the puzzle that ends the search in minutes. It is important to remember that decisions are made with the information we have at that time. For this reason, it is crucial for all information collected to be properly maintained, cross-checked, and utilized the entire time crews are searching.

## Witnessed Drowning

Witnessed drownings can range from the simple 30-second dive where a bystander is literally pointing to, standing at, or floating above the spot where the victim is, to recovery operations that take days or weeks. Water depth, clarity, current, and contamination can all be "wrenches in the gears" when it comes to public safety diving. Most rescuers would think a witnessed submersion would be simple, and sometimes it is. However, there are many missions that present with acres of water to search, because the witness was across the lake or on another boat going in the other direction at 60 mph. Often weather conditions can cause havoc, when rescuers must deal with currents or surface winds.

When facing a scene such as this, get reinforcements started toward the scene, including the tools that might help narrow down a large area, such as sonar units and/or aerial and land search crews. If you want to beat the golden hour, a lot of information must be processed, and quickly. On a true rescue mission, several rescuers may have to get involved with witnesses, and a quick agreement of the PLS after the witness interviews may be the best way to get started. Witnesses should be taken to the place on the water where they watched the event occur, even if it means taking them out in a boat. The point of view from two places around or on the water can be very different. Two witnesses might indicate the same general area,

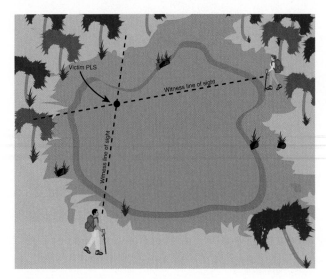

**FIGURE 14-5** Triangulation is using witnesses at different vantage points to narrow down the area where the victim was last seen. Remember to triangulate from where the witness observed the event happen.
© Jones & Bartlett Learning.

but rescuers can narrow it down a bit more using triangulation. **Triangulation** is the term for determining the PLS in an area narrowed down by witness statements (**FIGURE 14-5**). It is simply using intersecting lines of sight from your witnesses to provide a PLS. The PLS is examined in more detail later in this section. With a quick triangulation, you can at least start to set things up, for example, entry and exit points for the divers, EMS staging, etc.

## Witness Interview Techniques

The world is made up of many different people and personalities, and as interviewers, rescuers must take that into consideration. Interviewers have to be part coach, part psychologist, part interpreter, and part listener. People react in different ways to different situations, and screaming, eerie calmness, rambling, and sometimes just "checking out" are all natural reactions that people can have. Emotion plays a huge part in witnesses' behavior, and rescuers should be conscious of not letting what *they* think happened sway the story they are trying to bring forth from the person they are interviewing.

Witnesses should be interviewed separately, but at the place where they saw the event occur (**FIGURE 14-6**). Keep in mind that the witnesses may not have seen the actual submersion, but they might have been the last ones to have seen a missing person alive in the area. Asking questions quietly and listening intently are the two most positive traits an interviewer can have. Be careful about how questions are phrased. People

**FIGURE 14-6** Witness interviews should be done from the place the witness watched the submersion.
Courtesy of Steve Treinish.

under stress may be led into statements that are not entirely accurate. For example, some rescuers are so impatient to get the mission moving that they point to a spot and say something along the lines of, "over there, in line with that big tree on the bank?" They may not even realize they are leading the witness with that question. The witness may not agree with the information suggested by the interviewer, but desperately wanting to help, may be swayed.

Likewise, in a group scenario, the most dominant personality may change another person's statement before the other person gets a chance to speak, even though that person might be able to give a better location. The person who stands up and rattles off a loud and strong line of sight might convince others that they are right, even though others think they saw something entirely different. Let the witness explain what happened in his or her own words, and take your time, especially on missions where a recovery has already been deemed necessary. A witness will often be asked to reenact a story several times over the course of the recovery in the hope that he or she will remember something else that may be helpful to the mission. On recovery missions it is acceptable to do a longer investigation, and perhaps run a sonar search, long before the first diver ever suits up. Remember, the less the rescuer enters the environment, the less chance he or she has of being injured. By using technology, it may be possible to enable dive durations of minutes, rather than hours or days.

At a minimum, interviewers must find out the following information:

- Who is the victim? Name, age, what clothes they had on, whether they are able to swim, etc. Get very specific here. Stories may not agree, and investigators may use this information later in more detailed interviews.

- Where did the scene action take place? This is obviously a question that everyone wants answered, but it may be asked several times to the same witness. Ask your witness to go back and vocalize or recreate exactly what they remember. Retracing steps is a good way to help the witness remember details, even days after the submersion. Water current will change people's views. Remember the adage, "the river never rests?" The same is true in this case. While the witness is watching all this unfold, the current is still moving the victim. Often the victim is farther downstream than the witness realizes. On the other side of that coin, people can overestimate the effect of current on a victim, and place them too far downstream.

- Was the victim alone? Did anyone else try to enter the water to save the victim, and, if so, where is that person?

- What did the victim eat or drink recently? Certain foods or liquids can play into float time later in the mission.

- Was the victim injured, or did he or she appear injured? Even good swimmers can be hampered by injuries.

Keep your witness at the scene! If this is not possible, get a driver's license or identification and a method to contact them at a later time. Most witnesses will bend over backward to cooperate, but people do have lives they must attend to. If a witness will not give an identity or acts suspicious, have a description of that person ready for law enforcement. It is not unheard of to interrupt someone placing a body in the water, and that person then becomes a "witness" when they have to think fast. These people may exhibit a very strong urge to exit the area, and this alone is fairly suspicious.

Children, for the most part, can be solid witnesses, but might need to be handled with care. Children generally do not add to the story or try to fill in details they do not know. They may need some convincing that they are not in trouble, or that it is OK for them to speak. They also can become hesitant if many rescuers or personnel ambush them with a thousand questions. If need be, the child's mother or father can stay with the child, but don't let the adult tell the story. Children can be very smart and mature for their age, and with an effort not to embarrass or pressure them, they can be a great resource.

## Point Last Seen

All the information you gather and process will lead the team to the "starting blocks" of the dive mission, or the PLS. The PLS is usually determined by witness statements and witness demonstrations, and it becomes

# Voice of Experience

My county team was called across the state to help recover a teenager who had fallen into a fairly fast-moving river. The teams operating there were convinced he had been trapped underwater for two days in a large swirling current at a river bend. This bend created a large, circulating current, and was by far the most prevalent feature in this area of the river. It simply attracted your eye, and also held a moderate amount of floating debris. Because it had been two days since the drowning and we were fresh in the mix, I wanted to re-interview the best friend (whom I will call Cory, not his real name) who was at the water's edge when it happened. My request was not meant to disparage the crews that had been working hard at the scene, it was merely intended to bring some fresh perspective into the operation. My request was met with resistance from the teams on scene; however, because these teams and crewmembers insisted they had talked to Cory and were absolutely sure the point last seen their interviews had developed was spot on.

I noticed that it seemed these rescuers had quickly interviewed him in a "gang" fashion from above the swirling current. Multiple rescuers, while meaning well and knowing time was of the essence, had, to an extent, ambushed this boy with rapid-fire and intense questioning. After this quick interview, Cory was asked to leave with family. I also got the feeling that the river current had already been deemed as a point last seen regardless of what Cory may have seen or done when his friend went underwater. I suspected that the style of interview might have left some clues to be gained by a different approach.

The crews on scene did not want to take the time to bring Cory back on scene, and they were frustrated when I insisted. The local sheriff brought Cory to me, and it was obvious he was very shy and in much emotional pain. Riding in a law enforcement vehicle did not help his mental state. Cory's best friend had died just a few feet away from him, and he was sent to summer school the next morning. I spent about 10 minutes just talking to him and expressing my condolences and sorrow that he was going through this. Rather than have my team or other members assist in re-interviewing, I chose to talk to him alone, and I asked most of the bystanders to move to an area farther away. I wanted him to feel more comfortable by removing some of the folks he was uneasy around.

I took Cory down to the riverbank, and we recreated the event. We went very slowly, with no other people around to pressure or tell him what they thought. After several minutes, he gave up one important clue: When the victim entered the water, Cory stated that he started screaming for another friend who was playing on a swing downstream, around a sharp bend in the river. I asked him why he yelled, and he replied the downstream friend couldn't see them. The whirlpool was in plain sight of the swing the downstream friend was on, which led me to realize the victim actually went underwater a good deal upstream. The friend I was interviewing had to scream to get the attention, and then help, from the other buddy downstream.

*(continued)*

# Voice of Experience

I did not immediately react to this because I didn't want to distract him. We talked a few minutes more, and I again had him physically move us through the entire story, and he confirmed the entry point a second time. He looked his friend in the eyes as he submerged, but Cory was not able to swim. He could only scream for help and show me the exact area his friend went under. The boy had never been interviewed well enough to pass on this information, and was never asked to do more than point a finger from an elevated bank to "about" where it happened.

My team immediately started a walking search of the shallows around this area, and located a hole in the river that was about six to seven feet deep. We were in the process of setting up a dive search when the victimwas pulled from the river just down-stream from the bend. Even though he was recovered after floating, it was a good lesson in how taking the time to let a witness slowly tell or walk through the event can reveal crucial information that can quickly end a rescue or recovery.

**Steve Treinish**
Fairfield County Special Operations
Columbus, Ohio

the focal point for later operations and investigations. Finding and marking the PLS is therefore one of the most important aspects of any dive mission. Sometimes it is an easy call—for example, right off a dock or a seawall—and sometimes it is nearly impossible to find a PLS until later in the mission when new information becomes available. Chapter 1, *Understanding and Managing Water Rescue Incidents*, provides further discussion of determining the PLS.

**FIGURE 14-7** PLS markers should be different than other markers being used, and should not be removed until the victim is recovered. Beware of the entanglement hazard the buoy can cause!
Courtesy of Steve Treinish.

Once all of this information is obtained and triangulated, it can be plugged into the prevailing environment. Water temperature, currents, and obvious hazards play into the decision to dive or not to dive, but at least a PLS can be determined. Once a PLS is formulated, it must be marked (**FIGURE 14-7**). This mark becomes a reference point for future operations.

With global positioning system (GPS) programs, the PLS can be programmed as a set of coordinates, and mapping software is an invaluable tool for scene markings of all kinds. Smaller markers, such as pelican buoys, can be used, but pelican-type buoys are also used in other situations on the water, such as marking sonar hits, evidence, or bodies. Larger and different buoys can be more effective. The PLS marking system must not be pulled until the victim is recovered or the search is called off. If other teams are inbound to do a body recovery or further investigation, they can pick up the buoy or line and return it later. All operations will likely be based off this marking or location from this point on. Rescuers can place the buoy after deciding the best triangulation or by a boat or even a rescuer in the water. When the rescuer gets to the location where the witness believes the victim to be, the marker is deployed.

## Sketching and Recording

The interviewers' best friends are the plain old paper and pen. **Scene sketching** is recording on paper all mapping and physical scene information for later use, and these documents become legal evidence in any later proceedings. Any details, whether the interviewer thinks them pertinent or not, must be recorded. A sketch of the water area can provide a place to record line-of-sight information, landmarks, and obstructions in the water, and anything else that may be important at the time. Many agencies have a dedicated form to record witness information, which eliminates a lot of confusion; the information is merely transferred onto the paper. No matter how

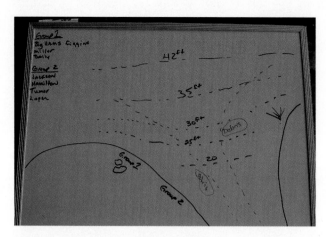

**FIGURE 14-8** A progress board showing diver progress, hazards, approximate depths, and other information can be kept during a mission.
Courtesy of Steve Treinish.

data is recorded, there should be a scene sketch, a place for any other information, and any other remarks the interviewer deems appropriate. Again, too much information will not be a problem in these missions. Use a new sheet for each interview if possible, but retain them for future comparison. Some witnesses end up recreating the scene two or three days in a row, and these sheets can all be compared to cross-check information. Maybe something will stick out on one sheet that will lead to a major break in operations. Remember, this information is now evidence and must be processed as such.

Many teams keep a large dry-erase board either in a vehicle or somewhere close to the scene to keep track of vital information. A large sketch of the water area can be used to show triangulation, search progress, and features. As certain areas are cleared, they can be shaded or marked in some way (**FIGURE 14-8**).

**LISTEN UP!**

A large dry-erase board can be outfitted with suction cups or large magnets and used on the side of a truck, a wall, or anywhere needed. It can also be used to supply a larger water sketch, team or agency contacts, or any other pertinent information.

Many rescue teams keep a small digital camera on the rescue vehicle to document scenes or information. Put that camera to work. Pictures are truly worth a thousand words, and digital evidence is admissible in court. If there is a true concern about evidence and digital pictures, consult with local law enforcement, as the memory card will likely be processed as evidence and sealed.

The standard in emergency medical services (EMS) documentation is that if something was not written down it did not happen. That also is entirely true with water rescue. Document everything the team does,

and why, as well as the victim's information and all witness information.

**LISTEN UP!**

Many rescuers carry smart phones, and snapping a few pictures may be a quick way to obtain scene documentation. Keep in mind, however, that if a phone is used to photograph a scene it can be considered evidence, and even if personally owned, it may be subpoenaed for evidence. Ensure the authority having jurisdiction (AHJ) has procedures in place for the use of personal phones.

## Dive Versus No Dive

The last decision to be made in the scene investigation is the go/no-go decision. There is no concrete way to decide whether a team will commit divers to any particular mission, unless obvious parameters—such as swift currents, confined space dives, or contaminated water—are present, or these types of dives are previously disallowed in team operating procedures. Simply put, the divers must be trained and equipped to perform these advanced dives. There are many variables that affect the go/no-go decision when teams are presented with dive runs, but safety must be your biggest concern. A great slogan has made its way into the fire service:

> We will risk our lives a lot, in a calculated manner, to save SAVABLE lives.
>
> We will risk our lives a little, in a calculated manner, to save SAVABLE property.
>
> We will not risk our lives at all for lives or property that are already lost.

This slogan speaks volumes. Dive rescue poses inherent risk to life, just as recreational diving does. Additionally, evidence collection, such as guns, cars, safes, and other items are a large part of many public safety dive (PSD) teams' mission capability. The idea that PSD does present some risk even by just submerging should always be considered.

**LISTEN UP!**

Imagine a family pet is stranded in a hole in the ice. Properly equipped, manned, and trained, would you commit surface ice rescuers? Many would, and make it a training evolution, because the public may attempt a rescue if we don't.

The same argument can be made for some types of public service runs. Divers often tire of the same old training patterns. Offering to help another fire company find a lost item such as a pike pole or hand tool from previous runs or trainings, diving for items such as a lost wallet, or even doing underwater cleanups at local areas such as boat ramps can be great training, too. Positive public perception can be huge doing this, but it must be done safely.

A small child fallen through the ice in calm water would elicit a very quick "go" decision from most teams capable of this, as it should. Put that small child in a swift current under ice, and now that dive is a whole new and very ugly monster. Current diving under an ice shelf without visibility is one of the most—if not the most—hazardous dives that can come up. If a team is not trained in these evolutions, they should stay topside. A mission with a victim submerged near a boil of a raging low head dam would probably be deemed unsafe to dive very quickly, and rightfully so. The chances of this person living are simply not worth the hazard of that dive. However, even in the same low head dam, low water conditions can often make a hazard much tamer, and provide a "bread and butter" effort for rescuers (**FIGURE 14-9**). If a team operates with safety in mind, and can plug its capabilities into the mix, decisions should be able to be made fairly quickly.

Those decisions may or may not be popular, even amongst the team, but we must be able to send our people home at the end of the day.

**KNOWLEDGE CHECK**

Which item is the most critical in establishing a point last scene?

**a.** Estimated time of submersion

**b.** Availability of a witness to the submersion

**c.** Availability of underwater scanning devices

**d.** Currents and weather conditions

*Access your Navigate eBook for more Knowledge Check questions and answers.*

# Dive Planning

Even with a good PLS, some parameters must be considered before the dive is deemed safe. The crews on scene must examine the dive environment to ensure safety. This can be done quickly, such as when an entire crew responds to a small pond, or may take some investigation of the type and location of entrapment, and sometimes an honest self-examination of the capability of the dive crew present. Experienced dive rescue crews can determine when a dive should be declined due to safety factors.

## Dive Environments

Although many drownings occur in open or recreational waters, this is not always the case. Industrial water intakes or discharges are found in many areas, and can present hazards to a diver in the form of differential pressures or underwater currents. Preplanning these areas is paramount, and divers who find these intakes should be cautious working around them. A good tactic when diving around these facilities might be to find the maintenance staff and cut the power to the pumps or utilities that supply them. Many retention ponds have fountain pumps or aerators to reduce aquatic growth and appearance factors, but these can easily injure or trap divers at the intake suction area, or electrocute divers when the power supply is not disconnected and locked out.

Another electric risk overlooked by many rescuers is a marina dock. The high-use area of a marina can easily have divers in the water for life rescue, but it also provides a good training area to search for lost items or even "treasure hunt" under or around the docks. However, many docks offer electric service for boaters, and people are electrocuted each year when the electrical equipment falls into the water, electrifying the water around the dock.

**A**

**B**

**FIGURE 14-9** This is the same low head dam, showing benign conditions at low water **(A)**, but deadly conditions in higher water **(B)**. Water environments should change the go/no-go decision!

Courtesy of Steve Treinish.

**LISTEN UP!**

Teams should carry a physical lock out/tag out kit to provide safety from inadvertent re-energizing of power supplies while rescuers are near or in the hazard zone. Do not rely on other people to keep the power turned off. A padlock can be used to lock the power out, and the key placed with the dive supervisor or dive safety officer, or command staff.

Boat traffic on larger water areas may cause safety issues when divers are underwater, especially when multiple agencies are present. Communication may not be possible across radio channels, so a strong perimeter of protection should be in place. Divers and crews should all know the procedures for keeping a diver on the bottom and ascending when needed.

Many underwater tasks and inspections require the use of commercial divers. Pipeline repair, valve replacements, bridge inspection and repair, and underwater utility work are all examples of jobs performed by commercial dive crews around the world. Commercial diving contractors often use surface supplied air, and they are frequently subjected to industrial regulations governing the safety of the commercial dive. When a rescue crew is called out for a trapped commercial diver, the best resource is often the dive supervisor in charge of the commercial crew. He or she can best advise on the needs of the rescue, and will likely have a plan in place to extricate the trapped diver.

Commercial divers can also provide great information on how to handle or look for hazards before a dive, or assist in preplanning water areas. For areas with a high chance of hazardous material releases, commercial dive companies might offer a great resource of equipment and manpower if needed. A true contaminated-water dive should be undertaken with surface supplied air and a drysuit system designed for protection from hazardous materials. While the basic encapsulation equipment suggested by NFPA 1006 is a great ensemble of PPE for most diving areas, it is not infallible. The typical dive drysuit system and full-face mask are similar to structural firefighting gear and some hazardous materials gear. Although firefighting PPE and SCBA provide a level of protection in some hazardous material responses, there is no substitute for mission- and material-specific PPE rated for that task, and substances that require a higher level of protection or a longer duration of protection need more specialized PPE.

## Dive Physiology

Any SCUBA diving, whether recreational or for public safety, presents some medical concerns and potential illnesses based entirely on the water depth and the duration of the dive. Simply put, the deeper a diver goes, the more pressure the body experiences. Modern dive equipment automatically provides the volume of gas needed to maintain enough pressure to match the surrounding water. The diver is required to breathe more air by volume. Breathing this extra gas causes more nitrogen to be stored in the body, and reducing pressure too quickly can cause a variety of maladies. **Barotraumas** are injuries caused by pressure changes. *Decompression illness* is the term used in the diving industry. Decompression illness is labeled as Type 1 and Type 2, with Type 2 being more involved and a potentially injurious decompression illness. Before further examining the decompression illnesses a diver may experience, one must understand how and why these illnesses can happen.

## Boyle's Law

Although you might not have thought you would be reading about physics in a water rescue textbook, it plays greatly into dive planning. **Boyle's law** states: If the temperature stays the same, the volume of a gas will get smaller at the same rate the surrounding pressure increases and density increases (**TABLE 14-1**). Due to Boyle's law, any volume of air will be reduced in size as surrounding pressure increases, and it will enlarge as surrounding pressure decreases, if the temperature remains constant.

When a diver descends in water, for every foot of depth the diver sinks, the water pressure exerted on him or her will increase by 0.432 psi in fresh water. Salt water increases this pressure to 0.445 psi, but for this text, we will assume fresh water dives. This does not sound like a lot of pressure, but in 4 feet (1.2 m) of

| **TABLE 14-1** Boyle's Law in SCUBA Diving | | |
|---|---|---|
| **Water Depth in Feet (m)** | **Atmospheres of Pressure (ATA)** | **Lung Size** |
| 0 (surface) | 1 | Full volume |
| 33 (10.0) | 2 | ½ volume |
| 66 (20.1) | 3 | ⅓ volume |
| 99 (30.2) | 4 | ¼ volume |
| 132 (40.2) | 5 | ⅕ volume |

water, there is enough pressure on your chest to prevent sucking air through a garden hose. When divers take and hold a breath of air at the surface, their lungs shrink in size as they swim down but return to normal when they return to the surface. However, divers cannot do this—lungs must stay properly inflated to function, and modern SCUBA regulators achieve this for them automatically.

The problem that divers have to deal with is Boyle's law. Normal atmospheric pressure exerts 14.7 psi on us. The atmosphere, per square inch, weighs 14.7 pounds, and this is the environment in which our bodies are meant to function. To keep it simple for the sake of this text, every 33 feet (10 m) of water depth produces a pressure increase equivalent to one more atmosphere, or 14.7 *more* psi. Therefore, at 33 feet (10 m) in water, a diver is at 29.4 psi and needs twice the volume of air to keep his or her lungs properly inflated. The diver will not feel this pressure or increase in gas volume, but when ascending, the diver's lungs will try to expand to twice their size if the diver does not exhale and if he or she does not control the rate of ascent. Bubbles in the bloodstream will also expand. In a normal ascent, these bubbles simply make their way to the diver's lungs, and are exhaled through normal breathing. These bubbles in the body can do harm to the diver if they end up in other areas of the body.

## Excess Nitrogen in the Body

When divers breathe the extra gas needed to conquer Boyle's law, they pay a price in also taking in the additional nitrogen found in that extra gas. Air is 78% nitrogen and 21% oxygen. (There are also some trace gases, but we will only discuss the two major ones here.) Oxygen is metabolized during the dive, but our bodies don't use the nitrogen. The excess has to go somewhere, so once again a physics law is used in SCUBA. Henry's law states that as pressure is increased, a gas will turn into a liquid solution. This means that all the extra nitrogen a diver is breathing turns into a liquid and is stored in tissues in the body. The deeper divers descend, or the longer they stay at depth, the faster the nitrogen piles in. This requires decompression stops to be made more frequently. **Decompression stops** are simply stopping at a certain point in the water column, for a certain amount of time, to allow excess nitrogen to come out of a liquid solution, into gas, and out of the body through exhaling. "Deco" diving can be very technical, must be planned beforehand, and also can present a problem with gas supply for the diver.

Every diver uses **dive tables** during basic recreational training that figure out the bottom times

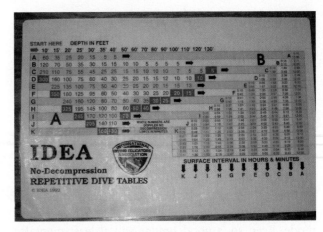

**FIGURE 14-10** Recreational dive tables help divers plan underwater times and depths and time spent topside to reduce the chances of dive injury.
Courtesy of Steve Treinish.

(**FIGURE 14-10**). The recreational SCUBA industry has taken the U.S. Navy dive tables and modified them for recreational, sport diver use. These tables place limits on dives to a maximum of 130 feet (39.6 m). Each depth has a maximum bottom time to stay within decompression limits. Staying after that time will result in the diver needing to do a decompression stop.

Divers also get credit for surface time, when the excess nitrogen comes out of the body and is exhaled normally. Deeper depths of 120 to 130 feet (36.6 to 39.6 m) may allow only five-minute dives, and this includes the time it takes to get there. Shallower depths allow quite a bit of dive time, often to the point of an air supply being exhausted. Repetitive dives can be done in many cases, but the repetitive dive must have the nitrogen absorbed on the preceding dive factored in, and the tables also help to determine these times. A qualified SCUBA instructor can explain how to use dive tables, and all SCUBA divers should know how to use them.

## Dressing a Diver

Many divers are surprised at the amount of work that is involved in just putting the equipment on, especially with a drysuit. Many teams dive with varying levels of PPE, but the NFPA standards now address diving PPE by recommending that divers use encapsulation gear, which consists of a drysuit with a latex hood, dry gloves, team buoyancy control device (BCD), and a full-face mask (FFM). The diver wears all of these while operating underwater on a dive rescue mission. Recovery dives go the same way, but do not have such a rapid time factor.
Dive gear is

discussed in more depth in Chapter 13, *Dive Equipment and Use*.

## Medical Checks

Before the diver ever hits the water, a medical check should be done as a baseline, just as is done prior to a hazardous materials run. In fact, encapsulated divers are wearing the equivalent to a Level B hazardous materials suit underwater. For rescue purposes, though, this may not be feasible if the dive must take place rapidly, which usually happens in true rescue mode. Even so, a good **tender** will at least mentally watch the diver. Does the diver appear to be OK? Some nerves may arise, but a diver acting totally out of touch might not be ready to go into the water under stress. On longer runs to a scene, a diver who can drink a quick bottle of water will be slightly ahead of the hydration curve, which may help ward off issues like decompression illness and muscle cramps.

> ### LISTEN UP!
> Fire fighter health has been in the forefront of many departments in the last few years. Cancer, heart health, and eating right are all things we try to control and can save lives. A daily check of vitals for divers, or for everyone, might help track general health issues.

> ### SAFETY TIP
> Another aspect of dive safety is flying after diving. In the heat of a rescue, a diver may forget about flying the next day. Flying commercially within 24 hours after diving can be hazardous and cause decompression illness. One exception to this rule may be travel in a medical helicopter, which normally flies at lower altitudes.

## Suiting Up

After the medical check, the drysuit is donned. The diver steps into the feet and legs and, with assistance from the tender, puts on the suit. Drysuits have latex seals at the wrists and neck to keep water out and warm air in. The seal should be stretched a few times to warm it up, and then walked over the neck or wrists, not pulled on like a sweatshirt. Pulling the seal over the wrist or neck can rip it very quickly, making the suit unusable until repairs are done. Also, make sure necklaces, watches and bracelets are removed, as these may rip the latex seals. Suits with quick-change seals are now available, but it will take time to change

**FIGURE 14-11** When zipping a drysuit, use care to pull the zipper by the loop, and brace the suit with the other hand.
Courtesy of Steve Treinish.

to those, as the cost can add up. Pure talc should be used to powder both sides of the suit seals to make the seals slide easier; this also will prolong the life of the seal. An old tube sock filled with a few ounces of pure talc can be used to pat the talc onto the seals. Zipper lubricants can be used on drysuits, but they should be applied normally after a dive to save time during rescue dressing.

Once the seals are in place, the dry zipper must be closed. Dry zippers give years of service with a little care. When pulling the zipper, grab the loop, grab the zipper behind it, and pull firmly. Zip the suit shut all the way to the stop, and tug a couple of times to make sure it is tight. Unzipping is done in reverse, but always using the off hand to grasp the suit (**FIGURE 14-11**). The angle of the pull should be parallel with the zipper, and not out from the suit. This puts less stress on the zipper and also pulls much easier. With longer response times, consider leaving the zipper undone and the diver partially dressed until you get a little closer. On hot days, the diver may be getting warmer and warmer, so have a towel and an extra water bottle handy.

> ### SAFETY TIP
> Any gear in a moving truck should be stowed safely and securely inside, and only if your department approves it. Trying to don a loose SCUBA setup while moving in a vehicle presents a fair risk, and safety belts must be worn when the truck is moving.

## Donning the SCUBA Equipment

Donning a mounted SCUBA setup in a vehicle is not bad; it is similar to an SCBA. Many teams will mount the entire setup in a harness holder, just like on the fire apparatus.

The tender can make life easier on the diver, though, by releasing the shoulder straps on the buoyancy control device (BCD), opening up the **cummerbund** (elastic belt and buckle at the belly), and letting the diver move backwards into the BCD. Then the tender buckles the diver's straps instead of the diver throwing both arms up and around. Depending on which weight system and drysuit insulation is being used, the weight belt may go on first, the harness may go on first, or the BCD might hold all the needed weight. Get the weight on the diver, but handle it carefully. It can easily break toes if dropped. The diver gets adjusted in the SCUBA equipment, and there are a couple of options to look at next.

---

### SAFETY TIP

Be aware of divers that tighten the BCD straps too tightly. Tightening BCD straps before air is added can cause such tightness that that diver cannot physically expand the lungs enough to breathe properly. When in a hurry, many divers mistakenly tighten the BCD straps as if they are wearing a structural SCBA.

---

## Fins

Fins should be kept with the SCUBA equipment, ready to go when the diver needs them. They do not have to go on until the diver reaches the water's edge, and instead of the diver trying to do it, the tender, once again, can make life easier. Usually, at the water's edge, the tender can wet the foot pocket of the fin, making donning them easier. Simply dunk it and place the fin on the ground with the strap under the fin foot pocket. Facing the diver, step on the blade of the fin, let the diver ram his foot into it, wiggle it to make sure it is in, and have the diver slightly pick up his heel. The strap can now be slid up and over the boot onto the Achilles tendon area. Consider the use of stainless steel spring-type fin straps, as they make doffing and donning easier, and also, properly adjusted, will provide a nice tension on the foot. Repeat for the other foot, and now the diver is dressed and ready to go.

## Minimum Reserve Pressure

Any diving done using cylinder-supplied breathing gas should have a specified minimum reserve pressure planned before the dive commences. Reserve pressure is just as it sounds: a volume of air, expressed in psi,

that remains for use in case of trouble. This pressure is what the diver should have in the cylinder upon surfacing, and tenders, divers, and supervisors should all know this target pressure before any diver submerges. The recreational dive industry has adopted 500 psi as the consensus standard, but this is based on fairly benign conditions and diving with a buddy. Many technical divers, including under-ice and cave divers, will plan to be on the surface and positively buoyant with one-third of the original gas supply remaining in the cylinder. Other agencies will use a cubic foot formula, taking into account a redundant air supply or the use of double cylinders. Using the cubic foot philosophy, some teams will allow a diver to surface with 500 psi, with the idea that the pony cylinder carried is part of the remaining reserve gas. Other agencies and teams recommend any diver surface with 1000 psi of gas. Arriving at the planned reserve pressure can be somewhat controversial, but is a good example of the need to plan operations based on crew ability, training, and conditions.

## Hot Exit

If the information being relayed to the truck is leading to a confirmed, close-to-shore submersion, the diver can **hot exit**, or exit the truck ready to have fins placed on his feet and enter the water (**FIGURE 14-12**). It is entirely possible to have a diver in the water and running searches within a few seconds of reaching the shoreline. Remember, every second is critical. This is the same type of situation as a fully dressed fire fighter coming off a truck with an SCBA mask and gloves in place. One thing to be aware of is fogging of the mask lens. This isn't a problem in and of itself, other than if

**FIGURE 14-12** In a confirmed, close-to-shore submersion, the diver can hot exit, or exit the truck ready to have fins placed on his feet and enter the water.
Courtesy of Steve Treinish.

**FIGURE 14-13** The tender can help guide the diver by keeping a hand on the yoke of the diver's tank.
Courtesy of Steve Treinish.

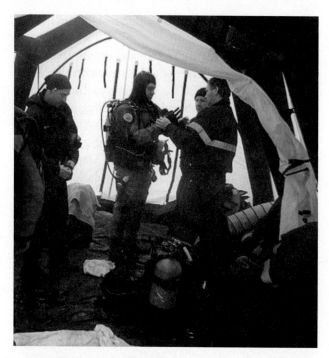

**FIGURE 14-14** Staging the diver somewhere until it is determined when and where to enter the water may keep divers more comfortable before a dive.
Courtesy of Steve Treinish.

a diver cannot see very well through the lens, he or she can easily trip or fall on the way to the entry point. A fix for this is to walk with the diver with a hand on the yoke of the tank (**FIGURE 14-13**). If a diver trips, the extra 100 pounds or so will likely put the diver face first into whatever he or she is walking on. Slippery rocks, pavement, grass, it doesn't matter—on the way out of the water, the diver will make everything wet and even more slippery. Tenders should walk their diver to the spot where undressing occurs and not let go until the diver is out of the equipment. Most divers are greatly appreciative of the help from the tender, and they are more than ready to get out of their rig and take a break.

## SAFETY TIP

It is a hard habit to break, but divers should think twice about carrying gear on their backs while on shore. Take the following example: A diver is wearing 80 pounds of SCUBA gear. He also is carrying his fins and a digital camera that he used to take pictures of a gun underwater. He is walking up a 20-degree boat ramp and slips. Do you think he will get his hands in front of him to break the fall? Even if he does, a Colles-style wrist fracture is very possible. A broken nose or broken facial bones also would not be out of the question. Injuries such as these are, for the most part, preventable. On a normal response, there are many untrained rescuers who can assist. Consider putting them to work carrying gear.

## Staging

The other option is to stage the diver somewhere until the location and the timing of the entry into the water is determined. In extreme weather environments it may mean sitting inside an air-conditioned vehicle or inside a portable shelter that can be heated. If this is the case, the diver can be staged on shore until diving plans are decided (**FIGURE 14-14**). When the time arrives to dive, bring the latex hood up and over the diver's head, with a skullcap for insulation if needed. Then the FFM goes on. Check to make sure the ambient air valve, if the mask is so equipped, is on so that the diver can breathe outside air, and then assist with or adjust the head straps. Many masks and straps may need to be placed as far down as possible on the face and head to eliminate the problem of the mask slipping up the diver's face. Do not overtighten the straps, as jaw fatigue and a headache are likely to occur if the straps are too tight. If using electronic communication, adjust the speaker to the diver's ear, and proceed to the dry gloves.

Another thing to consider when staging a diver is the outside air temperature. Cold weather in a drysuit with the proper thermal layers is not too bad, but allowing gear and divers to sit out in cold weather will chill both, and once the body core temperature goes down, it can be hard to bring it back up without starting from scratch. Conversely, a hot sun and no breeze can be problematic, with hyperthermia showing before the dive is made. Keep the diver cool, shaded, and hydrated. A quick blast of air from a drysuit inflator and out the neck can cool the diver's body. According to the U.S. Navy dive manual, there is a potential problem of divers developing hyperthermia at air temperatures above 90°F (32.2°C), or water temperatures above 82°F (27.8°C). And don't forget, a diver sitting in hot sun will be roasting in the drysuit, much like sitting on hot pavement. If a recovery decision is made rapidly, this can keep divers from dressing at all until needed.

## Staging Location

Perhaps the team has elected to postpone any active diving until investigations and interviews and/or a sonar search are completed. Divers also can simply be placed near the dressing point to wait in comfort. Thermal comfort such as a heated truck or a shaded spot, a windbreak, and a view of the water that may be searched are things all appreciated by divers. Seeing the water offers a chance to mentally prepare for the potential hazards of the dive, and allows the divers to get their game face on. Divers often will discuss water features, potential problems, and solutions when seeing where they may operate. Keep proper fluids available to drink at all times regardless of the temperature.

---

### SAFETY TIP

Diver hydration can help prevent dive-related illnesses and can help with muscle cramps during and after the dive. However, in cold weather, coffee or hot chocolate are frequently offered. Caffeinated drinks are diuretic, and they may cause the diver to lose hydration through urination. Typically, plain, cool (not cold) drinking water is the best fluid.

---

## Dive Checklist

There is a movement in the fire service to use checklists. We have guides and checklists for incident command, swiftwater rescue, urban search and rescue, and more. Pilots have used a written checklist for years in the cockpit. Why? There are steps and checks so crucial to the mission that omitting or forgetting one item could result in a major life-altering catastrophe. Believe it or not, the same holds true for the diver. The diver is about to enter an environment where he or she most likely will be alone, totally dependent on life support equipment, and likely battling a lack of visibility. Missing certain steps may result in injury or death, or may simply make the mission more difficult.

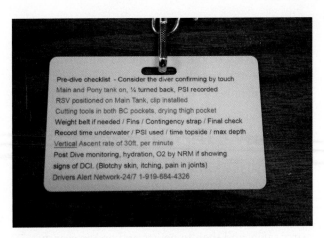

**FIGURE 14-15** A laminated checklist worn necklace-style.
Courtesy of Steve Treinish.

Every team should have a pre-deployment **diver checklist** not just on the truck, but also on the tender. Many teams use a paper sheet and a clipboard, and some teams use a small laminated card worn necklace-style. The paper form can actually be checked off line-by-line, which gives it the advantage, as a verbal checklist item is more likely to be skipped over by accident in a rescue rush. Some teams even have the diver or the tender read the list out loud to the other, and the member will give a yes, no, or condition reply (**FIGURE 14-15**). For example, when turning on the diver's air, the tender will not only say "Air on, primary tank" but verbalize the tank being on and the tank pressure, and make a note of it on the dive sheet. The diver does this while physically checking and turning the valve, then turning it back a quarter turn. (This is controversial in some agencies, but assures anyone else that the valve is indeed on, and does not need Hercules himself to open it.) Teams also can use technology in the form of laptop computers or tablets. EMS have been using them for years, and a good, user-friendly tracking log can be quickly filled in electronically at the water's edge and offers a great way to store, print, and document dives.

So, the gas is turned on in the primary and the pony tank, the manifold block is checked, weights are checked, carabiners and harnesses are checked, and

---

### SAFETY TIP

At some point, a diver will get into the water in the heat of a rescue and then remember that he or she had planned to fly the next day. The Divers Alert Network (DAN) recommends a 24-hour interval between a dive and flying to minimize decompression illness risks due to cabin pressures in the aircraft. A diver who has plans to fly must realize that diving will affect his or her travel plans, and the team will have to plan accordingly.

the diver's seals and mask fit are checked. Next, the diver should be able to grab each cutting tool, weight release buckles, or pouches, and verbalize what he or she is grabbing. Any diver having to make several grabs for equipment needs more muscle memory training.

# Diver Deployment

The normal **blackwater** public safety diver has to rely on the tender for many things. The tender is a coach, doctor, psychologist, and mechanic all rolled into one. The first order of business is to get the diver in the water safely and then to get started on the search.

## Diver Streamlining

The archenemy of public safety divers is entanglement. Once all the gear is put on, the diver has very little vision other than straight ahead and becomes much more dependent on feel. Rescuers should prioritize **streamlining** gear setups, making them free of any extra gear that may cause entanglement issues, both on the diver's body and on the SCUBA equipment. Anything that can be secured inside the BCD pockets should be. Dive knives strapped on the leg seem to pick up every piece of fishing line the diver gets close to. Divers who work missions with many gadgets hanging off them can get into problems, as all those gadgets make entanglements worse (**FIGURE 14-16**).

**FIGURE 14-16** The fewer gadgets attached to the diver, the more streamlined the diver and the lower the risk of entanglement. The diver shown here is carrying too much clutter.
Courtesy of Steve Treinish.

> ### SAFETY TIP
>
> Some recreational BCDs have pockets sewn into the material on the outside of the air bladder. This is not a big deal in the clear water, but if a diver keeps cutting tools in these pockets, there is a chance that with the air added for buoyancy control, the pockets could be squeezed so tight that it becomes almost impossible to remove the tool for use. If your team is buying BCDs, look for pockets usable at all levels of bladder inflation.

> ### SAFETY TIP
>
> At the beginning of a dive, the diver should be positively buoyant to prevent submersion before the diver is ready. Divers should enter the water with the BCD inflated enough to float the diver and all gear. When the diver is ready, the air in the BCD can be expelled and the dive begun.

## Water Entry and Exit

The diver should be escorted everywhere on land, and tenders might have to get partially wet while getting the diver into the water and moving without problems, and also assisting the diver from the water to land. Exiting the water can be especially problematic with water, mud, and silt contributing to divers slipping when trying to get topside. These slips and slides can plague a diver to the point where crawling out may be the best option, and even then it can be hard work (**FIGURE 14-17**).

Some heavily silted ponds may even require a diver to crawl in and out of the water. A set of hip waders or ¾ length rubber firefighting boots will go a long way toward diver safety, as they will enable the tender (with proper PPE, of course) to walk down most shorelines or gently sloping banks to the point where the diver can operate. A dive injury at the water's edge because someone didn't want to get his or her boots dirty is inexcusable. The diver who was so carefully helped to the water could have both feet stuck in shin-deep mud and end up falling and striking his or her head on a rock.

**FIGURE 14-17** Entering or exiting the water can be especially problematic with water, mud, and silt inhibiting the diver's movements.
Courtesy of Steve Treinish.

**FIGURE 14-18** A simple hauling system can facilitate diver entry and exit.
Courtesy of Steve Treinish.

Many people get injured each year by jumping or, even worse, diving headfirst into water that is too cloudy or stirred up to see through. Never let a diver jump into water that cannot be verified as having a depth sufficient to prevent injury. Take a dive weight, tie it to a small line, and sound the bottom, if possible. Sounding a bottom with a weight is similar to sounding a roof before walking on it; it is checking out the condition and depth of the water and bottom, and even then, it will not clear the area 100%. (Think vertical rebar.) Using a ladder or the tender rope to support the diver while entering or exiting the water can make operations easier for everyone (**FIGURE 14-18**). Even tenders can be supported using the back clip of a harness if the bank is steep enough, but make arrangements for this before the mission. Tying a support line at the water's edge up to an object at a higher point on the bank can be helpful. This line can give a handhold to help rescuers brace or pull themselves up. In extreme cases, a

**KNOWLEDGE CHECK**

To minimize risk of decompression illness, a minimum _____-hour interval between a dive and flying is recommended.

**a.** 12
**b.** 24
**c.** 36
**d.** 48

*Access your Navigate eBook for more Knowledge Check questions and answers.*

non–life safety lowering system will make deployment easier, but only if the diver's feet never leave the ground.

# Dive Signals

Most dive supervisors relax when a diver breaks the surface of the water, signals he or she is OK, and is positively buoyant on the surface. On the surface, the OK signal is indicated by patting a fist on the top of the head with either arm, or forming a large circle above the head with the arms (**FIGURE 14-19**). This exaggerated movement is done to discern between a rapid movement that a troubled diver might make and a diver truly giving the OK signal. Once the diver surfaces and is positively buoyant, he or she should signal an OK, even if he or she does not have the tender in sight. This tells the tender

**LISTEN UP!**

A ladder can be helpful to divers going into and coming out of the water up a steep, muddy bank. The ladder should be secure and lying on the ground, ensuring it will not drift or slide down the bank while the diver is on it.

**A**

**B**

**FIGURE 14-19 A**. OK signal with one arm. **B**. OK signal with two arms.
Courtesy of Steve Treinish.

that the diver has made a normal ascent and, at the moment, is fine. This hand and arm movement should be done even if using voice communications. If a diver seems to be having trouble on the surface, the tender can use this signal to ask the diver if he or she is OK, and should get the signal back if the diver is OK. This normally would be used with the diver adjusting something when the tender thought a descent was about to take place. During night operations, the OK signal can be given using a flashlight to draw a circle in the air. To the diver this looks like an "O." The same reply should be given or, in case of trouble, a left-to-right sweeping motion is used. It should be noted that some divers make this action into a habit, even to the point of signaling it without thinking when they are actually in trouble. The signal for "trouble" or "send help," is usually a wave of one arm or both arms back and forth, which is pretty much what people in trouble do.

A few other signals are sometimes used to shave seconds off a response. They are used when the diver needs to swim to a certain location in the water before submerging. Many new dive crewmembers do a witness interview, determine where they want the diver offshore, and then tell the diver this. In practice, the diver now must spin around and try to see this spot, but the diver really doesn't have to know it in any case. The diver swims best on the surface on his or her back. Use this to your advantage and direct the diver left, right, or straight with your extended arms. The diver swims or turns in the direction you are pointing and continues straight when you signal that. When the diver arrives at the point of submergence, give an OK, get an OK, and the dive can commence. Very speedy! This can be done with voice communication systems, but both the surface directions and the OK signal should be done to ingrain good habits in case of communication system failure.

**SAFETY TIP**

Vertical lowering and lifting of divers off bridges is possible, but requires specialized equipment and training before attempting. Most BCDs with nylon webbing crotch straps are not designed to support this type of load and may fail, causing injury or death.

## Rope Tug Signals

**Rope tug signals** are a set of tug pulls on a line between a diver and tender that signal different things.

The number of tugs equals a piece of information. For example, two tugs given to a diver underwater may mean "Stop, tighten the slack, and change directions." Four tugs given to a tender may mean, "I am entangled and need help." Keeping it simple may help a stressed diver underwater. Pull signals also can easily be printed on a small card and laminated for the crewmembers to wear. A great place for this information is on the back of a dive checklist card. There are many different rope tug signals—some are simple, and some are complex. A phrase many fire fighters like to use is to "Keep it simple, Simon!" and that holds true in public safety dives. While there have been serious debates among divers, teams, and agencies, the rope tugs that are chosen will boil down to what works for your team and what will be remembered under stress.

Planning for underwater emergencies directly relates to diver safety, and divers and crewmembers of any team should be well versed in using line pulls in case of communication failure. The use of electronic voice communication systems is becoming more and more common, but there are two factors that require rope signals to be learned and practiced.

| TABLE 14-2 Typical Rope Tug Pulls | |
|---|---|
| **Tender to Diver** | **Diver to Tender** |
| One pull on the line: OK, OK? | One pull on the line: OK, OK? |
| Two pulls: Stop, take out slack, reverse direction | Two pulls: Give line |
| Three pulls: Come to the surface | Three pulls: Object found |
| Four or more pulls: Stop, don't move (there could be danger ahead or a boat entering the search area) | Four or more pulls: Assistance needed |

First, electronic voice systems are fairly expensive; microphones and speakers must be purchased for the diver and the tender, and special communication rope must be used with hardwire communications. Some teams may not be able to afford these systems.

Second, even if a team has an electronic communication system, that system could fail. If a communication system fails in the middle of a dive for a child in shallow, cold water, should the dive be aborted? Not likely. There is no reason to discontinue that dive, provided known signals and a plan are in place before the dive starts. Many major public safety underwater training agencies operate using similar rope tug pulls (**TABLE 14-2**).

Many agencies and dive teams utilize voice communication systems but keep the rope tugs in operation while diving. Keeping these signals fresh in your mind makes them easier to remember if the electronics fail during a mission.

It is up to the team or team leader to decide which signals will be used for every response. Signals should be the same from the day they are decided; they should not be changed on the fly. A diver in distress is not likely to remember a tug pull adjustment when things go bad. How you give and answer a signal is more concrete. Rope tugs should be firm. The diver should send firm signals, as he or she may lose some power when pulling underwater. The distance from tender to diver will affect this—at a length of 20 feet (6.1 m) it does not take a great deal of force to send a tug, but tugging over 100 feet (30.5 m) of rope is a different story.

One thing that never should be done (unless the diver asks for it) is to pull too hard on the rope. If a diver is working on an entanglement with, say, fishing line, and the tender rope is pulled too tight, all the fishing line the diver is dealing with could be cinched down onto the diver, making the entanglement worse.

### SAFETY TIP

Rope tug pulls are effective only if the rope between the diver and tender does not bend around or become hung up on features underwater. Ropes catching on debris or bending around objects may impede rope tugs in both directions, and the tender cannot truly feel the diver operate underwater. Tenders may have to adjust their body position to keep the rope unobstructed. Keep in mind that this movement may slightly alter the search pattern, affecting its effectiveness.

## Electronic Voice Communications

Electronic voice communication systems have greatly increased safety in public safety diving, and they are required by the National Incident Management System (NIMS) for an agency to be listed as a "typed" team. These systems greatly increase safety by allowing the different crewmembers to converse while the diver or divers are underwater. Take, for example, a plain rope signal from a diver that he or she is hung up. The tender topside has no idea if that diver is hung up on one piece of rebar or if he or she swam into a commercial fishing net strung in a tree. Voice communications let the diver explain what is actually happening. Another advantage is that the diver can be spoken to calmly during stress or a case of nerves if it arises. With the proper setup, the backup diver can be placed into the system, and the two divers can talk while mitigating the situation. A backup diver listening to the conversation between the primary diver and tender will already have a head start on the problem he or she is sent to fix.

### Hardwire Systems

Hardwire systems use a kernmantle rope with a small communication wire strung down the center. For most public safety use, the diver uses a microphone and speaker system connected into the mask. The microphone has a short piece of wire coming from the mask as a pigtail, and the communication rope attaches to this. Usually, a figure eight knot is used to attach the rope to the diver's harness to prevent a pull directly on the connector. The tender, who wears the small belt pack and headset required for operation, holds the rope (**FIGURE 14-20**). Some systems use a larger box that features speakers and a hand-held microphone. The headset units give the tender more mobility on the shore when running dive patterns, and more

**FIGURE 14-20** Hardwire systems use a kernmantle rope with a small communication wire strung down the center.
Courtesy of Steve Treinish.

A

privacy if the diver finds something of a sensitive nature. Hardwire systems are not affected by water turbidity or line of communication problems, but can cost more because specialized rope must be used.

## Wireless Systems

Wireless systems broadcast signals from a transceiver on the diver to a base station or another diver. They can be used with any rope or even no rope, and are becoming more popular in the recreational arena. Some attach directly to the diver's head strap and use a speaker/microphone setup similar to the hardwire unit, and some higher power units have the transceiver in a box that the diver has to mount on gear or carry underwater. Shore personnel use a transceiver suspended in the water column to both talk and listen to anyone using that frequency (**FIGURE 14-21**). Wireless systems can be subject to transmit or receive problems when large objects, such as cars or boulders, are between the diver and the tender, and they can be affected by the amount of particles in the water. Another potential problem with some wireless systems is that thermoclines can cause degradation of the communication.

## Search Patterns

Now that the diver is dressed, checked, and ready to go, there must be a way for the diver to search underwater—but how can that be done with low or no visibility? There are many **search patterns** used out in the field, and they are simply ways to keep the diver covering new ground during the search, mostly by utilizing a rope system. Some are universal and some are

B

**FIGURE 14-21** Wireless communication systems connect the tender **(A)** to the diver **(B)**.
Courtesy of Steve Treinish.

**LISTEN UP!**

Keep in mind that patterns may have more than one name. Be sure that you understand and are trained in the search patterns used by your team.

developed and used by teams that dive in specific places and need to accomplish a specific task. We will cover just a few basic patterns commonly used in rescue dives. Other patterns are used for recoveries, but require intense training and are more complex.

In some situations a search can be accomplished in clear water, but this is done only if divers have a dive buddy with them underwater, have good visibility, and they can be tracked from shore by tenders. Wireless communication stations that allow tenders to talk to and listen to each diver are highly recommended. Rescuers who undertake underwater searches must also have near-perfect buoyancy skills, as an entanglement on a silty bottom will almost certainly destroy visibility.

## Search Pattern Selection

An initial search pattern may be changed to another pattern if conditions and feedback from the diver necessitate it. Often, smaller bodies of water hold trash, trees, and other debris, and a quick change of search patterns will greatly increase the speed and effectiveness of the search. Along the same lines, the bank can play a large part in pattern selection. Land points that jut out in the water lend themselves to a large, semicircle sweep pattern, and a straight-line wall or pier may be easier searched using the "walk the dog" pattern.

## Semi-Circle Sweep

The semi-circle sweep (also referred to as an *arc pattern* or *circle sweep*) is the most common pattern in public safety diving (**FIGURE 14-22**). It can be done off a dock, shore, or boat; is simple; provides good diver

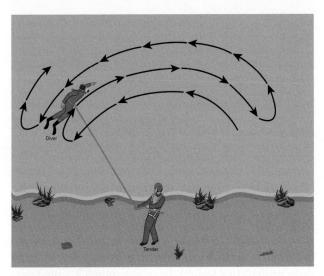

**FIGURE 14-22** Semi-circle sweep, also known as the arc pattern.

© Jones & Bartlett Learning

communication; and is easy to tend. Two dive crews working side by side must be able to stay untangled, but it can be done with practice and by being aware of the length of the diver's lines and the overlap that may occur.

For this pattern to be effective, the tender *must* stay in a fixed spot while the diver travels. The diver will use tension on the rope to pull/swim himself in an arc in the water. If the tender moves, the arc will move, and bottom areas will be missed. The tender is responsible for controlling the amount of rope in the pattern and when to have the diver change direction, and the diver must control the slack by swimming slightly away from the rope. The diver will be swimming at a slight angle away from the tender to do this. If the diver lets the rope slacken, the tender must stop the diver and fix it, either by tugs or voice. Missing a four-foot swath in the water may cause a diver to miss the tiny child he or she is trying to rescue. Slack rope is most often a training problem, but it can happen during true missions. Another problem is the rope snagging on obstructions as the diver goes deeper. A float can be attached to the tether line to hold it up and off the bottom to make things easier, but care must be used not to affect the amount of ground the diver is working.

### SAFETY TIP

One major rule of engagement is that no matter the pattern, a solo diver must be in direct contact with the tender. Diving without a direct link to the surface tender can be dangerous and is a major cause of public safety diving death and injury. A direct link is the backup diver's only route straight to the primary diver, and therefore is the only way a primary diver can receive help in a timely manner underwater.

Normally, the diver will swim this pattern away from the shore. Or, a diver can be placed at the farthest point offshore and work inward. This puts the diver back onshore as fatigue increases, and it also lets the diver work toward shore, creating a better feeling of getting closer to the finish. However, there are some concerns with using this method. The problem is that if there is anything on the bottom that will snag a rope, it will, because the diver is starting at the farthest point out from shore using the rope at its longest point. If a diver deploys with a 100-foot rope and there is a tree at 45 feet (13.7 m) out, the tether rope could get tangled in it.

Another concern is that bodies do not weigh a lot underwater, and a very hard swimmer on a larger rope could move a body across the bottom and out of the search area much more easily than most rescuers would think. The other kink is silting or stirring up visibility.

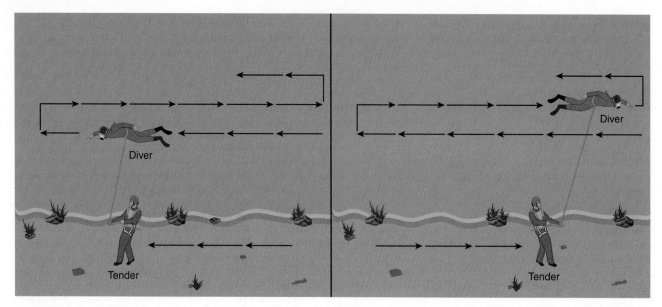

**FIGURE 14-23** Walking the dog.
© Jones & Bartlett Learning.

Remember, the diver is swimming away from shore to maintain tension on the rope, so the fins are stirring up the bottom the diver is swimming into. Three feet of visibility can be nice in the right conditions, and could also help provide pictures later in the dive if the object is found.

> **LISTEN UP!**
>
> In areas with heavy vegetation, the tether rope will often collect weeds and algae, making it impossible for the diver to effectively swim or even properly tension the tether rope. The addition of a large, inflatable buoy can help hold the rope above the weeds, making searches much more effective.

## Parallel Shore Pattern (Walking the Dog)

Walking the dog gets its name from the way it looks when it is in use (**FIGURE 14-23**). It works well for longer, straighter runs along a bank, dock, or walkway, and can keep two divers clearing ground at the same time. Two divers swimming at the same time are possible, but the bottom must be fairly clear. In this evolution, the diver swims away from the bank to maintain slack, but the tender walks with him, close to the water's edge. They walk and swim as a pair, resembling a human walking a dog that is trying to pull the leash to one side. When they arrive at a predetermined point, the "change direction" signal is given. The diver continues in the other direction to the other end of the search pattern, and so on. This is especially effective where people are reported to have fallen from a bank, because normally they will not be found far offshore.

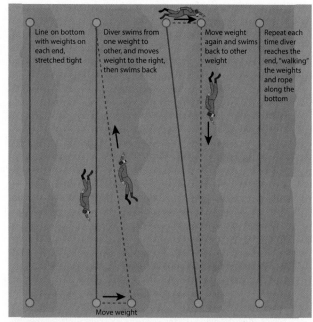

**FIGURE 14-24** Jackstay search pattern.
© Jones & Bartlett Learning.

## Jackstay/Running Weight Search

Jackstay patterns are not the first choice for rescue, because they can take longer to set up than other methods, which is time away from your victim. It is also harder to physically tend a diver using this pattern (**FIGURE 14-24**). The jackstay search pattern entails running a tight line between two weights, and having the diver swim the line between the two. Each time the diver returns to a weight, he or she moves it a predetermined distance in one direction. The weights are then "walked" across the bottom. It is a good pattern for smaller objects, because there is some double coverage of the bottom. Divers can run this pattern clipped onto the rope using a carabiner and short tether. While not physically connected to the jackstay rope, they still have hard contact with it and can be found on it. Wireless communications make jackstay patterns much easier, because a communication rope is not used.

## Jackstay Variation

In a variation on the jackstay pattern the tender becomes one of the anchors, either on shore or on a boat. The diver works the pattern the same way, but the jackstay line is run through a carabiner up to a float on the surface. The diver clips onto the rope with a carabiner and, instead of moving two ends, moves just one. The tender moves the other. The diver is still capable of feeling rope tugs, and is attached to the rope. The disadvantage is that tension must be held, and the line may rise up from the bottom slightly if bottom terrain is not smooth.

### SAFETY TIP

Many divers run jackstays simply holding onto the rope with their fingers. This is easier to manage, but will not allow a backup diver to find the primary if contact is lost. Clipping onto the jackstay rope guarantees the primary diver is still physically connected to the rope. Other teams run the jackstay with a diver on hardwire rope. If this is done, care must be used to keep the ropes from getting tangled underwater.

## Untended Patterns

Are there untended patterns? Yes, but only if certain safety parameters are met. A search may not entail using a rope, but may be done untended, as long as visibility in the water makes conditions safe enough to dive this way, and a route can be planned that does not need a rope to cover bottom surfaces effectively. Consider the base of a cliff that has 30 feet (9 m) of water below it, is 200 feet (60 m) long, and has 20 foot (6 m) visibility in the water, as you have preplanned it and previously conducted dives there. It normally presents pretty nice conditions for inland water. You are called in because someone found a fishing pole and a cold can of beer at the edge and a car parked close by with fishing gear in it. Could you run a pair of divers along the bottom and have them follow the cliff without using a rope? Sure!

This dive will be easy, quick, and under those conditions, quite safe. The buddy system provides dive safety, and other than looking for a human, this is no different than a typical recreational dive. This is one place where a wireless system with a surface unit would be attractive. Two divers can stay in visual contact during the dive, but have a margin of built-in safety and awareness when both are able to speak to the topside crew. These divers also must have a plan in place for if they lose visibility.

## Boat-Based Patterns

Some dives, such as those far enough offshore to prevent shore-based tending, or those at the base of a cliff, require the dive crew to operate from a boat. It is possible to search from a boat using a circular pattern, but these patterns generally present a sizable entanglement risk, and the time to set up these patterns generally precludes them from being used for rescue operations. With a good PLS, however, a boat can be used as a platform for dive operations. The goal with any type of usage like this is to get the boat stable in the water, resistant to a diver pulling it around while swimming a pattern. When close to bridge structures, cliffs, or walls, it is possible to tie up the boat to these objects and create a good, stationary platform. In open water, using three anchors helps stabilize the boat in the water. Using a single anchor allows the boat to be pulled in an arc and may affect the pattern and the diver's endurance.

### SAFETY TIP

Anytime anchors or other ropes are in the water with divers, the risk of entanglement increases greatly. All efforts should be made to place the search area away from the anchor lines, and ensure pattern selection does not place the diver near them.

## Monitoring the Diver

The most important job of the tender is to know exactly where, what, and how the diver is doing underwater. In true blackwater, on a rope with no voice communications, the tender is the diver's only link to the surface, and the diver depends on the tender for his or her well-being.

### Bubble Watching

Not to be confused with tourists "bubble watching" on a dive boat, the presence of exhalation bubbles are the first sign that the diver is exchanging air underwater, which is what everyone really likes. The ability to physically see a diver's bubbles is a major factor in the go/no-go decision, as tenders who cannot see bubbles cannot truly tell if a diver is still breathing. The tender must put himself or herself in a position to see the bubbles, and must focus on them at all times, no matter what. Yes, with electronic communication systems, we can hear the diver breathing, but if the system fails, then what? Your eyes must assume the role of seeing the diver breathing.

### SAFETY TIP

The development of sonar and underwater electronics has greatly reduced the need for divers to search vast underwater areas. This equipment is highly recommended, but keep in mind that a sonar unit requires a well-practiced operator and a boat.

A trained tender can glean a lot of information from the bubbles. Respirations should be calculated every five minutes at a minimum, and this will give the tender a baseline as to how the diver is handling the mission underwater. The five-minute mark is also a great time to have the diver check his or her remaining air pressure, so the tender can figure air consumption rates. Newer or less experienced divers may increase their breathing with stress or nerves, and a diver hyperventilating underwater is a hazard to himself or herself. More experienced divers have enough bottom time that their respirations can show up much more rhythmically, controlled, and slower. This would show the tender the diver is relaxed and dealing with the mission parameters well.

A regulator free flow will be obvious by the amount of air showing up at the surface. This is not a problem in and of itself, as the diver should be trained to handle this, but it is a reason to abort the dive immediately. Entanglements, finding bodies, or other stressors also can be discovered by a quick increase of the respiration rate by the diver, and a true sense of the diver's nerves is how fast he or she recovers.

### SAFETY TIP

Modern regulators are designed to flow incredible amounts of air, so if a free flow occurs, divers must realize they may have only seconds of air left to make a safe ascent. Additionally, if a free flow depletes the gas supply, the diver will have to manually inflate his or her BCD, drop weight, or both at the surface.

## Gas Supply

The diver's breathing gas supply is crucial to dive planning because the deeper the diver goes in the water column, the more air he or she needs to keep the lungs inflated and working properly. Whereas many typical dive rescue setups utilize an 80-cubic-foot cylinder, some deeper dives require the use of twin cylinders of more than 100 cubic feet each, and some AHJs deploy divers in a compact rescue pack with only 19 or 30 cubic feet of gas.

### LISTEN UP!

A small rapid-deployment dive setup that uses a pony cylinder is typically utilized by highly skilled divers in shallow waters offering good visibility. Rescuers using this equipment should be aware of its limitations concerning gas supply at depth, especially if emergencies or entanglements are considered.

At 33 feet (10 m), the diver consumes gas twice as quickly as at the surface, and at 132 feet (40.2 m), which is 5 ATA, or 73.5 psi, the diver will go through his or her air supply five times faster. This can lead to trouble, especially in the form of the 80-cubic-foot tanks. It is common for agencies to "cap," or limit, the depth divers may operate on a standard SCUBA system, depending on the conditions found at the scene. Deep dives are possible, especially in clearer water, but in blackwater, they can be treacherous. Entanglements at that depth increase breathing rates, and at five times the gas con-

### LISTEN UP!

Divers should calculate their SAC in full dive gear, reflective of what they will be using on missions. If they use a smaller, lighter, recreational SCUBA setup, it will not reflect the added air consumption of swimming with the heavier, bulkier gear that may be used in rescue or recovery operations. As an example, consider a diver with a SAC rate of 15 psi/min. At 33 feet, or two atmospheres, that diver will be using twice the air, meaning 30 psi/min. A 15-minute dive would be 15 mins × 30 psi, resulting in a need for approximately 450 psi of air. The same 15 minutes at 99 feet would be four atmospheres, or 120 psi/min, requiring 1800 psi of air supply. Take the same diver and use a SAC of 100 psi/min. At 33 feet, or two atmospheres, that 15-minute dive now uses 3000 psi of gas. A common standard for public safety divers working in low visibility water is to surface with no less than 1000 psi in the tank, so obviously there is concern with the SAC, the amount of time underwater, and the depth of the dive.

sumption, supply becomes a problem for both the diver and the backup.

**Surface air consumption (SAC) rates** are calculated for divers on the surface, and they are used to estimate the amount of air available for each diver at specific depths. SACs are commonly obtained by running a diver in shallow water, (less than five feet [1.5 m]), for a period of 10 minutes. The amount of air used (in psi) is divided by 10, giving the diver his or her own rate of air consumption expressed in psi per minute. Individual divers may have large differences in SACs, and this information should be readily known by the diver and recorded in team records for further reference. It is not unusual for fully rigged public safety divers to use 80 to 120 psi/minute. Remember, the diver will use more gas the deeper he or she dives.

Regardless of the SAC calculator used, the same basic principle applies. The SAC is plugged into the planned depth and the cylinder pressure, and an estimated time of the dive is shown (**FIGURE 14-25**). Many recreational divers are shocked to see the SAC rates of public safety divers, and public safety trainees should be made aware of the differences in SAC rates between public safety divers and recreational divers. Another option is to limit the time of the dive to result in the diver returning with no less than 1000 psi. Do not get complacent and think that shallow dives mean air calculations are not needed. Remember, a trapped diver with only 1000 psi may be in serious trouble, especially at deeper depths. Some divers will argue that the descent and ascent do not use the same air as the deeper depths, but calculating a total time at the maximum depth at which the diver is working is a good way to build in a safety margin.

## Rope Monitoring

Teams using electronic communication have a huge advantage in tracking their diver's well-being; they can simply ask. This is likely the largest advantage with these systems. Tenders can listen to the diver breathing on voice-activated systems, or ask the diver on push-to-talk setups. Divers may talk to themselves, and while not having to answer directly, tenders know exactly what is happening underwater. One point to remember with voice communications is that talking takes air. A controlled, relaxed breathing rate will use less air than talking needlessly. Tenders with voice communications also can be the first to break the circle of panic. The typical circle of panic starts with "uh-oh, I am in trouble!" and is quickly followed by an increase in vital signs and adrenaline output, and a decrease in fine motor skills. This makes things worse, and the diver now thinks, "uh-oh, I am *really* in trouble." A diver's motor skills can degrade in a panic to the point where only limb movements are happening. A calm, reassuring voice talking to the diver in a manner to break this circle will go further than anything else at that point. It may take a few minutes, and even a firm tone, but it must be done, to protect both the diver in trouble and the backup diver trying to help.

In the case of rope-only systems, the tender can still use the rope to ascertain if the diver is moving, by movement or tugging of the line. Obviously, a diver making steady progress while making steady bubbles does not need much input from a tender, but a diver constantly stopping or asking for an OK may need to be brought in and calmed down, or even replaced for the mission.

### LISTEN UP!

Never underestimate the power of humor to calm a diver down. A new diver wanted to get his first recovery done—just to get it over with. He assisted another team from an adjacent county on a recovery of a young adult victim in a small pond. After a 15-minute search, the diver found the victim by swimming right onto his torso. The tender, wearing a hardwire headset, was sitting in a boat, with the rest of the dive crew and some fire fighters within earshot. The tender was doing a great job coaching the diver, when he suddenly told the joke about the horse walking into a bar, where the barkeep asks "Why the long face?" The diver had gotten anxious when finding the body, and asked the tender to tell him a joke. It worked great, but definitely raised some eyebrows until later explained. Jokes are something you don't normally hear on dive missions, but in this case, it served its purpose.

## Diver Location

At any given time, the tender should be able to answer the following questions:

- Where is the diver?
- How do I think the diver is doing?

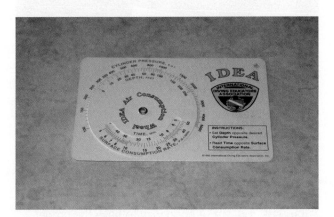

**FIGURE 14-25** SAC calculator.
Courtesy of Steve Treinish.

- Is the operation going well?
- How long has the diver been in?
- How much air does the diver have left?

While the diver is running the pattern, the scene sketch and clipboard should be used to provide triangulation for the areas searched. If the tender is turning (changing direction) the diver at the large oak tree across the pond, chart it. Keep track of how much rope the diver has used running the pattern. Ropes can be marked with a tape system, cable ties, or permanent markers (**FIGURE 14-26**). Knots can be used as rope length markers, but this uses up a lot of line, and can be confusing if a knot is quickly tied to mark a distance. However, knots are commonly used to mark a specific distance on the line if needed. If a diver has a major problem, or asks you to note something on the chart, you should be able to provide a good idea of where it is later. In some cases, a compass heading can be obtained from a fixed point on the shore, and this is used to pinpoint a location. Global positioning systems (GPS) are a great tool for obtaining and logging a location. Many sonar units are now equipped with onboard GPS, so even offshore points can be specifically marked. Finding a body on a rescue or a recovery will result in someone asking "where?" The more detailed your description, the easier it will be on the team later during debriefing and investigations.

## Diver Distress

A key point to remember with divers operating underwater is that as long as they have air in their tanks, they can work through almost anything that is thrown at them, short of a true medical emergency. A calm tender will go a long way in maintaining a calm diver, and remember: panic breeds panic.

The first line of defense is to know when a diver might be uptight about something. A sudden increase in respiration rate could mean a diver is stressed, and something as simple as a one-tug OK signal can calm a diver. Just knowing that the tender is truly watching out for the diver will help the diver's mental state. Voice communication systems allow coaching and calming of divers, both on the surface and underwater. All divers should be trained to drop their lead and manually inflate their BCD on the surface in an out-of-air emergency. Manual inflation is a skill learned in recreational SCUBA classes. Unfortunately, more often than not divers rarely, if ever, practice this skill. A vocal reminder may get the diver into that mode and take care of a problem at that moment.

If a diver cannot help himself or herself, surface tenders must have capabilities at the ready to retrieve the diver, whether by rope, boat, or other means. Most often, the rope will be used to pull a diver in on the water surface, but surface swimmers can quickly approach a diver in distress and start treatment sooner.

A good way to have the proper tools at the water's edge during a dive is to assemble a **shore box**, or similar tool kit. This kit contains, at a minimum, a throw bag, air horn or whistle, an oxygen cylinder and nonrebreather mask, some basic first aid supplies, and a flashlight (**FIGURE 14-27**). More equipment—such as extra carabiners, window punches, or cutting tools—can be added to tailor the mission, but at least the essentials would be in place. Endotracheal tubes or other airway equipment can be added if the rescuers have the capability to use them. A team able to administer aid right at the water's edge will give the diver every chance possible if the unthinkable happens, instead of running to the truck, or worse, calling for EMS and waiting.

**FIGURE 14-26** Ropes can be marked with a tape system, cable ties, or permanent markers.

Courtesy of Steve Treinish.

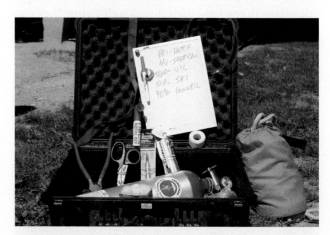

**FIGURE 14-27** A good way to have the proper tools at the water's edge during a dive is to assemble a "shore box," or similar tool kit.

Courtesy of Steve Treinish.

## Diver Medical Records

An emergency medical form may be filled out by the divers on the team and be kept on the truck in case of a medical emergency in which the diver is unable to provide crucial information. Allergies, previous medical history, prescriptions, etc. can be listed and documented to provide to EMS or the hospital in case of diver injury or illness. There can be privacy issues, but voluntary information can be used, and if properly stored and sealed, the only time this information is used is on a true emergency.

### SAFETY TIP

As a team leader, the issue of privacy must be weighed against being unable to provide medical information to an EMS crew. Having answers to basic questions might help alleviate part of any dive emergency. An EMS crew transporting a diver would be glad to have pertinent information while giving care.

# Decompression Illness

As discussed previously, the term **decompression illness** encompasses many different maladies that can strike a diver. Even a diver sticking to the depth and time parameters that are considered "safe" to dive may be stricken with decompression illness. Due to the workload and temperatures that many rescue divers may face, it is safe to say public safety divers face more of a risk than their civilian counterparts. Decompression illness is generally broken into two categories: Type 1 and Type 2. Decompression illness signs and symptoms may be felt within a few minutes of surfacing or perhaps as long as 24 hours after the dive.

## Type 1 Decompression Illness

Type 1 sickness generally involves the skin and muscular tissues of the body. When excess nitrogen is stored in the body, it is stored as a liquid in the fat and muscles. When the pressure is reduced, this liquid changes back to a gaseous state. Picture a soda bottle sitting on the shelf, under pressure. Not an issue at all, until the cap is released, and the pressure blows off quickly. Then bubbles seemingly form from everywhere. In a diver, some of this gas may end up in the tissues around the body, especially around the joints. Muscular or joint pain may be present, and may increase with time. Movement of the affected area does not normally cause

a difference in the pain. A red rash and/or itching can also be a sign of Type 1 decompression illness.

### SAFETY TIP

It is interesting to note that according to the Divers Alert Network, two of the main risk factors for decompression illness are cold temperatures and an increased workload while diving—both of which easily can fit a public safety diver's mission profile.

## Type 2 Decompression Illness

Type 2 decompression illness is thought to be more severe, and it affects the neurological and pulmonary areas of the body. Signs and symptoms of Type 2 decompression illness include weakness or paralysis, difficulty walking or urinating, confusion, chest pain, nausea, or ringing in the ears. Symptoms of Type 2 decompression illness can present quickly or slowly after the dive, with general fatigue often being attributed to the dive activity. This can let decompression illness worsen with denial or a lack of knowledge, with the diver attributing symptoms to other causes.

## Arterial Gas Embolisms

If the bubbles in the bloodstream become too big before exhalation expels them, they may lodge in the arterial blood vessels as they travel with the blood. This is called an **arterial gas embolism (AGE)**, and it is the most severe form of decompression illness (**FIGURE 14-28**). Because arteries in the body become progressively narrower, a blockage may form when a bubble gets stuck in

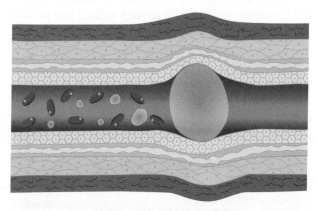

**FIGURE 14-28** An arterial gas embolism is similar to a stroke. As the diver ascends, pressure is reduced and the size of the gas bubbles increases, potentially blocking the artery.
© Jones & Bartlett Learning.

them. AGE is very harmful to divers, especially in the brain vessels. The typical brain AGE will have the same signs and symptoms as a cerebral stroke, and a diver will usually show an AGE quickly, by either lapsing into unconsciousness during the ascent or a few minutes after making the surface or shore. AGE may also result in frothy spit, confusion, tingling, or chest pain.

The alveoli in the lungs can also expand and burst if the diver holds his or her breath during the ascent. A breath-hold ascent of as little as 4 feet (1.2 m) after breathing air while underwater can result in lung injury. At 66 feet (20.1 m), the diver is now taking in three times the gas to inflate the lungs, and the alveoli expand to triple their size when coming back up to the surface. Normal exhalation lets this excess gas vent off. This is why the cardinal rule of diving, taught from day one for every diver, is to never, ever hold your breath.

## Decompression Illness Treatment

A suspicion of any type of decompression illness should be treated with high-flow oxygen immediately, and the diver should be transported to an emergency room for evaluation. Many divers do not admit to problems, however, or attribute their signs or symptoms to other issues, such as prior activity. Even if the signs and symptoms subside on a stricken diver, decompression illness could still be a problem and may not get better without treatment. According to DAN, transport to the closest emergency room is recommended, as is a call to the DAN hotline. DAN can assist in locating the proper facility to treat decompression illness, and that phone call may include a recommendation to divert transport to a location that has a **recompression chamber**, where a recompression of the diver may occur in a controlled environment.

Recompression chambers may be large enough to hold many people or small enough to hold one person lying down. The diver's ambient pressure is then increased; compressing the bubbles in the body, then slowly reduced under observation. The process includes a Doppler ultrasound to see the bubbles in the bloodstream. Many divers can recover from decompression illness and be released to dive again another day. In-water recompression is not recommended, as gas

**LISTEN UP!**

Many teams ask a diver/paramedic to specialize in the recognition and treatment of dive-related maladies to help educate the rest of the team. Many facilities with chambers will gladly educate and provide tours to teams who ask. "Chamber dives" are a great way to experience pressure safely, and are another perk of a chamber tour.

supply, hypothermia, and a lack of direction as to how long and how deep of a dive should be commenced are concerns. If proper medical protocol exists, a diver showing minor effects of decompression illness might be transported directly to a facility that operates a chamber, although passing a higher level of care must be planned before the dive, and is not endorsed without proper medical clearance from the AHJ. Aerial evacuations can be undertaken, but the cabin pressure should be maintained at 800 to 1000 feet (244 to 305 m) above sea level if possible. This all goes back to preplanning for medical considerations, as NFPA 1006 mentions.

**LISTEN UP!**

Many dive-related cases are transported to area emergency rooms where physicians may not be familiar with barotrauma as it relates to diving or how to recognize it. Barotrauma does not happen in many parts of the country with any regularity. The season may affect a doctor's diagnosis, because many people think wintertime and diving do not go together. Divers Alert Network (DAN) is a nonprofit organization dedicated to dive injury research, prevention, and response. DAN's expert medical staff and physician referral network assist divers in need 24 hours per day.

## Minimizing Decompression Illness Risk

While decompression illness can occur in any diver during or after any dive, there are some procedures and behaviors both divers and tenders can adopt to minimize the risk of a decompression illness occurrence. The greatest method of decompression illness risk reduction is to plan the dives, and dive the plans. Planning ahead on the shore to place a time and depth limit to a dive, and sticking to a proper ascent rate, are the two best ways to keep divers healthy.

### Ascent Rates

An ascent rate of 30 ft/min (9 m/min) will keep most divers safe. Ascending faster than this may not let the expanding gas in the body be expelled through the pulmonary systems. Nitrogen typically comes out of the liquid solution to form bubbles, which travel through the bloodstream to the lungs, and are simply exhaled.

Normally, in clear water, a diver will ascend while watching his or her smallest exhalation bubbles, or holding a line, or even monitoring a dive computer, all of which can offer cues to slow the ascent rate. Public safety divers dealing with no-visibility water, however, might not have outside references. Once again, a well-trained tender can guide the diver by helping to control the rate of ascent. Divers can be trained to ascend blind, but with a lack of sensory input, it can be very hard to do. If the diver comes up too quickly, he or she really can't tell, and if the diver is slightly negatively buoyant, it will cause him or her to settle back to the bottom after working hard while thinking he or she is headed up to the surface.

Divers can ascend on an **upline**, which is a line they deploy to hold onto while ascending. But in blackwater this can lead to entanglements, and while holding a victim, tending to their own BCD, and doing the ascent, divers can end up out of control. Divers coming up a slope can ascend under control by slowly swimming upslope. This allows the diver to vent the expanding air from the BCD and/or drysuit as needed, and also allows the diver to manage any other issues he or she may encounter. The tender merely keeps the rope slack controlled. It is not recommended to let the diver swim up his or her own tether without the slack being controlled because the diver may encounter loops or snags, and a snug tether line at all times makes it quicker for the backup to deploy to the primary diver faster, if needed.

When considering that during an ascent the diver must manage his or her buoyancy and ascent rate, while holding onto a victim and dealing with possible entanglements, tenders can elect to assist a diver up. Tenders must be aware of two things: the actual ascent rate and the slope of the bottom. Bringing a diver straight up off the bottom vertically must be done around 30 ft/min (9 m/min). This would be one foot every two seconds. On a gently sloping bottom, the actual vertical rise would be less, so a slight increase could be allowed, but why chance it? We don't want to injure a diver at any cost, even if he or she has the victim we are looking for.

Other problems the diver may face are entanglements or reverse squeeze, so the diver should have a way to tell the tender to stop the ascent. Unconscious victims are considered unable to hold their breath, and a diver equipped in a positive pressure FFM could still breathe on his own, even unconscious in the water. Unless the diver is absolutely known not to be breathing, his or her ascent rate should still be 30 ft/min (9 m/min).

Ascents are normally signaled to the diver, and he or she can assist or swim for the ascent, but may choose to let the tender control it with the tender line. It is reassuring to the diver that a competent and knowledgeable tender is bringing him or her up at the correct rate. The diver has enough to manage with a victim, his or her own gear, and the possibility of entanglements as it is.

## The Cardinal Rule of Diving

The cardinal rule of diving, learned from day one, is to never, ever hold your breath underwater. You must keep air moving through the lungs to prevent overexpansion injury. Many dive injuries are incurred when a panicked diver succumbs to the instinct of not being able to breathe underwater, throws away a perfectly functioning regulator, and holds their breath and bolts to the surface. Keep your divers breathing normally at all times.

## Dive Tables

Developed years ago by the U.S. Navy, dive tables are a quick way to plug in an expected time and an expected depth, and to get a glance of how much nitrogen the diver will absorb. These tables can give a maximum time for depths of up to 130 feet (39.6 m), and can also help plan concurrent dives and surface time a diver may have to spend off-gassing nitrogen from his or her body. Planning using dive table times can shave minutes off a dive to maintain a safety factor when dealing with decompression issues. However, a qualified dive instructor should explain their use, as they can be confusing to new or non-divers.

## Computer Planning

While generally not used for shallower dives of less than 10 to 20 feet (3 to 6 m), dive computers can help plan whether and for how long a diver may dive again. A dive computer computes and shows the time that can be spent at various depths at the time the user is observing the readout. As the surface interval time (the time a diver is topside above water) increases, the computer reflects the change in the diver's time and depth parameters. Applications for smartphones or tablets also are available to serve this purpose.

# Retrieving the Diver

Diver retrieval could be as easy as walking the diver up the steps of a dock, or could be as difficult as trying to get a diver into a johnboat in a strong current in a snowstorm. Operations-level dive rescuers can make a

**FIGURE 14-29** Tenders can play a huge role in reducing risks faced by divers entering and exiting the water.
Courtesy of Steve Treinish.

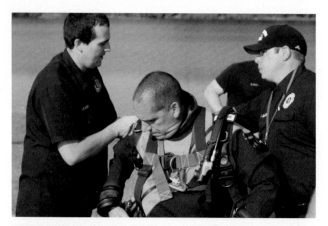

**FIGURE 14-30** With the diver seated, the tender can disconnect the straps, drysuit inflator, cummerbund, and other gear while the diver relaxes.
Courtesy of Steve Treinish.

huge difference in the ease of getting the diver out of all the gear worn for the dive. Often, divers find themselves crawling out of the water rather than trying to walk, to avoid sinking in mud or silt or tripping over rocks (**FIGURE 14-29**).

## Water Exits and Divers

Remember reading earlier in the chapter that a tender should properly assist a diver into the water? On exit, the diver is usually glad to be out of the dark, sometimes cold water, but is also fatigued, often to the point of just wanting to get the gear off and sit down. Granted, most dive gear becomes somewhat weightless to the diver underwater, but it is still work for the diver to move his or her body and the equipment though the water, and diving can be tiring mentally. Divers operating in cold water may have digits or limbs slightly impaired by the cold temperature, and may need some assistance getting up a bank. Regardless of any water condition, once again, the tender or a

shore support person should be holding the cylinder valve and regulator for support. It might be advantageous to sit the diver down at the water's edge, assist him or her in doffing the gear, and have helpers carry the gear back to the staging area.

When doffing a diver's SCUBA setup, turn off the air for the moment, sit the diver down or, if no seating is available, take the weight of the rig off the diver by lifting the tank. Sitting is preferred, as the tender can have the diver hold his or her arms out, shoulder level, and the tender can disconnect the straps, drysuit inflator, cummerbund, and other gear (**FIGURE 14-30**). Picnic tables or the floor of many EMS trucks can be used. The diver should not have to twist or strain to undo his or her straps, but should be allowed to simply sit and relax. As soon as the gear is loose, the diver simply stands up and takes a step away, and the gear is properly secured, cleaned, filled, and returned to service. Don't forget to lay down the setup so it doesn't fall over and injure anyone, and secure the weight belts. A weight belt falling off a bench onto a foot is indeed a toe breaker, and a completely preventable injury.

### SAFETY TIP

Many tenders loosen the diver's shoulder straps on the BCD and then have the diver pull his or her arms out. The diver has to rotate his or her arms out at a bad angle, so if the BCD is equipped with them, use the quick-release buckles instead.

## Watercraft Entry and Exit

Dive operation in watercraft ranges from easy to extremely difficult. Boats with dive steps or platforms can make entries and exits easy, and inflatables or johnboats

### LISTEN UP!

On almost every dive scene, there are responders on scene who have no water rescue training or who have training but simply don't have a position in the active crew. Instead of staging them at the truck to get bored, use them to carry gear! They don't have to know what it is or how it works, only that the pile of gear has to get from here to there. This can be applied across the board on all rescues, provided these responders are kept clear of the hot zone of the incident.

may be harder to use. Regardless, preplanning with the craft that will be used will go a long way toward proficiency.

One potential problem is the diver doffing gear in the water. This in itself is not a problem, as most recreational SCUBA divers learn this skill in early classes. However, if the diver is wearing a belt- or harness-style weight belt, taking off the BCD may result in lacking the buoyancy needed to stay afloat. Once the flotation of the BCD is removed, a diver may be pulled under by the extra lead, with no air to breathe. Removing a diver's weight system before the BCD can prevent this problem, but doing so also allows the diver to keep the mask on (if an FFM is being used), giving the diver air to breathe if he or she starts to sink or gets splashed. Be sure to provide proper buoyancy to the rescuer in the water. Once the buoyancy is in place, the diver can be extricated by whatever means is normally used for the watercraft. Chapter 15, *Dive Rescue Technician*, includes a National Institute for Occupational Safety and Health (NIOSH) report examining this problem.

## After-Dive Briefing and Reassignment

After checking through the medical precautions and undressing with the tender's help, the diver may be asked to submit to a quick interview with the dive supervisor, to quickly brief the diver on the progress made, bottom conditions, unusual hazards, or equipment concerns. The diver should be in clothing that fits the environment, whether warm or cool clothing, and should be getting the needed hydration, or possibly a snack. Dive gear should be checked, tanks filled, and equipment rinsed or dried as needed. If the diver clears a medical, has no problems, and the surface interval is sufficient to allow another dive, the diver may be rotated into a new task, usually acting as tender, then as backup or safety diver, then as primary once again.

Dive tables are used to determine how much nitrogen a diver will have in his or her system after a dive, and how long of a surface interval they require before another dive is permissible. Each diver must know his or her profile and whether he or she is clear to dive again. Each dive should have enough information listed on the tender sheet to make this a quick and conscious fact.

## Post-Dive Monitoring

Divers should be checked after exiting the water and after every dive, no matter how shallow or short in duration. The greatest change of pressure and hence, the greatest problem area in the water column is from the surface to 33 feet (10 m), and within that distance the last 10 or 12 feet (3 to 3.5 m) is the greatest change of pressure, meaning that even shallow dives can present risks.

Divers should be checked for signs of decompression illnesses with a field neurological exam, or field check. A **field neurological exam** is a quick way to ascertain how the diver is reacting to the pressure reduction he or she faced when coming up to the surface, and to the extra nitrogen onboard turning back to gas that might be causing issues. Each diver should expect these checks and, at a minimum, the check should include asking the diver the date, his or her name, and how he or she feels. How the diver feels should get a specific phrase in reply, such as "I feel fine." The *F* sound is easily slurred in case of problems, and gives the monitor something to look for. A diver slurring the *F* could be showing signs of a problem.

The monitor should also ask the diver to squeeze the monitor's fingers, and at the same time, ask the diver to resist the movement of each outstretched arm of the diver being pulled downward. This shows equal performance on each side, with no deficit. Any diver showing any signs of distress should be placed on 100% oxygen and transported to the closest proper facility. An emergency room lacking a hyperbaric chamber may be bypassed for another location that has one if medical clearance is given or approved beforehand. Helicopter transport is possible when dealing with long transport distances, but must be considered before the dive takes place. Generally, if the helicopter flight does not result in a change of 1000 feet (305 m) altitude, the flight might be the better alternative to long transport times on the ground. If altitudes higher than 1000 feet (305 m) are expected, ground transport may be a better choice. A quick informational call to DAN would be a good way to explore what might work best for a specific location.

Medical monitors or team members also should be quietly watching the diver for signs of mental stress. A heavy entanglement, or a victim similar to a diver's own family, or other factors can weigh heavy on the mind, and a tactic such as more rehab time, or being dismissed for the day may prevent a problem if the diver goes back underwater. All efforts should be made to mention anything like this discreetly, as embarrassment will only breed resentment.

## Diver Dehydration

All SCUBA dives cause some dehydration in divers. The air from the compressor or cascade fill system is dry to eliminate moisture issues in the regulators.

When breathed, this dry air wicks moisture away from the diver through the lungs. Even recreational divers experience dry mouths. Compound this with a rubber drysuit in high water temperatures (or even worse, high air temperatures) in the case of the safety diver, and sweat alone can cause severe dehydration. Make sure the divers drink plenty of fluids. Plain old water works well, as do sports drinks. Stay away from diuretic beverages, such as soda or coffee, and a good general rule is that if the divers do not urinate at least once an hour, they could use more liquids. Dehydration is considered to be a risk factor of decompression illness, and it also can lead to muscle cramps. A diver underwater experiencing cramps will not be able to perform to the best of his or her ability.

## Other Physiological Issues

Other problems that can afflict divers include ear infections, sinus irritation, toothaches, and eardrum ruptures. Each of these problems can occur in dives innocently enough, and some, such as ear infections, are common, even though they may not show up for a few days. Over-the-counter ear drying solution can be used to help eliminate ear infection risk, and proper handwashing technique can help eliminate exposure to various contaminants in the water, and hence on the dive gear. Treatment should follow the local medical protocols under which the dive team, AHJ, or local EMS is operating.

# Other Operations-Level Duties

It isn't all glory and medals at the water's edge. There are a host of other jobs that an operations-level rescuer may be asked to perform. They are all necessary, but nonetheless tough. A dive operation is continually evolving and can last for hours, if not days or weeks. Dive operations can be just as gear intensive as any other discipline, and in deeper waters or unwitnessed searches or recoveries, much more so. Let's take a look at some of the peripheral jobs an operations-level rescuer might have.

## Diver Medical Monitor

This position should be filled by an operations-level rescuer trained in oxygen therapy as it relates to diving, at the bare minimum. Preferably, a full advanced life support (ALS) crew will be standing by, but someone should be watching the divers, primarily for signs of

decompression illness. Decompression illness symptoms will normally show up within one hour of ascent, but can be as late as 24 hours after the dive. The field neurological exam and a set of vital signs can be repeated; generally, after 30 minutes, a diver will be rested enough to be placed in a topside support position if needed. Some divers will not have had enough of a surface interval to dive again, and this must be taken into account. Other divers may be suffering from a case of nerves or stress and should not be cleared to dive again.

Any diver showing signs of any decompression illness should be placed on oxygen and transported to a facility for higher-level care.

## Accountability

Most rescuers, when they think accountability, jump on Velcro tags or key chains with each responder's name. The same holds true with dive operations, but whoever is responsible for accountability also must know who is on which team; who are the primary, backup, and safety divers; and who are the supervisors of each. They also must know where and who are the medical, equipment, and rehabilitation groups, and they must have a list of divers who are ready to dive, those who are waiting for a surface interval to elapse, and those who are done for the day. It can get confusing, and it all leads to a big accountability system. Some teams use two-colored reversible T-shirts, with one color being a diver ready and able to dive and the other color signaling a diver unable to dive for whatever reason. Compound this by adding different teams, and managing accountability can become a tough job.

## Equipment Specialist

Some teams will outfit each diver with his or her own personal gear, and some teams use gear used by all team divers. Whichever the case, the **equipment specialist** has to know where it all is, whether it is ready to use, how it goes together, and where to stage all gear properly. Simple tarps on the ground make this job much cleaner, and using different colored tarps can allow for staging equipment with different meanings (**FIGURE 14-31**). Another common trick is to stage all full bottles with the valve cap on and pointed at the water, while empties have the cap off and are pointed away. Using the "off" side of the truck or trailer can be done, too: all gear between the truck and the water is fair game, and placing gear on the off side of the truck signals it is out of service for whatever reason.

**FIGURE 14-31** An equipment specialist makes sure all equipment is staged in a helpful way.

Courtesy of Steve Treinish.

**SAFETY TIP**

More than one diver has had a tank rigged that came from the "good" pile that was less than full. Make sure to physically check the submersible pressure gauge (SPG) before committing the diver. If this happens, it may be a sign that something needs to be adjusted in the support areas.

## Family Liaison

The **family liaison** can be one of the hardest jobs in dive rescue, or water rescue in general. Each dive operation should have an experienced diver from the team assigned to the family for the duration of the incident, if possible. A family liaison who does not have any training in crisis counseling will have much better outcomes if they work with a critical incident stress management (CISM) trained counselor. Once a trained counselor is on scene, the diver can act as a technical advisor to the counselor and family.

Critical incident trained counselors, grief counselors, or psychologists should be another source of help for the family, and a savvy team will have a callout list preplanned and available for response. Trained pastors, ministers, or clergy can be used, if the family is open to this. "Your baby is with God now," while meant well, is not what many families want to hear at this point. These trained responders, although not divers, have a crucial role in the mission, and teams should consider making them a part of the callout on any mission.

On many recoveries, the amount of time needed to find someone underwater can be long, and the family liaison may spend an incredible amount of time with the family. The liaison acting alone makes it easier on the family, as they are not constantly bombarded with new faces or new information, but makes it harder on the liaison by having to do it all. It is emotionally draining to play this role, but it is crucial to the overall operation. Even harder to deal with are families who have no idea about nature's effect on their loved one as a body decomposes. Animal feeding, body decomposition, current forces, and muddy water, while happening naturally, will not usually allow a "clean" recovery, and the family needs to know this upfront, before the recovery is made. Most families have no idea why recoveries take so long, or what the divers actually do underwater, and the liaison needs to be able to explain these points, as well as answer any other questions the family might have.

The media has been known to go to great lengths to film grieving families, or worse, the victim being recovered. If the scene is properly secured, this should not be an issue, but nevertheless it often happens. Secure the family in a private area. The family will likely want to be within eyeshot of the water where the search is ongoing, and will be vigilant of that water until the victim is found. This is natural, and all efforts should be made to place the family in a position where they can watch, but not interfere with personnel or be hassled by the media. Another factor is the stress rescuers will be dealing with. Rescuers sometimes use gallows humor while a traumatic mission is ongoing as a way to deal with stress. Other rescuers understand this, but the family may not and may take it the wrong way. Make sure the family realizes the stress that rescuers may face as divers, and that the family being close to crews can make a hard run harder.

How does one deal with survivors during a recovery? According to many families, the single best thing that can be done is to be honest. Honesty can be hard to present, but it should be done, and done respectfully. The family members may not like what they hear, and they may display emotions or reactions such as screaming, fainting, or even destroying things, but remember, these are natural reactions. Likewise, when delivering the news that the victim has been found, remember the person being recovered is not a "victim" to the family. Using the victim's name and referring to him or her in the present tense may help. The family liaison will be holding information that will change their lives forever. Some people will collapse with grief, while others will hardly flinch. Do not let them hurt themselves, but give them time to vent and digest the information.

Many rescuers put themselves on a mental "autopilot" on missions, and that is fine, but a family waiting on rescuers to bring back their loved one has no such autopilot. Rescuers may see denial, rage, and despair in an absolute raw form. Finding family members who can hold themselves somewhat steady is an asset in all areas of the mission. Use this person to ask questions of the family if there are any, and ask who should be in the area when giving updates or information. The family can allow certain people, such as close friends or a boyfriend or girlfriend there, but not everyone. They decide who is in "the loop," and you must abide by that or else information can be mishandled.

### SAFETY TIP

Many of the safety tips in this text have been directed at rescuers. Do not forget to watch out for families having medical problems when the body is recovered. Stress like this can and does cause heart attacks, fainting, and even suicidal thoughts among the living.

Expect a completely unknown reaction when delivering the news of a recovery, and be wary of delivering that news regarding a victim who is not 100% identified. Identification is the coroner's job. It might be possible to identify clues on the body, such as tattoos, body piercing, or jewelry, and relay this information to the family. While the 100% positive identification has not been declared by you or the team, the family can start preparing themselves for the likelihood that their loved one has been found. This begins the closure process for the families.

### LISTEN UP!

While viewing items on the body may be a way to tentatively relay identity to the family, the body should never be physically touched or removed, unless ordered by someone higher up in the evidence custody chain. Removing items or even cleaning skin to gain a closer look could be considered tampering with evidence, and should be discussed with law or investigative agencies.

Team liaisons can try to persuade the family members to wait until the victim is cleaned and dried off before they view their loved one. A glimpse of someone decomposed and muddy may not be a good thing for some family members, and with a bit more patience on the family's behalf, a body may be made more viewable with a bit of work. One capability that can be offered to the families of drowning victims is

for the liaison to view the body at the morgue while the technicians are cleaning up the victim. This lets the liaison confer with the family as to the physical state of the body, and gives the family an idea of what they can expect. Not many teams have had this offer refused, and many, many families have been thankful for it. It also can offer a great education to the rescuer by seeing and hearing information from the morgue technicians or coroner.

## Public Information Officer

Many jurisdictions have a public information officer (PIO) who deals with media requests and interviews, but a dive-specific PIO, or a diver assisting the department PIO, is a valuable tool in a dive crew. A news crew has a job to do, and working with them can help them deliver the story with respect and consideration while keeping selected details or information private. A good explanation or even demonstration of the dive search and rescue can even form positive opinions from the viewers, which often happen to be the voters who ultimately approve or reject tax levies that fund the very organization or agency doing the diving. Even after the incident is over, consider creating contacts within the media for other positive stories later.

## Sonar and Remotely Operated Vehicle Operators

Many teams tasked with performing underwater rescues and recoveries are making good use of technology. Quite simply, the more time a diver spends out of the water, the safer he or she is. Just a few years ago, acres of water would have to be cleared by hand searches underwater. This was time and resource consuming, and it subjected divers to problems just through the amount of time they had to submerge. Now, with sonar units and remotely operated vehicles, many searches have the divers diving on targets only after they are found by crews topside.

## Sonar Operator

Years ago, the use of LCD screen fish finders started making its way into the dive rescue arena. Many fishermen were becoming adept at locating and defining structures on the bottom, and dive teams utilized this technique. With today's color screen sonar units and GPS, a rescuer can get a great view of the bottom, and also see where exactly he or she has been in the water. **Color sonar** shows the bottom contours and objects in

| Depth (ft) | |
|---|---|
| **19.0** | |
| Speed (mph) | |
| **3.1** | |
| Heading | |
| **96.4** | |
| GPS Lat. | |
| **39° 19.205N** | |

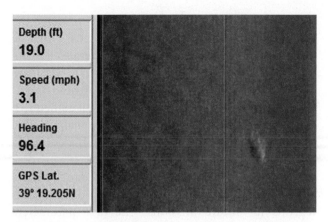

**FIGURE 14-32** Sonar can help limit the amount of time divers spend searching underwater. The shape on the screen is a victim lying on the bottom at 19 feet (5.8 m), with the head pointing in the 6 o'clock direction. The feet and legs can be seen pointing to the 12 o'clock position.

Courtesy of Steve Treinish.

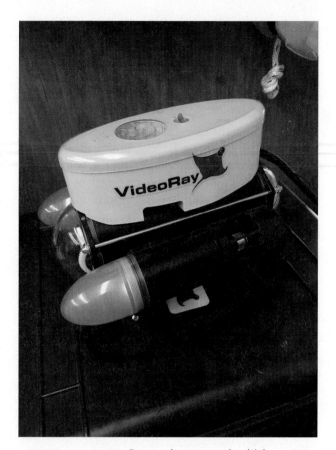

**FIGURE 14-33** Remotely operated vehicles can go places a diver cannot, namely to great depths. To offer a size comparison, the knot in the corner of this photo is from a throw bag.

Courtesy of VideoRay LLC.

color, and uses that color to show how dense the bottom is. Victims may show a different density, with bones and muscle tissue showing as different colors from the surrounding bottom. Color sonar has the disadvantage of being able to "see" only the bottom below or just to the side of the boat. Imagine the boat sitting on a traffic cone. The view from the bottom of the boat, through the traffic cone, is similar to how you would see the bottom using sonar. Sonar units are fairly inexpensive, and good units can be found for under $300.

**Side scan sonar** is rapidly becoming the choice of searchers, as it can shoot a signal out laterally from the boat, and the images it returns are incredibly detailed (**FIGURE 14-32**). Well-versed side scan operators on a high-end unit can see bodies easily, but they are more expensive, and can run upwards of $15,000 to $40,000. Smaller, fish finder–style units are on the market for around $2,000, and offer more use as the technology becomes more affordable. Even newer technology is the use of a drop-type, circular scanning sonar that is similar to radar. Deployed from the surface, the sonar unit sends signals back to the operator, who uses a laptop computer that can track not only the target, but also the diver moving to it in real time. These sonars resemble a radar unit shown on a laptop screen and have made for some quick dives.

### ROV Operator

**Remotely operated vehicles (ROVs)** can go places a diver cannot, namely to great depths (**FIGURE 14-33**).

True rescues of people submersed in hundreds of feet of water do not happen, as divers are limited to divable depths. Many trained, well-equipped technical divers can operate at depths of 200 feet (61.0 m) or so, but even then there is a decompression obligation to pay on the way up, and the setup time for the dive is simply too long. ROVs are small, swimming robots used in deep-sea exploration, and some rescue agencies have access to these units. They are very expensive, but they are gaining popularity, especially with deep water teams. They offer hope for returning loved ones who would have previously been deemed unrecoverable to families.

## Terminating a Dive Rescue Operation

As with all rescue incidents, the operations-level rescuer is responsible for terminating a dive rescue. Accountability is always checked, with all team members accounted for and safe. Sometimes, keeping the team at the ready can help other rescuers in boats or operating around the water's edge safer, or facilitate a recovery of a dropped tool or equipment. Once all other rescuers are clear of the hazard zone, the team can stand down.

Terminating a dive incident entails monitoring the divers' physical status, including hydration and any symptoms of decompression illness. A field neurological exam can be performed after the dive and again an hour or so later. Any exposure to the water or failure of PPE, especially dry hoods, must be noted, as ear infections can be common after diving. Handwashing is paramount after a dive, to ensure cleanliness of the hands that often wipe eyes, noses, and ears.

Gear should be returned to service and checked for safe and proper operation. Underwater mud and silt can find its way into every crack and crevice in the SCUBA system, including weights and belts. It may be necessary to disassemble the entire system to clean it thoroughly.

The task of restoring SCUBA equipment to operational readiness after cleaning should be assigned to a single individual. Often after a dive, too many people assist in gear assembly, and steps are omitted, creating issues with safe use later.

As always, an after-action review can be done before everyone leaves the scene, and is often advantageous for future missions. The dive parameters should be logged and stored however the AHJ chooses. Many teams now keep dive logs electronically. Dive scene sketches, witness statements, and other documentation should be processed as potential evidence in cooperation with the law enforcement assisting, if applicable.

# *After-Action* REVIEW

## IN SUMMARY

- Operations-level dive rescuers must be able to select proper personal protective equipment (PPE), assist the diver in dressing for the mission, select the proper search pattern and where to use it, and have a good understanding of drowning physiology and the human factors that will decide where to start a search.
- Operations-level rescuers in dive rescue must be able to perform witness interviews and develop a location to begin dive operations.
- It can be much harder to arrive at a point last seen (PLS) at an event with no one to give any location information, but using sound logic, reasoning, and good investigational tactics, it is possible to narrow things down enough to enable the go/no-go decision.
- Any SCUBA diving, whether recreational or for public safety, presents specific medical concerns and potential illnesses based entirely on the water depth and the duration of the dive.
- NFPA standards recommend each diver's PPE starts with encapsulation gear, which consists of a drysuit with a latex hood, dry gloves, team BCD, and a full-face mask.
- Every team should have a pre-deployment diver checklist not just on the truck, but also on the tender.
- Gear setups need to be streamlined, or free of any extra gear that may cause entanglement issues, anywhere on the diver's body and the SCUBA equipment worn.
- Depending on the circumstances, voice communication, hand signals, and rope tug signals can be used for communication between the diver and other responders on scene.
- An initial search pattern may be changed to another pattern if conditions and feedback from the diver necessitate it. Smaller bodies of water often hold trash, trees, and other debris, and a quick change of search patterns will greatly increase the speed and effectiveness of the search.
- The most important job of the tender is to know exactly where, what, and how the diver is doing underwater.
- Any signs and symptoms of decompression illness may be felt within a few minutes of surfacing, or perhaps as long as 24 hours after the dive.
- When doffing a diver's equipment, turn off the air for the moment, sit the diver down or, if no seating is available, take the weight of the entire setup off the diver by lifting the tank.
- Other operations-level positions that may be filled at a dive rescue or recovery include medical monitor, accountability, equipment specialist, family liaison, and sonar and ROV operators.
- Terminating a dive incident includes monitoring the divers' physical status, including hydration, any symptoms of decompression illness, and neurological status.

# KEY TERMS

*Access Navigate* **for flashcards to test your key term knowledge.**

**Arterial gas embolism (AGE)** A bubble in the bloodstream lodging in an artery, creating a blockage of blood flow to the surrounding tissues or organs.

**Barotrauma** Pressure-related injury. See *decompression illness.*

**Blackwater** Slang term used by many public safety divers to describe no-visibility water.

**Boyle's law** Law of physics that shows an airspace will decrease in volume as pressure increases and increase when pressure is reduced.

**Color sonar** Electronic unit using a transmitted beam aimed at the bottom surface to form pictures of the bottom density; similar to fish-finders found in many boats.

**Cummerbund** An elastic strap that holds the bottom of the BCD securely around the waist area.

**Decompression illness** A syndrome due to evolved gas in the tissues resulting from a reduction in ambient pressure. (NFPA 99)

**Decompression stops** Stopping an ascent from depth for a prescribed period of time to allow excess nitrogen to leave the diver's body. Decompression stops must be followed faithfully, or injury can occur.

**Dive tables** Tools used to calculate a diver's nitrogen loading based on depth, length of exposure to a hyperbaric environment, and intervals between exposures of an actual or a planned dive. (NFPA 1006)

**Diver checklist** A sheet the surface tender can use to check off diver preparations, thus making sure needed items are noted as ready to dive.

**Dry drowning** A death by suffocation, as the lungs are sealed off forcibly by a laryngospasm.

**Equipment specialist** Position that spearheads the status, field repair, and staging of all gear used in an operation.

**Family liaison** Position attached to the surviving family for channeling information between the family and the responders; usually filled by one person so the family does not have too many people to deal with.

**Field neurological exam** Rapid assessment of a diver's physical condition as it relates to pressure-related injuries; usually done on exit of the water or shortly thereafter.

**Go/no-go decision** The initial decision to deploy or hold personnel, after considering all aspects of the impending mission.

**Hot exit** Diver coming off the truck ready to dive.

**Laryngospasm** The involuntary reflex of the body forcibly clamping the airway shut in an effort to deny water entry.

**Most probable point (MPP)** In case of no witnesses, the location where the victim was likely to have been; determined by clues and logic.

**Negatively buoyant** The condition in which any object, including a person, sinks in the water column.

**Point last seen (PLS)** The last geographical location where the victim was physically witnessed to be; can be on the water or shore.

**Positively buoyant** The condition in which any object, including a person, floats in the water column.

**Recompression chamber** Small room or tube used to control pressures on the body; used for treating decompression illness.

**Remotely operated vehicles (ROVs)** Small, unmanned units used to survey areas underwater, controlled by an operator on the surface. Has advantages over divers in that depth, air, and decompression concerns are not issues to ROVs.

**Rope tug signals** A set of predetermined pulls on a line between a diver and tender signaling different communications.

**Scene sketching** Making a written record of the scene, conditions, witness information, and any other pertinent activity.

**Search patterns** Different methods of ensuring the diver covers bottom areas thoroughly during a search.

**Shore box** A case containing the essentials of dive safety, usually containing oxygen, throw bags, signaling or recall device, and a flashlight, at the least.

**Side scan sonar** Electronic unit similar to color sonar, but the beam is directed out to the side of the boat, giving lateral pictures from the boat position.

**Streamlining** Ensuring each diver does not present entanglement issues, such as tucking in or securing dangling objects, and taping or securing straps.

**Surface air consumption (SAC) rates** The rates at which divers use their gas supply at the surface of the water. This rate can be used to calculate approximate times a known gas supply should last underwater at different depths.

**Tender** A member of the dive team responsible for assisting divers with assembly and donning of equipment, communicating with divers, tracking the diver's status and location, and managing subsurface search operations, and trained to meet all the job performance requirements of operations-level diver rescue. (NFPA 1006)

**Triangulation** Using different points of view to produce a narrowed down point last seen. Intersecting lines of sight from these views may provide the PLS.

**Upline** A line divers deploy to hold onto while ascending.

**Wet drownings** Asphyxiation in which the lungs of the victim fill with fluids.

# On Scene

After the sixth drowning at a low-head dam near downtown, your department is mulling over forming a dive rescue team. You are tasked with providing the chief officers a list of risks and benefits, including gear concerns, diver selection, SOPs, and anything else dive related. Your department has a good water rescue response system, and there are a few recreational divers in your department who can offer advice. Seemingly everyone has opinions, including three dive retail centers that are now trying to convince you that their way is the right way.

**1.** From what three main problems does encapsulation gear protect divers?

**A.** Temperature extremes, biological hazards, and chemical hazards

**B.** Bodily fluids, mud and silt, and water temperatures

**C.** Water contamination, mud and silt, and water temperatures

**D.** In-water chemicals, biological fluids, and punctures

**2.** If a dive mission is deemed too dangerous to dive, what else can be done?

**A.** Dispatch watercraft capable of pulling drag hooks and dragging the water first.

**B.** Establish a PLS, perform a sonar grid search, and bring in other resources more equipped to handle this particular dive.

**C.** Call in the law enforcement team best equipped to handle a crime scene mission.

**D.** Wait for the body to decompose, causing it to float.

**3.** How much pressure is a diver subjected to at 33 feet (10 m) of depth?

**A.** 14.7 psi

**B.** 47.7 psi

**C.** 29.4 psi

**D.** 25.4 psi

**4.** Which pattern is most widely used by divers?

**A.** Walk the dog

**B.** Jackstay

**C.** Jackstay variation

**D.** Semi-circle sweep

**5.** When a diver ascends, all gas bubbles in his or her system will _____, and _____ can result.

**A.** stay the same; fatigue

**B.** increase in size; decompression illness or AGE

**C.** decrease in size; dehydration

**D.** increase in size; dehydration

*Access Navigate to find answers to this On Scene, along with other resources such as an audiobook and TestPrep.*

# CHAPTER 15

## Technician Level

# Dive Rescue Technician

## KNOWLEDGE OBJECTIVES

After studying this chapter, you should be able to:

- Identify and describe the certification levels for divers. (pp. 373–374)

- Describe public safety diving and how it differs from recreational diving. (pp. 374–375)

- Identify and describe common team positions in public safety diving. (**NFPA 1006: 18.3.1**, pp. 375–376)

- Describe the necessary components in starting new divers. (**NFPA 1006: 18.3.2, 18.3.3, 18.3.4, 18.3.6, 18.3.7**, pp. 376–380)

- Explain how to rescue divers safely and efficiently. (**NFPA 1006: 18.3.1, 18.3.2, 18.3.3, 18.3.7, 18.3.8, 18.3.9**, pp. 380–387)

- Explain the importance and the components of dive planning. (**NFPA 1006: 18.3.1, 18.3.2, 18.3.3, 18.3.6, 18.3.7, 18.3.9**, pp. 387, 389–392)

- Describe the search process in public safety diving. (**NFPA 1006: 18.3.1, 18.3.10**, pp. 392–393)

- Identify specialty dive conditions and how best to work with them. (**NFPA 1006: 18.3.3, 18.3.6**, pp. 393–400)

- Explain the purpose and process of diver decontamination. (pp. 400–402)

- Identify key considerations in leading dive evolutions. (**NFPA 1006: 18.3.5**, pp. 402–403)

## SKILLS OBJECTIVES

After studying this chapter, you should be able to:

- Manage stressors commonly found in a life-threatening aquatic environment. (**NFPA 1006: 18.3.7, 18.3.9**, pp. 376–387)

- Assist a submerged diver in distress. (**NFPA 1006: 18.3.8**, pp. 380–387)

- Use dive rescue equipment that properly protects the user, perform a search of a body of water typical of one found in a local district. (**NFPA 1006: 18.3.10**, pp. 389–393)

- Supervise and coordinate search activities and team operations, with consideration given to the parameters of water conditions, crew safety, and victim information and care. (**NFPA 1006: 18.3.5**, pp. 402–403)

# Technical Rescue Incident

Your department's dive rescue team is well equipped, respected, and very active. The team works in conjunction with the local police dive team, and your aspect of public safety diving is focused solely on rescue. Eighty-five percent of the runs your team responds to are in retention ponds less than 15 feet deep and rarely farther offshore than 50 feet.

Right at roll call, you receive a call to a city reservoir on an unknown run. During the response, dispatch advises of a person in the water off a dam at the south end of the lake. You are familiar with the lake but not the dam. Your educated guess is that the water is 75–85 feet deep with no visibility, and you know from experience it will be totally black and cold. You arrive and see a frantic jogger who says she saw a man fishing along the rail. On her return trip, about 35 minutes ago, she saw him fall in.

This scenario seems possible, given the amount and type of recreation that occurs on the dam and around the area. It is also known that many fishermen like to have a few adult beverages while spending the day fishing. The dam is releasing a small amount of water to the discharge stream below. The news media already have a helicopter overhead, and you have to make the decision for your divers to go or stay topside.

**1.** Compared to conditions in which your team normally operates, does the depth you face change your operations?

**2.** What hazards to the divers may be found in this water, and how will conditions affect dive planning?

**3.** How much air can you expect a diver to use to perform a 10-minute dive if rescue is attempted?

 *Access Navigate for more practice activities.*

# Introduction

Dive rescue can be one of the most hazardous undertakings in the water environment. With the proper precautions and equipment it can be relatively safe, but all diving, whether recreational or public safety, poses some inherent risk just by submerging. This chapter focuses on steps and procedures that can be taken to minimize that risk. All dive instruction should be undertaken only with a competent, experienced public safety dive instructor and with proper gear and precautions. Be wary of recreational instructors or dive shops with little or no public safety experience that claim to train dive teams. The National Fire Protection Association (NFPA) recommends that all public safety dive candidates possess at least a recreational-based advanced open water certification before starting public safety dive training, and that they fulfill the dive rescue job performance requirements (JPRs) in NFPA 1006, *Standard for Technical Rescue Personnel Professional Qualifications*.

# Certification Levels

All dive instruction should be done through an experienced instructor working under the guidelines of a nationally recognized dive-training agency. The World Recreational Scuba Training Council (WRSTC) comprises several dive-training agencies that develop and govern the knowledge and skills new divers are required to master before certification. Each agency offers its own spin on dive education, but in the end they all meet requirements set by the WRSTC.

## Open Water Dive Certification

Considered to be base knowledge and skills for any beginning diver, open water students learn the basics of gear, assembly, how to manage equipment, basic physics, and the crucial skills to survive underwater. Many dive shops offer equipment sales, dive instruction, and even dive tourism trips. Dive instructors, both those employed by shops and those operating independently,

**FIGURE 15-1** Recreational divers decide when and where to dive, often enjoying ideal dive conditions.
Courtesy of Steve Treinish.

are certified and tested by governing agencies addressing recreational dive standards, and adhere to best practices set by the groups under which they operate. Millions of people around the world have been certified in open water diving (**FIGURE 15-1**). The basic limitations are that a diver with this certification can dive safely above 60 feet (18.3 m) of depth during daylight and in calm, friendly water. The buddy system is paramount in recreational diving, and most skills reflect the use of the dive buddy to help mitigate problems underwater. Many skills learned in these classes are relevant to both recreational and public safety diving, but are not practiced enough to become reactionary when they are needed. For this reason, many divers each year die with functioning equipment still in place.

## Basic Dive Skills

Some of these basic skills apply to any type of personal protective equipment (PPE) the authority having jurisdiction (AHJ) or diver chooses to use. **Clearing a mask** is the process of using exhaled air to displace water that has filled a mask. This applies to a standard dive mask

or a full-face mask (FFM). When water presses or is present against the nasal area, a diver who is unaccustomed to this feeling may have trouble differentiating between breathing from the mouth, which is dry, and the nose, which is covered in water. When this happens, it can easily cause choking or coughing.

**Purging a regulator** is required underwater because ambient water will fill the mouthpiece or face piece when it is not in use by the diver. The diver must be able to blow the water out in order to get dry air to breathe. Normally, exhaling and pressing the purge button on the regulator will flush any water out. Failure to purge a regulator underwater will likely result in aspiration and inhalation of the water in it, causing the user to cough or choke. Worse, it may result in the diver believing the regulator is not delivering air and making an emergency ascent to the surface.

## Advanced Open Water Certification

Advanced open water certification is the next step in learning for recreational divers, and it covers classroom and in-water skills such as night diving, underwater navigation, deep diving (but shallower than 130 feet [39.6 m]), and some basic search and recoveries of objects such as anchors and blocks. The same dive shops that teach open water classes offer advanced certification classes. No public safety skills are presented, as this is also a recreational-based class. Generally, most divers feel more comfortable in the water after the advanced classes.

## Rescue Diver Certification

The term *rescue diver* is frequently used in lieu of *public safety diver*. A rescue diving class is based in recreational SCUBA diving and teaches civilian divers the basics of saving another diver from the generally friendly water conditions that recreational divers frequent. Subjects taught include first aid, cardiopulmonary resuscitation (CPR), and oxygen therapy, along with some basic ascent and diver extrication technique; however, this class is not a true public safety diving class. Beware of divers who possess a Rescue Diver card who try to perform public safety tasks.

## Public Safety Diving

When deciding to take a public safety dive course, many divers do not fully understand the differences between recreational diving and public safety diving. Recreational divers usually decide when to dive, and they normally enjoy conditions that are not considered harsh. Recreational diving is built on the premise of diving with a partner in the water to assist with problems, and while some divers dive solo, these divers usually have the benefit of sight underwater.

Some public safety divers operate in oceans, lakes, and rivers that have areas of good visibility, but many in the field operate nearly or even totally blind from the first few inches of descent to the time they surface. Underwater visibility refers to the distance a diver can see around them. In the Florida Keys, a bad day's visibility may be 10 feet (3 m). In many inland lakes and rivers, a good day's visibility might be 6 inches (15.2 cm). Recent rains, currents, runoff, and boat traffic can all affect visibility, as can weather and seasonal conditions. The term *blackwater* diving comes from the fact that the diver sees only blackness during the dive. Many public safety divers operate in cold, dark, and dirty conditions, especially when diving in inland or confined waters. The stressors they deal with are compounded by the fact that they usually dive alone. There is the possibility of getting entangled, and sometimes they have no way to communicate other than using rope tug signals. Nerves can become a significant factor in navigating blackwater dive environments. Divers should remember that they do not operate alone, but as part of a team. They merely dive alone, and a host of positions must be put in place to ensure the diver returns topside safely.

Many teams require diver candidates to perform a skills test underwater before they even start training. There is nothing wrong with this, but the test must be passable. Many excellent public safety divers are developed even if their skills are sketchy at the beginning of their training. A good base of skills to examine public safety dive (PSD) candidates would be mask clearing, no-mask swimming and operating underwater, buoyancy skills, cutting lines and ropes, and doffing and donning SCUBA gear underwater. Underwater obstacle courses are frequently created to hone these skills. The diver normally has a mask that is blacked out or taped over, and follows a rope around a pool, finding, assessing, and defeating obstacles similar to those they may find in the water in their jurisdiction (**FIGURE 15-2**).

Most divers who are completely comfortable with these skills can be easily trained in no-visibility water, but divers lacking these skills may need to "tune up" and polish them before trying no-visibility diving. Some techniques and equipment for training and operating in the no-visibility environment are discussed later in this chapter.

**FIGURE 15-2** Taping a dive mask can help sharpen skills in no-visibility water. Note the large tab created to enable quick removal of the tape in case of trouble.
Courtesy of Steve Treinish.

### SAFETY TIP

Any diver operating blind underwater must be monitored for safety and problems while underwater. A fully trained public safety diver with sight must be close enough underwater to intervene in case of panic or problems. An emergency signal also must be in place to alert others of a problem. Any emergency underwater should end the entire evolution for everyone, as two emergencies underwater will likely be tough to handle.

## Team Positions

The makeup of dive crews varies from team to team. Some AHJs require that four or even five qualified members be present before starting a dive, whereas other true rescue units may dive with three divers on scene, with one filling a supervisory role. Still other teams will let a single diver dive, and will let that diver make the decision upon arrival. The typical incident management system used in many other fire service missions can be placed into a dive rescue, with roles being specific to the task. For example, the creation of an equipment specialist allows the dive supervisor to focus on the dive, lets the tender focus on the diver, and lets the diver focus on the dive at hand. All three now know the equipment is being assembled, staged, and checked for operation by another team member, taking one aspect of stress off their respective tables. Rather than a team trying to manage extrication of a victim in shallow water, appointing a member (who may not need to be dive trained) to the shore to head up the removal of the victim will accomplish the same task. It is this person's job to locate the proper equipment to safely remove the victim once the divers get to the shallow water where it can be harder for them to move around and operate.

There are specific terms regarding crewmember titles that are common to almost every team in the world. These are:

- **Primary diver:** The diver actively diving, or the first diver to be sent into the water.
- **Secondary, or backup diver:** A fully trained, qualified, and equipped diver standing by as close as possible, even in the water, and able to descend within four to five seconds. They should have fins on, mask in place, and have no duty or tasks other than to be ready to assist the primary diver.
- **90 percent diver, or safety diver:** Many teams will stage a diver close by, but not in the way of the primary diver. This diver can be staged with the mask off and fins off. If the backup deploys, the 90-percent-ready diver becomes the backup diver. This position also is known as a safety diver.
- **Tender:** The controller of the dive, as related to the diver underwater. The tender keeps track of the times, pressures, locations, search patterns, and obstacles in the water. The tender is a key person in the operation and does not simply "hold the rope."
- **Dive supervisor:** A fully trained and experienced diver who fills the role of company officer in most dive evolutions. The dive supervisor controls the dive/no-dive decision, approves the pattern selection, and works closely with the incident commander to see that all of the team's needs are met, including emergency medical services (EMS), gear, rehabilitation, and air fills.

Many teams rotate the diving positions during missions, giving each diver the chance to get underwater and to help fill the support roles. This also allows teams to meet surface interval requirements on missions requiring multiple or deeper dives. It may be advantageous, however, to keep the tender and supervisor intact as long as possible for fluidity. More positions are examined in Chapter 14, *Dive Rescue Operations*.

**KNOWLEDGE CHECK**

Which is the dominant factor that distinguishes recreational diving from public safety diving?

a. The ability to choose when and where you're going to dive
b. The depths typically encountered
c. The types and style of equipment used
d. The basic survival skills needed

*Access your Navigate eBook for more Knowledge Check questions and answers.*

# Starting New Divers

Many teams and divers incorrectly assume that anyone with some SCUBA experience can be a public safety diver. This is wrong, as some great recreational divers cannot handle the potential claustrophobia of an FFM and/or being submersed in total blackness. There is a fine line between a diver adjusting to the stress in stages and becoming a good public safety diver and someone getting a case of nerves early on in the training and quitting. It can take some divers many dives to become comfortable, and others may never be able to handle the stress conditions that this type of diving can create. These divers can, however, make great tenders, because they know firsthand what the diver may be going through under the surface.

## Swim and Skills Testing

NFPA 1006 requires a swim test for all water rescuers aspiring to be trained to the technician level. This is a good place to start new divers, and highlights the swimming skills that need to be considered. Quite honestly, divers who are grossly out of shape or not comfortable swimming in the water should not perform public safety diving. NFPA also recommends the swim test be at least as strenuous as a divemaster level swim test. A **divemaster** is a diver with enough training and experience to be able to supervise diving operations at a professional level. The International Association of Dive Rescue Specialists (IADRS) provides a good test. Many teams add items, such as an underwater swim and a weight retrieval, and make this test an annual recertification requirement (**FIGURE 15-3**).

## Recreational Skills

Some of the basic skills that divers learn in recreational open water classes most definitely apply to public safety training, and sadly, many of these skills are not practiced enough. Divers have died while under duress by failing to perform some of these basic skills.

Manual inflation is done with the buoyancy control device (BCD) during an out-of-air emergency. The BCD is normally filled with air from the cylinder the diver breathes from, but with no air, it cannot be filled from that cylinder. The diver must tread water at the surface and press the deflation button while blowing into the mouthpiece. Once a full breath is blown into the BCD, the button is released; then, another breath is taken and exhaled into the BCD. Once the BCD is inflated, the diver will float. To manually inflate a buoyancy control device, follow the steps in **SKILL DRILL 15-1.**

# I.A.D.R.S ANNUAL WATERMANSHIP TEST

### Evaluation parameters
There are five exercises that evaluate stamina and comfort in the water, each rated by points.
The diver must successfully complete all stations and score a minimum of 12 points to pass the test.
The test should be completed with not more than 15 minutes between exercises.

**Exercise 1:       500 yard swim**
The diver must swim 500 yards without stopping using a forward stroke and without using any swim aids such as
A dive mask, fins, snorkel, or flotation device. Stopping or standing up in the shallow end of the pool at any point
During this exercise will constitute a failure of this evaluation station.

| Time to complete | Point awarded |
| --- | --- |
| Under 10 minutes | 5 |
| 10-13  minutes | 4 |
| 13-16 minutes | 3 |
| 16-19 minutes | 2 |
| More than 19 minutes | 1 |
| Stopped or incomplete | Incomplete |

**Exercise 2:       15 minute tread**
Using no swim aids and wearing only a swimsuit the diver will stay afloat by treading water, drown proofing,
bobbing or floating for 15 minutes with hands only out of the water for the last 2 minutes.

| Performance criteria | Point awarded |
| --- | --- |
| Performed satisfactorily | 5 |
| Stayed afloat, hands not out of water for 2 minutes | 3 |
| Used side or bottom for support at any time | 1 |
| Used side or bottom for support > twice | Incomplete |

**Exercise 3:       800 yard snorkel swim**
Using a dive mask, fins, snorkel, and a swimsuit (no BCD or other flotation aid) and swimming the entire time with
the face in the water, the diver must swim non stop for 800 yards. The diver must not use arms to swim at any time.

| Performance criteria | Point awarded |
| --- | --- |
| Under 15 minutes | 5 |
| 15-17  minutes | 4 |
| 17-19 minutes | 3 |
| 19-21 minutes | 2 |
| More than 21 minutes | 1 |
| Stopped at any time | Incomplete |

**Exercise 4:       100 yard inert rescue tow**
The swimmer must push or tow an inert victim wearing appropriate PPE on the surface 100 yards non stop and without assistance.

| Performance criteria | Point awarded |
| --- | --- |
| Under 2 minutes | 5 |
| 2-3  minutes | 4 |
| 3-4 minutes | 3 |
| 4-5 minutes | 2 |
| More than 5 minutes | 1 |
| Stopped at any time | Incomplete |

**Exercise 5:       free dive to a depth of nine feet and retrieve an object**

| Performance criteria | Point awarded |
| --- | --- |
| Performed satisfactorily | Pass |
| Stopped or incomplete | Incomplete |

Additional copies available at no charge via the international association of dive rescue specialists webpage. Visit www.IADRS.org

**FIGURE 15-3** IADRS swim and skills test.

# SKILL DRILL 15-1
## Manually Inflating a Buoyancy Control Device

**1** Grip the BCD inflation/deflation control with the finger ready to depress the dump valve.

**2** Place the mouthpiece of the BCD inflator into the mouth.

**3** While depressing the deflation button, blow a breath into the BCD air bladder through the BCD buoyancy control tube.

**4** Release the deflation button before removing the mouthpiece, keeping the exhaled air in the bladder. These steps may need to be repeated several times to ensure positive buoyancy is achieved.

Removal of the weight system the diver is using is also a basic skill learned in open water classes. Divers must be able to remove their own weights or those of another diver. This skill is discussed in detail later in this chapter.

## Dive Rescue PPE

In this chapter, we will assume the diver-in-training is wearing the full complement of encapsulation gear, which keeps the diver warm and dry and protected from the water temperature and possible non–life-threatening contaminants. A drysuit is highly recommended, in conjunction with a latex dry hood, dry gloves, and an FFM. For water that is considered to contain high levels of contamination or worse, a true surface supplied air (SSA) system may be a better option, but these systems do not generally lend themselves to rescue work because of the longer time needed to set up. Dive rescue PPE is covered in more detail in the earlier dive rescue chapters.

### LISTEN UP!

Encapsulation diving will not protect the diver from obvious water contamination such as highly-concentrated raw sewage, petroleum, chemicals, and the like. For dives that include these types of hazards, the decision to rescue must be weighed against the protection of the dive rescuer. The typical drysuit/FFM setup is considered to be a good baseline for most waters public safety divers encounter.

## Drysuits and Full-Face Masks

Two major items of PPE are specific to diving. A diving drysuit is similar to a surface water drysuit, but it has valves that allow the user to inflate and deflate the suit underwater. As the diver descends, the inflation valve is pressed to add air to the suit. This helps control buoyancy and eliminates the squeeze felt as the material is compressed onto the body of the diver. The addition of a latex hood and dry gloves and thermal insulation keep the diver protected from cold water temperatures and the contamination often found in public safety dive environments. Upon ascending, the air that was added on the way down will expand and must be vented out, most often from a valve on the shoulder of the suit. This air must be purged from the suit, because the expanding air causes an increase in buoyancy that can cause the ascent rate to become too fast.

The FFM many divers choose to wear is similar to a structural firefighting mask in shape and feel, and many fire fighters adapt readily to it. One aspect

**FIGURE 15-4** This diver is pushing his nose into a nose block in order to equalize his ears.
Courtesy of Steve Treinish.

that may take practice and adaptation is the process of equalizing the ears. Because the face is totally covered, the nostrils cannot be pinched in order to blow against them and equalize (**FIGURE 15-4**). Divers who need to perform this maneuver must push their nostrils into a nose block to occlude the nose enough to gently blow and equalize pressure. Another aspect of FFM diving is switching to an alternate air source if an emergency arises. It is rare that a mask fails to deliver air, but in the event it does fail, the mask must be removed underwater. A typical recreational second-stage backup regulator is often the air supply that the diver will switch to, and the actual switch is no different than the procedure learned in recreational classes. What is different is that now there is no underwater vision. For this reason, many public safety divers carry a recreational mask in a pocket as a backup while diving.

## The First Training Dive, "Lights Out"

Many public safety divers can relate to the first time they were dropped into dark water and told to swim around. Many of these same divers can tell you the nerves were a lot to handle and that taking time to get used to these conditions is beneficial. Divers can get used to no-visibility conditions with two simple pieces of equipment: a roll of clear tape and a roll of duct tape (**FIGURE 15-5**). Just as new fire recruits do not try to extinguish a fully involved house fire without first learning how to handle the small fires, the public safety diver candidate can be trained in a similar step-wise fashion.

**FIGURE 15-5** It is important to conduct training with simulated hazards.
Courtesy of Steve Treinish.

**FIGURE 15-6** Many important skills can be practiced in a shallow pool or even in the station.
Courtesy of Steve Treinish.

### SAFETY TIP

Consider performing all initial evolutions in a confined, shallow pool. Once a diver can operate in the deep end from start to finish on a typical dive, cut and fix entanglement issues, and assist a fellow diver in trouble, then real-world conditions can be introduced. Another plus is the ability to provide safety to a diver operating with occluded vision underwater, with the instructors and assistants likely having better conditions for communications and control of the training.

Clear plastic tape can be placed on the mask to allow some light penetration but partially occlude sight. Students can wear team gear, but only after proper training in the use and setup of that gear. That is a great way to start building muscle memory, but on the first dive, be wary of putting a student in gear with which they are not familiar.

Toddler toys or homemade puzzles can be utilized for the new diver to work on underwater and build their confidence. Instructor presence should be close, and squeeze signals or even a hand on the diver's shoulder can calm the first-time nerves. Later evolutions can be done with duct tape or blackout covers, leaving the diver in total darkness. An instructor should be close enough to render assistance, but the touches should become more infrequent as the diver improves. Respiration rates via watching the exhalation bubbles are a good way to monitor diver nerves. If available, wireless transceivers may allow some voice contact, but the student needs to get used to not being able to talk to anyone.

## Confidence is Key

Doffing and donning, cutting tool practice, etc., can all be practiced in a shallow pool or even on the bay floor

(**FIGURE 15-6**). As a matter of fact, using a building-block system, where training is done in steps, will result in better divers. Divers should be run through drills that make them completely comfortable cutting themselves out of entanglements and running search patterns completely blind, under an instructor's supervision and control. The diver's confidence will soar, and with other students acting as backup divers and tenders, the roles they will have to assume in the field are quickly learned. Another advantage to pool training is the ability to let the other students watch. Placing team members on the surface with a mask and snorkel to watch other students' training evolutions can be a key learning tool. Many divers fail to realize how much better a slow, controlled, deliberate movement in the water is until they watch a cutting operation done too fast.

Divers learning to operate as backup divers also should be stressed as a function in these evolutions. They are a critical part of the dive crew, and their job is no less stressful than that of the primary diver. The primary diver learns to deal with the entanglement, but with major fouling underwater, the diver should not hesitate to call down the backup diver. The backup should be able to descend without hesitation and assist the primary until the problem is resolved. In later evolutions, as a diver's confidence builds, consider surprising the backup diver with a call for assistance when they are not expecting it. This is especially effective when a backup diver's attention might be drifting somewhere else.

Just as in structural firefighting, stress responses should be created and molded during controlled training, not in emergency situations. Asking other teams and agencies about prior close calls and working through those types of incidents underwater not only will help train divers, but will also change entire procedures within a team once problems or solutions are identified.

# Diver Rescue

No one wants to imagine a diver from their team drowning, but it happens. Certain equipment choices can tip the scales toward safer operations, and some diver actions can save lives. A huge advantage to any diver in the water is the presence of a trained backup diver (**FIGURE 15-7**). What constitutes *presence* is another question. Some teams allow operations with a backup diver available in a parking lot, even though the safer setup places the backup diver in the water, ready to submerge at a moment's notice. This seems like common sense, but some divers and teams have no issues with a single diver entering the water with no backup yet on scene. While it is commonly argued that the diver is justified doing this while operating in a true rescue mode, it also brings up the argument of whether a single diver response should be an option. Constantly asking yourself, "What will happen if this goes wrong?" will provide great guidance.

In theory, anyone on the team should be able to fill the backup diver position. However, theory and real world operations sometimes differ. It can be advantageous to use the more experienced diver on the team in the backup role. Many people would automatically place the most experienced, seasoned diver in the primary role, but if the stronger skilled diver is used as the primary diver, the backup may not have the presence of mind that experience brings to the table. This statement should not be construed as advocating putting lesser-trained divers in the water. In a perfect world, everyone would have the same training, experience, nerves, and ability, but this is not always the case. Experience

or team operations may play into diver selection during missions.

Although the objective in dive rescue is to rescue the victim, it is not unusual for the diver to require some form of rescue as well, whether in the form of a quick snip of fishing line in a location on his body he cannot physically reach, or being brought to the surface by another diver after suffering some sort of medical emergency. Divers and tenders should anticipate potential emergencies, know the response plan, and be prepared for the following situations.

## Entanglements

It is safe to say a huge problem public safety divers will get into is entanglements. Entanglements can range from a piece of rope snagging a fin strap; to becoming wrapped up in fish netting; to finding large sections of wire rope, fencing, or concrete and rebar. Believe it or not, many aquatic plants will foul a diver to the point that it becomes hard to move at all. Entanglements are intimidating, but they can be dealt with safely (**FIGURE 15-8**).

The selection and placement of a diver's cutting tools are critical. It does no good for a diver to have cutting tools strapped or located where they cannot be reached. Divers tend to put their tools in places that can be reached easily on land or when free of entanglements, but once the diver gets into the water, the accessibility of their tools can change. For example, the giant knife strapped to the diver's lower leg may look cool, but it is a magnet for strings and lines underwater, and if the diver's arms are tangled, it may not be reachable. Another factor that can ease an entanglement problem is muscle memory. Divers who learn

**FIGURE 15-7** The safest staging location for a backup diver is in or immediately beside the water, ready to submerge at a moment's notice.
Courtesy of Steve Treinish.

**FIGURE 15-8** Entanglements are intimidating, but they can be dealt with safely.
Courtesy of Steve Treinish.

where each tool is, each time, will have a much easier time finding them in real life missions.

The **golden triangle** of the body is where most cutting tools should be kept. Buoyancy control device (BCD) pockets are the best place to keep them, as placement here will allow a more streamlined diver profile underwater (**FIGURE 15-9**). Many BCD pockets can be equipped with retractors or small lanyards, but these may not be long enough to reach areas of entanglement underwater. Many divers strap cutting tools to BCD shoulder straps or waist straps. This is acceptable, but they should be mounted cleanly. Set up and practice your cutting tools with your off hand. Many right-handed divers have mastered the art of cutting things underwater with their right hand, but if the right arm is fouled, the left arm will need to perform the same cutting skills.

The number of cutting tools carried varies from diver to diver, but most teams require at least two tools to be kept in different areas on the diver. Many agencies allow divers on the team to carry whatever they want within reason, as long there are two sets of cutting shears and any tool carried is secure and streamlined.

A small, non-folding knife or seat belt cutter tucked in a BCD pocket may come in handy.

## Types of Cutting Tools

Trauma shears or dive scissors make the best cutting tools for many situations. They are safer to use underwater, as there are no exposed blades or points when they are closed; they are usually readily available to fire and EMS personnel; and they are inexpensive.

Wire cutters also can be carried, especially in places where stainless steel leaders or wire rope are plentiful. Dive knives are useful in some situations, but consider using a blunt-point knife instead of a drop point to decrease the chance of puncturing gear or the diver. Dive knives have one disadvantage in that they may require that there be tension or pull applied to the line being cut. Sweeping the blade of a knife out away from the body may not tighten the line enough for the blade to cut it easily. Shears or wire cutters can be operated using one hand (**FIGURE 15-10**). Whatever tool you choose to use, remember that getting it back in the sheath or pocket can result in a torn suit, punctured BCD, or even a laceration. It also can be tricky to sheath tools underwater without losing them. Some teams operate in throwaway mode, with the knife simply discarded after use, but why throw away a tool? What if the same tool is needed again if more entanglements are encountered?

Entanglements underwater can be made more difficult due to the weightlessness of the line in which a diver is entangled. Monofilament fishing line is seemingly everywhere in many places underwater, and Kevlar fishing lines, commonly termed "spiderwire," has quickly become many divers' nemesis. Monofilament lines also can be easily camouflaged in aquatic growth and become

**FIGURE 15-9** Keep most cutting tools on the golden triangle of the body.

Courtesy of Steve Treinish.

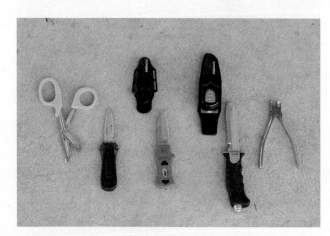

**FIGURE 15-10** Cutting tools can include shears, knives, and wire cutters.

Courtesy of Steve Treinish.

wrapped around arms, legs, or fins without the diver realizing it.

A cutting tool will cut whatever the user asks it to. Be careful and deliberate when using one blindly, especially with heavy insulation on your hands. SCUBA hoses can be mistaken for rope or tubing and be accidentally cut. If there is any doubt, trace the hose back to the first stage or the device it supplies. The last thing a diver wants to do is cut the primary air hose supplying a tangled team member.

Whatever tool is used, make sure that enough of the entanglement has been cut to get the diver free. While little pieces of line will not keep a diver entangled, longer ones may. Enough cutting should be done to completely free the diver, and remember, on the way up to the surface, the loops and bights in the entanglement may be pulled tight and cinched down again.

### LISTEN UP!

Expensive dry gloves can be ruined quickly when tangling with a piece of Kevlar fishing line, even in a recreational dive with great visibility. Simply pulling the line away from the body to see it better can slice gloves open—and also slice the hand underneath it.

### Entanglement Management

A general way to clear a diver of entanglement is to work from the head down. Clear the body of the tangled diver on all sides, pushing the lines down to the feet. Do not just cut and fling lines, strings, or ropes away, but try to control the pieces. Many times a student will rapidly flick or toss a piece of line away only to have it follow his hand back to the tangle with the water being pulled behind his hand. This might tangle the backup diver or re-tangle the primary diver. A few additional cuts to make the offending line smaller in length can make life easier on the way up.

The main cache of dive equipment on divers is generally found from the waist up, and these are also parts that frequently will tangle a diver badly. The fouled diver may be able to assist by pushing himself or herself off the bottom or slightly raising the legs or feet so the backup can get a good feel for the tangled part or slide an arm under the primary diver to feel for the issue. Remember, the tender can assist in this too. A tug

signal to take up slack may reveal the line that is keeping the diver attached to the bottom.

### SAFETY TIP

Some divers may elect to inflate the BCD enough to stretch the tangled lines up off the bottom, similar to a hot air balloon. If this diver cuts himself loose or the entanglement is freed, he may become positively buoyant and could be pulled to the surface too quickly. If a backup diver also gets caught in that situation, now two divers are at risk for ascent-related problems that may include arterial gas embolism. Divers should remember the flare position for an out-of-control ascent, which is basically a skydive position. Extend the legs and arms out to provide as much drag as possible in the water, thus slowing the ascent as much as possible.

## Underwater Communication and Problem Management

Being tangled to the point of needing help and having the backup find you is a great feeling, but how does communication take place? With electronic systems, it is easy for the primary to voice his or her problems to the backup, but for teams that cannot afford it, or in the event of a communication failure, the divers still need to "talk" underwater. Once again, a small amount of surface preplanning and practice will go a long way when things go bad.

Some teams simply wing it and hope for the best, but you can imagine what it would be like if no predetermined signals were in play with two people diving blindly underwater. Different teams and agencies use different methods. In a pinch, noisemakers or tank bangers can be used, although squeeze signals are generally used. For example, a signal of three blasts from an air-operated noisemaker could signal the three tugs, which may mean "bring me up." Three squeezes from the backup diver to the primary diver could also mean the backup believes he has the problem dealt with well enough that an ascent can be attempted. Squeeze signals and hand signals can be adapted to reflect the tug signals, but some differences also may arise, such as a way of communicating that a diver is tangled. Noisemakers also can be used in the water as a diver recall system, if used with a preplanned signal, like the three blasts mentioned previously. Rescuers must remember, however, that these air-operated devices operate from the same gas supply as the diver.

## Entanglement Communication

To begin, once the primary diver calls for the backup diver, there must be a meeting place where the backup can find the primary's hands. If the primary diver places his or her hands on the attachment point of the body harness (usually a carabiner or quick release), the backup will know on the way down where he or she should be able to start to "talk" to the primary. The first squeeze should be the backup signaling his or her arrival, usually an OK signal. The primary diver answers with one squeeze in return. This is the initial "I am OK, but the problem is...." After this initial contact, the primary can guide the backup's hands through some signal motions.

Generally, for entanglements, the primary will motion the backup diver's hand in whatever motion means "I am entangled," then place the backup's hand as close to the entanglement as possible. A common signal for this is the primary moving the backup's hand in a large circle, then pushing or guiding it to the place where the entanglement is felt. The backup will then deal with the issue with the primary lying as motionless as possible. Having the primary lay calmly lets the backup deal with the situation without having a thrashing diver bumping him and his cutting tools. It also keeps the water motion around the divers calmer. Lines or strings swirling around the area from limb movements can make things worse by moving entanglements elsewhere, or wreck what little visibility there might be. Once the entanglement is cut away or removed to the backup's satisfaction, he or she can then come back to the primary diver and signal that an ascent can take place. The backup diver then signals the tender to bring the pair in. If this seems like a complex and potentially frustrating situation, it is exactly that. Training for this situation is crucial to developing a team's reaction to duress, and it must be refreshed during subsequent trainings.

### KNOWLEDGE CHECK

Who makes the dive/no-dive decision?
a. Incident commander
b. Dive supervisor
c. Operations section chief
d. Safety officer

*Access your Navigate eBook for more Knowledge Check questions and answers.*

## Out-of-Air Emergencies

For a primary needing air, the backup's hands can be moved in another motion that signals "I need air," signaling the need for breathing gas. A common signal is tapping the backup's hand to the primary's mouth

area, or regulator. The backup can then use whatever method of air supply is utilized to deliver air to the trapped diver. Keep in mind, though, if donating a second stage regulator, that the backup can move only so far before the regulator will be pulled out of the primary diver's mouth. Once a signal is given, either way, the answer should be an "OK," or "understood" (one squeeze is a common signal for this).

The other problem that can arise is the backup needing to ascend to the surface for other supplies or more gas. Once again, a well-rehearsed system done on the surface is better than anything concocted underwater. One trick that can be done with confused divers is to literally use formed hand signals to get the message across. The backup mentioned here would certainly want the primary to know he or she is leaving, but may need to form the primary's hand into a pointer, then use it to point at the backup, then signal "ascent," then signal "air." The primary should answer with the OK signal. "Conversations" like this can be practiced anywhere, anytime.

## Out-of-Air on the Surface

Most every diver—recreational or public safety—becomes proficient at using the power inflator to fill the BCD with air and create positive buoyancy at the surface. However, a diver who has exhausted the gas supply must now orally inflate the BCD. This is like other recreational skills that get practiced infrequently in training or drills. It is imperative for divers to be able to inflate their own or others' BCDs manually (as shown

### SAFETY TIP

It may be somewhat controversial, but instructors might consider requiring public safety diving students to physically breathe a tank down to empty in the shallow end of a pool. Do not try this anywhere other than a pool or other safe water area in which the student can immediately stand up. A safe way to do this is to let the student lie down in three to four feet (0.9 to 1.2 m) of water, with an instructor standing beside the diver. This way, the diver simply stands up when the gas supply is exhausted. This may not be taught universally, but it is helpful for students to feel what a regulator change feels like when the gas supply goes away. The regulator will breathe harder for the last few breaths. It is better to let the diver experience this in a pool, rather than in blackwater where he or she might be wondering what is happening with the regulator. The last few breaths could mean the difference between having to do a true emergency ascent and being able to take a few more seconds to perform the ascent more slowly, and therefore, more safely.

in Skill Drill 15-1). An inflated BCD and ditched weight system will drastically, if not totally, eliminate the chances of a diver slipping back underwater. To remove a weight system, follow the steps in **SKILL DRILL 15-2.**

## Medical Emergencies

Medical emergencies can be signaled, but in a water environment, there is not much that can be done other than to make a slow, controlled ascent. Many medical signals involve the primary tapping the backup's hand to the primary's chest, then to the body area needing attention. Bringing up an unconscious or disoriented diver can be extremely difficult, and can tax even the best divers. Two key aspects of ascending with divers requiring assistance are the rate of ascent and making the diver positively buoyant on the surface. Many dive fatalities have occurred from divers being pulled back underwater by weights that were unable to be dropped.

### Ascents During Rescue

One of the most important aims in this text is decreasing the possibility of rescuers being so pumped up and excited that they injure themselves or each other. Bringing a fellow diver up safely can be one of the hardest things to temper in any diver, but that is exactly what we need to do. Barotrauma is a result of simple physics; it does not matter if your buddy was tangled and is now unconscious or if a diver was underwater for a few seconds. A careful, safe ascent is required to get someone to the surface. If the bottom is clear enough, or if the divers are coming straight up a tended line, the tender can play a key part in the ascent by pulling the diver or divers in at a slow enough rate that the vertical rise is within 30 to 40 feet per second. This way the diver making the rescue can concentrate on keeping the regulator in place if required, expelling the expanding air from drysuits and BCDs, and keeping a firm grasp on the diver being rescued.

If, for some reason, divers are coming up without a tended line, keeping the ascent rate slow enough now becomes the rescue diver's job. It sounds easy enough to most trained public safety divers, but remember, you might be bringing up a diver who has been trapped, tangled, or out of air, and as of that moment, he or she likely only wants to get up and get out. The urge to get out of the water at that point can be powerful. The rescuing diver must remember this and must also remember the body positioning needed to effectively vent expanding air from BCDs and drysuits. A vertical ascent allows air to rise to the top of the

BCD and drysuit, where the exhaust, or "dump" valve is located. Most drysuits are equipped with automatic dump valves, and it may need to be turned fully counterclockwise. Raising the valve by raising the diver's elbow, or simply pressing the valve, allows it to vent. It is true multitasking to do all this. Teams need to train in pools to prepare for this. Presence of mind and correct training and practice once again become a diver's key ally.

> **LISTEN UP!**
>
> It is extremely hard for a diver in no-visibility conditions to judge ascent rates. Divers are taught in open water classes to ascend by sight of their exhalation bubbles, or by watching a screen on the dive computer. This does not happen in blackwater. By staying slightly negatively buoyant, a diver can let the tender assist in maintaining a proper ascent rate. The tender can bring the diver up at the proper rate, using the tether. The tether will stay slightly taut in this fashion. If the tender feels slack develop, he or she can verbally instruct the diver to dump some air from the buoyancy system.

### Unconscious Diver Rescue

Generally, when bringing up a fellow diver from behind, grasping the diver at the tank yoke is preferable. In this position, the rescuing diver has good access to his or her own BCD controls, the BCD control of the stricken diver, and both drysuit exhaust valves. This situation is where the automatic exhaust valve on many drysuits shines, as simply raising an elbow above the shoulder will let the valve release air.

As rudimentary as it seems, airway control should be practiced often. Rescuing a diver creates high stress levels. FFMs offer good airway preservation with no outside moves or positioning by the rescuing diver, and even for a nonbreathing diver, will keep the airway dry, hopefully increasing chances for resuscitation. The FFM should be left in place with the ambient air valve closed (if equipped) until the diver is safely on shore in the hands of EMS care.

### Weight Removal

Once at the surface, the most important issue is to keep the troubled diver positively buoyant, no matter what. This means dropping the weights and inflating the BCD, either with the power inflator or manually if the gas supply is exhausted. Dropping weight belts or pouches is a reaction that should be absolutely ingrained in any diver. It must be a reaction the entire

# SKILL DRILL 15-2
## Removing a Weight System

**1** For this skill drill, the integrated weight pockets of the BCD are shown. The pocket and the weight inside will both be removed. Place the hands on both sides around the rib area.

**2** Find the pocket releases, using the hands on the sides of the BCD.

**3** Gripping the releases, pull the quick release buckles forward. The weight pocket and weight will start to pull forward.

**4** Hold both pockets and the weights contained in them away from the body and drop them. Ensure the pockets clear the body and the weight is not caught on the body or dive equipment and inadvertently retained.

Courtesy of Trevor Treinish.

team is familiar with, whether they use a harness, an integrated system, or a plain weight belt. One way to ensure these muscle memories is to have exiting divers remove their weight system every time they exit the water, before any other gear is removed. Ordering a diver to ditch the weights unannounced on the surface or the shore provides the element of surprise. Remember, divers will react how they train.

Consider the following scenario taken from the National Institute of Occupational Safety and Health (NIOSH) Report #98-F16:

On May 19, 1998, a diver (the victim) assigned to the fire department's diving squad drowned during a river search in an attempt to recover the bodies of two civilians. The victim and his SCUBA partner, both fully equipped with diving gear and a rope, entered the swift, murky river to assist another team's divers, who were already in the river. The victim and his partner entered the river at the location where the two civilians were reported to have gone down to perform an independent sweep search pattern. The search lasted approximately 10 to 15 minutes at an approximate depth of 25 to 30 feet. Due to zero visibility and the underwater current, the victim and his partner decided to surface and return to the staging area where they changed over to their underwater communication masks and received further instructions from the dive supervisor.

Once they returned to the staging area, the dive tender changed their tanks, assisted with the removal of their gear, provided them with Gatorade to drink, and placed a 50-foot long, four-inch round air float (rubber-jacketed fire hose) from shore to the U.S. Coast Guard cutter that had just arrived. After a brief conversation with the dive supervisor, the divers decided to remove their SCUBA gear and free float to the Coast Guard cutter using the four-inch float as a guide and flotation device, determining this would be the easiest way to enter the boat since it did not have a swim platform.

The victim, wearing his weight belt, began his free float to the boat, holding on to his buoyancy control device (BCD), tank, and the four-inch air float as flotation devices. The weight belt consisted of three 10-pound lead weights secured around his waist. As the victim was approaching the boat, he lost grip of the flotation devices and instantly went under the water due to the 30-pound weight belt that he did not release. His partner immediately went down after him, free diving with just his wet suit, which created a buoyancy problem and limited his dive depth. After two attempts to reach the victim, he surfaced and called for assistance from the other divers on scene.

One diver from another team descended to the area where the victim went down and located him. As the victim was pulled close to the water surface, the victim's partner grabbed him. The rescuing diver lost his grip on the victim while adjusting his own equipment, and because of the 30-pound weight belt around the victim's waist, the victim's partner was unable to hold onto him, and the victim descended for a second time. The victim was located and pulled from the water approximately 10 to 15 minutes later by the police rescue divers. The victim received immediate medical attention on shore before being loaded into the medical helicopter, which transported him to an area hospital where he was pronounced dead.

NIOSH investigators concluded that, to prevent similar incidents, fire departments should:

- Ensure that whenever divers remove their diving gear, the first piece of equipment to be removed is their weight belt.
- Ensure that whenever a dive boat is being used it is equipped with an adequate diving ladder or platform for the specific operation.

*(Full report available online)*

When a troubled diver is brought to the surface, he or she should not be allowed to sink again. Ditch the weights, regardless of the type of system! Any lead in place on a diver must have additional flotation in the form of air that is pumped or blown into the BCD. The BCD may be inflated to assist flotation, but be aware that as the BCD is inflated, the straps on the diver may tighten to the point of preventing the chest from expanding enough to get a good breath. Loosen the straps and check vital signs. If the diver is breathing, efforts must be made to protect the airway. FFMs are kept in place by the head strap and will continue to deliver air. Recreational regulators need to be kept in the mouth, if possible. The diver may also vomit. CPR is generally not recommended in the water; as effective as CPR is, it is hard, if not impossible, to do on the surface. It is better to get the diver to shore where effective EMS measures can start. EMS gear, such as a backboard and supplemental oxygen, should be staged as close as possible to the water.

## Removal to Shore

Most divers prefer the **push method** of transporting divers across open water (**FIGURE 15-11A**). In this method, the unconscious diver is on his or her back, with the rescuing diver on his belly, pushing the troubled diver's feet with his shoulders. This enables the rescuer to kick effectively with his fins, and steering the unconscious diver becomes much easier. This method is much faster than towing a diver; however, towing stricken divers may be preferred (**FIGURE 15-11B**).

Another method is to use the tether line. When using this method, the diver's head must be protected from bumps and submersion in the water. The rescuer can keep the diver hooked up and grab the line and the unconscious diver's tank valve. In this fashion, both divers can be pulled in. The tether line also can be easily pulled down and held at or with the tank valve or yoke so that the pull is from the rear of the unconscious diver, not the front. This positions the stricken diver with the head in a better position for keeping the airway face up and drier. The tender must remember not to pull too hard, or both divers could be pulled underwater.

**A**

**B**

**FIGURE 15-11** Most divers prefer the push method of transporting divers across open water (**A**) rather than towing (**B**) but both are acceptable.
Courtesy of Steve Treinish.

When the diver reaches the shore, all dive gear must be stripped off before removing the diver to the bank. Adding a few people at the water's edge in proper PPE helps, as does a blanket or bariatric-style moving tarp. Consider removing the dive rig, but don't forget to disconnect the drysuit inflator and crotch straps if they are in place. All dive gear should be quarantined for further examination and investigation.

At this point, it becomes a standard EMS run, and local protocols and medical direction should be followed, including a call to the Divers Alert Network (DAN).

# Dive Planning

Once a student diver is trained, confident, and properly equipped, it is time to get diving! However, before any rescue commences, some type of plan must be formulated, starting with the go/no-go (or dive/no-dive) decision. The most basic risk-versus-benefit decision is, "Can I handle this dive with my training and equipment, and do I have a need to deploy it?" Also consider the following questions:

- Is there a viable victim?
- Has the timeline been checked?
- Does the evidence support the story thus far?
- Can the rescuers be protected from the possible hazards, both topside and underwater?
- Are the divers OK with making this dive?

Many teams allow an individual diver to abort or decline a dive. If conditions or the tasks are more than what a diver feels they can do safely, remove them from the water. This is breaking the first link of the accident chain. Maybe the current they are feeling in the river is stronger than they have trained in, or the dive is beyond their skill level, or the diver woke up with a sinus infection that morning and doesn't think he can equalize his ears. Don't start a dive with one strike in the count before the diver even lands on the bottom.

## Depth and Time

Most agencies are currently using at least 60 minutes from the time of submersion for a window of rescue versus recovery time, and in colder water or under ice, some agencies have increased this window to 90 minutes. However, some conditions may preclude dive operations and result in searchers immediately declining the dive in the interest of safety. Water currents, the type and condition of the water, overhead environments, and the ability of the divers on scene can affect the search. Some groups give the on-scene dive supervisor some leeway to make this decision on scene on a case-by-case basis.

# Voice of Experience

On November 12, 2011, my department was alerted to assist with a dive operation in a neighboring jurisdiction. I had recently been asked to be the dive team leader, and I was out of the area at an event and could not respond to this call. Two divers went down to assist with the operation, but the victim was not located. The divers searched in cold, black water with a significant tidal shift, and the search operation went into the night-time hours before it was called off for the day.

After mustering more divers the following day, we returned to the site and continued to search the area. On this day, we enlisted the help of a neighboring dive team, and they brought their boat that had side scan sonar capabilities. Personally, I conducted a couple dives that day. There were numerous overhead obstructions, the current was significant in the creek bed, and the water was black and cold. Our search lines often got tangled in the obstructions, and our divers were quickly fatigued due to the temperature of the water, the air, and the less-than-desirable conditions. After numerous dives, and a tremendous amount of effort by the divers, it was decided to call off the search operations for the day.

On November 14, we received assistance from other area dive teams, but the results were the same. The body was never located on our searches; it surfaced a few days later and was recovered.

This was a very sobering operation for the divers, as well as me as a new dive team leader. At this time, our dive team had been around for a number of years and was well-respected in the region; however, there wasn't a defined progression when it came to training. While there were no injuries during this operation, several of us (myself included) were diving well beyond our scope and certification level. This weighed heavily on my mind as a new dive team leader, and this issue needed to be addressed. It became apparent to me that the key to a safe dive operation begins with proper training and not diving beyond the diver's scope. The progression for my team is now: Basic Open Water; Advanced Open Water; Full Face; Drysuit; Rescue Diver; Public Safety Diver; Underwater Criminal Investigator; and Master Underwater Criminal Investigator. All divers must also pass an annual swim test. Furthermore, we improved our record-keeping methods to ensure that supervisors understood the certification of all the divers and that personnel were not to dive beyond their scope.

Being a dive rescue technician is not as simple as having a couple of recreational certifications and equipment and being part of a public safety organization. You and your agency must ensure that *all* the pieces of the puzzle are in place (i.e., trained personnel, equipment, on-going training, and maintenance of equipment). There is a tremendous amount of training required to remain proficient at your skills, and personnel must ensure they are training in the environments in which they'll dive. They must test their comfort levels in these environments to answer the call when it comes in. Most importantly, personnel must show restraint and not dive beyond their training level. Being a public safety diver is very physically and mentally demanding, and you dive in some of the harshest environments. The key to being a safe public safety diver is receiving the proper certifications along with an ongoing training program that pushes you to your limits.

**Jack McGovern**
Lieutenant
Dive Team Leader
Fredericksburg Fire Department
Fredericksburg, Virginia

Entry points and exit points, search patterns, and the all-important point last seen also come into play. For example, a single cell phone call with no witnesses making themselves known at a scene might be a good reason not to commit divers, especially in large water bodies or moving currents. There is not enough information to warrant placing a diver in possible danger. A victim being circulated in a low head dam boil may call for a boat-based recovery attempt on the water, but no rescuer should be placed in that water.

If there is evidence of a victim being in the water longer than the AHJ's benchmark, the mission will likely be deemed a recovery operation. Rough parameters exist regarding how long a body may remain submerged, but the biggest factor is that the dive operation timeline changes. No matter what, the victim is still dead, and we can now wait out conditions so that we can dive safely. Side-scan sonar and/or remotely operated vehicles (ROVs) can be considered, even if divers are not getting in the water. Sonar has revolutionized dive safety by limiting the amount of time the divers are exposed to harm underwater.

If a dive is warranted, there must be a plan. Tank pressures, gear checks, diver health, and the approximate depth and duration of the dive should be known. It is important to know the depth of the water before deploying divers. There is a limit to the amount of time they can spend underwater due to nitrogen intake, and in blackwater there may not be enough visibility to see a computer or depth gauge. Many teams set limits for the depths at which their divers can operate. Deeper (more than 30- to 50-foot [9.1- to 15.2-m]) dives are much trickier due to the increased air consumption of the divers, and teams are completely warranted to factor safety into the decision to slow down or even call in better resources for that level of dive. Emergency plans and equipment also must be ready. A tender checklist should be used to ensure items or checks are not omitted.

## Dive Tables

Dive tables are a quick way to plug in an expected time and an expected depth and get a glance into how much nitrogen the diver will absorb. These tables can give a maximum time for depths of up to 130 feet (39.6 m), and also can help plan concurrent dives and surface time a diver may have to spend off-gassing nitrogen from his body. Planning using dive table times can help make the decision to shave some minutes off a dive in order to maintain a safety factor when dealing with decompression issues. Dive tables are discussed further in Chapter 14.

**LISTEN UP!**

Dive tables and computers both figure nitrogen intake and residual nitrogen times in intervals of 10 feet (3.0 m). Planning a dive to a depth 10 feet deeper than what will be the maximum depth creates a more conservative dive profile, adding a margin of safety for dives with a heavy workload or very cold water temperatures.

**SAFETY TIP**

A problem that has arisen on more than one team is a diver entering the water in the heat of a rescue, but then remembering that he or she is supposed to fly the next day. DAN recommends a 24-hour interval between diving and flying to minimize risk of decompression illness due to cabin pressures in the aircraft. A diver who has plans to fly must realize that diving will affect travel plans, and the team will have to plan accordingly.

## Search Pattern Selection

An initial search pattern may be changed to another pattern if conditions fed back by the diver necessitate it. Many times, smaller bodies of water will hold trash, trees, and other debris, and a quick change of search patterns will greatly increase the speed and effectiveness of the search. Along the same lines, the bank can play a large part in pattern selection. Land points that jut out into the water lend themselves to a large, semi-circle sweep pattern, and a straight-line wall or pier may be more easily searched using the "walk the dog" pattern.

Another aspect of pattern selection is the object of the search. A search pattern underwater must have a small amount of overlap on each pass in order to ensure that all areas are searched. Tenders must know how much slack to give the diver at each directional change. Small children may need slack of just 2 or 3 feet (0.6 or 0.9 m), while a search for an automobile can let a tender hand out 5 or 6 feet (1.5 or 1.8 m). Similarly, a jumbled bottom surface may reduce the pattern slack to ensure good searches, whereas a sandy, firm bottom with no entanglements may let the diver truly reach out to both sides and increase speed.

## Minimum Reserve Pressure

Any diving done using cylinder-supplied breathing gas should have a specified minimum reserve pressure planned before the dive commences. Reserve pressure is just as it sounds—a volume of air, expressed in psi, that remains for use in case of trouble. This pressure is what the diver should have in the cylinder upon surfacing, and tenders, divers, and supervisors should all know this target pressure before any diver submerges.

The recreational dive industry has adopted 500 psi as the consensus standard, but this is based on fairly benign conditions and diving with a buddy. Many technical divers, including under-ice and cave divers, plan to be on the surface and positively buoyant with one-third of the original gas supply remaining in the cylinder.

Other agencies use a cubic foot formula, taking into account a redundant air supply or the use of double cylinders. Using the cubic foot philosophy, some teams allow a diver to surface with 500 psi, with the idea that the pony cylinder carried is part of the remaining reserve gas. Other agencies and teams recommend any diver surface with 1000 psi of gas. Arriving at the planned reserve pressure can be somewhat controversial, but is a good example of the need to plan operations based on crew ability, training, and conditions.

## Redundant Gas Supplies

Redundant gas supplies are a stand-alone, completely separate breathing gas supply that the diver carries in case of a failure of equipment or in the event their primary supply is exhausted.

### Pony Bottles and Regulators

The addition of a pony bottle and regulator gives divers two options underwater. They can use it themselves if they need breathing gas or, if equipped properly, they can donate it to another diver in need of breathing gas. Pony bottles can be easily set up on a quick-release system that attaches to the main SCUBA tank, or some BCDs have pockets sewn into the rear on either side of the main tank. Either way, the general objective is to get the pony bottle free from the diver. The diver can then hand it to the diver needing additional air (**FIGURE 15-12**).

Divers using typical second-stage regulators only need to move the second stage regulator from their pony to their mouth. This exposes the diver to any ambient water contamination; however, air takes priority in an emergency. Divers or teams using a manifold block can switch the gas supplies using the block and a quick-connect fitting somewhere in the system.

Another aspect of using pony bottles is to make sure the diver has the bottle secured before letting go. It would not be a good feeling to have your precious air drop between your legs and out of reach. Snap clips, carabiners, or even webbing can be used to secure the pony to the diver who is out of air.

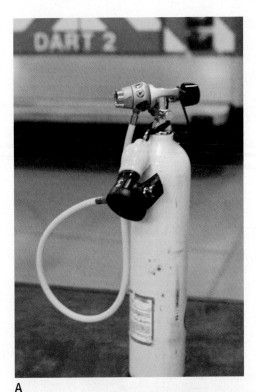

**A**

**B**

**FIGURE 15-12** A pony bottle can be used by the diver carrying it or by another diver who needs it. **A**. Stand-alone pony bottle. **B**. Plumbed pony bottle.
Courtesy of Steve Treinish.

### Contingency Cylinders

**Contingency cylinders**, or emergency air supplies, are simply SCUBA tanks that are either equipped with first- and second-stage regulators or a quick-connect end that will connect into a manifold block and a way to secure the cylinder to the diver (**FIGURE 15-13**). They are delivered and used the same way as the pony bottles described above, but are usually a 60- to 80-cubic-foot tank, thus providing much more air. There is no depth gauge or inflator hose needed, as this tank is used only for diver emergencies. A small, nut-style submersible pressure gauge (SPG) that installs in the high-pressure

**FIGURE 15-13** Contingency cylinders provide more air than pony bottles and are used only for diver emergencies.
Courtesy of Steve Treinish.

regulator port can be used, or a hose and full-size SPG can be attached. If a hosed SPG is used, secure it to the cylinder with webbing or bungee cord. It should not become a part of the tangle underwater.

There are few underwater problems divers cannot deal with as long as they have air to breathe. Once a backup or safety diver deploys this system, a window of time must be charted and followed. The amount of time the cylinder may last varies, but remember—a diver trapped at deeper depths will use more air and drain their gas supply faster (Boyle's law). No matter what the problem, a new, full cylinder can be delivered to the trapped diver while the problem is being worked out. It is extremely rare to deploy a contingency cylinder outside of training, but the skill will be there if needed.

### SAFETY TIP

A diver can deliver a SCUBA cylinder anywhere underwater, but if it is low or empty, it will do no good to anyone, and in the time it takes to get another one, the diver may well drown. Make sure this setup is full and ready to go when you are!

## Surface Supplied Air

Although true surface supplied air can be used for public safety diving, it is generally considered to be too bulky and complicated to set up quickly. Surface supply diving has the user attached to an umbilical hose, which supplies gas from a topside bank system. While it is standard equipment for hazardous material diving, it is somewhat cumbersome and the general thought is that a drysuit, equipped with dry gloves and a dry hood, and coupled with an FFM offers the best standard of protection for rescue work. There are systems on the market that some teams use with success, with an FFM plumbed into an automatic switching valve that lets the diver breathe from a surface supplied setup or a twin tank system carried on the diver. It can be used either as a standalone SCUBA setup or as an SSA-type system. The big advantage is that the air supply tubing, tether, and communication wires are all integrated into one umbilical cord, yet are small, making it easier to use as a tether (**FIGURE 15-14**).

## Entry and Exit Points

Entering and exiting the water can be problematic for divers at some locations. Slippery slopes, rocky or boulder-covered shores, deep silt and mud, and drop-offs can hinder or completely block water entries and exits. Crews must have a way to quickly extricate a diver who experiences trouble in or under water. In extreme cases, small watercraft or rope systems may need to be set up and staffed before dive operations commence. Even the addition of a small folding attic ladder on a bank can give a diver much better footing to enter and exit the water.

**FIGURE 15-14** Some surface supplied units combine air supply, communications, and a tether to the diver, placing the air supply and monitoring onshore.
Courtesy of Dive Rescue International, and Todd Rishling of Elk Grove Village Fire.

# The Search

Once the diver is trained and comfortable working blind, cutting himself or herself from entanglements, and calling for and operating as a backup diver, he or she can be considered ready for field training in "live," or open water.

For whatever reason, the idea of being in no-visibility water as opposed to a pool can bring out nerves in divers. Perhaps it is the knowledge that they will be alone, with no one watching; however, more likely it is the idea of bumping into something unknown. It may be easier to transition student divers to true blackwater with a quick jump back to the puzzles underwater. Sand beaches can be a good transition too, as the hard bottom and usually gentler slopes allow for easier diving.

## Vision Beyond Sight

There is a saying that flying an airplane on instruments alone is learning to trust vision beyond sight. This idea holds true for no- or limited-visibility diving, too. Whereas your eyes cannot see, your mind can, and you have great sensory input from touch. Straining your eyes to try to see in no-visibility water is useless, and in fact, may cause problems. Because your mind wants to see things so badly, it can and will often create things, including things that can startle you. Consider diving with your eyes closed whenever you are diving with limited or no visibility and searching for objects. Close your eyes before you submerge, and do not open them

until the tender tells you that you are topside. Once the object is located, go ahead and see if there is some sight there, but until that point, keep the mind's tricks at bay by closing your eyes and concentrating on the signals your body is giving you. You may well bump your victim with a fin or a brush of the fingertips. This is actually beneficial, because it takes some of the surprise factor away from finding a victim underwater.

It is highly advantageous to remember that divers are entering the home of a host of aquatic creatures. They will not bite, but a four-foot-long carp flipping around in front of a dive mask will certainly release adrenaline, and most likely also release an excited vocal response. It is OK to be startled, but not spooked. The fish are just busy being fish, and you are still at the top of the food chain.

One major way a diver can stack the deck favorably is to learn a few ways to control stress. Motor skills are lost during stress or panic, starting with the finer parts like the fingers. The greater the panic, the more skills are lost and eventually a diver may not be able to use anything but gross motor skills. Taking slow, deep breaths using the diaphragm muscles can calm a diver and help refocus mental abilities. Sometimes rescuers will find themselves in a bad situation, and their mental autopilot kicks in, but that autopilot must be programmed through proper training and then reinforced by constant practice. Keeping the mind sharp will help keep the body in check. A calm voice from the tender can do an incredible job calming, coaching, and reassuring a diver, but this too must be a practiced response. Stressing tenders in training can be a valuable way to protect the entire team from stress reactions.

## During the Search

For the typical public safety dive, it is assumed that the diver is using an FFM and an electronic communication system, and a tender and backup diver are in play. Upon submersion, the diver often must balance a line between excessive talking and feeding information to the topside.

**FIGURE 15-15** A competent tender is a diver's best friend during a rescue or recovery operation.
Courtesy of Steve Treinish.

Many divers and teams have verbal benchmarks the diver uses in order to let everyone on the dive communication system know certain things are happening. For example, "Diver on bottom" is self-explanatory and lets the tender know the diver is on the bottom and setting up to search. A few seconds to ensure the tether is placed properly, buoyancy is adjusted, and the diver is mentally ready may pay off during the dive. Verbalizing each change of direction and the amount of slack given when turning helps many divers visualize what they cannot see and helps to keep things sorted underwater.

Some agencies do not believe public safety divers need good buoyancy skills underwater; the diver just crawls/swims/scoots across the bottom anyway. This is not true, and good buoyancy skills can reduce the workload, and hence the air consumption, of the diver. Tall or thick vegetation, heavy mud, or deep silt can hinder forward swimming ability, slowing the search and causing the diver to start using his or her hands to crawl or pull their way along the bottom. This reduces the width of the search pattern. Adjusting buoyancy so that the body glides just over these surfaces allows both hands to sweep on either side of the diver, increasing effectiveness.

Verbally relaying the bottom conditions and obstacles and objects found helps the diver imagine their surroundings, and lets the tender and backup diver know conditions and potential issues before they appear.

Once a victim is found underwater, the diver should notify the tender, and make a quick note regarding the body position underwater for further investigation. A quick sweep of the body is not out of the question to check for entanglements. Some divers carry a piece of webbing to cinch onto the victim to make extrication easier, and some find it is easier to simply grab around the torso of the victim and let the tender bring them both up. Regardless of how a victim is brought to the surface, a note should be made on the scene sketch as to the location of the victim and any entanglements found. A knot in the tether rope can be measured after the rescue to provide a more detailed location for investigations later in the incident.

## On the Surface

Although the relatively light weight of a body underwater makes grasping a victim fairly easy, once the rescuer and the victim are back on the surface, it can be troublesome to swim with the victim when the weight and mobility of the dive gear being worn is considered. For this reason, staging other rescuers in the water in surface water PPE can be advantageous for both open water and shore rescue. Transferring the victim to these rescuers makes a much quicker and more efficient move to a boat or shore and EMS care. Once the victim is in EMS care, the same rescuers can assist the diver with gear or exiting the water. In ice dives, the diver can even submerge to provide more room in an ice hole and to help push the victim up onto the ice shelf.

# Specialty Dives

Public safety diving is inherently dangerous in calm conditions, but adding current, ice shelves, or automobiles to the mix can multiply the risks for divers. This section examines these types of dives; however, it is not to be construed as instructional material.

## Night Dives

Night diving in the recreational world is often thought of as "spooky" and mysterious, but offers a glimpse into a fascinating marine world. Public safety night diving in limited or no-visibility water really isn't any different for the diver. Flashlights may be carried, but generally do not improve conditions, because the light cannot penetrate the mud and silt frequently found. For divers working in water that may offer some visibility, two flashlights may be carried for redundancy. Regardless of the water conditions, divers should be marked with some type of light on the regulator area. Six-inch (15.2-cm) chemical snap stick lighting works extremely well for this, and different colors can be used on different dive teams or crews.

## Night Signals

Night diving is problematic in that tenders and shore crews may not be able to directly see the diver or exhalation bubbles on the surface. Even with advances in scene lighting, operations may be remote or start before the scene is lit up properly. Divers can signal they are OK to shore or each other by using dive lights. The diver or the tender can ask or reply "OK" by moving the light in a large circle. Waving the light back and forth is generally thought of as a distress signal. If visibility allows, lights can be used to illuminate a diver's hands underwater, and normal hand signals can be used. Dive lights will activate the luminescent gauges similar to SCBA gauges in smoke-filled environments.

## Overhead Environment Diving

A few underwater environments are found that may restrict a direct ascent to the surface, making these dives extremely hazardous. These dives may be considered confined-space dives, and teams and divers should be trained by qualified instructors and equipped with gear sufficient to allow the diver an escape procedure. Automobiles, buses, ice canopies, ship keels, and even odd occurrences such as construction accidents may necessitate that a diver operate without a direct escape upward. If the overhead obstruction is taken out of the equation, the actual dive is not much different for the public safety diver. He or she is still in a harness, still blind, and still attached to the rope. However, the line is now the diver's entire life. If the line breaks or the diver comes off the tether in poor or no-visibility water, he or she may not find the entry point. The diver is now effectively lost, with no easy way to return to the atmospheric air above him or her, and on a limited gas supply (**FIGURE 15-16**).

## Ice Dives

Public safety divers may encounter the need to dive under an ice surface for rescue. Ice fishing, snowmobiles, iceboats, all-terrain vehicles (ATVs), and even street legal vehicles can all be operated on ice, and because ice can fail, these items and the people operating them may

**FIGURE 15-16** Although not a dive many would plan, diving under a collapsed indoor pool ceiling may be a response faced by divers, and presents an overhead risk.
Courtesy of Steve Treinish.

submerge. It can be very dangerous under the ice. Even recreational **ice diving** is considered technical diving, as the ice shelf may block a direct ascent to the surface. While not much different underwater than a typical rescue dive, getting to the entry point and protecting the tenders can be a challenge.

**SAFETY TIP**

Tenders must be able to survive an immersion into the icy water. Flotation and thermal protection equipment must be worn. Placing the tender on an inflatable is an excellent way to keep tenders and gear dry, afloat, and much warmer (**FIGURE 15-17**).

The divers are now under an overhead environment, and a diver must be able to come out of the same hole he or she entered (**FIGURE 15-18**). Even a thin layer of ice could be unbreakable to a diver below it, due to the lack of leverage available. The worst-case scenario is a diver disconnected from the tether who does not know or see directions back to the entry hole. For this reason, the primary diver operates on a rope that is a predetermined length, and the diver goes no farther than that length, under any circumstance. The backup diver is on a rope that is generally 20 to 40 feet (6.1 m to 12.2 m) longer (**FIGURE 15-19**). The idea is that if the primary becomes disconnected from his or her rope, he or she ascends to the ice shelf and stays there.

Dropping weights also should be considered, because this diver does not want to go back down. The tender should have a general idea of where the diver was and how far out he or she is. The backup diver deploys to one side of the last known point of the primary and

**FIGURE 15-17** Tenders must be prepared to survive an immersion into the icy water.
Courtesy of Steve Treinish.

**FIGURE 15-18** In ice diving, the divers are under an overhead environment, and a diver must be able to come out of the same hole he or she entered.
Courtesy of Steve Treinish.

**FIGURE 15-19** The backup diver is on a rope that is generally 20 to 40 feet (6.1 m to 12.2 m) longer than the primary's rope.
Courtesy of Steve Treinish.

deploys the entire length of the rope. The catch here is that the backup stays up at the ice shelf, where the primary should be located. Once deployed, the backup starts a slow, controlled sweep, and should run his or her line across the trapped primary. If the primary is conscious, he or she can signal the backup or the tender, and the primary then stays on the rope. As the backup swims back to the hole, he or she should find the primary, and together they can retreat to the entry point.

Contingency straps can be stored in a BCD pocket and used to clip onto the backup's line, thus securing the primary onto the line that is now the route back to the topside. Unconscious divers can be treated as any other lost diver, and the backup deployed down the primary diver's rope.

Wagon wheel lines also can be shoveled in the snow onto the ice surface. To do this, small paths are cleared on the ice leading back to the entry/exit hole and, from underwater, they will appear as lighted lines in the shape of wheel spokes. But remember, we may be dealing with blackwater. Try it, but do not depend on it. Ice surfaces with no snow on them can be physically walked in search of a disconnected diver and, in clear conditions, the diver can actually see the person walking on the ice, or the walker might spot the diver. Another thing to watch for is bubbles moving around under the ice shelf. The walker may then be able to guide the diver to the hole. Presence of mind will go a long way in diver self-rescue.

## Regulator Freeze-Ups

The cold water also presents the problem of regulator freeze-ups. Environmentally sealed regulators use silicone to insulate the inner workings of the first stage and prevent frost from forming inside of it. Remember, when air is decompressed it cools and, under heavy use from rapid or heavy breathing, a regulator will be subject to these cooler temperatures. Freeze-ups are not a life-threatening problem per se, but the possibility that the gas supply may be exhausted is. Modern first stage regulators are designed to fail "downstream," or "open", and supplying air. The solution for any freeze-up to is to ascend—in a safe, controlled manner—and get positively buoyant once on top. In the event of an FFM free-flow, leave the mask in place. In fact, an effort may be needed to let some of the extra air out. Many divers do not realize how violently the mask will shake with the release of such a volume of air. Pulling the mask slightly away from the face may make a free-flow much easier to deal with.

Recreational primary second stages may be disregarded and the octopus switched, too; however, because the primary is already flowing, that air might as well be used. The diver may have to hold the regulator in place with a hand, as the force from the flowing air can make it harder to keep the mouthpiece in place. There is a time concern because many high-performance regulators flow enough air to drain a full SCUBA tank in a matter of minutes or even seconds, but a safe, proper rate of ascent is still crucial to the diver's well-being.

## Automobile Extrication

Many chances for dive rescue stem from a quick call to authorities after an automobile ends up in the water (**FIGURE 15-20**). Many civilians do not understand pressure issues and opening doors underwater, and will thus end up trapped in the car as it sinks underwater.

**FIGURE 15-20** Many chances for dive rescue stem from a quick call to authorities after an automobile ends up in the water.
Courtesy of Steve Treinish.

Couple this with cold water, and now the victim is relatively easy to find and in cold enough water to provoke the mammalian diving reflex. This adds up to one of the better scenarios in terms of chances of a successful save.

The first thing that must be understood is what the water pressure does to a car, as it relates to opening doors. Many people die each year thinking they can simply open the door and escape. Not true. The surrounding water pressure exerts so much force on the sides of the car that the doors are next to impossible to force by human muscle. The other problem is that most people will keep the windows up in an effort to keep air in. The only way to open a door easily underwater is to flood the passenger compartment, thus equalizing the pressures inside and out—then, the door can be opened.

Vehicles in water deeper than 10 to 15 feet normally sink motor first. This will likely put the car upside down on the bottom. It is also likely the trunk or rear hatch will pop open with the force of the air trying to escape. A panicky driver who does not flood the car to escape may find himself on the bottom, upside down, and surrounded by silt or mud.

When dive teams arrive at a car-in-the-water mission, the first thing they must do is to locate the car. How? There are many ways, including tracks to the water, air bubbles escaping from the car, gasoline or petroleum sheen on the surface, and bottom disturbance. Pick an entry point close to any signs like this, but remember to try to preserve the tracks. They might play a large part in the accident investigation later. Finding a car underwater can be done pretty quickly, as the search patterns can run with a larger sweep length. Once the car is found, expect it to be turtled, or upside down. More modern cars may float on the surface for a longer period of time, because interior compartments are more airtight, and this may result in the car being blown or carried by wind or water farther offshore than rescuers may think possible.

A fast response to a submerging car may present a victim in a viable airspace in a vehicle. Although this is a long shot for rescuers, larger airspaces such as in minivans or SUVs could potentially trap enough air to provide a few breaths until rescue. Because opening the doors or windows by any means may flood a car, a victim thus trapped could potentially need quick instruction on how to breathe from a regulator and not to hold their breath while the rescuer attempts to swim them to the surface. Remember, positive flotation will need to be provided at the surface for the victim, likely in the form of an inflated BCD. The rescuer should leave his or her air supply in place in case the victim forces the rescuer underwater.

## Automobile Searches

Even with the great chances of a rescue that cars underwater can offer, this evolution can still trap and kill divers not operating properly around them. A few critical aspects of operating around motor vehicles are discussed next.

## Vehicle Stability

One aspect of underwater work with any vehicle or watercraft is stability, especially in moving water. Being downstream or downslope of any vehicle is hazardous, because currents may carve out holes that the vehicle settles into, or they may simply slide farther down the sloping bottom. For vehicles close to shore, it may be advantageous to attach a cable to the upstream or up-slope area to try and stabilize it. Even a rope system from shore attached to a vehicle will provide some stability, but the best solution is to stay clear of the danger zones.

## Oriented Searches

Although it is difficult to get oriented on a car upside down in the water, if the diver can orient himself or herself to the car, it becomes easier for the diver and tender to both "see" where the diver is operating. When dealing with a car on its wheels, the diver can go to the roof, keeping the tether rope minded, and feel for the windows. Wipers, mirrors, and radio antennas can all be of help ascertaining the position of the car under the diver. Once the diver knows, it is a matter of reaching down and checking windows, then window condition can be verbalized to the tender. In this manner, the diver can move to the driver's side, reach in, clear it, and return to the roof. This can help keep the tether line from dropping down and becoming pinned under wheels, frames, or the car body. A diver who circles the car has a great chance of the tether becoming wedged underneath the vehicle.

## Gaining Access

While opening doors and sweeping is possible, many times the vehicle will sink in mud or silt and the doors may not be opened easily. The diver must have the ability to shatter glass to gain interior access. A window punch works best, as divers do not have to swing it to use it. Rescue hammers and even Halligan bars can work, but there is normally not a lot of leverage underwater to get a good swing at the glass, and these tools are heavy and extremely negatively buoyant.

Longer screwdrivers also can be used to shatter glass by prying between the glass and the door panel,

but it is much easier to use punches. Once the glass is broken, be aware of air pockets draining and the possibility of victims moving around in the interior compartments, especially if they are positively buoyant. Seat belt cutters are a handy tool at this point, and remember, cuts may be needed at both the lap belt and the shoulder belt.

## Compartment Penetration

Many divers will try to penetrate a car compartment, but most victims can be removed by reaching in no farther than the shoulder. Penetration into a car can be dangerous not only because of the seat belts, but from all of the other paraphernalia motorists might have in the car. For example, many construction contractors carry extension cords, ropes, air hoses, etc.; however, other objects may be loosened and make things hazardous. Imagine a child's car seat in the rear of the car (remember it might not be secured, or may not be in use) getting moved around the car by the diver's motions, and a diver who goes in head first will likely have to back out. Now, the same car seat has been moved into a position at the diver's ribs, below the arms, which are likely reaching forward doing a sweep. When the diver backs out, the car seat can jam between the window posts or doorframe and wedge the diver in badly enough that he or she cannot get out easily. If you think this can't happen, look at a modern car seat, and all of the places a SCUBA hose or regulator can snag into it, or all the strapping that might catch a diver's rig.

Additionally, penetrating a car or van is dangerous because once you are inside you cannot immediately surface in case of trouble, and you are now, technically, in a confined space. Making matters worse is that most car interiors are small spaces. Some subcompact cars may not even have enough room for a diver to bring his arms past his shoulders, so head entanglements become much harder to deal with. Picture a primary and backup diver working an entanglement in a Mini Cooper. Yes, it can be that bad.

Consider using spring-style window punches that are pulled back and allowed to snap back onto the glass. They have no moving parts. The more common punch that is pushed into the glass to break the glass has a tendency to rust if not dried or maintained properly or if it is kept in a BCD (**FIGURE 15-21**).

**FIGURE 15-21** Spring-style window punches are reliable, easy to use, and have no moving parts.

Courtesy of Steve Treinish.

**FIGURE 15-22** Winching an auto is a specialized evolution that can bring a vehicle to shallower water when victims are trapped or cannot be removed underwater.

Courtesy of Steve Treinish.

Divers should be aware of the potential for victims to be floating at the rear of the car. Victims will often climb to the back to try to keep their face in a pocket of air, and with a car sinking motor first, this air is often in the rear compartment. Victims who dry drown, suffer from laryngospasm, or are wearing clothing that can trap air, can float at the roof level.

## Searching the Surrounding Area

After the interior compartment is cleared, remember to search the area around the car, and also the route back to shore. The impact with water can be great, and people may be ejected through the windshield or side windows via rollover accidents. Perhaps an occupant escaped the car, tried to get to the closest shore, and didn't make it. A dive team proclaiming an area clear then finding a body between the shore and car would certainly have their methods questioned.

If an occupant is pinned in the car, it will usually be easier to remove the car, then extricate, rather than try to use heavy hydraulic tools underwater. They will function underwater, but they require each tool to have its own lift bag, and the combination of blackwater, lack of visibility, and tools that can easily injure the car's occupants make this a very specialized and training-intensive evolution.

## Winching Operations

Occasionally, teams may be presented with multiple victims in a vehicle, or victims who cannot be removed through windows. When dealing with bariatric victims, windows with sheet tinting, or cars sunk too far in the mud to break windows or open doors, it may be easiest and fastest to hook the car and winch it to shallower water. In rescue, the main goal is to get the car to a shallow enough depth that surface rescuers can extricate victims. This is normally done in conjunction with either personnel trained in winch operations with proper winch systems or tow truck drivers (**FIGURE 15-22**).

**FIGURE 15-23** Winching a vehicle to shallow water can allow other rescuers access.
Courtesy of Steve Treinish.

**FIGURE 15-24** Consider using a $^9/_{16}$ inch kernmantle rope doubled to have a "J" hook tied at one end for winching operations.
Courtesy of Steve Treinish.

The general rule is similar to that for diving around autos in current: stay clear of areas downslope of the car, and divers and tenders must be mindful of the winch cable and tether line. A steel "J" hook can be used to hook the vehicle at the axle, springs, suspension, or frame. This winching operation is used only to drag the vehicle to shallower water in which surface rescuers can operate, not for lifting. Using surface rescuers to open the vehicle, extricate, and remove the victims to shore is much more effective than having divers try to do the same with the weight of the dive gear (**FIGURE 15-23**). In water shallow enough for extrication, the divers frequently will sink in mud and silt to their knees, and mobility will suffer greatly.

**LISTEN UP!**

Rather than having a diver try to pull a heavy steel cable, consider the use of a $^9/_{16}$ inch (14-mm) kernmantle rope doubled to have a "J" hook tied at one end (**FIGURE 15-24**). Using cable ties to keep the loop together produces a light, mobile line that divers can move around much more easily. The winch hook used can be brought as far as possible into the water, a figure eight placed in the rope, and the winch hook attached to the knot. Once everyone is clear, the winch can be used to pull the vehicle into shore. A $^9/_{16}$ inch kernmantle rope has a breaking strength of around 20,000 pounds; this will easily handle most cars.

**SAFETY TIP**

Dive rescuers must be keenly aware of the risks of dragging a winch cable underwater to a vehicle. It is easy to flip the cable over the tether line, preventing a direct contact to shore or from the backup diver. Entanglement potential is high in this evolution.

Divers should swim the route back to the shore, checking for large obstructions that may hinder the winching operations. Finding the large boulder that would block progress could easily be overcome by simply moving the winching vehicle a few feet. No winching operation should take place until the divers are clear of the water, no matter how hectic the rescue scene is. All personnel must be clear of the winch path in case of breakage. Once the car is in shallow water and victims are extricated, it becomes a simple removal operation and is no longer a rescue.

## Legal Issues

While removing an occupant from a vehicle underwater in rescue mode, teams and divers must remember that they are dealing with an active crime scene, both underwater and on the shore. Law enforcement and accident investigators still have to do their jobs. Divers can make this easier for them by documenting which windows they shattered, if any, and any damage felt on the car during the search. It is also helpful for the divers to protect the car's entry point into the water. Other debris floating on the surface may interest investigators, as criminal evidence can be kicked from the car on the way into the water. Wood chunks, floating beach toys, and balloons all can be used to pin accelerator pedals to power the auto into the water. If the car is removed from the water, keep the doors shut if possible to allow the car to drain slowly. A lot of evidence can be washed right out of the car if it is allowed to drain rapidly, and a slower draining action can preserve things the way they were when the car went into the water. Search the trunk, if possible, but document your actions as you do so.

If a victim is beyond the scope of rescue, it is out of our hands as rescue personnel. The scene should be locked down, protected, and turned over to the law enforcement agency doing the investigation. Many times the entire vehicle will be taken into custody as evidence, to the point of removing the victim to a lab or morgue. All actions on the car by rescue personnel should be documented and turned over to the investigators. Preplanning and working in conjunction with local law enforcement agencies is a good way to make this type of run go smoothly.

## Watercraft Searches

Just as vehicles experience crashes on land, watercraft accidents also may result in dive rescues or searches. Although the use of personal flotation is encouraged, it is often not used, especially on larger craft. Finding the actual point of the crash may require some investigation, because currents and wind or wave action may move floating debris away from where victims might be submerged. Using floating debris fields or petroleum sheen on the surface may help point searchers to potential locations underwater; however, keep in mind that, just as with automobiles, the area between the suspected crash site and the present location of the debris should be searched in case someone floated for a time before submerging.

Other aspects of underwater searches of watercraft are the boat's construction and the typical load out. Fiberglass material is sharp, and shards of fiberglass can puncture or tear drysuits or skin. Divers searching around a damaged boat should use caution. Fishing line, ropes, and anchor lines are commonly found on underwater wrecks, and they may be draped or distributed over the wreck, causing severe entanglements.

## Other Overhead Dives

As most technical rescuers know, in any discipline, there are always some incidents that may not have ever crossed the minds of the responders. Cisterns, industrial tanks, water intakes, and even building collapses can present environments that divers may have to search. These are extremely hazardous dives, and should be treated as such.

### Current Diving

NFPA 1006 defines swiftwater as water moving faster than one knot. As you might guess, **current diving** is diving in water moving faster than one knot. That does not sound like much, but it can fatigue a diver,

putting large amounts of force on a diver's body. The first problem to face is being able to swim in it. The diver must be able to search the area thoroughly, without being pushed around by the current. The possibility that a diver may be pushed into an underwater strainer downstream is a significant risk in this sort of diving. Current diving is often considered to be the most dangerous undertaking in public safety diving, especially in no-visibility conditions.

There are some ways to use the moving water to your advantage, but they can be manpower intensive and do not help with a quick search, especially in blackwater. An effective way to deploy a diver in current is with a movable control point, controlling them from the topside (**FIGURE 15-25**). This entails setting a static line across the river and using a movable control point to control the divers or the boat from which they are diving. The divers do not have to fight current; they are tethered to a rope system that assumes the current force. If a diver falls off, the downstream area must provide some ability to pick up the diver, and the area downstream must be clear of hazards. Movable control point operations are discussed in more detail in Chapter 6, *Swiftwater Rescue Operations*.

Operating just upstream of a strong low-head dam is one instance of choosing the no-dive option when considering the overall scene. Upstream hazards also must be considered, even if this simply means putting an observer upstream to watch for things coming downstream. Divers who choose to dive directly tethered in current should consider using a quick-release snap shackle, which gives them the option of pulling a small cord and releasing themselves from the tender rope in case of entanglement (**FIGURE 15-26**). No-visibility currents of more than three knots generally are considered not diveable, and generally warrant waiting conditions out. Some agencies can successfully dive in faster currents, but they are highly trained and equipped. Currents that are moving that fast do not generally lend themselves to true rescue operations.

# Decontamination

If we consider the water in which we dive dirty, it only makes sense to wash it off when we get out. It makes no sense to bring that dirty water back to the truck, or let the tender splash it all over when he or she helps us undress.

## Gross Decontamination

**Gross decontamination** is simply rinsing the diver, in full gear, with large amounts of clean water. Engine companies, garden hoses, or freshwater showers can be

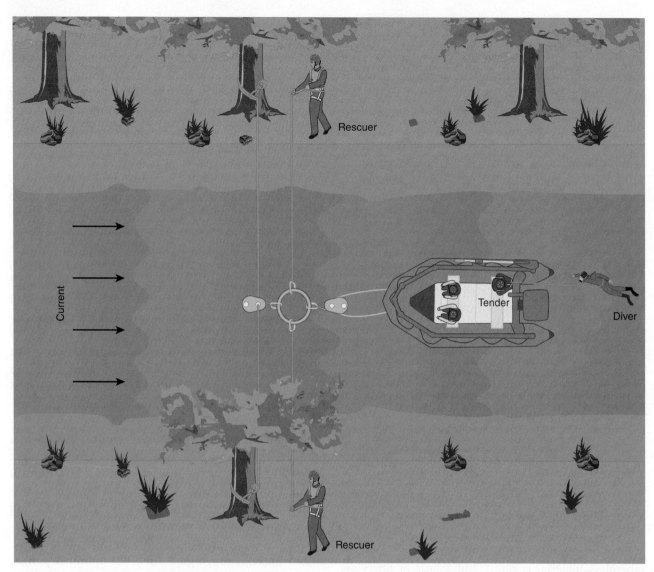

**FIGURE 15-25** A movable control point offers an effective way to deploy a diver in current.
© Jones & Bartlett Learning.

used. Remember, this is not a true hazardous material decontamination, so runoff is generally not a problem. If the runoff is considered problematic, consider using small children's pools to contain it, and rescuers must be certain that the gear worn truly will protect the diver. Consider that if runoff needs to be contained, an SSA system with an encapsulating drysuit probably should be in use.

A better way to clean divers is with a soap or detergent solution. A nontoxic cleaner, such as a 10% solution of Simple Green, scrubbed on with a long-handled brush and rinsed off will take off most nasty stuff (**FIGURE 15-27**). In the event of biohazard, a 10% bleach solution could be used, but it must sit on the equipment for 7 to 10 minutes to work effectively. After that, clean water can be used to rinse. Again, this

is meant for dealing with everyday water we dive in, not a true hazardous materials environment, and in some areas washing divers after dives is a good idea but not absolutely critical. Good information can be found from the manufacturer of the dive PPE in use.

Another system, powered by compressed air from an SCBA cylinder, sprays solution on the diver before the PPE is doffed. This works well to get cleaning

**LISTEN UP!**

A portable water extinguisher is a good source of clean water in remote areas. It is not a lot—approximately two and a half gallons for most cans—but it is better than nothing. A pail, a brush, and a bottle of soap combined with the "can" makes for quick cleaning.

**FIGURE 15-26** A snap shackle connecting the diver to the tether can enable the diver to release the tether in current.
Courtesy of Steve Treinish.

**FIGURE 15-27** Gross decontamination.
Courtesy of Steve Treinish.

solution into every nook and cranny, and maintains the diver's ability to breathe tank air if needed (**FIGURE 15-28**).

Tenders also should be aware of water contamination, because their hands will be exposed to the wet

**FIGURE 15-28** Commercial decontamination agents are available for use if manufacturers approve their use.
Courtesy of Steve Treinish.

and dirty gear as they tend ropes or assist divers. Consider the use of EMS gloves at all times, and beware of wiping the eyes or nose before hands are washed.

# Leading Dive Evolutions

Just as with other disciplines of technical rescue, the leadership position of a dive rescue can be one of the most stressful and nerve wracking experiences in a career. Dive supervisors must ensure that the divers they serve are well trained, well equipped, and 100% confident in their ability to handle a search. The issuance of a dive certification does not mean a diver is qualified to dive. Continuing education is absolutely required in this discipline, and because the divers operate alone on most dives, the supervisor must be certain the diver is not "rusty." For this reason, most teams require a diver to train in public safety gear at least several times over the course of a year. Dive supervisory personnel must have the ability to forecast potential problems, and they must relay these to the crew. An honest look at the

**FIGURE 15-29** Dive supervisors must be able to address and provide safety and planning for all phases of a dive rescue, including removing divers stricken by injury or illness from areas such as this.

Courtesy of Steve Treinish.

ability and experience of the divers on a scene often will lead to good decisions, and insisting on proper setup, planning, and execution of a dive rescue is paramount (**FIGURE 15-29**). Any diver should have the ability to decline a dive if he or she feels their ability is outweighed by the dive conditions, and supervisors should listen if a diver elects to do so. A wealth of experience provides safety, as does the firmness and confidence to abort a dive that is deemed unsafe.

Checklists are invaluable to ensure that safety checks are not overlooked before a dive, and to provide written insurance that a dive setup is rigged and operational before the diver submerges. The general hazards noted on a scene should be relayed to the team before the dive, even if it costs a bit of time in a rescue. The dive area should also be well marked. A quick pre-dive briefing can occur at the water's edge, and with team members knowing their roles ahead of time, this briefing may last only a few seconds.

After the dive, supervisors must ensure that the diver is observed and medically evaluated for signs of decompression illness, proper hydration, and fitness to return to duty. Any actions undertaken or situations found by the diver should be documented in a written statement, and the entire team should be briefed regarding how the rescue progressed.

# *After-Action* REVIEW

## IN SUMMARY

- With the proper precautions and equipment, diving can be relatively safe, but all diving, whether recreational or public safety, poses some inherent risk just by submerging.
- All dive instruction should be done through an experienced instructor working under the guidelines of a nationally recognized dive-training agency.
- Some public safety divers operate in oceans, lakes, and rivers that have areas of good visibility, but many in the field operate nearly or even totally blind from the first few inches of descent to the time they surface.
- Some authorities having jurisdiction (AHJs) require that four or even five qualified members be present before starting a dive, whereas other true rescue units may dive with three divers on scene, with one filling a supervisory role. Still other teams will let a single diver dive, and will let that diver make the decision upon arrival.
- There is a fine line between a diver adjusting to the stress in stages and becoming a good public safety diver and someone getting a case of nerves early on in the training and quitting.
- A huge advantage to any diver in the water is the presence of a trained backup diver.
- The most basic risk-versus-benefit decision is: Can I handle this dive with my training and equipment, and do I have a need to deploy it?
- It may be easier to transition student divers to true blackwater through puzzles underwater or sand beaches where there is a hard bottom and usually gentler slopes that allow for easier diving.

- Public safety diving is inherently dangerous in calm conditions, but adding current, ice shelves, or automobiles to the mix can multiply the risks for divers.
- Automobiles, buses, ice canopies, ship keels, and even odd occurrences such as construction accidents may necessitate that a diver operate without a direct escape upward.
- If the runoff from gross contamination is considered problematic, consider using small children's pools to contain it, and rescuers must be certain that the gear worn truly will protect the diver.
- Dive supervisors must ensure that the divers they serve are well trained, well equipped, and 100% confident in their ability to handle a search.

## KEY TERMS

*Access Navigate* **for flashcards to test your key term knowledge.**

**Clearing a mask** Expelling water inside a dive mask while underwater in order to return it to a dry state.

**Contingency cylinders** Tanks only used for an out-of-air diver trapped underwater. Delivered by the backup diver, they are used to buy time while other problems are worked on.

**Current diving** Diving in water that is moving at more than 1 knot.

**Dive supervisor** The member of a dive team who has the authority and expertise to manage and direct all aspects of the diver operation and has been trained to meet all nondiving job performance requirements of technician-level dive rescue. (NFPA 1006)

**Divemaster** A professional level of dive rank, capable of supervising and leading safe dive operations and assisting instructors.

**Encapsulation gear** Dive PPE consisting of a drysuit, dry gloves, dry hood, and full face mask.

**Golden triangle** The area on a diver consisting of the neck to both hips. Cutting tools and the octopus should be carried in this area.

**Gross decontamination** The phase of the decontamination process during which the amount of surface contaminants is significantly reduced. (NFPA 472)

**Ice diving** Any SCUBA operation under an ice canopy that blocks a normal emergency ascent.

**90 percent diver, or safety diver** The backup's backup, dressed, but usually staged with the mask off.

**Primary diver** The diver who is first to dive, or the diver actually in the process of searching at that time.

**Purging a regulator** Expelling water inside a dive regulator in order to use the breathing gas supply underwater.

**Push method** A way to get a tired or injured diver to shore. The injured diver is made positively buoyant by ditching weight and inflating the BCD, and the pushing diver pushes the injured diver by the feet. This provides good swim ability for the pushing diver.

**Secondary, or backup diver** The diver standing by at the water's edge to assist the primary, or to take over the search pattern when the primary is relieved.

**Tender** A member of the dive team responsible for assisting divers with assembly and donning of equipment, communicating with divers, tracking the diver's status and location, and managing subsurface search operations, and trained to meet all the job performance requirements of operations-level diver rescue. (NFPA 1006)

# On Scene

You are proud to be a part of your agency's dive team, but you are nervous at the same time. You feel comfortable in the water knowing an instructor is there, but just like taking a line into a true working fire, you both want and dread the first one. Your instructor chuckled a bit when you asked on the last day of class when you could dive. He said it might be hours, months, or years. Today, less than 24 hours after the class, you get your wish. You are called to a car in the water in an apartment complex with reports of a woman and three kids in the car. Your supervisor on the team says there is nothing like trial by fire, and deems you the primary diver. Your pulse is racing as you dress, and you rehearse some of the training advice that you were given in the pool.

Upon arrival, you see numerous onlookers around a pond, with several in near panic. Witnesses are reporting bubbles, and the first-in ladder has found tire tracks leading into the water. While the water looks dark, you see oil sheen halfway across the pond and your supervisor believes this is where the car is located. Just as you mask up, the building maintenance person runs up and advises your crew that instead of the typical 15-feet-deep (4.6-m-deep) pond, this is one is different. Once the permit for the pond was approved, the contractors dug it out to almost 50-feet (15.2-m) deep and then used it as a construction dump. He believes there are stumps, shingles, electric wire, and vinyl siding present at the very least.

Your breathing quickens yet again when you realize how deep 50 feet can be. You deploy into the water, and start down the slope of the pond. Abruptly, the slope changes to a near vertical grade, and just past 16 feet (4.8 m), the water turns to complete no-visibility conditions. You hear the tender tell the supervisor that he believes you to be around 45 feet (13.7 m) deep. During the next pass you feel something you cannot identify, and you realize it is the car on its roof.

**1.** How can a safety factor be built into dive planning?

**A.** Reduce the time the diver is submerged.

**B.** Don't dive at all.

**C.** Attach another pony bottle to the dive rig.

**D.** Deploy the diver from a location that lets the diver search from uphill of the car.

**2.** You suddenly feel an unknown rope or line tangle your fin. How can you best reduce stress response?

**A.** Stop, take a couple of breaths, consider your entanglement, and alert the backup diver.

**B.** Stop, take a couple of breaths, consider your entanglement, and deploy the backup diver.

**C.** Use a cutting tool, and sweep behind you toward your fins.

**D.** Immediately pull on the snag progressively harder to pull or pop it off your fin.

# On Scene Continued

**3.** When the diver surfaces, what is the minimum decontamination recommended?

**A.** Gross decontamination using plenty of water

**B.** Soap and water on the diver's face and hands

**C.** Soap and water applied to the gear after the diver dresses down

**D.** Gross water decontamination, followed by soap and water solution and rinsing

**4.** What stressors can be found in the water that divers must contend with?

**A.** Marine life, entanglements, darkness

**B.** Physical workload, breathing resistance with depth

**C.** Nerves, marine life, risk of entanglements, loss of air

**D.** Bodies, chemicals, exposures, and loss of dexterity

*Access Navigate to find answers to this On Scene, along with other resources such as an audiobook and TestPrep.*

## Operations Level

# Helicopter Rescue Support

## KNOWLEDGE OBJECTIVES

After studying this chapter, you should be able to:

- Describe water rescue situations that might benefit from helicopter support. (**NFPA 1006: 16.2.9, 21.1.5**, p. 408)
- Describe the standard features of a helicopter. (**NFPA 1006: 16.2.9, 21.1.5**, p. 411)
- Identify and describe the components of helicopter landing. (**NFPA 1006: 16.2.9, 21.1.5**, pp. 411–412, 414–415)
- Identify different types of lifting devices trained crews can deploy for rescue use. (**NFPA 1006: 16.2.9**, pp. 415–417)
- Identify victim attachment devices used in helicopter assistance at rescue incidents. (**NFPA 1006: 16.2.9**, pp. 417–419)
- Explain how rescuers can facilitate helicopter operations as part of the aircrew. (**NFPA 1006: 16.2.9**, pp. 419–420)

## SKILLS OBJECTIVES

After studying this chapter, you should be able to:

- Set up a helicopter landing zone. (**NFPA 1006: 16.2.9, 21.1.5**, pp. 411–412)

# Technical Rescue Incident

With the small child still inside and the water still rising around the car, your rescue crew is quickly running out of options. The boat you usually run with is out of service, and because this flood has affected the entire area, other resources are worked over pretty well, too. The river is wide—really wide, almost 300 feet (91.4 m)—and somehow the car ended up almost in the middle of it. It is stable for now, but the rising waters are almost certain to push it off the gravel bar where it is currently sitting. The father of the child is frantic. He almost had his child out when the child panicked, and then the force of the river pulled the father away from the car. He is a strong swimmer and was able to swim to safety downstream.

The other crews at the water's edge saw the television helicopter holding overhead and waved it in. Because the news chopper had scanners that covered the fire band, the pilot could hear them but not respond. They instructed the pilot to come over and land. The plan was to hang a fire fighter from a rope tied to the landing skid so he could grab the child and fly to shore.

A $^9/_{16}$ inch (14.3 mm) life safety rescue rope was tied to the right-hand skid of the aircraft, and a length of about 100 feet (30.5 m) of rope was laid out beside the helicopter. At the end of this rope, a full-body harness was attached for the rescuer, and another smaller harness and attachment hardware were already in place and secure. The only thing the rescuer had to do was put the child in the harness while the helicopter hovered above him, and their planning was sure to keep it to a fast rescue.

As the improvised lift started, the fire fighter was understandably nervous but not scared. Heights were not a problem for him, and he had experience hanging from rope. What was lacking, however, was any lifting experience on the pilot's end. He had learned to fly as a private student, and he had minimal flying experience in this type of helicopter.

**1.** Should other peripheral roles in rescues be questioned during times of high urgency?

**2.** List some things that could go wrong with this rescue.

**3.** Does this evolution seem feasible, given the mission parameters above?

*Access Navigate for more practice activities.*

# Introduction

Helicopters are used in rescue work every day, worldwide. People are moved, evacuated, and rescued with helicopters in many locations, making helicopter rescue and transport familiar to most rescuers, especially in rural areas. Still more helicopters and crews assist in rescues indirectly by providing lighting, reconnaissance, and emergency medical services (EMS). How many times have we read about or packaged patients in car crashes who are flown to trauma centers? Dedicated-to-response aircrews have the equipment and the training to use aircraft safely in rescue situations (**FIGURE 16-1**). However, these crews may not have the ability or training to perform lifting or hoisting evolutions, and the rescue crews might be able to use aircraft only in a support role, such as providing aerial views or transportation.

Although some areas of the country use rescue helicopters for seemingly routine missions, many more places never use a helicopter directly except in the event of a catastrophic incident in which the military responds or a helicopter is "drafted" by the players in the mission. Picture hurricane Katrina and all the aerial shots of the military using helicopters to lift people from rooftops. NFPA 1006, *Standard for Technical Rescue Personnel Professional Qualifications*, in addition to having a dedicated helicopter rescue chapter,

**FIGURE 16-1** Dedicated helicopter crews offer excellent skills, training, and equipment to rescue teams.

Courtesy of Steve Treinish.

also acknowledges that helicopters are pressed into service by water rescue personnel, and it includes awareness-level job performance requirements (JPRs) for helicopter support in the watercraft chapters. While it may be tempting to ask media helicopter crews to assist in rescue operations, the use of rescuers, pilots, or flight crews who are not trained in emergency helicopter rescue operations is extremely dangerous. No matter how urgent the rescue need, air and rescue crews must ensure that the crews are knowledgeable in rigging, lifting, and lowering rescuers, victims, or litter baskets. Often, the training and capability levels of a media helicopter, or the crews operating with or under it, are not sufficient training for direct helicopter rescue. There is a real risk of a crash, injury, or death if a helicopter is drafted into service in this manner.

Although helicopter rescue operations are not very forgiving to untrained crews, helicopters can offer a multitude of support functions, especially in remote areas. Lighting, crew and victim transport, aerial reconnaissance, and even active search and rescue can be accomplished with aircraft, and mixing their equipment into our mission is certainly a possibility, if not a requirement (**FIGURE 16-2**).

# Before the Mission

Inaccessible or hard-to-access areas, offshore water, or fast-moving water can present opportunities for helicopters to shine; however, for many rescuers, helicopter operations should be pressed into service only when there is no other option at the time. Normally drafted in times of duress, helicopter rescue missions can present a great danger to the helicopter crew and to the rescue crew and victim, and they should be considered only in true life-or-death situations.

A high risk is presented when untrained crews and pilots attempt something that they have never tried or even thought of before that particular mission. Rescue crews and aircraft crews should make sure, if

**FIGURE 16-2** Helicopters, even if not directly involved with rescue, can offer a multitude of support functions, including lighting, reconnaissance, transportation, and aerial views. Here, a levee is repaired by a helicopter crew dropping sandbags into a breach.

Photo by Jocelyn Augustino/FEMA.

attempting to use helicopters in such a way, that all other rescue options have been examined and discarded, and they should think things through critically before using helicopters in an unplanned evolution. Most agencies operating helicopters for support-type operations will not allow live hanging loads. Discussions with your local helicopter agencies can help ascertain exactly what the aircraft and crew can do in a rescue mission. Before the water rescue, the aircrew should know their role to assist rescuers, and the rescuers should know how to assist aircrews.

## SAFETY TIP

Helicopter use in water rescue is not something to be taken lightly. This chapter merely reflects and examines some of the capabilities and methods used by helicopter crews; it should not be considered comprehensive training material.

# Helicopter Features

All helicopters incorporate certain features and hazards. All choppers are not created equal, and as is the case with firefighting apparatus, there are different vehicles for different tasks.

## Basic Helicopter Layout

Helicopters are complex pieces of machinery. Many parts must endure high speeds, high temperatures, and high stress. Some parts are operating at a speed that makes them hard, if not impossible to see with the naked eye. The main parts and features of a typical helicopter include the following (**FIGURE 16-3**):

- **Main rotor.** The main rotor spins on the top of the fuselage, driven by the power plant, or engine. The rotor is what gives the helicopter the edge for rescue work, including the ability to hover and to take off and lift loads vertically.
- **Tail rotors.** Tail rotors provide the pilot a means of equalizing torque forces from the main rotor.
- **Landing skids (or landing gear).** The landing skids or gear of the helicopter are typically steel tubing, but some larger helicopters may be equipped with wheels or floats.
- **Fuselage.** The body of the aircraft.
- **Fuselage doors.** The entrance to the cargo area, fuselage doors are usually found on the side of the craft, though some are in the rear.
- **Cargo area.** Extra seats, litters, cots, or equipment can be fastened in and transported in the cargo area. Some larger helicopters have winches (e.g., U.S. Coast Guard).
- **Control seats.** Where the actual flying and navigating takes place.

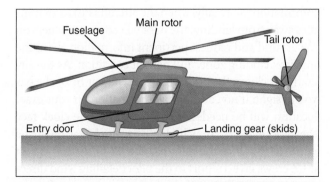

**FIGURE 16-3** The major parts of the helicopter are the fuselage, main rotor, tail rotor, landing gear or skids, and entry doors.

© Jones & Bartlett Learning.

# Helicopter Safety

Safe operations are paramount around helicopters. The rotors spin at incredibly high speeds, making injury or death highly likely for anyone who gets in the way of one. Another problem with rotors is debris being kicked up from the downdraft. When helicopters fly, the air passes through the main rotors downward. This gives the helicopter the lift it needs to fly, but also forces the air down underneath it. Helicopters can stir up a huge amount of dust and debris, and forceful air can blow debris around on scene.

## Communication

Most medical and law enforcement helicopters have the ability to transmit and receive on channels often used by rescuers in the fire service. However, trying to inform a pilot of the hazards around a landing area is no time to be competing for airtime; a channel that handles only helicopter communication is preferable. In the event of a non–public safety helicopter being pressed into service, standard aviation radios can be used. A phone call to a nearby major airport may be helpful, or the helicopter's operator may provide information regarding which aviation radio channel to use. A lack of communication to any aircraft can result in frustration or risk if hand signals are attempted.

**SAFETY TIP**

Approaching a helicopter after landing should be done only after the pilot gives permission to do so. The pilot must be aware of your presence and intention. If not, the aircraft may be moved or rotated, possibly striking someone with the rotors.

## The Clock Face

Pilots always refer to the aircraft nose as 12 o'clock, regardless of whether it is a helicopter or a fixed-wing aircraft (**FIGURE 16-4**). This facilitates directional spotting or communicating directions. These directions are always based on where the aircraft is flying, not on your location.

# Helicopter Landing

Landing zones (LZs) are simply areas that are being used for the helicopter as takeoff or landing sites. They can be visually selected by the pilot or scouted from the ground by rescuers. Often the rescuers will receive advance notice of a landing and have the time to scout

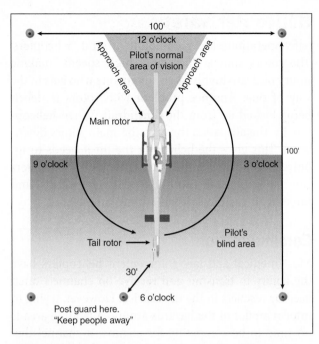

100'
12 o'clock
Approach area
Pilot's normal
area of vision
Approach area
Main rotor
100'
9 o'clock
3 o'clock
Tail rotor
Pilot's
blind area
30'
6 o'clock
Post guard here.
"Keep people away"

**FIGURE 16-4** A clock face is used to differentiate the sides, front, and rear of a helicopter.

© Jones & Bartlett Learning.

out and examine an LZ. Solid, level ground is preferred, and hard surfaces, such as pavement or concrete, are even better. Debris that can be stirred up should be removed, and overhead obstructions should be relayed to the helicopter crew. Power lines are a major problem for helicopter pilots, and so are loose gravel and water. Helicopter engines, especially turbines, can suck trash or debris into the intake port of the motor. Just one piece of trash or debris that is sucked into the intake port of the motor can destroy an engine in seconds and may result in an aircraft crash. The pilot may approach from an angle or hover for a few seconds before setting down in an LZ. This lets the downdraft of the rotor blow any debris out and away before landing.

Pilots prefer to land and take off into the prevailing wind if possible, rather than take off or land vertically. Placing the LZ in an area where this is possible helps the aircraft and the pilot.

LZs should be as large as possible, and an LZ of 100 feet × 100 feet (30.5 m × 30.5 m) should suffice for most landings (**FIGURE 16-5**). Smaller helicopters may be comfortable with as little as 75 feet × 75 feet (22.7 m × 22.7 m), but the pilot will use his or her discretion. LZs should be marked on each corner.

Commercially made devices are available for marking the LZ, but it also can be done with orange traffic cones during the daytime. The cones should be weighted or heavy enough not to become projectiles from the downwash. A nice

touch is to lay the cones down with the tip of the cone pointing inward toward the LZ. Rocks or other weights can then be placed inside them, making them more resistant to blowing around. Placing a flashlight in these cones at night helps illuminate them, without the brightness of LED apparatus lights. Pilots are susceptible to night blindness from high-powered lights, and modern night vision devices only make this worse. Marking an LZ at night with apparatus is possible, provided the other trucks' lights are turned off to prevent confusion and the pilot has confirmed this. Strobes can be used at night, but should be turned off at the request of the pilot.

NFPA 1006 mentions the presence of fire suppression systems during emergencies. Because flight emergencies can occur in the takeoff or landing phase of flight, basic fire protection should be in place if possible. This can be something as simple as having all-purpose extinguishers manned at the LZ, to using a hand line off of one of the fire apparatus marking the LZ in case of fire. Either way, a plan should be worked out before the aircraft is on approach.

**SAFETY TIP**

Be aware of creating an inadvertent hazard when turning off emergency lighting, especially in roadways. Motorists may be confused about what they see, or they may be distracted by the scene. If a helicopter is used, the area should be clear of civilian foot and road traffic.

## Location of the Landing Zone

The location of the landing zone should be communicated to the aircrew as soon as possible. GPS coordinates, addresses, cross streets, and directions from a town or pertinent landmark can suffice and provide additional information. Normally, when the aircraft is 10 to 15 minutes away from landing, the pilot will contact rescuers on the ground and request pertinent information about the mission, landing zone, and patient condition. As the aircraft flies closer, the pilot will relay whether he or she has you in sight. If not, additional information on your exact location will be needed, and once again, the clock face the pilot refers to can help describe which way he or she needs to fly. Once the pilot has the LZ in sight, it is best if LZ crews give a report to the pilot detailing wind direction, hazards, and exactly what the helicopter will be landing on. In the event of an EMS flight, patient condition should be given first.

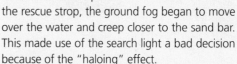

# Voice of Experience

During an early nighttime training mission, Naval Air Station Whidbey Island Search and Rescue (NASWI SAR) was notified by a Skagit sheriff of a swiftwater rescue involving three survivors in a small boat. NASWI SAR's helicopter, Rescue 58, was approximately 20 miles to the west, operating in a local training area for monthly rappel currency. Once air traffic control notified the crew, Rescue 58 departed to the datum that was passed from the sheriff.

While transiting out of the training area, the rescue swimmer began to dress out in his swimmer gear. Meanwhile, the SAR Mission Commander (SMC) and Co-Pilot (CP) were discussing the area with the Crew Chief (CC). The SAR Paramedic/Med Tech (SMT) works with the SMC in identifying local hospitals based on potential injuries, on-scene time based on distance, and fuel arrangements.

Within 5 miles of the location, Rescue 58 began slowing our air speed to allow for an acceptable search speed of 60–70 knots, allowing for safe search/cruising speed over the winding river, as well as airspeed for emergencies. Radio communications were established with the sheriff and a rescue boat, and Rescue 58 learned that the rescue boat was running aground roughly 500 meters away from the survivor's boat, which had run aground on some rocks in the middle of the river following a blind bend.

As Rescue 58 arrived in a 100-ft hover, a small ground fog was noticed in a field to the right of the river. The three survivors were standing in waist-deep water inside the boat. Once in an established hover, the CC worked with the SMT to get into a position for the CP to get a decent visual on and under night vision. The CC stayed in the gunner's window clearing trees, as the closest tree in the hover was within 4 ft of the rotor arch. The SMT took over as the host operator and hoisted the swimmer down to the boat. Once on the boat, the rescue swimmer detached from the hook and determined the boat was on a sand bar with large rocks, with knee-deep water over the sand bar and raging swift-water off the sand bar. As the swimmer sent the first survivor up in the rescue strop, the ground fog began to move over the water and creep closer to the sand bar. This made use of the search light a bad decision because of the "haloing" effect.

With the ground fog moving in, the SMT rigged the rescue basket to allow for quicker survivor packaging. The second and third survivors were put in the rescue basket without complications. Once all the survivors were in the basket, the swimmer turned off the emergency radio/beacon on the boat and was hoisted to the aircraft. With the main cabin door closed, the SMT passed verbal control of the helicopter to the CC and began treatment of the survivors. There were no noted injuries on the patients; all three were cut free of their wet clothing, which they had been wearing for 4 hours in the water. Active warming measures were taken by turning on the Environmental Control System (ECS) heat and supplying survival blankets. One of the patients was a Type II diabetic but had missed his medication and had not eaten. The local hospital was a 6-minute flight from the area on the river. The sheriff notified the local hospital of the survivors' arrival.

Turnover went smoothly. In the post-incident debriefing, it was noted that the ground fog made the CP want to have better visuals while in the hover. The coordination of verbal calls from the front to the back, and back to front, allowed for close hovering to terrain to allow the successful rescue of the three survivors from the river.

**Wayne Papalski, BS, NR-P/FP-C/TP-C**
Leading Flight Paramedic
Naval Air Station Whidbey Island Search and Rescue
Oak Harbor, Washington

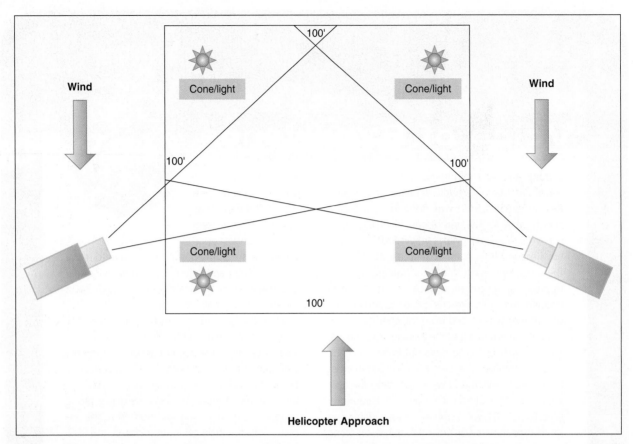

**FIGURE 16-5** Landing zones should be 100 feet × 100 feet (30.5 m × 30.5 m) if possible, marked on all four corners, and set so the aircraft can approach and depart into the wind. Loose debris should be removed.
© Jones & Bartlett Learning.

## Approaching Helicopters

Want to make a pilot nervous? Run out toward a running helicopter without clearance from the pilot. There is a certain way to approach helicopters with rotors spinning, and doing it the correct way could save a life.

The safest place to be when any rotor is spinning is outside the LZ. Limit the number of rescuers approaching the LZ. There is no reason to put too many people in that area, and the helicopter crews appreciate that. Helicopter crews have more training on operations around a helicopter that is "hot," or whose rotor is still turning, and many prefer to do loading or unloading themselves.

### Approach Zones

There are specific **approach zones** that you should be aware of when approaching a hot aircraft (see Figure 16-4). Think of the helicopter as a clock, with the center of the windshield at 12 o'clock and the tail at 6 o'clock. For most helicopters, approaching on the ground is done from either the 9 o'clock point or the 3 o'clock point, provided two things are done: First, make eye contact with the pilot or copilot on the side from which you are approaching, and proceed only when you are cleared. Second, stay low. That may

sound ridiculous, but in the heat of the mission, a rotor that is spinning quickly may not be seen, and standing up can literally take off your head. It should be somewhat intimidating to walk under a rapidly spinning rotor, even if it provides clearance. An LZ that is sloped makes this problem worse. Hilly terrain and circumstances may entail a pilot setting down in a small, flat area with a slope on either side. The uphill side will have less vertical distance between the rotor and the ground, and should be avoided at all costs (**FIGURE 16-6**).

### SAFETY TIP

Appointing a safety person to monitor the tail rotor and rescuer traffic around the helicopter may increase safety by ensuring rescuers approach the aircraft from proper directions. This rescuer must stay in place the entire operation, and be able to signal other rescuers to immediately stop.

From 7 o'clock to 8 o'clock and from 3 o'clock to 4 o'clock, it is possible to approach the craft, but extra caution should be used because eye contact with the pilot can be reduced, and that places a rescuer closer to

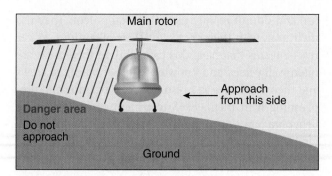

**FIGURE 16-6** Approaching spinning rotors should be done only on the downslope side of a helicopter, because there will be reduced vertical clearance on the upslope side between the rotor and the ground.
© Jones & Bartlett Learning.

**FIGURE 16-7** One-skid landings require a high degree of skill, and if not done correctly, quickly, and smoothly may injure or displace the victim or the rescuer.
Courtesy of Aiut Alpin Dolomites.

the tail rotor. Tail rotors can pass within two to three feet (0.6 to 0.9 m) of the ground, so although the main rotor may be seven feet (2.1 m) up, the tail rotor is not.

From 10 o'clock to 2 o'clock, you may be able to approach, but the main rotors can tilt downward in the front of certain helicopters, and clearance must be obtained from the pilot before approaching.

From 5 o'clock to 7 o'clock, you are in danger of contacting the tail rotor. *Stay out of this area!*

---

**KNOWLEDGE CHECK**

In helicopter terminology, the clock face applies to:
a. in-flight navigation.
b. landing zone orientation.
c. aerial search orientation.
d. the nose, sides, and rear of the aircraft.

*Access your Navigate eBook for more Knowledge Check questions and answers.*

---

## One-Skid Landings

Another option for skilled helicopter pilots is a **one-skid landing**, in which the pilot places the aircraft near the victim (i.e., the victim can reach out and climb aboard) but does not actually land the aircraft (**FIGURE 16-7**). If we could land it, we would probably not need it! Good pilots can hover like this, but it takes an experienced pilot and a steady hand. Another problem is the closeness of the entire aircraft to the rescue operation. There is no cushion of vertical distance as with a haul or hoist system, and the rotors are close to the action. Similar to approaching a boil line in a hydraulic, working underneath a helicopter that is fully throttled and hovering should definitely make you swallow and glance upward.

If this is the only option, so be it. When the aircraft is close enough, get the victim aboard. Beware of

letting the victim or other rescuers simply hold onto a skid, or hang on with their arms. Likewise, do not give the order to ascend or start moving until you are sure the victim is secure. This is not a dress rehearsal, and you may get only one chance.

## Helicopter Lifting

While seemingly routine for the crews that are trained to perform helicopter rescue missions, the fact of the matter is that, for the most part, many water rescuers will use helicopters only as a last-ditch effort to save a viable victim. An untrained rescuer who simply ties himself or herself to a skid and hopes for the best is an accident waiting to happen. Another issue is pilots who do not have experience moving or hovering in a precise location. Just keeping the helicopter in one place can be a challenge with wind currents, weight shifts, and rescuer activity below. Placing a rescue line becomes a challenge for the pilot, and pilots who do not lift or move at the right time can injure rescuers and victims.

It is a good idea to place a rescuer, preferably with radio contact, in eyesight of the pilot as a safety spotter. The pilot may not be able to see everything below him, and a set of eyes at another vantage point can help. Ideally, the pilot should have contact with the crew in the fuselage, but in a case of a helicopter being "drafted" into service this is not likely. Most helicopter lifting consists of either fixed position hoisting or static line lifts.

## Weights and Balances

The biggest problem with hanging off any helicopter on one skid is the potential upset of the **weight and balance** in the aircraft. Helicopters are susceptible to being jostled out of control when weight is suddenly removed or loaded on one side, and the pilot does not have optimal control if the lateral balance is upset. Because the skids are somewhat long, attaching a load too far forward or rearward can upset the balance longitudinally. Helicopters like to be balanced side-to-side and front-to-back.

Engine power is a factor as well. The motor must be able to lift, hover, and transport all the weight in or attached to the helicopter; current or water forces pulling on the rescuer and victim will add to this total weight. The time of year and weather conditions also can factor into lifting, as hot and humid conditions, as well as altitude, will affect the amount of lift the air provides.

## Fixed Position Hoisting

**Hoisting** is accomplished with the aircraft hovering in place while the victim is lifted up to and inside the aircraft fuselage. The aircraft is equipped with an onboard winch in a fixed position, most often above a side door (**FIGURE 16-8**). The winch is rated to pick up the victim and the rescuer, while providing safety margins. Picture the U.S. Coast Guard performing offshore ocean rescues, and you have a good example of this style of hoisting. The victim cannot be slung under the helicopter and transported any distance to speak of. Hanging off a line for extended periods of time in a rescue sling likely will result in injury, and hypothermia in wet patients is a real concern. Remember, the victim is under the rotor wash and moving laterally, so windchill can play into a lift.

Hoist operations are used by the military, but are not done as frequently by public safety teams. The teams that operate military helicopters are proficient in their

**FIGURE 16-8** Winches attached to the outside of the aircraft can lift directly up to the fuselage and can lift while the helicopter is hovering stationary.

U.S. Department of Defense. The appearance of U.S. Department of Defense (DoD) visual information does not imply or constitute DoD endorsement.

use and do not readily use other equipment. They use their gear, with their crews, for their operations. Water rescue teams that have this capability in their areas are often well versed and knowledgeable about these crews and their capacities as a result of cross training and pre-planning with them. EMS treatment can begin in the helicopter cargo or treatment area, thus stabilizing the victim. Just starting the warming process during the trip to the eventual destination is helpful.

## Fixed Line Lifting

Fixed line, or static line, hauling is done using the helicopter as a vertical lifting unit to pick up a fixed length of rope or cable. This is also called a short haul lift. The victim (and the rescuer) must be transported by the pilot to shore or another safe location while hanging under the aircraft.

Many helicopters are equipped with cargo hooks on the belly of the aircraft that allow both centering of the hanging load and cutting it loose in case of a control failure. Even if the hooks are rated for the lift, having a way to control the length of the rope and a safety backup system can still be an advantage. Lifting the victim and rescuer is not much of an option, but lowering is. Don't look at it as lowering per se, but rigging this way provides the advantage of being able to give slack to the rescuer.

There are multiple ways to tie off a **fixed line haul**, but one thing to remember is that the pilot would like the load centered under the aircraft if possible. The use of a 10- to 20-pound weight bag at the bottom of the static line helps keep the line under the helicopter instead of having it blowing around where it could be sucked into the rotors and/or the engine intakes.

### SAFETY TIP

Whichever method is used to haul, hoist, or sling a victim, it is important to let the end of the line or rope touch the ground before you do. A huge amount of static electricity is built up under the spinning rotors of the aircraft; allowing the line or rope to touch the ground first lets the charge dissipate to the ground.

## Tying Off Fixed Lines

The use of twin ropes for short hauls is strongly recommended as a safety. There is no reason not to double up, and your rescuer hanging off the helicopter will likely appreciate it. With a cargo hook under the helicopter, one way to secure these lines is to use figure eight knots and carabiners to attach the ropes to the

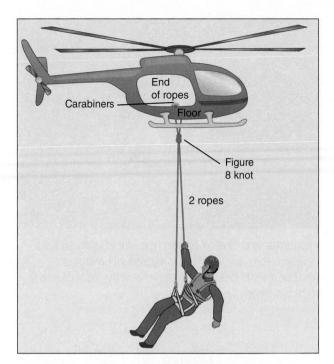

**FIGURE 16-9** A typical fixed-line rigging, using twin lines attached to the cargo hook, with the tails of the line fastened together after routing them through each door. This provides a backup to the hook.
© Jones & Bartlett Learning.

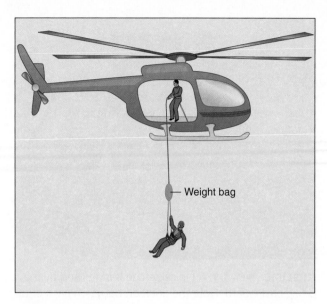

**FIGURE 16-10** A belay system with the rescuer in the helicopter adjusting slack through a belay device. The lowered rescuer may be harnessed in or use a short tether and carabiner to attach to a figure eight or another carabiner. Note that the belaying member is tied in for safety.
© Jones & Bartlett Learning.

hook, but leave long enough tails to run up and inside the fuselage (**FIGURE 16-9**). These tails can then be clipped together. If the hook fails, the loop inside the cabin will then grab the load.

Another option is to set up these lines by running them through a belay device anchored to a center ring, if the helicopter is equipped with one (**FIGURE 16-10**). This gives the rescuer manning the belay the option to extend the rope but still lock it off to perform the haul. Remember, many helicopters are not set up to do this, and the pilot may not be able to see much underneath the aircraft.

---

### SAFETY TIP

Even if you are not the one hanging from the rope underneath the aircraft, don't forget that working around open doors can be just as fatal. The helicopter will likely be moving around in the air, and one slip from a rescuer who is not tied in can result in injury or death.

---

## Rescuer Deployment

One method of deploying rescuers is to give the rescuer the ability to harness in, or "package," a victim while the helicopter stages somewhere close by, but in a better place—namely higher. The rescuer, wearing a harness, has the ability to disconnect from the static

line or hoist and work with a victim, or, in some cases, even travel in moving water with them. When the rescuer is ready, signals are given to the pilot, and the helicopter is brought back into position.

The line from the aircraft is equipped with a figure eight belay device steel ring, or similar equipment, and the rescuer clips his or her carabiner into the system. The lift can then proceed as planned.

---

### SAFETY TIP

Many agencies and crews recommend spring-locking carabiners. Locking carabiners that twist a knurled knob to lock may vibrate loose with the vibration of the aircraft.

---

## Victim Attachments

Getting the victim from the hazard to the safe area can be a chore, even with a properly trained and equipped helicopter crew. Fortunately for most rescuers, if the helicopter crew has approval to perform a rescue, it probably is staffed with qualified personnel. However, it is worth examining the process they may elect to use.

### Basket-Type Equipment

Most people have seen video of victims being placed in a **rescue basket**. These baskets are equipped for one person, offer flotation and protection from striking

**FIGURE 16-11** A typical rescue basket used by the U.S. Coast Guard. Note the flotation buoys on each end.

Photo by Petty Officer 2nd Class Adam Eggers-U.S. Coast Guard.

**FIGURE 16-12** A basket-type litter used to haul incapacitated, injured, or uninjured victims. It is equipped with flotation, a cervical spine board, and blankets, ready for use.

U.S. Navy photo by Mass Communications Specialist 2nd Class Nathan Burke/Released.

other objects, and are very strong (**FIGURE 16-11**). Even though some models can fold for storage, they are fully capable of lifting and protecting the human cargo. Also in this category is the Billy Pugh net, which is basically made of ropes and a hard seat that resembles a bird cage.

## Stokes Baskets or Litters

Many helicopters employ some kind of steel or aluminum Stokes basket, used for injured or ill victims (**FIGURE 16-12**). They are similar to Stokes baskets found on many rescue vehicles, although solid plastic Stokes litters may spin more when exposed to rotor wash under the aircraft. The victim is packaged in with strapping, a bridle of some kind is fastened to the haul line or hoist line of the helicopter, and away they go. Other rescuers or swimmers may be called on to ride the basket, clipped in at the haul line and riding in a harness. This may offer the victim ongoing care or, in the case of a panicky victim, a nearby calm, reassuring voice. Without that reassurance, panicky victims may get so upset that they wiggle out of the strapping and fall out.

## Sling-Type Harnesses

Trained rescuers have the ability to hang from a harness or a cinch sling, directly under the helicopter. This eliminates the bulk of a basket or cage. Victims may not have the ability or the strength to enter the device, and even then, most will not know how to properly secure themselves in it, which is a recipe for disaster. Often, the rescuers will descend to the victim. Slings are a good choice only when the victim is conscious and capable of helping himself or herself. Victims who are unconscious, severely hypothermic, or unable to follow instructions may be better off in cages or Stokes baskets.

**FIGURE 16-13** A screamer sling, also known as a diaper harness.

Courtesy of Steve Treinish.

**Screamer Slings.** Screamer slings, or "diapers" as some rescuers like to call them, are bag-style slings in which the victim sits or lies (**FIGURE 16-13**). They are somewhat easy to manage and load, but the victim may feel a bit loose in them. They fold down into a small area, making them easy to store or carry.

**Cinch Straps.** Cinch straps are used as a quick sling for a victim. The idea with any strapping or slings is that the duration of the haul is kept short, and the victim is moved only as far as is necessary to a safer area. Comfort is not high on the priority list. Because cinch straps self-tighten, good grip on the victim is maintained (**FIGURE 16-14**). Rescuers can

**FIGURE 16-14** Close-up view of a cinch-type harness.
Courtesy of Casey Ping.

**FIGURE 16-15** Jungle penetrators have been used heavily in combat overseas, and they can be used in heavy brush or trees. A rescuer and victim can both sit on the fold-down seats.
U.S. Department of Defense.

wear body harnesses and work hands-free to get a sling on a victim.

**Jungle Penetrators.** Jungle penetrators are used to get through heavy brush or trees, and were adapted from the military (**FIGURE 16-15**). A heavy, slim spike is used to get down to the victim. When ready, a barb-like hook can be folded down from the spike, providing a place to hook up or sit.

---

**KNOWLEDGE CHECK**

Which is a common component of a jungle penetrator?
a. GPS beacon
b. Fold-down seat
c. Basic survival gear
d. Victim cage

*Access your Navigate eBook for more Knowledge Check questions and answers.*

---

# Emergency Flying Considerations

While not specifically trained to act as a crewmember, it would not be out of the realm of possibility for a rescue team member to have to travel in a

helicopter. One of the uses for rescuers is to provide "eyes in the sky" during times of emergency. Putting a rescuer in a seat to help search or to get the big picture is a possibility. Water rescuers should be guided by the pilot's or aircrew's commands, and what they say goes. Seat belts, helmets, and emergency procedures are for your safety.

## Sterile Cockpit

Pilots are trained in basic flight operations to use a sterile cockpit to avoid distractions that may lead to crashes or missing important checks during takeoff and landings. Pilots often refer to a typical flight as hours of boredom ended by a few minutes of nerves as the landing occurs. All talk not related to the flight or mission should be curtailed, unless approved by the pilot or crew. During takeoff and landing or emergencies, the best advice is to stay quiet and not undo any safety equipment until ordered to do so.

## Crew Instruction

If you are in a helicopter, the biggest help you can provide is to listen to the crew. When they give you instructions on how the doors operate, how to operate the seat belt, or when to talk, pay attention.

Typically the pilot or crew will explain how to exit the aircraft, where to go afterward, and what to do next,

**FIGURE 16-16** This police pilot successfully autorotated to an emergency landing after engine failure. No injuries occurred, and the helicopter did not tip until the skid sank in the soft mud at the water's edge.
Courtesy of Steve Treinish.

**FIGURE 16-17** The Helicopter Emergency Escape Device (HEED) is a small can of compressed air for use when landing in the water.
Photo by Staff Sgt. Mike Alberts 25th Combat Aviation Brigade Public Affairs.

if it should become necessary. Helicopters can autorotate in case of engine failure. Autorotation is using the wind speed to keep the rotors turning, thus maintaining some lift and making an emergency landing possible (**FIGURE 16-16**). This also means the helicopter will not be able to travel far laterally, so when the pilot decides to set down, it will happen quickly.

## HEED

The Helicopter Emergency Egress Device (HEED) is a small, hand-held bottle of compressed air that crews can use or even strap to their flight suits (**FIGURE 16-17**). It lasts for only a few breaths, but because helicopters have a tendency to invert when in the water, many crews require them to be onboard when flying over water.

# *After-Action* REVIEW

## IN SUMMARY

- Although some areas of the country use rescue helicopters for seemingly routine missions, many more places never use a helicopter directly except in the event of a catastrophic incident in which the military responds, or a helicopter is "drafted" by the players in the mission.
- Rescue crews and aircraft crews should make sure, if attempting to use helicopters in such a way, that all other options have been examined and discarded, and they should think things through critically before using helicopters in an unplanned evolution.
- All helicopters incorporate certain features and hazards. All choppers are not created equal, and as is the case with firefighting apparatus, there are different vehicles for different tasks.
- Just keeping the helicopter in one place can be a challenge with wind currents, weight shifts, and rescuer activity below. Placing a rescue line becomes a challenge for the pilot, and pilots who do not lift or move at the right time can injure rescuers and victims.
- Getting the victim from the hazard to the safe area can be difficult, but if a helicopter crew has approval to perform a rescue, it is probably staffed with qualified personnel.
- Water rescuers should be guided by the pilot's or aircrew's commands. When assisting as aircrew, seat belts, helmets, and emergency procedures are for your safety.
- While not specifically trained to act as a crewmember, it would not be out of the realm of possibility for a rescue team member to have to travel in a helicopter. Basic considerations include understanding the importance of a sterile cockpit, listening to the crew, and understanding emergency devices.
- Helicopters can offer a multitude of support functions, including lighting, crew and victim transport, aerial reconnaissance, and active search and rescue.

## KEY TERMS

*Access Navigate for flashcards to test your key term knowledge.*

**Approach zones** With the nose of the aircraft at the 12 o'clock position, the sides of the helicopter.

**Cargo area** The section behind the control seats, carrying crew, cargo, or patients.

**Cinch straps** A quick, self-tightening sling to provide a good grip on the victim.

**Control seats** The front seats of the aircraft, where the pilots sit.

**Fixed line haul** Using the helicopter to provide a lift for a line attached to the aircraft, with no hoisting capability.

**Fuselage** The main body of an aircraft. (NFPA 403)

**Fuselage doors** The entrance to the cargo area.

**Hoisting** Using a winch onboard the aircraft to pick up the rescuer and/or victim, placing them into the fuselage.

**Jungle penetrator** A steel, hook-like device lowered through tree canopies, then unfolded to provide a seat for the rescuer and victim.

**Landing skids (or landing gear)** Steel feet the helicopter sits on when on the ground.

**Main rotor** The top blades on the helicopter that provide lift and some forward thrust.

**One-skid landings** Placing one skid at a distance so that rescuers or victims can get onto the skid or into the fuselage, but not fully landing the aircraft.

**Rescue basket** A stainless steel floating cage in which the victim sits while being lifted to the helicopter.

**Screamer slings** A diaper-like bag the victim sits in to be lifted.

**Tail rotors** The smaller blades at the tail.

**Weight and balance** The figurement of the helicopter's total lift capabilities, and the weighting of all loads as not to upset the attitude of the aircraft in flight.

# On Scene

Two canoeists could not wait for the water to warm up before getting a good paddle in, and they failed to realize the stream height. After one dropped a paddle, the remaining paddler could not control the canoe, and it capsized and wrapped around a tree in the middle of a stream with water moving at almost 5 knots.

One victim is clinging to a tree, using his legs and arms to hug the tree, with one foot barely on the canoe supporting his weight. It is truly a dramatic rescue. The man's girlfriend, who was wearing a personal flotation device (PFD) is missing. There is a real possibility that she is simply downstream in a wooded area, but with the spring foliage, spotting is hard to do. The police helicopter is flying around overhead, and you decide a quick rescue could be done using your kernmantle rope tied to a skid and moving a rescuer into position to do a pick-off rescue, saving time and resources for the search for the missing woman.

You dial up the law enforcement helicopter on a mutual-use radio channel and order it to descend and land in a nearby field. After landing, the pilot is adamant she will not use the helicopter to perform your lift. She will support the search in any way she can, but other methods will have to be used to rescue the man. A medical helicopter is inbound for patient care for the man trapped.

**1.** You are asked to set up an LZ for an incoming medical helicopter that is providing aerial search for this incident. These duties can include:

**A.** sweeping for debris before landing, securing any loose articles, and advising of LZ conditions when hailed by the pilot.

**B.** marking the LZ with traffic cones and ensuring they are weighted to prevent becoming projectiles.

**C.** providing basic fire suppression in the form of an extinguisher or hand line from an engine.

**D.** All of the above

**2.** The helicopter crew advises a rescuer may ride in the rear seat to help look for the victim. You assume a sterile cockpit environment. What does this mean?

**A.** No hazardous materials risk will be brought into the aircraft.

**B.** No medical or biological materials can be brought on board.

**C.** You help declare hazards the pilot cannot see in the blind spots.

**D.** Talking is severely restricted during takeoff, landing, or high-stress times.

**3.** A pilot declines to use a fixed line to haul a victim to safety, citing a lack of centrally located hooks on her aircraft. While not very popular with command, this is because:

**A.** the jet engine may not have enough power to pick up the person.

**B.** the added weight on one side may unbalance the forces of the rotors, causing a control failure.

**C.** her supervisor at the hanger was not able to get permission from the flight director.

**D.** she does not trust the rescue rope supplied by rescue crews.

*Access Navigate to find answers to this On Scene, along with other resources such as an audiobook and TestPrep.*

# Appendix A

## Ropes and Knots in Water Rescue

## Introduction

Although NFPA 1006, *Standard for Technical Rescue Personnel Professional Qualifications*, requires swift-water operations-level rescuers to have rope training as a prerequisite before taking a class, the fact is that most of the disciplines in water rescue require the use of basic ropes and knots. Any of the water rescue technician levels may require rescuers to perform tethered rescues, and any operations-level rescuer should be able to tie off and manage the technician doing the "go" rescue. This appendix is a guide to the basics of the knots and systems that are most commonly used in water rescue. This information is for reference only and will not meet requirements for rope rescue training. Rope rescue training should be taught by a qualified rope rescue instructor and should reflect NFPA 1006, Chapter 5: *Rope Rescue*.

## Rope Types and Construction

Ropes have come a long way in the last 40 years. Science has developed fibers that are strong, resistant to rot, and light enough that we can carry a lot of it with us. Water rescue rope also has evolved into a specific-use rope, in that it is lighter than most rescue rope and floats on water. Many agencies use their rappelling or lifting rope for water rescue work, but that can make for some tough operations, because true nylon kernmantle will sink when it gets wet. Standard kernmantle rope systems are used for many functions, such as a highline, which is basically a trolley system that is used to lower rescuers in harnesses from above when water conditions are too dangerous to enter.

Sinking ropes, however, add weight in the water when it is not needed, and they present a safety hazard if the rope sinks and they get pulled under. While the weight of the rope alone probably is not enough

to pull the rescuer under, if it sinks and gets wrapped around something underwater, then what? The bank crews may pull too hard, and rescuers will follow what they are attached to.

Most water rescue rope is made from polypropylene, which gives it flotation, but also provides good working strength. Rescuers don't need rope with a 9000-pound (4082-kg) breaking strength when they are not lifting vertically. Different ropes for different tasks call for smaller lines. A good example of this is a throw bag. Many throw bags are stuffed with 0.5-inch (12.7-mm) or even larger-diameter rope, but some are filled with smaller-diameter rope, such as 0.28-inch (7-mm) diameter rope (**FIGURE A-1**). With less drag, the 7-mm rope can easily travel farther through the air, and we need a rope that will get there. Even taking into consideration a victim weight of 350 pounds (158.8 kg), the minimum breaking strength of 7-mm water rescue rope is just over 2900 pounds (1315 kg), which constitutes a significant safety margin. In a pinch, water rescue throw bags can be used for non–life safety raising or lowering use, such as tethering or creating a tension diagonal across the water. This type of rope also works with common rope hardware, such as Prusiks and pulleys, and can be integrated into water rescue work if required.

## Webbing

Tubular webbing is often seen in rope rescue and water rescue, most often to build anchor systems. It is inexpensive, easy to use and clean, and provides good strength; but in terms of true breaking strength, it is often the weakest link in the system. It can range in width from three-quarter inch (19.1 mm) up to three to four inches (76.2 to 101.6 mm) for specially made anchor straps, with one-inch (25.4-mm) webbing being most common. One-inch webbing also happens to fit most rope hardware nicely.

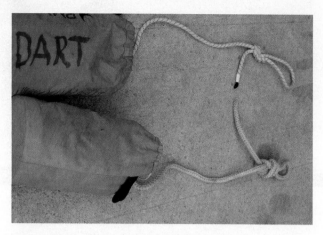

**FIGURE A-1** Smaller-diameter, floating polypropylene water rescue rope is best for throw bags, as the smaller-diameter line allows longer throws and still provides a good working strength.

Courtesy of Steve Treinish.

> ### LISTEN UP!
>
> One way to extend the life of webbing and provide a great abrasion guard is to run it through an old fire hose. Small sections of fire hose make great edge protection and can be left on the webbing ready to go.

# Basic Knots for Water Rescue

Carabiners and webbing are found on most fire apparatus these days, and common knots used in water rescue include the following:

- Figure eight on a bight
- Figure eight follow through
- Grapevine knot
- Bowline
- Water bend

Of course, there are many other knots with many other uses, but a good working knowledge of the knots listed here will allow most rescue evolutions to be done strongly and safely. These knots can be practiced on a short length of rope kept in a locker or perhaps with turnout gear. Often, fire fighters gather after dinner on the front or back ramp to chat, and it can easily lead to a good opportunity to practice tying knots and even rigging rope systems.

All knots should be dressed, which simply means that all the pathways forming the knot are easy to see and correctly routed. Dressing the knot ensures the knot is as strong as it can be and helps identify the knot at a glance.

## Figure Eight on a Bight

The figure eight on a bight is one of the easiest and quickest knots. It provides good hold when it is tied into a line, is easy to untie after being loaded, and can be left in place and used with a carabiner for great speed in hooking up systems. It is a great basis for rescue work in general. If you learn only three or four knots, make this one number one. To tie a figure eight on a bight, follow the steps in **SKILL DRILL A-1**.

## Figure Eight Follow Through

A variation of the figure eight knot is the figure eight follow through. This knot is used for tying around solid objects for anchoring or lifting, and creates a nice loop for attaching carabiners. To tie a figure eight follow through, follow the steps in **SKILL DRILL A-2**.

## Water Knot

Water knots are used to make webbing into loops and are most commonly used for anchors and safety systems. They are easy to tie, but they must be tied properly. To tie a water knot, follow the steps in **SKILL DRILL A-3**. For clarity, the illustrations in this skill drill show two pieces of webbing in different colors. A single piece of webbing would be tied at the two ends in the same fashion. Make sure the webbing is not twisted before tying the knot.

## Grapevine Knot

The grapevine knot is used to tie two ends of a Prusik line together, forming a Prusik. The Prusik is used as a safety device or to hold a carabiner on a larger rope. This knot ties fairly easily, but must be practiced. To tie a grapevine knot, follow the steps in **SKILL DRILL A-4**. The grapevine knot can be made into a triple bend easily.

## Prusik Bend

Properly wrapping a Prusik onto a rope can be confusing. The Prusik bend is a knot that is easy to learn, yet just as easy to forget. It is used for attaching pulleys to larger ropes or providing a grip on the rope, most

# SKILL DRILL A-1
## Tying a Figure Eight on a Bight

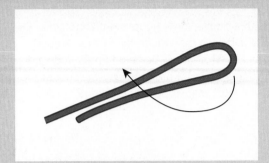

**1** Form a bight at one end of the rope, allowing a couple feet for the tail.

**2** Wrap the tail around the mainline once, then again.

**3** Bring the tail back up through the loop created at the beginning.

**4** Tighten the knot.

# SKILL DRILL A-2
## Tying a Figure Eight Follow Through

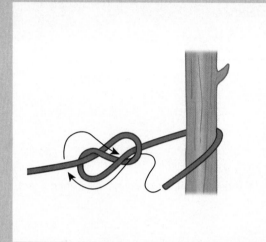

**1** Start with a simple, loose figure eight knot tied in the mainline, allowing for enough tail to wrap around the object to which it will be tied. Run the rope around the anchor, and start it back through the knot at the point where the main rope exits the knot.

**2** Keep the tail following the path of the knot around the loops of the original figure eight.

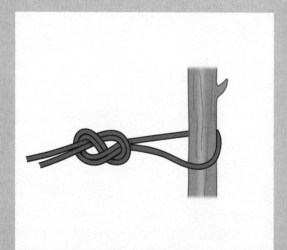

**3** Tighten the knot down, making sure the ropes are not twisted over each other and the knot is "clean."

# SKILL DRILL A-3
## Tying a Water Knot

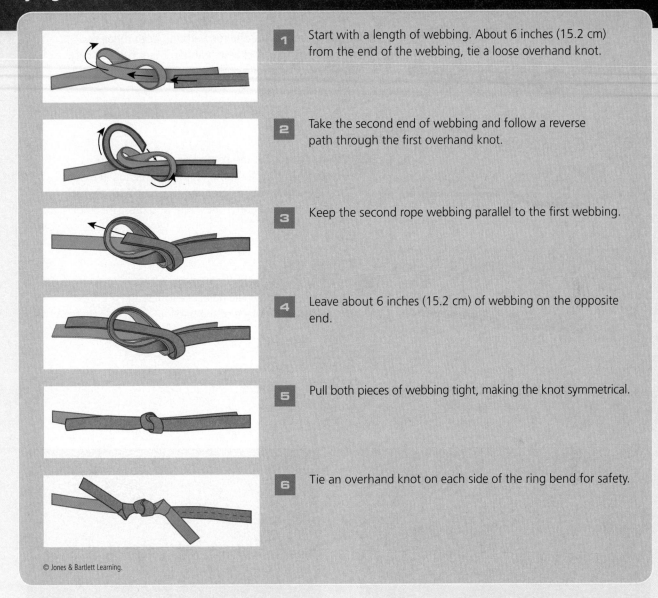

**1** Start with a length of webbing. About 6 inches (15.2 cm) from the end of the webbing, tie a loose overhand knot.

**2** Take the second end of webbing and follow a reverse path through the first overhand knot.

**3** Keep the second rope webbing parallel to the first webbing.

**4** Leave about 6 inches (15.2 cm) of webbing on the opposite end.

**5** Pull both pieces of webbing tight, making the knot symmetrical.

**6** Tie an overhand knot on each side of the ring bend for safety.

© Jones & Bartlett Learning.

often a safety backup. Climbers use Prusik wraps to provide a safety belay while rappelling or to tie into a rope to gain a foothold. A general rule is that the diameter of the Prusik lines used must be two-thirds to three-quarters of the diameter of the larger rope they are attaching to. Prusik bends rely on friction and binding to function properly, and may present slippage problems on frozen ropes. To tie a Prusik bend, follow the steps in Skill Drill A-4. To clear up any confusion, a grapevine knot forms a Prusik loop, and a Prusik bend is what is tied with a Prusik loop. This holds onto the larger rope via friction.

A properly tied and tensioned Prusik bend will grab the larger rope tightly when tensioned, but can be easily moved when the Prusik is loosened (**FIGURE A-2**). Many rescuers will try to run the wraps through quickly, and the barrel of the Prusik flipping over the larger rope will cause tangles. This will not provide the needed friction to properly grab and hold the rope when tension is applied (**FIGURE A-3**).

# SKILL DRILL A-4
## Tying the Grapevine Knot

**1** Wrap the end of one rope around both ropes two full turns in a clockwise manner. Then thread it back through the inside of the two turns.

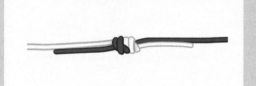

**2** Wrap the end of the second rope two full turns around both ropes in the opposite direction in a counterclockwise manner and thread it back through both turns. Pull both ropes tight and dress the knot.

**3** When tied correctly, the two turns from each half of the knot should lie parallel and flat against one another on one face of the knot.

© Jones & Bartlett Learning.

**FIGURE A-2** A properly dressed Prusik wrap will grab the rope it is tied to using friction from each of the wraps.
Courtesy of Steve Treinish.

**FIGURE A-3** A Prusik wrap that is not cleaned up. The wraps are on top of each other and will not provide a good friction grab. It will likely fail when loaded.
Courtesy of Steve Treinish.

# Common Rope Hardware

Rope systems require hardware. Agencies tend to have preferences for certain brands and types of hardware, but whichever equipment gets the nod from your authority having jurisdiction (AHJ), it is important to keep it maintained and to handle it with care. Dropping it may cause microscopic cracks that can lead to failure later.

# Carabiners

Carabiners may be of the locking or non-locking type, and each style has its merits. Pulling "biners" across ice or rocks may open the gates on non-locking carabiners, but water in locking carabiners may allow them to freeze shut if they get cold enough. Helicopter evolutions are naturally heavy with vibration, so automatic spring-loaded carabiners are recommended there.

Steel carabiners can be used, as can aluminum, but both materials have one weakness: side loading. Side loading places a load on the longer sides of the carabiner, which is not how carabiners are designed to be used. They are engineered to have loads placed on the long axis, and can fail if side loaded. That is one thing to check when tension checking loads. Tension checking puts a small amount of load on the system and allows rescuers to check for potential problems before the full load is applied.

# Descenders

Descending devices, or descenders, can be used to control a load as it is lowered. These devices are used to belay or to control a load at the end of a rope. Water rescue work generally uses three types: figure eight plate (no relation to the knot), bar racks, and automatic belay devices. Many teams also use descending devices as a stop or safety while raising a load, but this can present problems. If the rope is not kept tight while traveling through the descending device, a fall may not be properly arrested.

Figure eight plates are used with the rope run through them and held stationary by an anchor (**FIGURE A-4**). Some have "ears" to limit the tendency of the rope to slide completely off the device.

**FIGURE A-4** Figure eight plates offer a quick setup and small packaging, but do not allow a large amount of friction adjustment.

Courtesy of Steve Treinish.

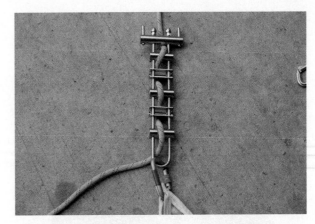

**FIGURE A-5** Brake bars offer good friction and the ability to tailor the needed friction to the device; however, they take a bit longer to set up than a figure eight plate and are a bit larger to carry.

Courtesy of Steve Treinish.

Steel and aluminum are both used to make figure eight plates, and each offers different friction. As with any friction, heat is generated when the rope moves through the device, but in water rescue operations ropes normally do not move fast enough for this to be an issue.

Brake bars, also known as bar racks, also attach to an anchor and control the rope by running it through small bars that connect perpendicularly to a *U*-shaped bar (**FIGURE A-5**). These bars can be added or removed from the system to add or take away friction as needed. An advantage to the bar rack is that bars can be added while a rope is in place. These devices are available in aluminum or steel.

Automatic belay devices are rigged and operated as a safety device, and as the name implies, will automatically lock onto a rope and arrest an uncontrolled descent (**FIGURE A-6**). Once it is locked onto a rope, a lever can be used to reset it and continue. Proper rope rigging and using the diameter of rope for which the belay device is rated are crucial for proper operation.

## SAFETY TIP

Rescuers must be well versed in the rigging and operation of automatic belay devices or multipurpose devices for load arresting, and they must also consider the condition of the rope used. Dirty, older rope may cause frequent lock-ups of these devices.

# Ascenders

Ascenders are used to grab the rope in a manner similar to a Prusik wrap, but they are somewhat easier to move up the rope, because they use a cam system to

**FIGURE A-6** Automatic belay devices automatically lock onto a rope and arrest an uncontrolled descent.
Courtesy of Steve Hudson.

**FIGURE A-7** Ascenders use a cam to grab the rope when it is pulled in one direction, but slide freely over the rope when pulled in the opposite direction.
Courtesy of Pigeon Mountain Industries.

**LISTEN UP!**

Rope hardware can be expensive. Some automatic belay devices can cost upwards of $500, and an entire hardware setup can cost thousands of dollars for the entire cache. An advantage of using basic rope gear is lower replacement cost if equipment is lost in the water.

grab the rope. They are specific to the job, and while there are concerns that putting too much of a load on an ascender too quickly may damage the rope,

ascenders are available that have less aggressive teeth on the cams and thus present a lower chance of damaging the rope (**FIGURE A-7**).

# Rope Systems

True vertical rope work is rare in water rescue, except in victim removal and highline evolutions. Even then, rope rescue technicians often are working with or cross-trained in that discipline, so it is not a huge issue with proper preplanning. The National Fire Protection Association (NFPA), in NFPA 1006, specifies that technician-level swiftwater rescuers must be able to control and haul a load safely, without stressing the system or causing danger to anyone involved. Here is a review of some rope systems, but for further education, or to read about highlines, refer to the text *High-Angle Rope Rescue Techniques.*

## Anchors

Any anchor used in any system should be heavy enough to handle the load (**FIGURE A-8**). Trees are often used around the water, but larger trees should be sought out, and they need to be solid. It does no good to anchor to a tree only to have it snap because it is dead and rotting. Anchoring to a tree farther from the water's edge will typically provide a better anchor. The root system is farther from the bank, which might have been washed out, exposing the roots. If the tree's roots are not in dirt, they will not hold much at all. Anchoring farther from the shore also allows the rescuers to spread out and not congregate at the water's edge, increasing working room and safety.

**FIGURE A-8** Basic anchors should be stout enough to handle any load likely to be placed on the system. Wrapping trees with tubular webbing is common in water rescue.
Courtesy of Steve Treinish.

**FIGURE A-9** Pickets can be used in areas with no anchors available.
© Jones & Bartlett Learning.

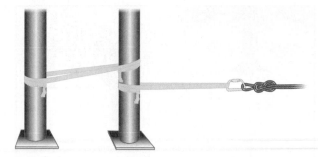

**FIGURE A-10** A back tie uses a connection to combine the two anchors.
© Jones & Bartlett Learning.

**FIGURE A-11** Consider using a vehicle as an anchor.
Courtesy of Steve Treinish.

Some river areas have concrete or steel retaining walls, and many of these walls have anchor rings or steel anchors that offer easy attachment for webbing or rope and also offer great strength. If there is any doubt about the strength or the ability of the anchor to hold the load, find another anchor. But remember, most waterborne rescue rope systems do not suspend live loads and do not generate high forces on anchor systems.

Multiple or load-equalizing anchors can be used, and they provide a measure of safety. If one fails, the others take the load and keep the system intact. Back tying an anchor is merely using a suitable anchor behind another anchor to provide a safety. This is commonly done with webbing or another rope system. The two anchors tied together should be in line with the direction of pull.

Pickets are large, steel spikes that can be driven into the ground, with the area left above the terrain used as an anchor. Multiple pickets should be used if used as an anchor, and you must consider the soil type. As you may imagine, sandy soil will not hold as well as hard clay. Pickets should be driven at least 3 feet (0.9 m) into the ground, at an angle away from the load. Once driven, the pickets should be lashed together to equalize the force across all of them. This back ties the entire picket line together (**FIGURE A-9**).

Multipoint anchors, or load-sharing anchors, utilize two or more separate anchors rigged together (**FIGURE A-10**). In the water environment, trees can often be used to create an anchor that spreads the force of the rope system to multiple trees, using only webbing and carabiners.

Vehicles can be used for anchor systems, and in the rural or wilderness environment can be helpful by literally driving to the point where the anchor is needed. Most fire apparatus is heavy enough to provide good holding power, and steel construction of the main features such as framing, axles, or wheels provide easy attachment points (**FIGURE A-11**).

When using a vehicle as an anchor, remember some key rules:

1. Anchor as low as possible, preferably straight to the frame or to an anchor point if it is properly rated. Ensure webbing does not lie on sharp edges that may cut or fray the webbing. Chain can be used if needed and available.
2. Ensure the object being used as the attachment point is actually attached to a substantial component of the vehicle. Many bumpers have tie-off points, but some may not be attached directly to the frame of the vehicle.
3. Pull to the sides of the truck. If the brakes fail, the truck will not move easily sideways. If the truck is pulled straight back or forward, in line with the wheels, it is much easier to roll, which may cause problems.
4. Ensure the vehicle is not started or moved, either by staging a safety rescuer at the cab or with a lock out/tag out system.

## Change of Direction

A change of direction is simply changing the lane of the pull or lower. Using an anchored pulley, the rope can be pulled at a different angle (**FIGURE A-12**). Consider pulling up a hill with a fence at the top.

**FIGURE A-12** A basic change of direction only changes the direction from which the pull or lower is being applied. It does not change the mechanical advantage of any rope system by itself.

© Jones & Bartlett Learning.

Rescuers might run out of room when pulling up, but if a pulley is installed before the fence, and the area to the left or right is clear, the rope can then be pulled laterally while still delivering the same pull up the hill.

---

**SAFETY TIP**

A good practice is to "dead man" the rope, or to provide some way to stop it from pulling completely through the pulley should it slip. Simply putting a figure eight knot in the end of the rope might stop it at the pulley should a slip occur. Another option is to tie off and anchor the rope somewhere past the farthest point that is expected to be used in the operation.

---

## Lowering Systems

Lowering systems are often used in water rescue when lowering boats downstream. Don't rule out lowering a boat down a bank if that is the best way to get it there; just remember that you must be able to control it.

Most lowering systems are built with an anchor and friction device controlling the rope and the load. In water rescue, we can get even simpler than that by lowering through a change of direction pulley and using muscle, provided we give ourselves an out and use proper safety techniques.

For larger loads, such as the boat mentioned earlier, the friction device provides a way to control the load. A suitable anchor is found, wrapped, and the device is attached to the anchor. The load is attached to the rope, the rope is run through the pulley, and the rope is then held by manpower. A safety just ahead of the pulley should be installed and manned. To bring the load back up, add a piggyback Z-rig on the rope at the load, then use the safety to hold and reset the system.

## Safeties

Any rope pulling on anything should be equipped with a safety device. If rescuers lose their grip on the rope, the safety will do two things: It will keep the victim from falling back to wherever he or she was at the start of the rescue, and it will let rescuers grab and continue the pull without losing too much ground. Highly complex safety systems can be installed, but this text will briefly touch on only the Prusik systems, and will assume a victim is being transported up a low-angle slope. True vertical, live-load rope systems require much more work to set up and operate. This is not that type of work.

Prusik loops are smaller-diameter ropes tied in a 2- to 5-foot-long (0.6- to 1.5-m) loop with a grapevine knot. The Prusik loop becomes a "grab" on the mainline if the pull from the rescuers slip. The Prusik also needs to be anchored to something substantial enough to hold the entire load, and using two of them side by side increases the safety margin. With two, if one fails, the other will assume the load, and this is a highly recommended practice.

To start, the first small Prusik loop is draped over the mainline. The Prusik is then wrapped around the mainline twice or even three times, back through the first loop that was created when the Prusik was laid over the mainline. This forms the Prusik wrap described earlier. The knot is cleaned up, and each loop will tighten into a row of rope wrapped around the mainline. If one of these loops is on the other and not lying flat, the Prusik will slip. Make sure the grapevine knot is not interfering with the friction loops the Prusik is providing. These loops, when tightened down, will bind and clamp the rope, allowing the Prusik to assume the load. To safety a mainline, the Prusik must be kept tight against the anchor, and the mainline rope fed through it. To do this, simply twist and loosen the wraps slightly. The Prusik loop will now slide easily over the mainline, but grab it if allowed to tighten up.

The Prusik loop (or loops if using two) are then attached to a rigged anchor. The Prusik system is now ready to assume the load if something slips, and it must be minded or slid over the traveling mainline by a rescuer. This rescuer instantly lets go of the Prusik if something or somebody slips, and the load then transfers to the Prusik loop. To keep shock loads to the system should something fail, no more than 12 inches (30.5 cm) of slack should ever be found in a Prusik system. It must be kept tight to work properly.

## Prusik Minding Pulleys

One way to install a safety system directly at the pulley is to use a Prusik minding pulley (**FIGURE A-13**). A Prusik wrap is installed onto the rope, and the loop

**FIGURE A-13** Prusik minding pulley.
Courtesy of Pigeon Mountain Industries.

end is attached to the carabiner. The shape of the pulley lets the Prusik slide over the rope as it travels through the pulley, but if direction is reversed, the Prusik grabs the rope.

## Basic Advantage Systems

Commonly known as a Z-rig, the 3:1 advantage system uses a mechanical advantage built with pulleys, Prusiks, and carabiners. It is most commonly used in water rescue to tension a static line stretched across a body of water. Caution should be used when tensioning any rope, as pulling forces (including rescuers) in the system can quickly add up. To rig a static line and apply tension to it using a 3:1 system, follow the steps in **SKILL DRILL A-5**.

There are a few things to remember when using a 3:1 advantage system. First, the entire system is only as strong as its weakest link, and proper safeties can save the static line if something breaks or slips in the rigging. Second, because it is a 3:1 system, every 3 feet pulled by rescuers will pull the mainline only 1 foot. A change of direction pulley for the pull might come in handy. This system can place huge amounts of stress on the rope hardware used to build it, so rescuers must ensure that the hardware used is rated for the task. An NFPA-certified pulley should cover these concerns. Consider using a maximum of three to four rescuers to pull the rope tight. This will further limit the amount of tension placed on the system. Two entire fire companies pulling as tightly as possible can quickly provide too much pull and possibly fail the system.

What if crews run out of room? A long enough pull might end up with the traveling pulley meeting the anchored pulley, or space to pull might be a problem. Use the safety Prusik to grab and hold the mainline as far out as possible. This lets the safety assume the load, and the traveling pulley can then be slid back from the anchored pulley and reset. Now, rescuers can start pulling again until the desired tension is applied (**FIGURE A-14**).

## Complex Rope Systems

When faced with steep terrain and current too fast or dangerous for swimming or watercraft use, rescuers may need to use complex, technician-level rope

# SKILL DRILL A-5
## Rigging a 3:1 Mechanical Advantage System

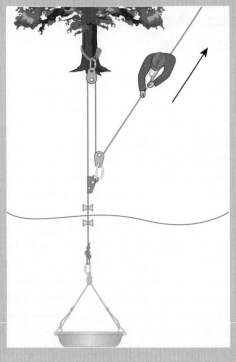

**1** Select two anchors. Anchor the rope end to one of them (e.g., with a tensionless hitch). To the other anchor, normally the river bank with the most room to work, wrap the anchor and attach a pulley with webbing. Run the line through that pulley and back toward the water.

**2** Before the water's edge, attach a Prusik wrap or an ascender and pulley to the main rope coming from the opposite bank. Moving this as close as possible to the shore allows a longer pull without resetting the system. The pull rope will travel 3 feet (0.9 m) for every 1 foot (0.3 m) the main rope travels.

**3** Apply tension by pulling the rope. As rescuers pull the rope, the pulley attached to the rope travels backward with the pull, allowing good tension to be applied. Add a safety device or Prusik wrap ahead of the hardware, and finish by controlling the pulled rope to hold the tension, usually by half hitching the anchor.

© Jones & Bartlett Learning.

systems to lower rescuers down to the victim's location, package or place the victim in a harness or Stokes litter, and be raised and pulled to either shore. This is true rope technician work, and while not covered directly in this text, the basic layout is shown here, as water rescuers might face a situation in which this is the only option.

A highline, or Tyrolean, system is an entire rope system rigged and operated to pull rescuers over a location, lower and raise a victim, then return the victim to shore, all via rope.

The highline system is similar to the MCP in that a static line is used as a track line. Highlines are also commonly called English reeve systems or Tyrolean systems. For this text, we assume a single rescuer will be lowered and raised from the water. Access to lower this rescuer down is gained only from two cliffs that

line the river. The track line is the rope used to support the rescuer, who is riding under a pulley. The pulley is pulled back and forth between the two sides of the cliff. These are called tag lines, and are used to control the side-to-side axis as it travels on the track line.

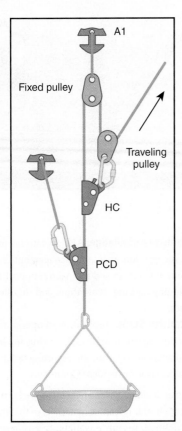

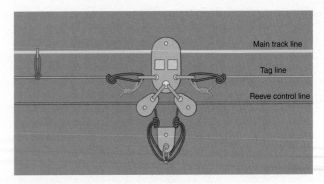

**FIGURE A-15** A typical highline system.
© Jones & Bartlett Learning.

**FIGURE A-14** This figure shows a typical Z-rig, although the finished system is often spaced farther apart. At the middle is the safety, or progress capture system, which assumes the load while the system is manually reset for another pull, or assumes the load if any part of the mechanical advantage system fails.
© Jones & Bartlett Learning.

Because the rescuer must be both lowered and raised, another line—the reeve control—is rigged under the traveling pulley. This lets the same crews on the cliff position move the rescuer across the canyon, then lower him or her to the victim. Stokes-style litters also can be moved in this fashion. Once the rescuer is ready to be hauled up, the reeve line is tightened, the rescuer is raised, and then the tag lines are used to pull the rescuer and victim back to the cliff. Reeve lines are often rigged with a 3:1 mechanical advantage on each end, given the weight of the load that may be faced. It is a complicated evolution, and it can require ingenuity in getting the track line across the distance spanned and to set up the rope systems in areas to allow good communication between crews. **FIGURE A-15** shows the rigging of a typical highline system.

# Appendix B

**An Extract from NFPA 1006, *Standard for Technical Rescue Personnel Professional Qualifications*, 2017 Edition**

## Chapter 16 Surface Water Rescue

**16.1\* Awareness Level.** The job performance requirements defined in 16.1.1 through 16.1.4 shall be met prior to awareness-level qualification in surface water rescue.

**16.1.1** Recognize the need for technical rescue resources at an incident, given AHJ guidelines, an operations- or technician-level incident, so that the need for additional resources is identified, the response system is initiated, the scene is secured and rendered safe until additional resources arrive, and awareness-level personnel are incorporated into the operational plan.

**(A) Requisite Knowledge.** Operational protocols, specific planning forms, types of incidents common to the AHJ, hazards, incident support operations and resources, and safety measures.

**(B) Requisite Skills.** The ability to apply operational protocols, select specific planning forms based on the types of incidents, identify and evaluate various types of hazards, request support and resources, and determine the required safety measures.

**16.1.2** Establish scene safety zones, given an incident, scene security barriers, incident location, incident information, and personal protective equipment (PPE), so that safety zones are designated, zone perimeters are consistent with incident requirements, perimeter markings can be recognized and understood by others, zone boundaries are communicated to Incident Command, and only authorized personnel are allowed access to the scene.

**(A) Requisite Knowledge.** Use and selection of PPE, zone or area control flow and concepts, types of control devices and tools, types of existing and potential hazards, methods of hazard mitigation, organizational standard operating procedure, and staffing requirements.

**(B) Requisite Skills.** The ability to select and use PPE, apply crowd control concepts, position zone control devices, identify and mitigate existing or potential hazards, and personal safety techniques.

**16.1.3** Identify and support an operations- or technician-level incident, given an incident, an assignment, incident action plan, and resources from the tool kit, so that the assignment is carried out, progress is reported to command, environmental concerns are managed, personnel rehabilitation is facilitated, and the incident action plan is supported.

**(A) Requisite Knowledge.** AHJ operational protocols, hazard recognition, incident management, PPE selection, resource selection and use, scene support requirements including lighting, ventilation, and monitoring hazards zones.

**(B) Requisite Skills.** Application of operational protocols, functioning within an IMS, following and implementing an incident action plan, and reporting task progress status to supervisor or Incident Command.

**16.1.4** Size up an incident, given an incident, background information, and applicable reference materials, so that the operational mode is defined, resource availability and response time is considered, types of rescues are determined, the number of victims are identified, the last reported location of all victims is established, witnesses and reporting parties are identified and interviewed, resource needs are assessed, search parameters are identified, and information required to develop an incident action plan is obtained.

**(A) Requisite Knowledge.** Types of reference materials and their uses, risk/benefit assessment, availability and capability of resources, elements of an action plan and related information, relationship of size-up to the incident management system, and information gathering techniques and how that information is used in the size-up process.

**(B) Requisite Skills.** The ability to read specific rescue reference materials, interview people, gather information, relay information, manage witnesses, and use information sources.

**16.2\* Operations Level.** The job performance requirements defined in Section 16.1 and 16.2.1 through 16.2.14 shall be met prior to operations-level qualification in surface water rescue.

**16.2.1\*** Develop a site survey for an existing water hazard, given historical data, specific PPE for conducting site inspections, flood insurance rate maps, tide tables, and meteorological projections, so that life safety hazards are anticipated, risk/benefit analysis is included, site inspections are completed, water conditions are projected, site-specific hazards are identified, routes of access and egress are identified, boat ramps (put-in and take-out points) are identified, method of entrapment is considered, and areas with high probability for victim location are determined.

**(A) Requisite Knowledge.** Requisite contents of a site survey; types, sources, and information provided by reference materials; hydrology and influence of hydrology on rescues; types of hazards associated with water rescue practices scenarios, inspections practices, and considerations techniques; risk/benefit analysis; identification of hazard-specific PPE; factors influencing access and egress routes; behavioral patterns of victims; and environmental conditions that influence victim location.

**(B) Requisite Skills.** The ability to interpret reference materials, perform a scene assessment, evaluate site conditions, complete risk/benefit analysis, and select and use necessary PPE.

**16.2.2\*** Select water rescue PPE, given a water rescue assignment and assorted items of personal protective and life-support equipment, so that the rescuer is protected from temperature extremes and environmental hazards, correct buoyancy is maintained, AHJ protocols are complied with, swimming ability is maximized, routine and emergency communications are established between components of the team, self-rescue needs have been evaluated and provided for, and preoperation safety checks have been conducted.

**(A) Requisite Knowledge.** Manufacturer's recommendations; standard operating procedures; basic signals and communications techniques; selection criteria of insulating garments; buoyancy characteristics; personal escape techniques; applications for and capabilities of personal escape equipment; hazard assessment; AHJ protocols for equipment positioning; classes of personal flotation devices; selection criteria for personal protective clothing, personal flotation devices, and water rescue helmets; personal escape techniques; applications for and capabilities of personal escape equipment; and equipment and procedures for signaling distress.

**(B)\* Requisite Skills.** The ability to use PPE according to the manufacturer's directions, proficiency in emergency escape procedures, proficiency in communications, don and doff equipment in an expedient manner, use preoperation checklists, select personal flotation devices, don and doff personal flotation devices, select water rescue helmets, don and doff water rescue helmets, select personal protective clothing and equipment, don and doff in-water insulating garments, proficiency in emergency escape procedures, and proficiency in communicating distress signals.

**16.2.3\*** Define search parameters for a water rescue incident, given topographical maps of a search area, descriptions of all missing persons and incident history, hydrologic data including speed and direction of current or tides, so that areas with high probability of detection are differentiated from other areas, witnesses are interviewed, critical interview information is recorded, passive and active search tactics are implemented, personnel resources are considered and used, and search parameters are communicated.

**(A) Requisite Knowledge.** Topographical map components, hydrologic factors and wave heights, methods to determine high probability of detection areas, critical interview questions and practices, methods to identify track traps, ways to identify spotter areas and purposes for spotters, personnel available and effects on parameter definition, the effect of search strategy defining parameters, communication methods, and reporting requirements.

**(B) Requisite Skills.** The ability to interpret and correlate reference and size-up information; evaluate site conditions; complete risk/benefit analysis; apply safety, communications, and operational protocols; specify PPE requirements; and determine rescue personnel requirements.

**16.2.4** Develop an action plan for a shore-based rescue of a single or multiple waterbound victim(s), given an operational plan and a water rescue tool kit, so that all information is factored, risk/benefit analysis is conducted, protocols are followed, hazards are identified and minimized, personnel and equipment resources will not be exceeded, assignments are defined, consideration is given to evaluating changing conditions, and the selected strategy and tactics fit the conditions.

**(A) Requisite Knowledge.** Elements of an action plan; types of information provided by reference materials and size-up; hydrology; types of hazards associated with water rescue practices; risk/benefit analysis; identification of hazard-specific PPE; factors influencing access and egress routes; behavioral patterns of victims; environmental conditions that influence victim location; safety, communications, and operational protocols; and resource capability and availability.

**(B) Requisite Skills.** The ability to interpret and correlate reference and size-up information; evaluate site conditions; complete risk/benefit analysis; apply safety, communications, and operational protocols; specify PPE requirements; and determine rescue personnel requirements.

**16.2.5** Conduct a witness interview, given witnesses and checklists, so that witnesses are secured, information is gathered, last seen point can be determined, last known activity can be determined, procedures to re-contact the witnesses are established, and reference objects can be utilized.

**(A) Requisite Knowledge.** Elements of an action plan; types of and information provided by reference materials and size-up; hydrology; types of hazards associated with water rescue practices; risk/benefit analysis; identification of hazard-specific PPE; factors influencing access and egress routes; behavioral patterns of victims; environmental conditions that influence victim location; safety, communications, and operational protocols; and resource capability and availability.

**(B) Requisite Skills.** The ability to interpret and correlate reference and size-up information; evaluate site conditions; complete risk/benefit analysis; apply safety, communications, and operational protocols; specify PPE requirements; and determine rescue personnel requirements.

**16.2.6\*** Deploy a water rescue reach device to a waterbound victim, given required equipment and PPE so that the deployed equipment reaches the victim(s), the rescue equipment does not slip through the rescuer's hands, the victim is moved to the rescuer's shoreline, the victim is not pulled beneath the surface by rescuer efforts, the rescuer is not pulled into the water by the victim, and neither the rescuer nor the victim is tied to or entangled in the device.

**(A) Requisite Knowledge.** Types and capabilities of PPE, effects of hydrodynamic forces on rescuers and victims, physiological effects of immersion, hydrology and characteristics of water, behaviors of waterbound victims, water rescue rope-handling techniques, incident-specific hazard identification, criteria for selecting victim retrieval locations based on water environment and conditions, hazards and limitations of shore-based rescue, local policies and procedures for rescue team activation, and information on local water environments.

**(B) Requisite Skills.** The ability to select PPE specific to the water environment, don PPE, identify water hazards (i.e., upstream or downstream, current or tides), identify hazards directly related to the specific rescue, and demonstrate appropriate shore-based victim removal techniques.

**16.2.7\*** Deploy a water rescue rope to a waterbound victim, given a water rescue rope in a throw bag, a coiled water rescue rope 50 ft to 75 ft (15.24 m to 22.86 m) in length, and PPE, so that the deployed rope lands within reach of the victim, the rescue rope does not slip through the rescuer's hands, the victim is moved to the rescuer's shoreline, the victim is not pulled beneath the surface by rescuer efforts, the rescuer is not pulled into the water by the victim, and neither the rescuer nor the victim is tied to or entangled in the throw line.

**(A) Requisite Knowledge.** Types and capabilities of PPE, effects of hydrodynamic forces on rescuers and victims, hydrology and characteristics of water, behaviors of waterbound victims, water rescue rope-handling techniques, incident-specific hazard identification, criteria for selecting victim retrieval locations based on water environment and conditions, hazards and limitations of shore-based rescue, local policies and procedures for rescue team activation, and information on local water environments.

**(B) Requisite Skills.** The ability to deploy both a water rescue rope bag and a coiled water rescue rope, select PPE specific to the water environment, don PPE, identify water hazards (e.g., upstream or downstream, current or tides), identify hazards directly related to the specific rescue, and demonstrate appropriate shore-based victim removal techniques.

**16.2.8\*** Develop and implement an action plan for the use of watercraft to support the rescue of a single or multiple waterbound victims, given watercraft, trained operator(s), and policies and procedures used by the AHJ, so that watercraft predeployment checks are completed, watercraft launch or recovery is achieved, rescuers are deployed and recovered, both onboard and rescue operations conform with watercraft operational protocols and capabilities, communications are clear and concise, and the candidate is familiar with watercraft nomenclature, operational protocols, design limitations, and launch/recovery site issues.

**(A) Requisite Knowledge.** Entry/exit procedures, communications techniques, boat operation techniques, design limitations, climactic conditions, tides, and currents.

**(B) Requisite Skills.** Implement entry and exit procedures and communications with watercraft crew, use emergency/safety equipment, identify hazards, and operate within the rescue environment.

**16.2.9\*** Define procedures to provide support for helicopter water rescue operations within the area of responsibility for the AHJ, given a helicopter service, operational protocols, helicopter capabilities and limitations, water rescue procedures, and risk factors influencing helicopter operations, so that air-to-ground communications are established and maintained, applications are within the capabilities and skill levels of the helicopter service, the applications facilitate victim extraction from water hazards that are representative of the bodies of water existing or anticipated within the geographic confines of the AHJ, air crew and ground personnel safety are not compromised, landing zones are designated and secured, and fire suppression resources are available at the landing zone.

**(A) Requisite Knowledge.** Local aircraft capabilities and limitations, landing zone requirements, hazards to aircraft, local protocols, procedures for operating around aircraft, dynamics of rescue options, crash survival principles, PPE limitations and selection criteria, ancillary helicopter rescue equipment, and helicopter surf rescue procedures.

**(B) Requisite Skills.** The ability to determine applicability of air operations, establish and control landing zones, assess fire protection needs, communicate with air crews, identify hazards, rig aircraft for anticipated rescue procedures, apply crash survival procedures, select and use PPE, and work with air crews to rescue a victim from the water.

**16.2.10\*** Implement procedures for performing watercraft-based rescue of an incapacitated waterbound victim, as a member of a team, given a water hazard that is representative of the anticipated rescue environment watercraft that is available to the team (if applicable), designated victim packaging and management equipment, and water rescue PPE, so that the control and stability of the watercraft is maintained, risks to the victim and rescuers are minimized, and the victim is removed from the hazard.

**(A) Requisite Knowledge.** Limitations and uses of available watercraft, local environmental entry and exit procedures, parbuckling (rollup) techniques, dynamics of moving water and its effects on watercraft handling, conditional requirements for PPE, and effects of extrication on watercraft handling and stability.

**(B) Requisite Skills.** The ability to move about in a designated watercraft in conditions representative of the anticipated rescue environment while managing the movement of a waterbound victim using techniques identified by the AHJ.

**16.2.11** Demonstrate fundamental survival swimming and self-rescue skills, given safety equipment, props, and a controlled setting representative of the anticipated rescue environment, so that the risk of injury is minimized, flotation is maintained, available PPE is utilized, and egress is accomplished.

**(A) Requisite Knowledge.** Basic forward stroke swimming theory (surface skills).

**(B) Requisite Skills.** Basic swimming skills, including the ability to swim and float in different water conditions with and without flotation aids or swimming aids as required and apply water survival skills; don and doff PPE; select and use PPE, flotation aids, and swim aids; use communications systems; and evaluate water conditions to identify entry points and hazards.

**16.2.12** Identify procedures for operation of rope systems particular to the water rescue needs of the AHJ, given rescue personnel, an established rope system, a load to be moved, and PPE, so that the movement is controlled, the load is held in place when needed, and operating methods do not stress the system.

**(A) Requisite Knowledge.** Ways to determine incident needs as related to the operation of rope systems, capabilities and limitations of various rope systems, incident site evaluation as related to interference concerns and obstacle negotiation, system safety check protocol, procedures to evaluate system components for compromised integrity, common personnel assignments and duties, assignment considerations, common and critical operational commands, common rope system problems and ways to minimize or manage them, and ways to increase the efficiency of load movement.

**(B) Requisite Skills.** The ability to determine incident needs, complete a system safety check, evaluate system components for compromised integrity, select personnel, communicate with personnel, manage movement of the load, and evaluate for potential problems.

**16.2.13** Support operations, given a designated mission, safety equipment, props, and water body, so that skills are demonstrated in a controlled environment, performance parameters are achieved, hazards are continually assessed, correct buoyancy control is maintained, and emergency procedures are demonstrated.

**(A) Requisite Knowledge.** Support procedures, including search patterns, operation support equipment, and communications issues.

**(B) Requisite Skills.** Basic support skills, including the ability to assist technicians in different water conditions including ice, surf, swiftwater conditions, and so forth.

**16.2.14\*** Terminate an incident, given PPE specific to the incident, isolation barriers, and tool kit, so that rescuers and bystanders are protected and accounted for during termination operations; the party responsible is notified of any modification or damage created during the operational period; documentation of loss or material use is accounted for, scene documentation is performed, scene control is transferred to a responsible party; potential or existing hazards are communicated to that responsible party; debriefing and post-incident analysis and critique are considered, and command is terminated.

**(A) Requisite Knowledge.** PPE characteristics, hazard and risk identification, isolation techniques, statutory requirements identifying responsible parties, accountability system use, reporting methods, post incident analysis techniques.

**(B) Requisite Skills.** Selection and use of task and hazard specific PPE, decontamination, use of barrier protection techniques, data collection and recordkeeping/reporting protocols, postincident analysis activities.

**16.3\* Technician Level.** The job performance requirements defined in Section 16.2 and 16.3.1 through 16.3.4 shall be met prior to technician-level qualification in surface water rescue.

**16.3.1\*** Swim a designated water course, given a course designated by the AHJ as demonstrating the capabilities necessary to operate in the anticipated rescue environment, water rescue PPE, and swim aids as required, so that the specified objective is reached, all performance parameters are achieved, movement is controlled, hazards are continually assessed, distress signals are communicated, and rapid intervention for the rescuer has been staged for deployment.

**(A) Requisite Knowledge.** Hydrology and specific hazards anticipated for representative water rescue environments (shoreline, in-water, and climatic), selection criteria for water rescue PPE and swim aids for anticipated water conditions and hazards, and swimming techniques for representative body of water.

**(B) Requisite Skills.** The ability to swim and float over the required distances and necessary duration as outlined in the watermanship test found in Annex K with and without flotation aids or swim aids, apply water survival skills, don and doff PPE, select and use swim aids, use communications systems, and evaluate water conditions to identify entry points and hazards.

**16.3.2\*** Perform a swimming surface water rescue, given a simulated victim, water rescue PPE, conditions representative of the anticipated rescue environment, swim aids as required, flotation aids for victims, and reach/extension devices, so that victim contact is maintained, the rescuer maintains control of the victim, the rescuer and the victim reach safety at a predetermined area, and medical conditions and treatment options are considered.

**(A) Requisite Knowledge.** Hydrology and specific hazards anticipated for representative water rescue environment (shoreline, in-water, and climatic); victim behavior patterns; emergency countermeasures for combative victims; selection criteria for water rescue PPE, swim aids and flotation aids for anticipated water conditions; victim abilities and hazards; swimming techniques for representative bodies of water; and signs, symptoms, and treatment of aquatic medical emergencies.

**(B) Requisite Skills.** The ability to swim and float in different water conditions with and without flotation aids or swim aids; apply water survival skills; manage combative waterbound victims; don and doff PPE; select and use PPE, flotation aids, and swim aids; utilize communications systems; select equipment and techniques for treatment of aquatic medical emergencies; and evaluate water conditions to identify entry points and hazards.

**16.3.3** Demonstrate defensive tactics in the water rescue environment, given a waterbound victim in a stressed or panicked situation so that the rescuer can maintain separation from the victim to create or maintain personal safety and can perform self-defense techniques to prevent rescuer submersion if direct contact is made between a panicked victim and the rescuer.

(A) **Requisite Knowledge.** Basic emergency procedures for applicable environments and situations with stressed or panicked victims at water rescues.

(B) **Requisite Skills.** The ability to release oneself effectively from the grasp of a panicked victim, including blocks, releases, and escapes.

**16.3.4** Supervise, coordinate, and lead rescue teams during operations, given incident checklists, maps, topographic surveys, and charts, so that teams are managed, personnel are supervised, hazards are assessed and identified, safety and health of team is ensured, qualifications/abilities of rescuers are verified, pre-entry briefing is conducted, and debriefing is performed.

(A) **Requisite Knowledge.** Supervisory practices, emergency procedures, communications procedures, local protocols, and safety checks.

(B) **Requisite Skills.** The ability to implement emergency procedures, communications procedures, and leadership/management skills.

## Chapter 17 Swiftwater Rescue

**17.1 Awareness Level.** The job performance requirements defined in 17.1.1 through 17.1.4 shall be met prior to awareness-level qualification in swiftwater rescue.

**17.1.1** Recognize the need for technical rescue resources at an incident, given AHJ guidelines, an operations- or technician-level incident, so that the need for additional resources is identified, the response system is initiated, the scene is secured and rendered safe until additional resources arrive, and awareness-level personnel are incorporated into the operational plan.

(A) **Requisite Knowledge.** Operational protocols, specific planning forms, types of incidents common to the AHJ, hazards, incident support operations and resources, and safety measures.

(B) **Requisite Skills.** The ability to apply operational protocols, select specific planning forms based on the types of incidents, identify and evaluate various types of hazards within the AHJ, request support and resources, and determine the required safety measures.

**17.1.2** Establish scene safety zones, given an incident, scene security barriers, incident location, incident information, and personal protective equipment (PPE), so that safety zones are designated, zone perimeters are consistent with incident requirements, perimeter markings can be recognized and understood by others, zone boundaries are communicated to incident command, and only authorized personnel are allowed access to the scene.

(A) **Requisite Knowledge.** Use and selection of PPE, zone or area control flow and concepts, types of control devices and tools, types of existing and potential hazards, methods of hazard mitigation, organizational standard operating procedure, and staffing requirements.

(B) **Requisite Skills.** The ability to select and use PPE, apply crowd control concepts, position zone control devices, identify and mitigate existing or potential hazards, and personal safety techniques.

**17.1.3** Identify and support an operations- or technician-level incident, given an incident, an assignment, incident action plan, and resources from the tool kit, so that the assignment is carried out, progress is reported to command, environmental concerns are managed, personnel rehabilitation is facilitated, and the incident action plan is supported.

(A) **Requisite Knowledge.** AHJ operational protocols, hazard recognition, incident management, PPE selection, resource selection and use, scene support requirements including lighting, ventilation, and monitoring hazards zones.

(B) **Requisite Skills.** Apply operational protocols, function within an IMS, follow and implement an incident action plan, and report task progress status to supervisor or Incident Command.

**17.1.4** Size up an incident, given an incident, background information, and applicable reference materials, so that the operational mode is defined, resource availability and response time, types of rescues are determined, the number of victims are identified, the last reported locations of all victims are established, witnesses and reporting parties are identified and interviewed, resource needs are assessed, search parameters are identified, and information required to develop an incident action plan is obtained.

(A) **Requisite Knowledge.** Types of reference materials and their uses, risk/benefit assessment, availability and capability of the resources, elements of an action plan and related information, relationship of size-up to the incident management system, and information gathering techniques and how that information is used in the size-up process.

(B) **Requisite Skills.** The ability to read specific rescue reference materials, interview people, gather information, relay information, manage witnesses, and use information sources.

**17.2 Operations Level.** The job performance requirements defined in Section 10.1, 10.2.1 through 10.2.5, Section 16.1 and 16.2.1 through 16.2.13 shall be met prior to operations-level qualification in swiftwater rescue.

**17.2.1** Construct rope systems particular to the swiftwater rescue needs of the AHJ, given rescue personnel, rope equipment, a load to be moved, and PPE, so that the movement is controlled, the load is held in place when needed, and operating methods do not stress the system.

(A) **Requisite Knowledge.** Rope systems specific to the swiftwater environment, capabilities, and limitations of various rope systems, incident site evaluation as related to interference concerns and obstacle negotiation, system safety check protocol, procedures to evaluate system components for compromised integrity, common personnel assignments and duties, common and critical operational commands, and methods to increase the efficiency of load movement.

**(B) Requisite Skills.** The ability to determine incident needs, complete a system safety check, evaluate system components for compromised integrity, select personnel, communicate with personnel, manage movement of the load, and evaluate for potential problems.

**17.2.2** Support operations, given a designated mission, safety equipment, props, and water body, so that skills are demonstrated in a controlled environment, performance parameters are achieved, hazards are continually assessed, and emergency procedures are demonstrated.

**(A) Requisite Knowledge.** Support procedures, including search patterns, equipment setup, operation support equipment, and communications issues.

**(B) Requisite Skills.** Basic support skills, including the ability to serve as an upstream or downstream safety or spotter, and tend a "go" rescuer.

**17.2.3** Assess moving water conditions, characteristics, and features in terms of hazards to the rescuer and victims, given an incident scenario and swiftwater tool kit, so that flow and conditions are estimated accurately, mechanisms of entrapment are considered, hazards are assessed, depth and surrounding terrain are evaluated, and findings are documented.

**(A) Requisite Knowledge.** Flow calculation methods, map or chart reading, local water hazards and conditions, entrapment mechanisms, and human physiology and survival factors.

**(B) Requisite Skills.** Determination of flow and environmental factors, the effects on victims and rescuers, and interpretation of maps or charts.

**17.2.4\*** Perform a nonentry rescue in the swiftwater and flooding environment, given an incident scenario, PPE, and swiftwater rescue tool kit, so that rescue is accomplished, and adopted policies and safety procedures are followed.

**(A) Requisite Knowledge.** Types and capabilities of PPE, effects of hydrodynamic forces on rescuers and victims, hydrology and characteristics of water, behaviors of water-bound victims, water rescue rope-handling techniques, incident-specific hazard identification, criteria for selecting victim retrieval locations based on water environment and conditions, hazards and limitations of shore-based rescue, local policies/procedures for rescue team activation, and information on local water environments.

**(B) Requisite Skills.** Select PPE specific to the water environment, don PPE, identify water hazards (i.e., upstream or downstream, current or tides), identify hazards directly related to the specific rescue, and demonstrate appropriate shore-based victim removal techniques.

**17.2.5\*** Terminate an incident, given PPE specific to the incident, isolation barriers, and tool kit, so that rescuers and bystanders are protected and accounted for during termination operations; the party responsible is notified of any

modification or damage created during the operational period; documentation of loss or material use is accounted for, scene documentation is performed, and scene control is transferred to a responsible party; potential or existing hazards are communicated to that responsible party; debriefing and postincident analysis and critique are considered; and command is terminated.

**(A) Requisite Knowledge.** PPE characteristics, hazard and risk identification, isolation techniques, statutory requirements identifying responsible parties, accountability system use, reporting methods, postincident analysis techniques.

**(B) Requisite Skills.** Selection and use of task and hazard-specific PPE, decontamination, use of barrier protection techniques, data collection and recordkeeping/reporting protocols, postincident analysis activities.

**17.3 Technician Level.** The job performance requirements defined in Section 10.2, 10.3.1 through 10.3.3, and 16.3.1 through 16.3 shall be met prior to technician-level qualification in swiftwater rescue.

**17.3.1** Perform an entry rescue in the swiftwater and flooding environment, given an incident scenario, PPE, and swiftwater rescue tool kit, so that rescue is accomplished, and adopted policies and safety procedures are followed.

**(A) Requisite Knowledge.** Types and capabilities of PPE, effects of hydrodynamic forces on rescuers and victims, hydrology and characteristics of water, behaviors of water-bound victims, water rescue rope-handling techniques, incident-specific hazard identification, criteria for selecting victim retrieval locations based on water environment and conditions, hazards and limitations of shore-based rescue, local policies/procedures for rescue team activation, and information on local water environments.

**(B) Requisite Skills.** Select PPE specific to the water environment, don PPE, identify water hazards (i.e., upstream or downstream, current or tides), identify hazards directly related to the specific rescue, and demonstrate appropriate victim removal techniques.

**17.3.2** Negotiate a designated swiftwater course, given a course that is representative of the bodies of swiftwater existing or anticipated within the geographic confines of the AHJ, water rescue PPE, and swim aids as required, so that the specified objective is reached, all performance parameters are achieved, movement is controlled, hazards are continually assessed, distress signals are communicated, and rapid intervention for the rescuer has been staged for deployment.

**(A) Requisite Knowledge.** Hydrology and specific hazards anticipated for representative water rescue environments (shoreline, in-water, and climatic), selection criteria for water rescue PPE and swim aids for anticipated water conditions and hazards, and swimming techniques for representative body of water.

**(B) Requisite Skills.** The ability to swim and float in different water conditions with and without flotation aids or swim aids as required, apply water survival skills, don and doff PPE, select and use swim aids, utilize communications systems, and evaluate water conditions to identify entry points and hazards.

**17.3.3** Perform a swiftwater rescue from a rescue platform such as a vessel, boat, watercraft, or other waterborne transportation aid while negotiating a designated swiftwater course, given a course that is representative of the bodies of swiftwater existing or anticipated within the geographical confines of the AHJ, water rescue PPE, and swim aids as required, so that the specific objective is reached, all performance parameters are achieved, movement is controlled, hazards are continually assessed, distress signals are communicated, and rapid intervention for the rescuers has been staged for deployment.

**(A) Requisite Knowledge.** The operator and/or crew of any waterborne transportation aid must be knowledgeable in the application and safe operation of the waterborne transportation device and its limitations, and follow all manufacturers' recommendations. The operator and crew of the waterborne transportation aid must comply with all regulatory and applicable laws of safe water transportation according to the AHJ.

**(B) Requisite Skills.** The ability of the operator and crew to enter and exit the waterborne transportation device in a swiftwater condition, to correct a capsized waterborne transportation aid, to assist with safe waterborne transportation operations as members of a swiftwater rescue team on a vessel.

# Chapter 18 Dive Rescue

**18.1 Awareness Level.** The job performance requirements defined in 18.1.1 through 18.1.4 shall be met prior to awareness-level qualification in dive rescue.

**18.1.1** Recognize the need for technical rescue resources at an incident, given AHJ guidelines, an operations- or technician-level incident, so that the need for additional resources is identified, the response system is initiated, the scene is secured and rendered safe until additional resources arrive, and awareness-level personnel are incorporated into the operational plan.

**(A) Requisite Knowledge.** Operational protocols, specific planning forms, types of incidents common to the AHJ, hazards, incident support operations and resources, and safety measures.

**(B) Requisite Skills.** The ability to apply operational protocols, select specific planning forms based on the types of incidents, identify and evaluate various types of hazards within the AHJ, request support and resources, and determine the required safety measures.

**18.1.2** Establish scene safety zones, given an incident, scene security barriers, incident location, incident information, and personal protective equipment (PPE), so that safety zones are designated, zone perimeters are consistent with incident requirements, perimeter markings can be recognized and understood by others, zone boundaries are communicated to incident command, and only authorized personnel are allowed access to the scene.

**(A) Requisite Knowledge.** Use and selection of PPE, zone or area control flow and concepts, types of control devices

and tools, types of existing and potential hazards, methods of hazard mitigation, organizational standard operating procedure, and staffing requirements.

**(B) Requisite Skills.** The ability to select and use PPE, apply crowd control concepts, position zone control devices, identify and mitigate existing or potential hazards, and personal safety techniques.

**18.1.3** Identify and support an operations- or technician-level incident, given an incident, an assignment, incident action plan, and resources from the tool kit, so that the assignment is carried out, progress is reported to command, environmental concerns are managed, personnel rehabilitation is facilitated, and the incident action plan is supported.

**(A) Requisite Knowledge.** AHJ operational protocols, hazard recognition, incident management, PPE selection, resource selection and use, scene support requirements including lighting, ventilation, and monitoring hazards zones.

**(B) Requisite Skills.** Application of operational protocols, function within an incident management system, follow and implement an incident action plan, report task progress status to supervisor or Incident Command.

**18.1.4** Size up an incident, given an incident, background information and applicable reference materials, so that the operational mode is defined, resource availability and response time, types of rescues are determined, the number of victims is ascertained, the last reported location of all victims is established, witnesses and reporting parties are identified and interviewed, resource needs are assessed, search parameters are identified, and information required to develop an incident action plan is obtained.

**(A) Requisite Knowledge.** Types of reference materials and their uses, risk benefit assessment, availability and capability of the resources, elements of an action plan and related information, relationship of size-up to the incident management system, and information gathering techniques and how that information is used in the size-up process.

**(B) Requisite Skills.** The ability to read specific rescue reference materials, interview and gather information, relay information, manage witnesses, and use information sources.

**18.2 Operations Level.** The job performance requirements defined in Section 18.1 and 18.2.1 through 18.2.8 shall be met prior to operations-level qualification in dive rescue.

**18.2.1** Define search parameters for a dive rescue incident, given topographical maps of a search area, descriptions of all missing persons and incident history, and hydrologic data, including speed and direction of current or tides, so that areas likely to contain the subject are differentiated from other areas, witnesses are interviewed, critical interview information is recorded, passive (indirect) and active (direct) search tactics are implemented, personnel resources are considered and used, and search parameters are communicated.

**(A) Requisite Knowledge.** Criteria for determining rescue versus recovery modes, human physiology related to dive environment, re-float theory, topographical map components, hydrologic factors, methods to increase probability of detection, methods to determine areas likely to contain the subject, critical interview questions and practices, methods to identify track traps, ways to identify spotter areas and purposes for spotters, personnel available and effects on parameter definition, the effect of search strategy defining the parameter, communication methods, and reporting requirements.

**(B) Requisite Skills.** The ability to interpret reference materials, perform a scene assessment, evaluate site conditions, complete risk/benefit analysis, and select and use necessary PPE.

**18.2.2\*** Implement an action plan for a dive operation, given an operational plan and a dive rescue tool kit, so that all information is factored, risk/benefit analysis is conducted, protocols are followed, hazards are identified and minimized, personnel and equipment resources will not be exceeded, assignments are defined, consideration is given to evaluating changing conditions, and the selected strategy and tactics fit the conditions.

**(A) Requisite Knowledge.** Elements of an action plan; types of and information provided by reference materials and size-up; hydrology; types of hazards associated with dive rescue practices; risk/benefit analysis; identification of hazard-specific PPE; factors influencing access and egress routes; behavioral patterns of victims; environmental conditions that influence victim location; safety, communications, and operational protocols; and resource capability and availability.

**(B) Requisite Skills.** The ability to interpret and correlate reference and size-up information; evaluate site conditions; complete risk/benefit analysis; apply safety, communications, and operational protocols; specify PPE requirements; determine rescue personnel requirements; and monitor and record submerged diver location, breathing, and dive times.

**18.2.3\*** Implement procedures for use of watercraft in dive operations, given watercraft used by the AHJ, trained operator(s), and the agency's procedures so that watercraft pre-deployment checks are completed; watercraft launch or recovery is achieved as stipulated by AHJ operational protocols; divers are deployed recovered, and protected from harm; both onboard and dive rescue operations conform with watercraft operational protocols and capabilities; communications are clear and concise; and the candidate is familiar with watercraft nomenclature, operational protocols, design limitations, and launch/recovery site issues.

**(A) Requisite Knowledge.** Entry/exit procedures, communications techniques, boat anchoring procedures specific to dive operations, and boat diving operation techniques.

**(B) Requisite Skills.** The ability to implement entry/exit procedures and communications with watercraft crew and use emergency/safety equipment.

**18.2.4** Support entry-level dive rescue operations, given a designated mission, a dive plan, safety equipment, props, and conditions consistent with the anticipated rescue environment, so that communication is maintained with divers while they are on the surface and submerged; status of divers' bottom time, location, repetitive dive status, and the progress of subsurface search operations is tracked and documented; skills are demonstrated in a controlled environment; performance parameters are achieved; hazards are continually assessed; and emergency procedures are demonstrated.

**(A) Requisite Knowledge.** Support procedures, including search patterns, dive equipment setup, operation support equipment, air panels, and communications issues.

**(B) Requisite Skills.** Basic support skills, including the ability to assist technicians in different water conditions, use communication tools, read dive tables, and record necessary information.

**18.2.5\*** Secure the area as a potential crime scene and generate an accurate record of possible evidence and its environment, given paper and pencil, evidence tube or container, marker float, GPS, and last seen point, so that items are secured; possible evidence is preserved by taking notes on, documenting, making sketches of, photographing, or retrieving evidence; chain of custody and evidentiary nature is maintained; and information is passed to law enforcement.

**(A) Requisite Knowledge.** Understand and maintain the "chain of evidence," camera operations, scent article handling and preservation, clue awareness, and specific scene situation considerations (i.e., wreckage, bodies, injury, evidence).

**(B) Requisite Skills.** Interview skills of corroborating witnesses and basic drawing skills.

**18.2.6** Select and assemble PPE to assist rescue divers, given a subsurface mission and personal protective and life-support equipment, so that rescuer is protected from temperature extremes, correct buoyancy is maintained, AHJ protocols are complied with, swimming ability is maximized, routine and emergency communications are established between components of the team, self-rescue needs have been evaluated and provided for, and pre-dive safety checks have been conducted, to include complete encapsulation, including dry suit with attached hood, boots, and gloves and full facemask.

**(A) Requisite Knowledge.** Manufacturer's recommendations, standard operating procedures, basic signals and communications techniques, procedures for the use of electronic communications equipment, selection criteria of insulating garments, buoyancy characteristics, personal escape techniques, applications for and capabilities of personal escape equipment, hazard assessment, and AHJ protocols for equipment positioning.

**(B) Requisite Skills.** The ability to use PPE according to the manufacturer's directions, be proficient in emergency escape procedures, be proficient in communications, don and doff equipment in an expedient manner, and use pre-dive checklists.

**18.2.7\*** Assist a surfaced diver in distress, given safety equipment; PPE; water hazard; and a tired, entrapped, or stressed diver, so that the diver is rescued or assisted, and the victim is extricated from the environment.

**(A) Requisite Knowledge.** Techniques for approach and assistance of surfaced victims or divers, buoyancy control techniques, disentanglement procedures, and communication procedures.

**(B) Requisite Skills.** The ability to use PPE, flotation devices, techniques for rescue or assistance, swimming techniques, and panicked diver evasion techniques.

**18.2.8\*** Terminate an incident, given PPE specific to the incident, isolation barriers, and tool kit, so that rescuers and bystanders are protected and accounted for during termination operations; the party responsible is notified of any modification or damage created during the operational period; documentation of loss or material use is accounted for, scene documentation is performed and scene control is transferred to a responsible party; potential or existing hazards are communicated to that responsible party; debriefing and postincident analysis and critique are considered, and command is terminated.

**(A) Requisite Knowledge.** PPE characteristics, hazard and risk identification, isolation techniques, statutory requirements identifying responsible parties, accountability system use, reporting methods, postincident analysis techniques.

**(B) Requisite Skills.** Selection and use of task and hazard specific PPE, decontamination; use of barrier protection techniques, data collection, and recordkeeping/reporting protocols; postincident analysis activities.

**18.3\* Technician Level.** The job performance requirements defined in Section 18.2 and 18.3.1 through 18.3.10 shall be met prior to technician-level qualification in dive rescue.

**18.3.1** Develop a dive plan, including the projected dive profile, given a predive checklist, dive tables, and a subsurface mission so that elements of the plan, including maximum bottom time, depth limit, minimum reserve breathing air pressure, risk/benefit analysis, hazard-specific equipment, access/egress routes, type of search to be performed, and communication methods, are defined.

**(A) Requisite Knowledge.** Use of references; use of dive tables; searcher limitations; incident management systems resource capabilities; search technique and theory; SCUBA limitations/abilities; float/refloat theory; and movement of a body, or evidence on the surface, during descent, and once on the bottom in still water and, if applicable, in moving water.

**(B) Requisite Skills.** The ability to use dive tables; develop plan; implement incident management; read and interpret maps; interview witnesses; translate information given into a search plan; use communications equipment; define search parameters; determine hydrology, critical interview questions, spotter placement, and strategies; and evaluate bottom topography, composition, debris, water visibility, current, and diver/tender capabilities to determine the safest and most appropriate search pattern.

**18.3.2\*** Select and use PPE, given a subsurface mission and personal protective and life-support equipment, so that rescuer is protected from temperature extremes and environmental hazards, correct buoyancy is maintained, AHJ protocols are complied with, swimming ability is maximized, routine and emergency communications are established between components of the team, self-rescue needs have been evaluated and provided for, predive safety checks have been conducted, and the diver returns to the surface with no less than the minimum specified reserve primary air supply pressure.

**(A) Requisite Knowledge.** Manufacturer's recommendations, standard operating procedures, basic signals and communications techniques, selection criteria of insulating garments, buoyancy characteristics, personal escape techniques, applications for and capabilities of personal escape equipment, hazard assessment, and AHJ protocols for equipment positioning.

**(B) Requisite Skills.** The ability to use PPE according to the manufacturer's directions, be proficient in emergency escape procedures, be proficient in communications, don and doff equipment in an expedient manner, and use predive checklists.

**18.3.3\*** Select and use a standard or full-face mask, given a subsurface mission and personal protective and life-support equipment, so that rescuer is protected from temperature extremes and environmental hazards, correct buoyancy is maintained, AHJ protocols are complied with, swimming ability is maximized, routine and emergency communications are established between components of the team, self-rescue needs have been evaluated and provided for, and predive safety checks have been conducted.

**(A) Requisite Knowledge.** Manufacturer's recommendations, standard operating procedures, basic signals, communications techniques, selection criteria of insulating garments, buoyancy characteristics, personal escape techniques, applications for and capabilities of personal escape equipment, hazard assessment, and AHJ protocols for equipment positioning.

**(B) Requisite Skills.** The ability to use PPE according to the manufacturer's directions, be proficient in emergency escape procedures, be proficient in communications, don and doff equipment in an expedient manner, and use predive checklists.

**18.3.4\*** Negotiate a SCUBA water course, given a SCUBA dive designated course, safety equipment, props, and water body, so that skills are demonstrated in a controlled environment, performance parameters are achieved, hazards are continually assessed, correct buoyancy control is maintained, and emergency procedures are demonstrated.

**(A) Requisite Knowledge.** Basic SCUBA theory (subsurface skills).

**(B) Requisite Skills.** Basic SCUBA skills, including the ability to maneuver using SCUBA in different water conditions, including limited visibility, and apply water survival skills.

**18.3.5** Supervise, coordinate, and lead dive teams during operations, given incident checklists, dive checklists, maps, topographic surveys, charts, and predive/postdive medical evaluation checklist, so that teams are managed, personnel are supervised, hazards are assessed and identified, safety and health of team is ensured, qualifications/abilities of divers are verified, predive briefing is conducted, and postdive medical evaluation and briefing are performed.

(A) **Requisite Knowledge.** Divemaster-level knowledge; knowledge of supervisory practices, dive tables, emergency procedures, communications procedures, local protocols, and predive safety checks.

(B) **Requisite Skills.** The ability to use SCUBA, dive tables, emergency procedures, communication procedures, and leadership and management skills.

**18.3.6\*** Select and use dive rescue equipment, given a dive rescue assignment and assorted items of personal protective and life-support equipment, so that the rescuer is protected from temperature extremes, correct buoyancy is maintained, AHJ protocols are complied with, swimming ability is maximized, routine and emergency communications are established between components of the team, self-rescue needs have been evaluated and provided for, predive safety checks have been conducted, and the diver returns to the surface with no less than the minimum specified reserve primary air supply pressure.

(A) **Requisite Knowledge.** Manufacturer's recommendations, standard operating procedures, basic signals and communications techniques, selection criteria of PPE, including full-face masks, if applicable, and redundant air systems, buoyancy characteristics, personal escape techniques, applications for and capabilities of personal escape equipment, hazard assessment, AHJ protocols for equipment, personal escape techniques, applications for and capabilities of personal escape equipment, and equipment and procedures for signaling distress.

(B) **Requisite Skills.** The ability to use PPE, including full-face mask equipment and redundant air systems, according to the manufacturer's directions; proficiency in emergency escape procedures; proficiency in communications; can don and doff equipment in an expedient manner; use predive checklists; use water rescue PPE, so that the rescuer will be protected from temperature extremes and blunt trauma, the rescuer will have flotation for tasks to be performed, swimming ability will be maximized during rescue activities, self-rescue needs have been evaluated and provided for, and a means of summoning help has been provided; proficiency in emergency escape procedures; and communicating distress signals.

**18.3.7** Manage physiological and psychological stressors in the aquatic environment for the diver and surface support personnel, given a simulated life-threatening situation, so that problems are recognized, corrective actions are initiated, and the situation is stabilized.

(A) **Requisite Knowledge.** Hazard identification and management techniques specific to the stressors and problems present with the environment of public safety diving, and commonly encountered life-threatening problems in the underwater environment.

(B) **Requisite Skills.** Diver monitoring and observation, communication and intervention techniques, use of diver checklists, and diver recall procedure implementation.

**18.3.8\*** Assist a submerged diver in distress, given safety equipment; PPE; and an entrapped, tired, or distressed diver, so that the diver is rescued or assisted, and the victim is extricated from the environment.

(A) **Requisite Knowledge.** Techniques for approach and assistance of conscious and unconscious divers, buoyancy control techniques, out-of-air emergency procedures, use of secondary air systems, procedures for disentanglement, and communications procedures.

(B) **Requisite Skills.** The ability to use PPE, techniques for rescue or assistance of conscious and unconscious divers, buoyancy control devices, regulators, weight belt removal, communication via hand signals, and emergency ascents.

**18.3.9\*** Escape from simulated life-threatening situations, including out-of-air emergencies, entanglements, malfunction of primary air supply source, loss of buoyancy control and disorientation, given safety equipment, a pool or controlled water environment, SCUBA equipment, and props, so that hazards are recognized, emergency procedures are performed, diver escapes from situation to safety, and problems can be identified prior to work in a high-stress environment.

(A) **Requisite Knowledge.** Basic SCUBA emergency procedures for applicable environments and emergency medical treatment protocols for oxygen toxicity, bends, decompression injuries, and other dive-related injuries and illnesses.

(B) **Requisite Skills.** The ability to implement loss of communications procedures; regulator loss, failure, or out-of-air procedures; disentanglement and self-extrication procedures; severed or entangled umbilical or tag line procedures; equipment loss or failure procedures; and emergency treatment of injured divers.

**18.3.10** Perform environment-specific search of the water body, given search parameters for a dive rescue incident, hydrologic data (including speed and direction of current or tides), descriptions of missing persons and incident history, checklists, conditions affecting overlap, pattern selection, water body representative of the AHJ, and safety and SCUBA equipment, so that areas with high probability of detection are differentiated from other areas, witnesses are interviewed, critical interview information is recorded, personnel resources are considered, search parameters are communicated, search is performed, and object is found.

(A) **Requisite Knowledge.** Search theory, environmental considerations, procedures/protocols, hydrologic factors, methods to determine high probabilities of detection areas, and critical interview questions and practices.

(B) **Requisite Skills.** The ability to negotiate a body of water, use rope or items in search, and implement procedures for effective underwater communications.

# Chapter 19 Ice Rescue

**19.1 Awareness Level.** The job performance requirements defined in 19.1.1 through 19.1.4 shall be met prior to awareness-level qualification in ice rescue.

**19.1.1** Recognize the need for technical rescue resources at an incident, given AHJ guidelines, an operations- or technician-level incident, so that the need for additional resources is identified, the response system is initiated, the scene is secured and rendered safe until additional resources arrive, and awareness-level personnel are incorporated into the operational plan.

**(A) Requisite Knowledge.** Operational protocols, specific planning forms, types of incidents common to the AHJ, hazards, incident support operations and resources, and safety measures.

**(B) Requisite Skills.** The ability to apply operational protocols, select specific planning forms based on the types of incidents, identify and evaluate various types of hazards, request support and resources, and determine the required safety measures.

**19.1.2** Establish scene safety zones, given an incident, scene security barriers, incident location, incident information, and personal protective equipment (PPE), so that safety zones are designated, zone perimeters are consistent with incident requirements, perimeter markings can be recognized and understood by others, zone boundaries are communicated to incident command, and only authorized personnel are allowed access to the scene.

**(A) Requisite Knowledge.** Use and selection of PPE, zone or area control flow and concepts, types of control devices and tools, types of existing and potential hazards, methods of hazard mitigation, organizational standard operating procedure, and staffing requirements.

**(B) Requisite Skills.** The ability to select and use PPE, apply crowd control concepts, position zone control devices, identify and mitigate existing or potential hazards, and personal safety techniques.

**19.1.3** Identify and support an operations- or technician-level incident, given an incident, an assignment, incident action plan, and resources from the tool kit, so that the assignment is carried out, progress is reported to command, environmental concerns are managed, personnel rehabilitation is facilitated, and the incident action plan is supported.

**(A) Requisite Knowledge.** AHJ operational protocols, hazard recognition, incident management, PPE selection, resource selection and use, scene support requirements including lighting, ventilation, and monitoring hazards zones.

**(B) Requisite Skills.** Application of operational protocols, function within an IMS, follow and implement an incident action plan, report task progress status to supervisor or Incident Command.

**19.1.4** Size up an incident, given an incident, background information and applicable reference materials, so that the operational mode is defined, resource availability and response time, types of rescues are determined, the number of victims is identified, the last reported location of all victims is established, witnesses and reporting parties are identified and interviewed, resource needs are assessed, search parameters are identified, and information required to develop an incident action plan is obtained.

**(A) Requisite Knowledge.** Types of reference materials and their uses, risk benefit assessment, availability and capability of the resources, elements of an action plan and related information, relationship of size-up to the incident management system, and information gathering techniques and how that information is used in the size-up process.

**(B) Requisite Skills.** The ability to read specific rescue reference materials, interview and gather information, relay information, manage witnesses, and use information sources.

**19.2 Operations Level.** The job performance requirements defined in Section 19.1 and 19.2.1 through 19.2.4 shall be met prior to operations-level qualification in ice rescue.

**19.2.1** Support Level II operations, given a designated mission, safety equipment, props, and water body, so that skills are demonstrated in a controlled environment, performance parameters are achieved, hazards are assessed continually, and emergency procedures are demonstrated.

**(A) Requisite Knowledge.** Support procedures, including search patterns, equipment setup, operation support equipment, and communications issues.

**(B) Requisite Skills.** Basic support skills, including the ability to serve as an upstream or downstream safety or spotter and tend a "go" rescuer.

**19.2.2** Assess ice and water conditions, characteristics, and features in terms of hazards to the rescuer and victims, given an incident scenario and ice rescue tool kit, so that conditions are estimated accurately, mechanisms of entrapment are considered, hazards are assessed, depth and surrounding terrain are evaluated, and findings are documented.

**(A) Requisite Knowledge.** Ice assessment, flow calculation methods, map or chart reading, local water hazards and conditions, entrapment mechanisms, and human physiology and survival factors.

**(B) Requisite Skills.** Determination of flow and environmental factors and the effect on victims and rescuers, and interpretation of maps and charts.

**19.2.3** Perform a nonentry rescue in the ice rescue environment, given an incident scenario, PPE, and ice rescue tool kit, so that rescue is accomplished and adopted policies and safety procedures are followed.

**(A) Requisite Knowledge.** Types and capabilities of PPE, effects of hydrodynamic forces on rescuers and victims, physiological effects of immersion and cold water near-drowning, hydrology and characteristics of water/ice, behaviors of victims, water rescue rope-handling techniques, incident-specific hazard identification, criteria for selecting victim retrieval locations based on water/ice environment and conditions, hazards and limitations of shore-based rescue, local policies/procedures for rescue team activation, and information on local water environments.

**(B) Requisite Skills.** The ability to select PPE specific to the ice rescue environment, don PPE, identify water hazards (e.g., upstream or downstream, current or tides), identify hazards directly related to the specific rescue, and demonstrate appropriate shore-based victim removal techniques.

**19.2.4\*** Terminate an incident, given PPE specific to the incident, isolation barriers, and tool kit, so that rescuers and bystanders are protected and accounted for during termination operations; the party responsible is notified of any modification or damage created during the operational period; documentation of loss or material use is accounted for, scene documentation is performed, scene control is transferred to a responsible party; potential or existing hazards are communicated to that responsible party; debriefing and postincident analysis and critique are considered, and command is terminated.

**(A) Requisite Knowledge.** PPE characteristics, hazard and risk identification, isolation techniques, statutory requirements identifying responsible parties, accountability system use, reporting methods, postincident analysis techniques.

**(B) Requisite Skills.** Selection and use of task and hazard specific PPE, decontamination, use of barrier protection techniques, data collection and recordkeeping/reporting protocols, postincident analysis activities.

**19.3 Technician Level.** The job performance requirements defined in Section 19.2 and 19.3.1 through 19.3.2 shall be met prior to technician-level qualification in ice rescue.

**19.3.1** Demonstrate techniques for movement on ice, given an ice formation that is representative of the bodies of water and ice existing or anticipated within the geographic confines of the AHJ, ice rescue PPE, and swim aids as required, so that the specified objective is reached, all performance parameters are achieved, movement is controlled, hazards are continually assessed, distress signals are communicated, and rapid intervention for the rescuer has been staged for deployment.

**(A) Requisite Knowledge.** Hydrology and specific hazards anticipated for representative ice rescue environments (shoreline, in-water, and climatic), selection criteria for ice rescue PPE and swim aids for anticipated water conditions and hazards, and swimming techniques for representative body of water.

**(B) Requisite Skills.** The ability to swim and float in different water conditions with and without flotation aids or swim aids as required, apply water survival skills, self-rescue with and without use of grip aids in the event of breakthrough, don and doff PPE, select and use swim aids, utilize communications systems, use task-specific equipment, and evaluate water/ice conditions to identify entry points and hazards.

**19.3.2** Perform an entry rescue in the ice rescue environment, given an incident scenario, PPE, and ice rescue tool kit, so that independent positive buoyancy is established for the victim, rescue is accomplished, and adopted policies and safety procedures are followed.

**(A) Requisite Knowledge.** Types and capabilities of PPE, effects of hydrodynamic forces on rescuers and victims, hydrology and characteristics of water, behaviors of victims, physiological effects of immersion and cold water near-drowning, water rescue rope-handling techniques, incident-specific hazard identification, criteria for selecting victim retrieval locations based on water environment and conditions, hazards and limitations of entry rescue, local policies/procedures for rescue team activation, and information on local water environments.

**(B) Requisite Skills.** The ability to select PPE specific to the water/ice environment, don PPE, identify water/ice hazards (i.e., upstream or downstream, current or tides), identify hazards directly related to the specific rescue, and demonstrate appropriate victim removal techniques.

## Chapter 20 Surf Rescue

**20.1 Awareness Level.** The job performance requirements defined in 20.1.1 through 20.1.4 shall be met prior to awareness-level qualification in surf rescue.

**20.1.1** Recognize the need for technical rescue resources at an incident, given AHJ guidelines, an operations- or technician-level incident, so that the need for additional resources is identified, the response system is initiated, the scene is secured and rendered safe until additional resources arrive, and awareness-level personnel are incorporated into the operational plan.

**(A) Requisite Knowledge.** Operational protocols, specific planning forms, types of incidents common to the AHJ, hazards, incident support operations and resources, and safety measures.

**(B) Requisite Skills.** The ability to apply operational protocols, select specific planning forms based on the types of incidents, identify and evaluate various types of hazards within the AHJ, request support and resources, and determine the required safety measures.

**20.1.2** Establish scene safety zones, given an incident, scene security barriers, incident location, incident information, and personal protective equipment (PPE), so that safety zones are designated, zone perimeters are consistent with incident requirements, perimeter markings can be recognized and understood by others, zone boundaries are communicated to incident command, and only authorized personnel are allowed access to the scene.

**(A) Requisite Knowledge.** Use and selection of PPE, zone or area control flow and concepts, types of control devices and tools, types of existing and potential hazards, methods of hazard mitigation, organizational standard operating procedure, and staffing requirements.

**(B) Requisite Skills.** The ability to select and use PPE, apply crowd control concepts, position zone control devices, identify and mitigate existing or potential hazards, and personal safety techniques

**20.1.3** Identify and support an operations- or technician-level incident, given an incident, an assignment, incident action plan, and resources from the tool kit, so that the assignment is carried out, progress is reported to command, environmental concerns are managed, personnel rehabilitation is facilitated, and the incident action plan is supported.

**(A) Requisite Knowledge.** AHJ operational protocols, hazard recognition, incident management, PPE selection, resource selection and use, scene support requirements including lighting, ventilation, and monitoring hazards zones.

**(B) Requisite Skills.** Application of operational protocols, function within an IMS, follow and implement an incident action plan, report task progress status to supervisor or Incident Command.

**20.1.4** Size up an incident, given an incident, background information and applicable reference materials, so that the operational mode is defined, resource availability and response time, types of rescues are determined, the number of victims are identified, the last reported location of all victims are established, witnesses and reporting parties are identified and interviewed, resource needs are assessed, search parameters are identified, and information required to develop an incident action plan is obtained.

**(A) Requisite Knowledge.** Types of reference materials and their uses, risk benefit assessment, availability and capability of the resources, elements of an action plan and related information, relationship of size-up to the incident management system, and information gathering techniques and how that information is used in the size-up process.

**(B) Requisite Skills.** The ability to read specific rescue reference materials, interview and gather information, relay information, manage witnesses, and use information sources.

**20.2 Operations Level.** The job performance requirements defined in Section 20.1 and 20.2.1 through 20.2.5 shall be met prior to operations-level qualification in surf rescue.

**20.2.1** Develop a site survey for an existing surf site, given historical data, PPE for conducting site inspections, rescue equipment for effecting surf rescues, tide tables, currents, and wave heights and meteorological projections, so that life safety hazards are anticipated, risk/benefit analyses are included, site inspections are completed, ocean conditions are projected, site-specific hazards are identified, routes of access and egress are identified, boat ramps are identified, entry and exit points to surf sites are identified, methods of entrapment are considered, and areas with high probability for victim location are determined.

**(A) Requisite Knowledge.** Contents of a site survey; types, sources, and information provided by reference materials; hydrology and influence of hydrology on rescues; types of hazards associated with ocean rescue practice scenarios, inspection practices, and consideration techniques; risk/benefit analyses; identification of PPE; identification of rescue equipment for effecting surf rescues; factors influencing access and egress routes; behavioral patterns of victims; and environmental conditions that influence victim location.

**(B) Requisite Skills.** The ability to interpret reference materials, perform a scene assessment, evaluate site conditions, complete risk/benefit analyses, select and use necessary PPE, and select and use appropriate rescue equipment for effecting surf rescues.

**20.2.2\*** Demonstrate survival swimming skills in low-surf environment, given safety equipment and a water body with low surf, so that basic survival skills are demonstrated in a representative environment as found in the jurisdiction, performance parameters are achieved, and problems can be identified prior to working in a low-surf environment.

**(A) Requisite Knowledge.** Types of fundamental swimming skills to enter a surf zone, maneuver within the surf zone, and exit the surf zone, and fundamental surf knowledge that includes knowing how waves form, why waves are seasonal, how to judge wave heights, recognizing the difference between plunging and spilling waves, knowing the dynamics of surf-related currents such as long-shore and riparian (or rip) currents, and familiarity with the user groups in the surf zone and the types of equipment they use.

**(B) Requisite Skills.** The ability to perform fundamental swimming skills in a surf zone, including the ability to enter, maneuver in, and exit the surf zone; swim in different surf conditions with and without flotation aids or swim aids; apply water survival skills; complete a distance swim in any open body of water using any stroke and without the aid of any surf rescue equipment; identify different types of waves, different types of currents, and the user groups in the surf zone and the types of equipment they use.

**20.2.3\*** Deploy a nonmotorized watercraft and rescue a waterbound surf victim, given watercraft used by the AHJ, so that watercraft predeployment checks are completed, watercraft launch or recovery is achieved as stipulated by AHJ operational protocols, both onboard and surf rescue operations conform with watercraft operational protocols and capabilities, communications are clear and concise, and the candidate is familiar with watercraft nomenclature, operational protocols, design limitations, and launch/recovery site issues.

**(A) Requisite Knowledge.** Types of fundamental skills to enter a surf zone, maneuver within the surf zone, and exit the surf zone on a nonmotorized watercraft; fundamental surf knowledge that includes knowing how waves form, why waves are seasonal, how to judge wave heights, recognizing the difference between plunging and spilling waves, knowing the dynamics of surf-related currents such as longshore and rip currents, and familiarity with the user groups in the surf zone and the types of equipment they use; and victim retrieval and removal techniques.

**(B) Requisite Skills.** The ability to perform fundamental deployment and maneuvering skills in a surf zone on a nonmotorized watercraft, including the ability to enter, maneuver in, and exit the surf zone; perform in different surf conditions; negotiate a measured distance in any open body of water using the watercraft used by the AHJ;

identify different types of waves, different types of currents, and the user groups in the surf zone and the types of equipment they use; and maneuver in the surf zone after retrieving a victim and demonstrate appropriate victim removal techniques.

**20.2.4\*** Define procedures to provide support for surf rescue operations within the area of responsibility for the AHJ, given motorized watercraft used by the AHJ, protocols and procedures, boat-to-shore communication, extraction issues, and safety procedures, so that communications are clear and concise, and the candidate is familiar with boat nomenclature, operational protocols, and design limitations.

**(A) Requisite Knowledge.** Limitations and uses of available boats, dynamics of moving water and its effects on boat handling, launch and docking procedures, conditional requirements for PPE, applications for motorized and appropriate boats, operating hazards as related to conditions, and crew assignments and duties.

**(B) Requisite Skills.** The ability to ride the boat, evaluate conditions for launch, don water rescue PPE, utilize communications systems, and apply procedures for rescuing a victim in the surf zone, including assisting the victim into the boat.

**20.2.5\*** Terminate an incident, given PPE specific to the incident, isolation barriers, and tool kit, so that rescuers and bystanders are protected and accounted for during termination operations; the party responsible is notified of any modification or damage created during the operational period; documentation of loss or material use is accounted for, scene documentation is performed, scene control is transferred to a responsible party; potential or existing hazards are communicated to that responsible party; debriefing and postincident analysis and critique are considered, and command is terminated.

**(A) Requisite Knowledge.** PPE characteristics, hazard and risk identification, isolation techniques, statutory requirements identifying responsible parties, accountability system use, reporting methods, postincident analysis techniques.

**(B) Requisite Skills.** Selection and use of task and hazard-specific PPE, decontamination, use of barrier protection techniques, data collection and recordkeeping/reporting protocols, postincident analysis activities.

**20.3 Technician Level.** The job performance requirements defined in Section 20.2 and 20.3.1 through 20.3.3 shall be met prior to technician-level qualification in surf rescue.

**20.3.1** Demonstrate advanced swimming skills in the surf environment, given safety equipment and a water body with high surf, so that advanced skills are demonstrated in an environment representative of conditions experienced in the jurisdiction, performance parameters and objectives are achieved, and problems can be identified prior to working in a high surf environment.

**(A) Requisite Knowledge.** Types of fundamental swimming skills to enter, maneuver, and exit a surf zone; wave

formation theory; wave types; dynamics of surf-related currents such as longshore and rip currents; and familiarity with victim behavior and recreational equipment used.

**(B) Requisite Skills.** The ability to perform advanced swimming skills in the surf zone, including the ability to enter, maneuver in, and exit the surf zone; swim in different surf conditions with and without flotation aids or swim aids; apply water survival skills; complete a distance swim in any open body of water using any stroke without the aid of any surf rescue equipment; and identify wave types, current types, and potential victim behavior.

**20.3.2** Perform a swimming rescue for a waterbound surf victim, given PPE, including a pair of swimming fins and a surf rescue tube with a shoulder strap, safety equipment, and a water body with high surf representative of the jurisdiction's conditions, so that the victim is secured within the surf rescue tube and towed out of the surf impact zone to shore or to a surf-free zone for pickup by a watercraft, boat, or helicopter.

**(A) Requisite Knowledge.** Types of fundamental swimming skills to enter, maneuver, and exit a surf zone; wave formation theory; wave types; dynamics of surf-related currents such as longshore and rip currents; and familiarity with victim behavior and recreational equipment used.

**(B) Requisite Skills.** The ability to perform advanced swimming skills in the surf zone, including the ability to enter, maneuver in, and exit the surf zone; swim in different surf conditions with and without flotation aids or swim aids; apply water survival skills; complete a distance swim in any open body of water using any stroke without the aid of any surf rescue equipment; identify wave types, current types, and potential victim behavior; and maneuver in the surf zone with a surf rescue tube, tow a victim with the tube, and demonstrate victim removal techniques.

**20.3.3** Perform a subsurface retrieval of a submerged victim in a surf environment, given PPE; swimming fins, mask, and snorkel; and a water body with high surf representative of the jurisdiction's conditions, so that the victim is located and brought to the surface, removed out of the surf impact zone to shore or to a surf-free zone for pickup by a watercraft, boat, or helicopter.

**(A) Requisite Knowledge.** Types of fundamental swimming skills to enter, maneuver, and exit a surf zone; wave formation theory; wave types; dynamics of surf-related currents such as longshore and rip currents; and familiarity with victim behavior and recreational equipment used.

**(B) Requisite Skills.** The ability to perform free diving skills in the surf zone, including the ability to enter, maneuver in, and exit the surf zone while towing a victim; swim in different surf conditions with and without flotation aids or swim aids; apply water survival skills; complete a distance swim in any open body of water using any stroke without the aid of any surf rescue equipment; identify wave types, current types, and potential victim behavior; and maneuver in the surf zone with a surf rescue tube, tow a victim with the tube, and demonstrate victim removal techniques.

# Chapter 21 Watercraft Rescue

**21.1\* Awareness Level.** The job performance requirements defined in 21.1.1 through 21.1.8 shall be met prior to awareness level qualification in watercraft rescue.

**21.1.1** Recognize the need for support resources, given a specific type of rescue incident, so that a resource cache is managed, scene lighting is provided for the tasks to be undertaken, environmental concerns are managed, personnel rehabilitation is facilitated, and the support operation facilitates rescue operational objectives.

**(A) Requisite Knowledge.** Equipment organization and tracking methods, lighting resource type(s), shelter and thermal control options, and rehab criteria.

**(B) Requisite Skills.** The ability to track equipment inventory, identify lighting resources and structures for shelter and thermal protection, select rehab areas, and manage personnel rotations.

**21.1.2** Recognize incident hazards and initiate isolation procedures, given scene control barriers, personal protective equipment (PPE), requisite equipment, and available specialized resources, so that all hazards are identified, resource application fits the operational requirements, hazard isolation is considered, risks to rescuers and victims are minimized, and rescue time constraints are taken into account.

**(A) Requisite Knowledge.** Resource capabilities and limitations, types and nature of incident hazards, equipment types and their use, isolation terminology, methods, equipment and implementation, operational requirement concerns, common types of rescuer and victim risk, risk/benefit analysis methods and practices, and types of technical references.

**(B) Requisite Skills.** The ability to identify resource capabilities and limitations, identify incident hazards, assess victim viability (risk/benefit), utilize technical references, place scene control barriers, and operate control and mitigation equipment.

**21.1.3** Recognize needed resources for a rescue incident, given incident information, a means of communication, resources, tactical worksheets, personnel accountability protocol, applicable references, and standard operating procedures, so that references are utilized, personnel are accounted for, necessary resources are deployed to achieve desired objectives, incident actions are documented, rescue efforts are coordinated, the command structure is established, task assignments are communicated and monitored, and actions are consistent with applicable regulations.

**(A) Requisite Knowledge.** Incident management system (IMS); tactical worksheet application and purposes; accountability protocols; resource types and deployment methods; documentation methods and requirements; availability, capabilities, and limitations of rescuers and other resources; communication problems and needs; communications requirements, methods, and means; types of tasks and assignment responsibilities; policies and procedures of the agency; and technical references related to the type of rescue incident.

**(B) Requisite Skills.** The ability to implement an IMS, complete tactical worksheets, use reference materials, evaluate incident information, match resources to operational needs, operate communications equipment, manage incident communications, and communicate in a manner so that objectives are met.

**21.1.4** Initiate a discipline specific search, given hazard-specific PPE, equipment pertinent to search mission, an incident location, and victim investigative information, so that search parameters are established; the victim profile is established; the entry and exit of all people either involved in the search or already within the search area are questioned and the information is updated and relayed to command; the personnel assignments match their expertise; all victims are located as quickly as possible; applicable technical rescue concerns are managed; risks to searchers are minimized; and all searchers are accounted for.

**(A) Requisite Knowledge.** Local policies and procedures and how to operate in the site-specific search environment.

**(B) Requisite Skills.** The ability to enter, maneuver in, and exit the search environment and provide for and perform self-escape and self-rescue.

**21.1.5\*** Perform ground support operations for helicopter activities, given a rescue scenario/incident, helicopter, operational plans, PPE, requisite equipment, and available specialized resources, so that rescue personnel are aware of the operational characteristics of the aircraft and demonstrate operational proficiency in establishing and securing landing zones and communicating with aircraft personnel until the assignment is complete.

**(A) Requisite Knowledge.** Ground support operations relating to helicopter use and deployment, operation plans for helicopter service activities, type-specific PPE, aircraft familiarization and hazard areas specific to helicopters, scene control and landing zone requirements, aircraft safety systems, and communications protocols.

**(B) Requisite Skills.** The ability to provide ground support operations, review standard operating procedures for helicopter operations, use PPE, establish and control landing zones, and communicate with aircrews.

**21.1.6** Initiate triage of victims, given triage tags and local protocol, so that rescue versus recovery factors are assessed, triage decisions reflect resource capabilities, severity of injuries is determined, and victim care and rescue priorities are established in accordance with local protocol.

**(A) Requisite Knowledge.** Types and systems of triage according to local protocol, resource availability, methods to determine injury severity, ways to manage resources, and prioritization requirements.

**(B) Requisite Skills.** The ability to use triage materials, techniques, and resources and to categorize victims correctly.

**21.1.7** Select and don PPE, given including PFDs, helmets and exposure garments that are consistent with the needs of the incident and type of watercraft so that the wearer is protected from the effects accidental immersion, exposure to the elements and injury from unanticipated movement of the watercraft.

(A) **Requisite Knowledge.** Hazards present on and near the water and aboard watercraft used by the AHJ including those presented by weather, current, and water conditions and the capacities.

(B) **Requisite Skills.** Locate, identify, and don PPE and flotation devices.

**21.1.8** Board and exit a watercraft given a selected watercraft used by the AHJ so that the stability of the craft is not compromised, the possibility of a fall is minimized, and the rescuer is protected from harm.

(A) **Requisite Knowledge.** Elements that affect the stability of watercraft, including mass, center of gravity, weight distribution, impact loads, current, sail area, and wind and water conditions.

(B) **Requisite Skills.** Boarding a watercraft in a manner that prevents injury and minimizes the impact on the stability of the watercraft.

**21.2\* Operations Level.** The job performance requirements defined in Section 21.1 and 21.2.1 through 21.2.18 shall be met prior to operations level qualification in watercraft rescue.

**21.2.1** Identify the types of watercraft, given a list of watercraft used by the organization, so that their limitations, capabilities, load ratings, performance criteria, and considerations for their deployment and recovery in the intended environment are identified.

(A) **Requisite Knowledge.** Types of watercraft used by the organization and the qualities and attributes of each craft that affect how it is utilized in the intended environment; inherent conditions of the intended environment including wind, current, and water conditions that affect vessel selection and use; mission scope and tactical objectives that affect watercraft selection.

(B) **Requisite Skills.** Identify watercraft characteristics such as draft, sail area, method of propulsion, size, weight, method of deployment, and configuration that affect its selection for use in a specific environment for a specific mission.

**21.2.2** Identify the configuration of watercraft given a watercraft available to the agency so that the location of access and egress points, propulsion system components, steering controls, communication equipment, emergency equipment, through hull and deck fittings, portals, and fittings necessary for water- and weathertight integrity are located.

(A) **Requisite Knowledge.** Location of equipment, watercraft components, and configuration of watercraft.

(B) **Requisite Skills.** Identify fittings, portals, and other equipment.

**21.2.3** Use the available methods of communicating between the watercraft and other rescuers in the water, on shore, in other watercraft, and in aircraft as applicable given communication tools, so that routine mission-related information and emergency messages are communicated to the intended recipient.

(A) **Requisite Knowledge.** Methods of communication available to rescuer and their limitations given specific weather conditions, visibility, and distances from the intended recipient.

(B) **Requisite Skills.** Select utilize available communication tools such as radios, hand signals, lights, audible signals, and loud hailers for the specific environment to communicate information.

**21.2.4** Identify conditions that require the notification of local and federal authorities, given conditions that require their involvement, including vessels in distress, hazards to navigation, release of hazardous or toxic substances, and other conditions that affect the health and safety of those in and around navigable waters, so that the proper agency is notified and relevant information is communicated.

(A) **Requisite Knowledge.** Laws, regulations, and standards the identify conditions that require notification of outside agencies, the method of notification, and required other actions.

(B) **Requisite Skills.** Identification of specific conditions that require notification of outside agencies and perform methods for their notification.

**21.2.5** Identify and interpret navigational aids given marine lights, structures and markings on land, other vessels, or on the water, so that nautical landmarks and other vessels are identified, intended course is selected, and collisions are avoided.

(A) **Requisite Knowledge.** Appearance and color of visual aids and navigation markers and their meaning.

(B) **Requisite Skills.** Interpret markers, lights, and signals to determine a course that will avoid other vessels.

**21.2.6** Perform self-rescue and survival swimming skills so that flotation is maintained, body heat is conserved, and egress is accomplished.

(A) **Requisite Knowledge.** Effects of hypothermia and cold water immersion and survival skills.

(B) **Requisite Skills.** Performing techniques to float and move through the water to reach a point of egress or await rescue while conserving body heat.

**21.2.7** Use PPE according to the manufacturer's directions, be proficient in emergency escape procedures, be proficient in communications, don and doff equipment in an expedient manner, use preoperation checklists, select personal flotation devices, don and doff personal flotation devices, select water rescue helmets, don and doff water rescue helmets, select personal protective clothing and equipment, don and doff in-water insulating garments, be proficient in emergency escape procedures, and be proficient in communicating distress signals.

(A) **Requisite Knowledge.** The capabilities and limitations of PPE and personal flotation devices.

(B) **Requisite Skills.** Donning and doffing PPE and personal flotation devices.

**21.2.8** Navigate a watercraft as a helmsman given a watercraft, navigation tools, and a plotted course, so that the course is followed, obstacles and other vessels are avoided, wind and currents accounted for, awareness of position is maintained, and the destination is reached.

**(A) Requisite Knowledge.** Operation of the controls relevant to the watercraft and how they affect speed and direction of the vessel; how the associated navigational tools, such as compass and GPS devices, function and are interpreted; marking on charts or plotters and their meanings; and the effects of local water, wind, and weather conditions on the direction and speed of the watercraft.

**(B) Requisite Skills.** Operate the controls of the watercraft and use the navigational tools and indicators on the watercraft, and select a heading and speed for the vessel for the existing conditions so that it follows its intended course.

**21.2.9** Perform docking or watercraft recovery operations given a watercraft and an operator so that communication is maintained with the operator, current and wind are accounted for mooring lines and fenders are rigged, the dock or slip and watercraft are protected from impacts, and the vessel is positioned properly at the slip and secured from unintended movement.

**(A) Requisite Knowledge.** Methods for securing a watercraft and rigging fenders to prevent damage and minimize undesired movement of the watercraft; means of maneuvering a watercraft using lines or other external systems to position the watercraft as desired; how wind, weather, and water conditions affect watercraft movement as it approaches the slip and after being secured; and considerations for specialized tools or conveyances used to recover watercraft such as trailers, jet docks, and davits.

**(B) Requisite Skills.** Rigging lines and tying knots, bends, and hitches related to mooring; securing and maneuvering vessels into or at a moorage location or conveyance; predicting direction and speed of approach to a moorage conveyance based on the boat operator actions; and the effects of wind, weather, and wave actions.

**21.2.10** Launch or deploy a watercraft from a pier, dock, slip, trailer or other conveyance, given a watercraft and an operator so that communication is maintained with the operator, current and wind are accounted for, mooring lines are managed, and equipment is secured against unintended movement.

**(A) Requisite Knowledge.** Methods for launching or deploying a watercraft and rigging, and securing equipment to prevent damage and minimize undesired movement of the watercraft; means of maneuvering a watercraft using lines or other external systems to position the watercraft as desired; how wind, weather, and water conditions affect watercraft movement as it leaves the slip and after being deployed; and considerations for specialized tools or conveyances used to deploy watercraft such as trailers, jet docks, and davits.

**(B) Requisite Skills.** Rigging lines and tying knots, bends, and hitches related to mooring and maneuvering vessels into or at a moorage location or conveyance; predicting direction and speed of vessel departing from a moorage or conveyance based on the boat operator actions; and the effects of wind, weather, and wave action.

**21.2.11** Perform anchoring operations given a watercraft, an operator, and anchoring equipment, so that the anchor is deployed to prevent vessel movement; and anchor swing, weather, and current and tide change is are accounted for.

**(A) Requisite Knowledge.** Techniques for setting anchor, requirements for anchor size, line length for the vessel and weather conditions, and the effects of vessel movement while at anchor.

**(B) Requisite Skills.** Set an anchor to minimize the potential for drag, and pay out anchor line to ensure proper scope is achieved for weather and tide changes.

**21.2.12** Perform procedures for a crew overboard (COB) event so that the incident is communicated to the operator, visual location of the subject is maintained, the location is marked, and recovery of the subject is accomplished.

**(A) Requisite Knowledge.** Vessel procedures for man overboard, methods of communication of COB event to operator, crew tactics for marking location to assist with returning to location of event, and effects of immersion and hypothermia.

**(B) Requisite Skills.** Deploy a surface marker or utilize other methods for marking the location of the COB event, deploy flotation aid to the member, perform operations specific to maneuvering the vessel and preparing to recover the subject, and perform recovery operations.

**21.2.13** Perform procedures for launching and recovery of "go" rescuers given a watercraft available to the agency, "go" rescuers and a watercraft operator so that the watercraft is not broached, control of the watercraft is maintained so that the rescuers are deployed and recovered at the designated location and are protected from injury.

**(A) Requisite Knowledge.** Watercraft specific procedures for deploying and recovering rescuers, including methods for avoiding contact with propulsion elements of the watercraft and uncontrolled falls or potential for entanglement on entry or exit.

**(B) Requisite Skills.** Rig or configure elements of the watercraft required for launching or recovery of rescuers, manage the operation of propulsion systems and other mechanical elements of the watercraft, and coordinate vessel movement and location so the rescuers are deployed and recovered at the desired location.

**21.2.14** Perform a watercraft-based rescue of an incapacitated water bound victim given a watercraft that is available to the team, a water rescue tool kit, a means of securement, and water rescue PPE, so that the watercraft is not broached; control of the watercraft is maintained; risks to the victim and rescuers are minimized; and the victim is removed from the hazard.

**(A) Requisite Knowledge.** Watercraft-specific procedures for recovering victims, including methods for avoiding contact with propulsion elements of the watercraft and uncontrolled falls or potential for entanglement on recovery.

**(B) Requisite Skills.** Rig or configure elements of the watercraft required for recovery of a victim, manage the operation of propulsion systems and other mechanical elements of the watercraft, and coordinate vessel movement and location so the victim is recovered at the desired location.

**21.2.15** Perform procedures to take another watercraft under tow so that the relative sizes of both watercraft are considered; neither vessel is broached; wind, weather, and water conditions are accounted for; lines are connected between the vessels; maneuverability and control are maintained; and the watercraft is protected from damage.

**(A) Requisite Knowledge.** Watercraft-specific procedures for taking a vehicle under tow, including rigging methods, anchor locations, methods for chafe and impact protection; watercraft handling dynamics while towing; propulsion capacities and impact of wind, weather, and water conditions on combined mass and surface area of both vessels; limitations on size and weight of vessel being towed.

**(B) Requisite Skills.** Rig lines; impact and chafe protection; control movement and direction of watercraft under tow; monitor position and condition of vessel under tow; and communicate with watercraft operator to maneuver the watercraft.

**21.2.16** Perform emergency procedures for a watercraft, given a watercraft available to the agency and emergency equipment so that help is summoned, emergency actions are taken, and risks to the occupants of the vessel are minimized.

**(A) Requisite Knowledge.** Location of emergency equipment such as signaling devices, fire extinguishers, distress beacons, life rafts, PFDs, exposure suits, and other related equipment and how to operate and deploy them.

**(B) Requisite Skills.** Deploy and activate life safety and emergency equipment.

**21.2.17** Conduct dewatering operations, given a watercraft available to the jurisdiction and dewatering equipment so that undesired water is reduced or eliminated in the watercraft, vessel stability is maintained, and damage to the watercraft is prevented.

**(A) Requisite Knowledge.** Watercraft specific dewatering plan, operation of onboard dewatering equipment, and effects of excessive water on stability and seaworthiness of the watercraft.

**(B) Requisite Skills.** Operate onboard dewatering equipment to remove water from the vessel.

**21.2.18\*** Terminate an incident, given PPE specific to the incident, isolation barriers, and tool kit so that rescuers and bystanders are protected and accounted for during termination operations; the party responsible is notified of any modification or damage created during the operational period; documentation of loss or material use is accounted for, scene documentation is performed; scene control is transferred to a responsible party; potential or existing hazards are communicated to that responsible party; debriefing and postincident analysis and critique are considered, and command is terminated.

**(A) Requisite Knowledge.** PPE characteristics, hazard and risk identification, isolation techniques, statutory requirements identifying responsible parties, accountability system use, reporting methods, and postincident analysis techniques.

**(B) Requisite Skills.** Selection and use of task and hazard-specific PPE, decontamination, use of barrier protection techniques, data collection and recordkeeping/reporting protocols, and postincident analysis activities.

**21.3\* Technician Level.** The job performance requirements defined in Section 21.2 and 21.3.1 through 21.3.12 shall be met prior to technician level qualification in watercraft rescue.

**21.3.1** Prepare a watercraft to get underway, given a watercraft available to the agency so that preoperational checks are performed, systems are energized, propulsions systems started, functional checks are conducted, and the watercraft is ready to be deployed.

**(A) Requisite Knowledge.** Watercraft system operational procedures and readiness checks.

**(B) Requisite Skills.** Check proper fluid levels, charges, connections, and lubrication of systems and connections.

**21.3.2** Operate a watercraft to perform tasks typical of the mission defined by the AHJ in conditions representative of the waters and weather common to the jurisdiction, given a watercraft available to the agency so that the objectives are achieved, the occupants and crew are protected from harm, and damage to the watercraft is prevented.

**(A) Requisite Knowledge.** Vessel-specific policies and procedures for operating the watercraft; capabilities and limitations of the watercraft; common wind, weather, and water conditions for the jurisdiction.

**(B) Requisite Skills.** Operate the controls of the watercraft and maneuver to achieve the objective while preventing damage to the watercraft or other vessels.

**21.3.3** Plot a course given navigational tools and charts so that heading, speed, and course are determined and wind, weather, current, and water conditions are taken into account.

**(A) Requisite Knowledge.** How to operate conventional and electronic navigation tools used by the agency.

**(B) Requisite Skills.** Determine location, heading, and speed to achieve the desired course.

**21.3.4** Operate the watercraft while performing docking or watercraft recovery operations, given a watercraft and watercraft crewmember(s) so that communication is maintained with the crew, current and wind are accounted for, and the vessel is positioned properly at the slip and secured from unintended movement.

**(A) Requisite Knowledge.** Methods for maneuvering or approaching moorage; how wind, weather, and water conditions affect watercraft movement as it approaches the slip and after being secured; and considerations for specialized tools or conveyances used to recover watercraft such as trailers, jet docks, and davits.

(B) **Requisite Skills.** Control and maneuver vessel into or at a moorage location or conveyance, predicting direction and speed of approach to a moorage conveyance based on the boat operator actions and the effect of wind, weather, and wave actions.

**21.3.5** Operate a watercraft as it is launched or deployed from a pier, dock, slip, trailer, or other conveyance, given a watercraft and watercraft crewmember(s) so that communication is maintained with the crew, current and wind are accounted for, and damage to the watercraft is prevented.

(A) **Requisite Knowledge.** Methods for maneuvering and operating the watercraft while launching or deploying a watercraft so it is positioned as desired, how wind, weather, and water conditions affect watercraft movement as it leaves the slip and after being deployed; and considerations for specialized tools or conveyances used to deploy watercraft such as trailers, jet docks, and davits.

(B) **Requisite Skills.** Maneuvering while departing the moorage location, predicting direction and speed of departure from a moorage or conveyance based on conditions, the characteristics of the watercraft, and the effect of wind weather and wave action.

**21.3.6** Operate a watercraft to conduct anchoring operations, given a watercraft, watercraft crewmember(s), and anchoring equipment so that the anchor is deployed to prevent vessel movement; and anchor swing, weather, current, and tide change are accounted for.

(A) **Requisite Knowledge.** Techniques for setting anchor, requirements for anchor size and line length for the vessel, weather conditions, and the impact of vessel movement while at anchor.

(B) **Requisite Skills.** Set an anchor to minimize the potential for drag, and pay out anchor line to ensure proper scope is achieved for weather and tide changes.

**21.3.7** Operate a watercraft in response to a COB event, given a watercraft available to the agency and watercraft crewmember(s) so that the incident is communicated to the operator, visual location of the subject is maintained, the location is marked, and recovery of the subject is accomplished.

(A) **Requisite Knowledge.** Vessel procedures for man overboard, methods of communication of COB event to the crew and tactics for noting COB locations to assist with returning to location of event, methods of quickly maneuvering the watercraft back to the COB location.

(B) **Requisite Skills.** Note location of the COB event using traditional or electronic methods, maneuvering the vessel to return to the COB location, and approaching the target area to recover the subject.

**21.3.8** Operate a watercraft to deploy and recover "go" rescuers, given a watercraft available to the agency, "go" rescuers and watercraft crewmember(s) so that the watercraft is not broached; control of the watercraft is maintained so that the rescuers are deployed and recovered at the designated location and are protected from injury.

(A) **Requisite Knowledge.** Watercraft-specific procedures for deploying and recovering rescuers, including methods for avoiding contact with propulsion elements of the watercraft and uncontrolled falls or potential for entanglement on entry or exit.

(B) **Requisite Skills.** Maneuver and control a watercraft, manage the operation of propulsion systems and other mechanical elements of the watercraft, coordinate vessel movement and location so the rescuers are deployed and recovered at the desired location.

**21.3.9** Operate a watercraft to perform a rescue of an incapacitated waterbound victim, given a watercraft that is available to the team, a water rescue tool kit, watercraft crew member(s), a means of securement, and water rescue PPE, so that the watercraft is not broached; control of the watercraft is maintained; risks to the victim and rescuers are minimized; and the victim is removed from the hazard.

(A) **Requisite Knowledge.** Watercraft-specific procedures for recovering victims, including methods for approach, avoiding contact with propulsion elements of the watercraft, and communication methods with crew.

(B) **Requisite Skills.** Maneuver watercraft while approaching waterbound victim, manage the operation of propulsion systems and other mechanical elements of the watercraft, coordinate vessel movement and location so the victim is recovered at the desired location.

**21.3.10** Operate a watercraft with another watercraft under tow, given a watercraft available to the agency and water rescue crewmember(s) so that the relative sizes of both watercraft are considered, ; neither vessel is broached; wind, weather, and water conditions are accounted for; lines are connected between the vessels; maneuverability and control are maintained; and the watercraft is protected from damage.

(A) **Requisite Knowledge.** Watercraft-specific procedures for taking a vehicle under tow; watercraft handling dynamics while towing; propulsion capacities and impact of wind, weather, and water conditions on combined mass and surface area of both vessels; limitations on size and weight of vessel being towed.

(B) **Requisite Skills.** Control movement and direction of the watercraft and the watercraft under tow, monitor position and condition of vessel under tow, and communicate with crewmembers to maneuver the watercraft.

**21.3.11** Operate ancillary navigation and electronic systems given a watercraft available to the agency so that the objective is achieved and the desired information is obtained.

(A) **Requisite Knowledge.** Watercraft- and agency-specific procedures for the use of radar, plotters, and visual aids.

(B) **Requisite Skills.** Operate equipment specific to the watercraft and the agency such as radar, plotters, and forward looking infrared radar (FLIR).

**21.3.12** Shut down and secure a watercraft, given a watercraft available to the agency so that post-shutdown checks are conducted, and the craft is protected from damage and tampering.

**(A) Requisite Knowledge.** Agency-specific procedures for watercraft operations; shutdown procedures for propulsion and ancillary systems; methods of securing craft from unwanted movement, theft, and vandalism; connecting and ensuring shore systems are operational.

**(B) Requisite Skills.** Tie knots, bends, and hitches required to moor or secure craft for long-term storage; use conveyances such as trailers, davits, or jet docks that the agency might use for storing or securing the craft; activate or operate systems that control or maintain the environment inside the craft, such as climate control and bilge monitoring systems; and connect and verify operation of shore support systems such as ac power.

# Chapter 22 Floodwater Rescue

**22.1 Awareness Level.** The job performance requirements defined in 22.1.1 through 22.1.4 shall be met prior to awareness-level qualification in floodwater rescue.

**22.1.1** Recognize the need for technical rescue resources at an incident, given AHJ guidelines, an operations- or technician-level incident, so that the need for additional resources is identified, the response system is initiated, the scene is secured and rendered safe until additional resources arrive, and awareness-level personnel are incorporated into the operational plan.

**(A) Requisite Knowledge.** Operational protocols, specific planning forms, types of incidents common to the AHJ, hazards, incident support operations and resources, and safety measures.

**(B) Requisite Skills.** The ability to apply operational protocols, select specific planning forms based on the types of incidents, identify and evaluate various types of hazards within the response area, request support and resources, and determine the required safety measures.

**22.1.2** Establish scene safety zones, given an incident, scene security barriers, incident location, incident information, and personal protective equipment (PPE), so that safety zones are designated, zone perimeters are consistent with incident requirements, perimeter markings can be recognized and understood by others, zone boundaries are communicated to incident command, and only authorized personnel are allowed access to the scene.

**(A) Requisite Knowledge.** Use and selection of PPE, zone or area control flow and concepts, types of control devices and tools, types of existing and potential hazards, methods of hazard mitigation, organizational standard operating procedure, and staffing requirements.

**(B) Requisite Skills.** The ability to select and use PPE, apply crowd control concepts, position zone control devices, identify and mitigate existing or potential hazards, and personal safety techniques.

**22.1.3** Identify and support an operations- or technician-level incident, given an incident, an assignment, incident action plan, and resources from the tool kit, so that the assignment is carried out, progress is reported to command, environmental concerns are managed, personnel rehabilitation is facilitated, and the incident action plan is supported.

**(A) Requisite Knowledge.** AHJ operational protocols, hazard recognition, incident management, PPE selection, resource selection and use, scene support requirements including lighting, and monitoring hazards zones.

**(B) Requisite Skills.** Application of operational protocols, function within an IMS, follow and implement an incident action plan, report task progress status to supervisor or Incident Command.

**22.1.4** Size up an incident, given an incident, background information and applicable reference materials, so that the operational mode is defined, resource availability, response times, and types of rescues are determined, the number of victims is identified, the last reported locations of all victims are established, witnesses and reporting parties are identified and interviewed, resource needs are assessed, search parameters are identified, and information required to develop an incident action plan is obtained.

**(A) Requisite Knowledge.** Types of reference materials and their uses, risk/benefit assessment, availability and capability of the resources, elements of an action plan and related information, relationship of size-up to the incident management system, and information gathering techniques and how that information is used in the size-up process.

**(B) Requisite Skills.** The ability to read specific rescue reference materials, interview people, gather information, relay information, manage witnesses, and use information sources.

**22.2\* Operations Level.** The job performance requirements defined in Sections 16.1, 16.2, 17.1, 17.2, 22.1, and in 22.2.1 through 22.2.5 shall be met prior to operations-level qualification in floodwater rescue.

**22.2.1** Support technician-level floodwater rescue operations, given a designated mission, safety equipment, props, and water body, so that skills are demonstrated in a controlled environment, performance parameters are achieved, hazards are continually assessed, and emergency procedures are demonstrated.

**(A) Requisite Knowledge.** Support procedures, including search patterns, equipment setup, operating support equipment, and communications systems.

**(B) Requisite Skills.** Basic support skills, including the ability to serve as a safety or spotter and tend a "go" rescuer.

**22.2.2\*** Assess floodwater conditions, characteristics, and features in terms of hazards to the rescuer and victims, given an incident scenario and a floodwater tool kit, so that flow and conditions are estimated, depth and surrounding terrain are evaluated, and findings are documented.

**(A) Requisite Knowledge.** Flow calculation methods, characteristics of floodwater events, map reading, interpreting local terrain data, local water hazards and conditions, entrapment mechanisms, weather forecasts, human physiology and survival factors.

**(B) Requisite Skills.** Assessment of water flow and environmental factors, the ability to acquire and interpret weather forecasts and local terrain data, and evaluate their impact on victims and rescuers.

**22.2.3** Perform a nonentry rescue in the floodwater environment, given an incident scenario, PPE, and a floodwater rescue tool kit, so that rescue is accomplished, and adopted policies and safety procedures are followed.

**(A) Requisite Knowledge.** Types and capabilities of PPE, effects of hydrodynamic forces on rescuers and victims, hydrology and characteristics of water, behaviors of waterbound victims, water rescue rope-handling techniques, incident-specific hazard identification, criteria for selecting victim retrieval locations based on water environment and conditions, hazards and limitations of shore-based rescue, local policies/procedures for rescue team activation, and information on local water environments.

**(B) Requisite Skills.** Select PPE specific to the water environment, don PPE, identify water hazards (i.e., upstream or downstream, current or tides), identify hazards directly related to the specific rescue, and demonstrate appropriate shore-based victim removal techniques.

**22.2.4** Develop and implement an action plan for the use of watercraft to support floodwater search and rescue operations, given a watercraft, trained operator(s), policies, and procedures used by the AHJ, so that floodwater specific hazards are addressed, watercraft predeployment checks are completed, watercraft launch or recovery is achieved, rescuers are deployed and recovered, both onboard and rescue operations conform with watercraft operational protocols and capabilities, communications are clear and concise, and the candidate is familiar with watercraft nomenclature, operational protocols, design limitations, and launch/recovery site issues.

**(A) Requisite Knowledge.** Entry/exit procedures, communications techniques, boat operation techniques, design limitations, climactic conditions, characteristics of floodwater events, and specific hazards presented by floodwater events in the potential rescue environment.

**(B) Requisite Skills.** Implement entry and exit procedures and communications with watercraft crew, use emergency/safety equipment, identify hazards, and operate within the rescue environment.

**22.2.5** Implement an action plan for the use of air assets to support floodwater search and rescue operations, given an action plan, access to air assets, policies, and procedures used by the AHJ, so that floodwater specific hazards are addressed, rescuers are deployed and recovered as required, both onboard and rescue operations conform with aircraft operational protocols and capabilities, communications are clear and concise, and the candidate is familiar with aircraft nomenclature, operational protocols, and design limitations.

**(A) Requisite Knowledge.** Means of contacting and accessing agencies with air assets, the role of aircraft in the support of floodwater events, the limitations of the available aircraft in the conditions associated with the rescue environment, the role of the rescuer as part of an aviation team.

**(B) Requisite Skills.** Implement a notification plan to request air assets, develop a list of tactical objectives to be achieved the aircraft, communicate mission priorities with the aircrew or operator of the aircraft.

**22.2.6\*** Implement measures identified by the AHJ to limit exposure of victims and rescuers from potentially contaminated floodwater given a floodwater event, a flood rescue tool kit, protocols and practices identified by the AHJ, and access to the required engineering controls and decontamination tools so that the sources of potential contamination are identified, and its effects and those of cross contamination are minimized.

**(A) Requisite Knowledge.** Sources of contamination, indicators of the presence of contaminants, methods to limit exposure to contaminated water, and decontamination methods targeted at the potential specific contaminants.

**(B) Requisite Skills.** Use of related engineering controls and PPE and practices that limit an individual's likelihood of exposure to contaminants and implementing methods for the removal of potential contaminants or rendering them inert.

**22.2.7\*** Identify locations at a floodwater search and rescue incident that have a high probability of containing victims, given an incident consistent with the predicted rescue environment and a flood water search and rescue tool kit so that all accessible areas of the incident are surveyed and the victim locations are marked.

**(A) Requisite Knowledge.** Locations that are specifically associated with areas of entrapment or refuge during floodwater events, including the interior of vehicles and attic spaces of structures.

**(B) Requisite Skills.** Assessing and surveying a floodwater environment for potential locations where victims might be trapped or have taken shelter.

**22.2.8\*** Identify and manage the hazards unique to the terrain and environment when covered with floodwater or subject to differential pressures.

**(A) Requisite Knowledge.** Specific hazards that could be present in the floodwater environment that are hidden or covered by water.

**(B) Requisite Skills.** Ability to survey the rescue environment for indicators of potential hazards.

**22.2.9\*** Navigate terrain covered in floodwater given a floodwater incident, a floodwater rescue tool kit, and practices identified by the AHJ so that the positions of the rescuers are known, hazards are avoided, search progress is documented, and geographic baselines are established.

**(A) Requisite Knowledge.** The use and implementation of GPSs and alternate mapping techniques.

**(B) Requisite Skills.** The ability to establish a baseline location using a GPS or other improvised method from which to conduct a search or coordinate the movement of resources and use of methods to determine the location of submerged hazards and geographical features.

**22.2.10** Terminate an incident, given PPE specific to the incident, isolation barriers, and tool kit, so that rescuers and bystanders are protected and accounted for during termination operations; the party responsible is notified of any modification or damage created during the operational period; documentation of loss or material use is accounted for, scene documentation is performed, and scene control is transferred to a responsible party; potential or existing hazards are communicated to that responsible party; debriefing and postincident analysis and critique are considered, and command is terminated.

**(A) Requisite Knowledge.** PPE characteristics, hazard and risk identification, isolation techniques, statutory requirements identifying responsible parties, accountability system use, reporting methods, and postincident analysis techniques.

**(B) Requisite Skills.** Selection and use of task and hazard-specific PPE, decontamination, use of barrier protection techniques, data collection and recordkeeping/reporting protocols, postincide analysis activities.

**22.2.11** Perform a floodwater rescue from a rescue platform such as a vessel, boat, watercraft or other waterborne transportation aid, given a trained operator(s), a course representative of the anticipated rescue environment, water rescue PPE, floodwater rescue tool kit, so that the specific objective is reached, all performance parameters are achieved, movement is controlled, hazards are continually assessed, and any related distress signals are communicated.

**(A) Requisite Knowledge.** The operator and/or crew of any waterborne transportation aid must be knowledgeable in the application and safe operation of the waterborne transportation device and its limitations and follow all manufacturers' recommendations. The operator and crew of the waterborne transportation aid must comply with all regulatory and applicable laws of safe water transportation according to the AHJ.

**(B) (B) Requisite Skills.** The ability of the operator and crew to enter and exit the waterborne transportation device in a floodwater condition, to correct a capsized waterborne transportation aid, and to assist with safe waterborne transportation operations as members of a floodwater rescue team on a vessel.

**22.3 Technician Level.** The job performance requirements defined in Sections 16.3, 22.2, and 22.3.1 shall be met prior to technician-level qualification in floodwater rescue.

**22.3.1\*** Perform an entry rescue in the floodwater environment, given an incident scenario, PPE, and floodwater rescue tool kit, so that rescue is accomplished, and adopted policies and safety procedures are followed.

**(A) Requisite Knowledge.** Types and capabilities of PPE, effects of hydrodynamic forces on rescuers and victims, hydrology and characteristics of water, behaviors of water-bound victims, water rescue rope-handling techniques, incident-specific hazard identification, criteria for selecting victim retrieval locations based on water environment and conditions, hazards and limitations of shore-based rescue, local policies/procedures for rescue team activation, information on local water environments, and methods of breaching or defeating structural components of vehicle or structures.

**(B) Requisite Skills.** Select PPE specific to the water environment, don PPE, identify water hazards (i.e., upstream or downstream, current or tides), identify hazards directly related to the specific rescue, and demonstrate appropriate victim removal techniques.

---

**NFPA 1006 - 2017 Edition**
*Standard for Technical Rescue Personnel Professional Qualifications*
**TIA Log No.:** 1308
**Reference:** 17.2 and 17.3
**Comment Closing Date: October 19, 2017**
**Submitter:** Peter M. Schecter, Oakland Park, FL

*1. Revise 17.2 and 17.3 to read as follows:*

**17.2 Operations Level.** The job performance requirements defined in Section 16.1, 16.2.1 through 16.2.14, Section 17.1, and ~~10.1 10.2.1~~ 17.2.1 through 17.2.5 ~~10.2.5 Section 16.1 and 16.2.1 through 16.2.13~~ shall be met prior to operations-level qualification in swiftwater rescue.

**17.3 Technician Level.** The job performance requirements defined in 16.3.1 through 16.3.4, Section 17.2, and ~~10.2, 10.3.1~~, 17.3.1 through 17.3.3 ~~10.3.3 and 16.3.1 through 16.3.4~~ shall be met prior to technician-level qualification in swiftwater rescue.

**Substantiation.** This change is being made based on language submitted as final work product that never made it into the 2017 version of the document.

**Emergency Nature.** The standard contains an error or an omission that was overlooked during the regular revision process.
The proposed TIA corrects errors in the current published version of the document, which incorrectly refers to or identifies pre-requisites for this chapter. The TIA corrects these errors, and points the user to the correct pre-requisite and location.

**NFPA 1006 - 2017 Edition**
*Standard for Technical Rescue Personnel Professional Qualifications*
**TIA Log No.:** 1309
**Reference:** 22.2
**Comment Closing Date: October 19, 2017**
**Submitter:** Peter M. Schecter, Oakland Park, FL

*1. Revise 22.2 to read as follows:*

**22.2\* Operations Level.** The job performance requirements defined in Sections ~~16.1~~ 16.2, ~~17.1~~, 17.2, and 22.1 and 22.2.1 through ~~22.2.5~~ 22.2.11 shall be met prior to operations-level qualification in floodwater rescue.

**Substantiation.** This corrects an error in the published version of NFPA 1006 (2017) that is not reflective of the final work product of the committee.

**Emergency Nature.** The standard contains an error or an omission that was overlooked during the regular revision process.
The current version of the standard incorrectly identifies and references pre-requisites for this chapter. The proposed TIA corrects this error by referring users to the correct pre-requisites.

# Appendix C

An Extract from NFPA 1670, *Standard on Operations and Training for Technical Search and Rescue Incidents*, 2017 Edition

## Chapter 16 Surface Water Search and Rescue

**16.1 General Requirements.** Organizations operating at surface water search and rescue incidents shall meet the requirements specified in Chapter 4.

### 16.2 Awareness Level.

**16.2.1** Organizations operating at the awareness level at surface water search and rescue incidents shall meet the requirements specified in Section 16.2.

**16.2.2** Each member of an organization operating at the awareness level shall be a competent person as defined in 3.3.21.

**16.2.3** Organizations operating at the awareness level at surface water search and rescue incidents shall implement procedures for the following:
(1) Recognizing the need for surface water search and rescue
(2)* Implementing the assessment phase
(3)* Identifying the resources necessary to conduct safe and effective water operations
(4)* Implementing the emergency response system for surface water rescue incidents
(5)* Implementing site control and scene management
(6)* Recognizing general hazards associated with surface water search and rescue incidents and the procedures necessary to mitigate these hazards within the general search and rescue area
(7)* Determining rescue versus recovery

### 16.3 Operations Level.

**16.3.1** Organizations operating at the operations level at surface water search and rescue incidents shall meet the requirements specified in Section 16.2 and in 16.3.1 through 16.3.7.

**16.3.2** Any member of the organization who could be expected to perform at the operations level for surface water search and rescue shall be provided training to meet the job performance requirements for operations-level surface water rescue as defined in NFPA 1006.

**16.3.3*** Any member of the organization who could be expected to perform functions as a crewmember on a watercraft shall be provided training to meet the job performance requirements for operations-level watercraft operations as defined in NFPA 1006 for the types of watercraft used by the agency under conditions representative of those typically encountered in the work environment.

**16.3.4*** Any member of the organization who could be expected to perform functions as the operator of a watercraft shall be provided training to meet the job performance requirements for technician-level watercraft operations as defined in NFPA 1006 for the types of watercraft used by the agency under conditions representative of those typically encountered in the work environment.

**16.3.5** Organizations operating at the operations level at surface water search and rescue incidents shall develop and implement procedures for performing a risk benefit analysis that shall include the following:
(1)* A survival profile of the potential victim
(2)* A risk profile for the proposed rescue operation

**16.3.6*** Personnel operating in the hazard zone who are not expected to enter the water as part of the rescue plan shall be provided the following minimum PPE:
(1)* PFD or other PPE approved by the AHJ as designed to provide inherent or on-demand positive buoyancy to the user for the expected tasks and conditions encountered in the specific rescue environment
(2) Whistle or other audible signaling device
(3)* Visible signaling device

**16.3.7** Organizations operating at the operations level at surface water search and rescue incidents shall develop and implement procedures for performing nonentry rescue, including the following:
(1)* Initial and ongoing size-up of existing and potential conditions at incidents where surface water search and rescue training and operations will be performed
(2)* Ensuring personal safety at water operations
(3)* Assessing water conditions in terms of hazards to the victim and the rescuer
(4) Separating, isolating, securing, and interviewing witnesses
(5) Evaluating or assessing the potential rescue problems
(6)* Evaluating the progress of the planned response to ensure the objectives are being met
(7)* Conducting shore-based rescue operations

**(8)**\* Using throw bags and related retrieval tools

**(9)**\* Providing assistance to organizations operating at the technician level

**(10)**\* Intervention and self-rescue methods for rescuers who accidentally become immersed

**(11)**\* Identifying and managing heat and cold stress to the rescuer

**(12)** Using packaging devices identified by the AHJ to be employed for removal of water-bound patients

**(13)** Transferring victim information, including location, surroundings, condition when found, present condition, and other pertinent information, to emergency medical services personnel

**(14)**\* Using watercraft-assisted and watercraft-based operations if watercraft are used by the organization

**(15)** Planning to meet operational objectives

**(16)**\* Performing rapid extrication of accessible victims

**(17)** Performing search operations for missing subjects, which do not require the rescuer to enter the water but that identify areas of highest probability and track progress of the search

**(18)**\* Managing incidents that involve waterbound vehicles, vessels, structures, or other circumstances that pose additional challenges to the rescue operation

**(19)** Providing a method for accounting for the location of all responders at the scene and ensuring their welfare

### 16.4 Technician Level.

**16.4.1** Organizations operating at the technician level for surface water search and rescue shall meet the requirements in Sections 16.2, 16.3, and 16.4.

**16.4.2** Any member of the organization who could be expected to perform at the technician level for surface water search and rescue shall be provided training to meet the job performance requirements for technician-level surface water rescue as defined in NFPA 1006.

**16.4.3** Organizations operating at the technician level at surface water search and rescue incidents shall develop and implement the following procedures, which allow for deploying a rescuer or rescuer(s) into the water to conduct a search and rescue task:

**(1)** Performing a risk benefit analysis based on the victim's projected survival profile and the potential risks the operation poses to the responder

**(2)** Using a checklist or other method to ensure all required elements of the rescue plan are in place prior to deploying a rescuer into the water

**(3)**\* Providing an intervention plan with specific methods for rescue or removal of rescuers who become injured or fatigued while in the water

**(4)**\* Conducting a search for a missing victim(s) or victims so that the areas of highest probability are identified and progress of the search can be monitored and documented

**(5)** Methods for managing incidents that involve waterbound vehicles and vessels or other circumstances that pose multiple concurrent challenges to the rescue operation

**(6)** Providing a method to maintain communication or contact with a rescuer(s) in the water so that the rescuer's location is known and assistance can be summoned immediately

# Chapter 17 Swiftwater Search and Rescue

**17.1 General Requirements.** Organizations operating at swiftwater search and rescue incidents shall meet the requirements specified in Chapter 4.

### 17.2 Awareness Level.

**17.2.1** Organizations operating at the awareness level at swiftwater search and rescue incidents shall meet the requirements specified in Section 17.2.

**17.2.2** Each member of an organization operating at the awareness level shall be a competent person as defined in 3.3.21.

**17.2.3** Organizations operating at the awareness level at swiftwater search and rescue incidents shall implement procedures for the following:

**(1)** Recognizing the need for swiftwater search and rescue

**(2)**\* Implementing the assessment phase

**(3)**\* Identifying the resources necessary to conduct swiftwater search and rescue operations

**(4)**\* Implementing the emergency response system for swiftwater search and rescue incidents

**(5)**\* Implementing site control and scene management

**(6)**\* Recognizing general hazards associated with swiftwater search and rescue incidents and the procedures necessary to mitigate these hazards within the general search and rescue area

**(7)** Determining rescue versus recovery

### 17.3 Operations Level.

**17.3.1** Organizations operating at the operations level at swiftwater search and rescue incidents shall meet the requirements specified in Section 16.3 and 17.3.1 through 17.3.4.

**17.3.2** Any member of the organization who could be expected to perform at the operations level for swiftwater search and rescue shall be provided training to meet the job performance requirements for operations-level swiftwater rescue as defined in NFPA 1006.

**17.3.3** Organizations operating at the operations level for swiftwater rescue shall be capable of applying the requirements of Section 8.3 under conditions representative of the swiftwater environment.

**17.3.4** For personnel operating in the hazard zone at a swiftwater search and rescue incident, the minimum PPE provided shall include the following:

**(1)** Personal flotation device (PFD) intended for use in the swiftwater environment

**(2)** Thermal protection

**(3)**\* Helmet appropriate for swiftwater rescue

**(4)** Cutting device that is easily accessible and that will at a minimum cut the ropes and webbing used by the AHJ

**(5)** Whistle or audible signaling device

### 17.4 Technician Level.

**17.4.1** Organizations operating at the technician level for swiftwater search and rescue shall meet the requirements in Section 8.4, Section 17.3, and 17.4.1 through 17.4.5.

**17.4.2** Organizations operating at the technician level for swiftwater search and rescue shall apply the requirements of Section 8.4 under conditions representative of the swiftwater environment.

**17.4.3** Any member of the organization who could be expected to perform at the technician level for swiftwater search and rescue shall be provided training to meet the job performance requirements for technician-level swiftwater rescue as defined in NFPA 1006.

**17.4.4** Organizations operating at the technician level at swiftwater search and rescue incidents shall develop and implement procedures for applying rope rescue techniques in the swiftwater environment.

**17.4.5** Organizations operating at the technician level at swiftwater search and rescue incidents shall have the following capabilities:
- **(1)** Constructing and operating rope rescue system anchors and mechanical advantage systems as specified by the AHJ
- **(2)** Constructing a tension diagonal rope system
- **(3)** Constructing a highline system over water
- **(4)** Constructing and operating rope systems that position and move a tethered boat controlled by ropes

**17.4.6** Organizations operating human-powered watercraft in a swiftwater search and rescue environment shall develop and implement procedures for the use of human-powered watercraft in the swiftwater search and rescue environment.

**17.4.7** Organizations operating motorized watercraft in a swiftwater search and rescue environment shall develop and implement procedures for the use of motorized watercraft in the swiftwater search and rescue environment.

# Chapter 18 Dive Search and Rescue

### 18.1 General Requirements.
Organizations operating at dive search and rescue incidents shall meet the requirements specified in Chapter 4.

### 18.2 Awareness Level.

**18.2.1** Organizations operating at the awareness level at dive search and rescue incidents shall meet the requirements specified in Section 18.2.

**18.2.2** Each member of an organization operating at the awareness level shall be a competent person as defined in 3.3.21.

**18.2.3** Organizations operating at the awareness level at dive search and rescue incidents shall implement procedures for the following:
- **(1)** Recognizing the need for dive search and rescue
- **(2)\*** Implementing the assessment phase
- **(3)\*** Identifying the resources necessary to conduct dive rescue operations
- **(4)\*** Implementing the emergency response system for dive rescue incidents
- **(5)\*** Implementing site control and scene management
- **(6)\*** Recognizing general hazards associated with dive search and rescue incidents and the procedures necessary to mitigate these hazards within the general search and rescue area
- **(7)\*** Determining rescue versus recovery

### 18.3 Operations Level.

**18.3.1** Organizations operating at the operations level at dive search and rescue incidents shall meet the requirements specified in Section 8.3, Section 18.2, and 18.3.1 through 18.3.5.

**18.3.2** Any member of the organization who could be expected to perform at the operations level for dive search and rescue shall be provided training to meet the job performance requirements for operations-level dive rescue as defined in NFPA 1006.

**18.3.3** Organizations operating at the operations level for dive rescue shall be capable of applying the requirements of Section 8.3 under conditions representative of the dive rescue environment.

**18.3.4** For personnel operating in the hazard zone at a dive rescue incident, the minimum PPE provided shall include the following:
- **(1)\*** Personal flotation device (PFD) or other PPE approved by the AHJ as designed to provide inherent or on-demand positive buoyancy to the user for the expected tasks and conditions encountered in the specific rescue environment
- **(2)** Thermal protection
- **(3)** Whistle or audible signaling device
- **(4)** Cutting tool

**18.3.5** Organizations operating at the operations level for dive rescue shall develop and implement procedures for fulfilling the function of a dive tender at a dive rescue incident, including the following:
- **(1)\*** Recognizing the unique hazards associated with dive operations
- **(2)\*** Serving as surface support personnel, including obtaining and assembling the diver's gear, assisting with donning, and performing all pre-entry checks
- **(3)** Identifying water characteristics
- **(4)\*** Operating surface support equipment used in water operations
- **(5)** Procuring the necessary equipment to perform dive operations
- **(6)** Employing techniques for water access, entry, and egress for divers
- **(7)\*** Participating in dive operations at any time of day or in any climate the organization encounters
- **(8)** Recognizing conditions or situations where a diver might need assistance
- **(9)** Implementing standardized contingency procedures for dive-related emergencies, including a diver in distress, a missing or injured diver, and related medical emergencies

**(10)** Providing the necessary medical equipment at the designated egress point to manage medical emergencies commonly associated with compressed gas diving

**(11)** Tracking and documenting status of divers, including bottom time, location, repetitive dive status, and, when possible, depth

**(12)** Using standardized methods to communicate with divers while they are on the surface and while submerged

**(13)** Tracking and documenting the progress of subsurface search operations

## 18.4 Technician Level.

**18.4.1** Organizations operating at the technician level for dive search and rescue shall meet the requirements in Section 8.3 and 18.4.1 through 18.4.11.

**18.4.2** Organizations operating at the technician level for dive rescue shall apply the requirements of Section 16.4 (technician-level surface water rescue) in a manner consistent with the anticipated conditions of the rescue environment.

**18.4.3** Any member of the organization who is recognized as a diver shall be provided training to meet all the job performance requirements for technician-level dive rescue as defined in NFPA 1006.

**18.4.4*** The AHJ shall ensure that all members of the organization who are recognized as divers obtain and maintain current dive certification from an agency or organization recognized as providing a curriculum focused on public safety diving.

**18.4.5*** For all diving members of a technician-level organization, an annual fitness test and a watermanship/skills test and basic scuba skills evaluation supplied by the International Association of Dive Rescue Specialists (IADRS) shall be conducted to maintain public safety diver capability.

**18.4.6** Prior to engaging in subsurface operations, any organization operating at the technician level at dive rescue incidents shall make provisions for the following functions whenever divers are in the water, and these functions shall be exclusive of other duties such as supervision, surface support, and standby resources:

**(1)*** Designating an on-site dive supervisor who has the authority to manage all aspects of the dive operation and has been trained to meet all nondiving job performance requirements of technician-level dive rescue as defined in NFPA 1006

**(2)** Designating a dive tender who is responsible for assisting divers with assembly and donning of equipment, communicating with divers, tracking their location, and managing subsurface search operations and who has been trained to meet all the job performance requirements of operations-level dive rescue as defined in NFPA 1006

**(3)*** Designating a safety diver who is equipped and positioned to immediately submerge and lend assistance to a diver in distress or to engage in a search for a missing diver

**(4)*** Designating a 90 percent diver who is equipped and positioned to quickly enter the water and assume the role of safety diver if necessary

**18.4.7** The agency shall ensure that the following equipment is present at the dive site and readily available prior to engaging in subsurface activities:

**(1)** Medical oxygen and related delivery equipment

**(2)** Backboard or other device suitable for the movement of a nonambulatory diver

**(3)** Means of summoning aid without leaving the dive site

**(4)** A dive flag or float in areas subject to vessel traffic readily visible to vessels approaching the dive location

**(5)** Copy of the agency's dive emergency response plan

**(6)** Audible signaling device

**(7)** Means of immediately recording required information relating to each diver's status and dive profile in a manner that is readily communicated or transferred to other members of the team or medical professionals

**18.4.8** Organizations operating at the technician level at dive incidents shall develop and implement procedures for performing public safety scuba diving, including the following:

**(1)*** Managing a diver's breathing gas supply and bottom time so that on reaching the surface the diver has a minimum reserve pressure that reflects one third of the entire rated capacity of the total primary breathing gas available to the diver and in no case allowing the established minimum reserve pressure for the primary source of breathing gas to be less than 500 psi.

**(2)** Applying an understanding of physics and physiology as they relate to the diver, diver-related emergencies, and the underwater environment

**(3)*** Applying dive tables or other methods designated by the AHJ that use a diver's bottom time and depth to determine his/her level of hyperbaric exposure, including the use of letter group designators, any potential decompression obligation, and the ability to perform repetitive dives

**(4)** Identifying and evaluating underwater environments and conditions to which the public safety diver could be exposed

**(5)** Identifying and managing the hazards posed by underwater plants and animals

**(6)** Conducting and supervising dive operations, including planning a dive based on projected depths, bottom times, and available air supply for a particular mission

**(7)*** Identifying, selecting, and implementing standardized techniques to perform and track the progress of a search that is consistent with the mission of the agency and anticipated conditions that might be encountered in their response area

**(8)*** Using recognized tools, such as a field neurological exam, to identify divers who are experiencing dive-related maladies, including psychological and physiological stress, air embolism, and decompression sickness

**(9)** Recognizing and managing the impact of near-drowning in cold water

**(10)**\* Identifying, selecting, and implementing standardized methods of communicating between a submerged diver and the surface so that the diver can immediately summon help, be recalled to the surface, directed in a search pattern, and warned of imminent hazards

**(11)**\* Utilizing redundant and alternative air sources and techniques during low-air or out-of-air emergencies

**(12)**\* Using full-body encapsulation equipment, including dry suits, dry hoods, and dry gloves, with a full-face mask as required by the AHJ, to protect divers from cold or potentially contaminated water

**(13)**\* Rescuing an entangled diver

**(14)**\* Performing pre- and post-entry medical monitoring of divers

**(15)**\* Recovering evidence, including locating, securing, and packaging evidence, documenting and maintaining the chain of custody, and documenting the scene

**(16)**\* Implementing standardized contingency procedures from the agency's dive emergency response plan for rescue operations in the event of primary diver injury, entrapment, loss of communication, and/or disconnect

**(17)** Using positive connection systems such as chest harnesses and tending lines with quick-release connectors when the use of such systems does not compromise the safety of the diver

**(18)**\* Using standardized written checklists to verify the condition, proper configuration, and operation of a diver's equipment before he/she enters the water

**18.4.9\*** All diving members of the organization shall have a medical exam conducted by a physician with specific training in hyperbaric exposure and dive-related injuries before engaging in dive operations and annually thereafter.

**18.4.10** Organizations operating human-powered watercraft in a dive rescue environment shall develop and implement procedures for the use of human-powered watercraft in the dive rescue environment.

**18.4.11** Organizations operating motorized watercraft in a dive rescue environment shall develop and implement procedures for the use of motorized watercraft in the dive rescue environment.

**18.4.12\*** The AHJ shall ensure that all diving members of the organization complete a subsurface task utilizing tools and tactics identified by the AHJ as consistent with the mission of the team under conditions representative of the rescue environment no less than 4 times over 12 months.

# Chapter 19 Ice Search and Rescue

**19.1** General Requirements. Organizations operating at ice search and rescue incidents shall meet the requirements specified in Chapter 4.

## 19.2 Awareness Level.

**19.2.1** Organizations operating at the awareness level at ice search and rescue incidents shall meet the requirements specified in Section 19.2.

**19.2.2** Each member of an organization operating at the awareness level shall be a competent person as defined in 3.3.21.

**19.2.3** Organizations operating at the awareness level at ice search and rescue incidents shall implement procedures for the following:

**(1)** Recognizing the need for ice search and rescue

**(2)**\* Implementing the assessment phase

**(3)**\* Identifying the resources necessary to conduct ice rescue operations

**(4)**\* Implementing the emergency response system for ice rescue incidents

**(5)**\* Implementing site control and scene management

**(6)**\* Recognizing general hazards associated with ice search and rescue incidents and the procedures necessary to mitigate these hazards within the general search and rescue area

**(7)** Determining rescue versus recovery

## 19.3 Operations Level.

**19.3.1** Organizations operating at the operations level at ice search and rescue incidents shall meet the requirements specified in Section 8.3 and 19.3.1 through 19.3.4.

**19.3.2** Any member of the organization who could be expected to perform at the operations level for ice search and rescue shall be provided training to meet the job performance requirements for operations-level ice rescue as defined in NFPA 1006.

**19.3.3** Organizations operating at the operations level for ice search and rescue shall be capable of applying the requirements of Section 8.3 under conditions representative of the ice rescue environment.

**19.3.4** For personnel operating in the hazard zone at an ice search and rescue incident, the minimum PPE provided shall include the following:

**(1)** Personal flotation device (PFD) or other PPE designed with inherent buoyancy intended for use in the ice rescue environment

**(2)** Thermal protection

**(3)** Whistle or audible signaling device

**(4)** Ice awls/picks

**19.3.5** Organizations operating at the operations level for ice search and rescue incidents shall develop and implement procedures for evaluating ice strength and conditions.

## 19.4 Technician Level.

**19.4.1** Organizations operating at the technician level for ice search and rescue incidents shall meet the requirements in Section 8.4, Section 19.3, and 19.4.1 through 19.4.6.

**19.4.2** Organizations operating at the technician level for ice search and rescue shall apply the requirements of Section 8.4 under conditions representative of the ice rescue environment.

**19.4.3** Any member of the organization who could be expected to perform at the technician level for ice search and rescue shall be provided training to meet the job performance requirements for technician-level ice rescue as defined in NFPA 1006.

**19.4.4** Organizations operating at the technician level at ice search and rescue incidents shall develop and implement procedures for applying specialized tools and rescue techniques for the ice rescue environment.

**19.4.5** Organizations operating human-powered watercraft in an ice search and rescue environment shall develop and implement procedures for the use of human-powered watercraft in the ice search and rescue environment.

**19.4.6** Organizations operating motorized watercraft in an ice search and rescue environment shall develop and implement procedures for the use of motorized watercraft in the ice search and rescue environment.

# Chapter 20 Surf Search and Rescue

## 20.1 General Requirements.

**20.2 Awareness Level.** Organizations operating at surf search and rescue incidents shall meet the requirements specified in Chapter 4.

**20.2.1** Organizations operating at the awareness level at surf search and rescue incidents shall meet the requirements specified in Section 20.2.

**20.2.2** Each member of an organization operating at the awareness level shall be a competent person as defined in 3.3.21.

**20.2.3** Organizations operating at the awareness level at surf search and rescue incidents shall implement procedures for the following:

(1) Recognizing the need for surf search and rescue and conducting nonentry victim location and observation techniques
(2)* Conducting a dynamic size-up and hazard/risk assessment
(3)* Identifying the resources necessary to conduct surf search and rescue operations based on conditions observed
(4)* Implementing the emergency response system for surf search and rescue incidents
(5)* Implementing site control and scene management, including a personnel accountability system
(6)* Recognizing general hazards associated with surf search and rescue incidents and the procedures necessary to mitigate these hazards within the general search and rescue area
(7)* Determining rescue versus recovery

## 20.3 Operations Level.

**20.3.1** Organizations operating at the operations level at surf search and rescue incidents shall meet the requirements specified in Section 8.3 and 20.3.1 through 20.3.4.

**20.3.2** Any member of the organization who could be expected to perform at the operations level for surf search and rescue shall be provided training to meet the job performance requirements for operations-level surf rescue as defined in NFPA 1006.

**20.3.3** Organizations operating at the operations level for surf search and rescue shall be capable of applying the requirements of Section 8.3 under conditions representative of the surf search and rescue environment.

**20.3.4** For personnel operating in the hazard zone at a surf search and rescue incident, the minimum PPE provided shall include a personal flotation device (PFD) or other PPE designed with inherent buoyancy intended for use in the surf search and rescue environment.

**20.3.5** Organizations operating at the operations level for surf search and rescue shall develop and implement procedures for evaluating surf size, strength, and conditions.

## 20.4 Technician Level.

**20.4.1** Organizations operating at the technician level for surf search and rescue shall meet the requirements in Section 8.4, Section 20.3, and 20.4.1 through 20.4.6.

**20.4.2** Organizations operating at the technician level for surf search and rescue shall apply the requirements of Section 8.4 under conditions representative of the surf search and rescue environment.

**20.4.3** Any member of the organization who could be expected to perform at the technician level for surf search and rescue shall be provided training to meet the job performance requirements for technician-level surf rescue as defined in NFPA 1006.

**20.4.4** Organizations operating at the technician level at surf search and rescue incidents shall develop and implement procedures for applying specialized tools and rescue techniques for the surf search and rescue environment.

**20.4.5** Organizations operating human-powered watercraft in a surf search and rescue environment shall develop and implement procedures for the use of human-powered watercraft in the surf search and rescue environment.

**20.4.6** Organizations operating motorized watercraft in a surf search and rescue environment shall develop and implement procedures for the use of motorized watercraft in the surf search and rescue environment.

# Chapter 21 Watercraft Search and Rescue

## 21.1 General Requirements.

**21.1.1** Organizations operating watercraft at search and rescue incidents shall meet the requirements specified in Chapter 4.

**21.1.2*** This chapter outlines the requirements for use of both human-powered and motorized watercraft to perform search and rescue operations.

**21.1.3*** The AHJ shall ensure that the requirements of this section are met in a manner consistent with the water and weather conditions typically associated with the agency's projected mission.

**21.1.4** No part of this section shall be used to abridge or circumvent certifications or licenses legally required to operate specific watercraft in a particular region, city, or state.

## 21.2 Awareness Level.

**21.2.1** Organizations operating at the awareness level at watercraft search and rescue incidents shall meet the requirements specified in Section 21.2.

**21.2.2** Each member of an organization operating at the awareness level shall be a competent person as defined in 3.3.21.

**21.2.3** Organizations operating at the awareness level at watercraft search and rescue incidents shall implement procedures for the following:

**(1)** Recognizing the need for watercraft in a search and rescue operation

**(2)** Implementing the assessment phase

**(3)** Identifying the resources necessary to conduct watercraft search and rescue operations, including launching and recovery sites

**(4)** Implementing the emergency response system for mobilizing search and rescue watercraft

**(5)** Implementing site control and scene management

**(6)** Recognizing general hazards associated with watercraft search and rescue operations and the procedures necessary to mitigate these hazards within the general search and rescue area

**(7)** Determining rescue versus recovery, if possible

## 21.3 Operations Level.

**21.3.1** Organizations operating at the operations level at watercraft search and rescue incidents shall meet the requirements specified in Section 21.2 and 21.3.1 through 21.3.5.

**21.3.2** Any member of the organization who could be expected to perform at the operations level for watercraft search and rescue incidents shall be provided training to meet the job performance requirements for operations-level surface water rescue as defined in NFPA 1006.

**21.3.3\*** Any member of the organization who could be expected to perform functions as a crewmember on a watercraft shall be provided training to meet the job performance requirements for operations-level watercraft operations as defined in NFPA 1006 for the types of watercraft used by the agency under conditions representative of those typically encountered in the work environment.

**21.3.4\*** Personnel operating in or on watercraft who might be exposed to accidental immersion shall wear the following minimum PPE:

**(1)** Personal flotation device (PFD)

**(2)** Whistle or other audible signaling device

**(3)\*** Visible signaling device

**21.3.5\*** Organizations operating at the operations level at watercraft search and rescue incidents shall develop and implement procedures for using watercraft in search and rescue operations, including the following:

**(1)** Identifying the types of watercraft available to the agency and their capabilities, limitations, and any special considerations associated with each type of craft

**(2)** Identifying the roles of crewmembers for each type of watercraft available to the agency

**(3)** Providing for the safety of each crewmember and passenger on the watercraft, including methods for accountability and briefing passengers on emergency procedures

**(4)** Performing an ongoing size-up of existing and potential conditions where watercraft search and rescue operations and training will be performed

**(5)** Assessing water conditions in terms of hazards to the victim and the rescuer and the capability of the watercraft

**(6)\*** Communicating with other agencies or resources that might be part of a watercraft-based search and rescue operation

**(7)\*** Conducting operations to take a vessel under tow with motorized watercraft, if used by the AHJ

**(8)\*** Conducting watercraft-based operations for deploying and recovering rescuers from the water

**(9)\*** Conducting watercraft-based operations for rescuing and recovering both unconscious and conscious waterbound subjects

**(10)\*** Deploying and recovering any watercraft used by the organization

**(11)\*** Deploying crew overboard (COB) measures, including a U.S. Coast Guard–approved Type IV throwable PFD, water rescue throw bags, heaving lines, or similar devices, for passengers or crew who fall overboard

**(12)** Performing watercraft-based search operations that identify areas of highest probability and areas previously searched

**(13)** Managing incidents that involve operating around waterbound vehicles, other vessels, submerged hazards, or other circumstances that pose additional challenges to the rescue operation

**(14)** Identifying navigational aids, such as lights, symbols, or sounds that are used to identify other watercraft, navigational channels, waterway features, or hazards

**(15)** Identifying and utilizing audible and visual distress signals

**(16)** Identifying emergency conditions on the watercraft, such as fire or flooding, and implementing required actions

## 21.4 Technician Level.

**21.4.1** Organizations operating at the technician level at watercraft search and rescue incidents shall meet the requirements specified in Section 21.3 and 21.4.1 through 21.4.3.

**21.4.2\*** Any member of the organization who might be expected to perform functions as the operator of a watercraft shall be provided training to meet the job performance requirements for technician-level watercraft operations as defined in NFPA 1006 for the types of watercraft used by the agency under conditions representative of those typically encountered in the work environment.

**21.4.3\*** Organizations operating at the technician level at watercraft search and rescue incidents shall develop and implement procedures for operating watercraft in search and rescue operations, including the following:

**(1)*** Operating a motorized watercraft with a vessel under tow, if used by the AHJ

**(2)*** Operating a watercraft while deploying and recovering rescuers to and from the water

**(3)*** Operating a watercraft for recovering both unconscious and conscious waterbound subjects

**(4)** Using watercraft-specific navigational systems, tools, and techniques so that the position of the craft can be accurately determined and a desired destination reached

**(5)*** Operating a vessel or watercraft in response to a crew overboard (COB) event, which includes a U.S. Coast Guard–approved Type IV throwable PFD, water rescue throw bags, heaving lines, or similar devices

**(6)** Operating and navigating watercraft in a search operation that identifies areas of highest probability and documents areas previously searched

**(7)** Operating watercraft in environments that include waterbound vehicles, vessels, submerged objects, or other hazards that pose additional challenges to the rescue operation

**(8)** Incorporating the use of navigational aids, such as lights, symbols, or sounds, that are used to identify other watercraft, features, or hazards to reach the intended destination and avoid collisions and groundings

# Chapter 22 Flood Search and Rescue

**22.1 General Requirements.** The AHJ operating at flood search and rescue incidents shall meet the requirements specified in Chapter 4 and Chapter 8.

**22.1.1** The AHJ shall evaluate the need for a missing person search in flood incidents that might occur within its response area.

**22.1.2** The AHJ shall provide a search capability commensurate with the identified needs.

**22.2 Awareness Level.**

**22.2.1** Members of organizations at the awareness level shall be permitted to assist in support functions on a flood search and rescue operation but shall not be deployed into the floodwater-affected areas.

**22.2.2** Organizations operating at the awareness level at any flood search and rescue incident shall have the following capabilities:

**(1)** Recognizing the need for a flood search and rescue–type response

**(2)*** Initiating the emergency response system for flood search and rescue

**(3)*** Initiating incident management systems suitable to the scale and nature of the flood

**(4)*** Recognizing the hazards associated with flood search and rescue incidents

**(5)*** Recognizing the types of floods and the impact to the organization

**(6)*** Recognizing the different phases of a flood and the impact to the organization

**(7)** Recognizing the limitations of emergency response skills and equipment in the flood environments

**(8)** Initiating the collection and recording of information necessary to assist operational personnel in a flood search and rescue incident

**(9)*** Understanding the social, economic, and political issues associated with flood incidents

**(10)*** Recognizing and implementing a search marking system suitable for the flood environment

**22.3 Operations Level.**

**22.3.1** Organizations operating at the operations level at flood search and rescue incidents shall meet the requirements specified in Section 8.3 and 22.3.1 through 22.3.7.

**22.3.2** Any member of the organization who could be expected to perform at the operations level for flood search and rescue shall be provided training to meet the job performance requirements for operations-level flood rescue as defined in NFPA 1006.

**22.3.3** Organizations operating at the operations level for flood search and rescue shall be capable of applying the requirements of Section 8.3 under conditions representative of the flood environment.

**22.3.4** For personnel operating in the hazard zone at a flood search and rescue, the minimum PPE provided shall include the following:

**(1)** Personal flotation device (PFD) intended for use in the flood environment

**(2)** Thermal protection

**(3)** Cutting device that is easily accessible and that will at a minimum cut the ropes and webbing used by the AHJ

**(4)** Whistle or audible signaling device

**(5)** PPE consistent with expected contaminated water

**22.3.5*** Organizations operating at the operations level shall be capable of operating at flood incidents that are limited to requiring a response based on surface water search and rescue operations capabilities on and around flood-affected areas.

**22.3.6** Organizations at the operations level shall be permitted to support organizations operating at the technician level but shall not deploy into higher risk, difficult, or complex flood environments.

**22.3.7** Organizations operating at the operations level at flood search and rescue incidents shall develop and implement procedures for the following:

**(1)** Identifying flood characteristics specific to the cause of the flooding and the geographic area flooded

**(2)** Operating surface support equipment used in flood search and rescue operations

**(3)*** Identifying and operating watercraft appropriate for use in the flood environment

**(4)*** Navigating through the flood-affected area

**(5)** Identifying potential sources of floodwater contamination

**(6)** Implementing decontamination procedures for personnel, casualties, and equipment

## 22.4 Technician Level.

**22.4.1** Organizations operating at the technician level for flood search and rescue incidents shall meet the requirements in Section 8.4, Section 22.3, and 22.4.1 through 22.4.7.

**22.4.2** Organizations operating at the technician level for flood search and rescue incidents shall apply the requirements of Section 8.4 under conditions representative of the flood environment.

**22.4.3** Any member of the organization who might be expected to perform at the technician level for flood search and rescue shall be provided training to meet the job performance requirements for technician-level flood rescue as defined in NFPA 1006.

**22.4.4** Organizations operating at the technician level at flood search and rescue incidents shall be capable of operating in, on, and around higher risk, difficult, or complex flood environments and shall have the following capabilities:

**(1)** Recognizing higher risk, difficult, or complex flood environments, and implementing systems to maximize the safety of responders

**(2)** Conducting search operations in areas affected by flood waters, including building and structure entries, as required to support the task

**(3)** Performing extrication and rescue operations involving packaging, treating, and removing victims trapped by floodwaters

**(4)** Transporting victims to a location where they can be removed from the flood-affected area

**22.4.5** Organizations operating at the technician level at flood search and rescue incidents shall meet the requirements specified in this chapter, in the following chapters, and in NFPA 1006.

**(1)** Rescuers expected to enter the water in floodwater environments that present swiftwater hazards shall meet technician-level requirements of Chapter 10 of NFPA 1006.

**(2)** Organizations operating at the technician level at flood search and rescue incidents in areas where swiftwater hazards are present shall meet the technician-level requirements of Chapter 17, Swiftwater Search and Rescue.

**(3)** Organizations operating at the technician level at flood search and rescue incidents that operate with helicopters shall meet the requirements of Chapter 15, Helicopter Search and Rescue.

**22.4.6** Organizations operating human-powered watercraft in a flood search and rescue environment shall develop and implement procedures for the use of human-powered watercraft in the flood search and rescue environment.

**22.4.7** Organizations operating motorized watercraft in a flood search and rescue environment shall develop and implement procedures for the use of motorized watercraft in the flood search and rescue environment.

# Appendix D

## NFPA 1006 and 1670 Correlation Guide

| NFPA 1006, *Standard for Technical Rescue Personnel Professional Qualifications,* 2017 Edition | Corresponding Chapter(s) | Corresponding Page(s) |
|---|---|---|
| 16.1 | 1 | 5 |
| 16.1.1 | 1 | 5–6, 8–16, 21–26 |
| 16.1.1(A) | 1 | 5–6, 8–14, 21–26 |
| 16.1.1(B) | 1 | 5–6, 8–11, 21–26 |
| 16.1.2 | 1 | 5–6, 8–9, 13–14, 16–21 |
| 16.1.2(A) | 1 | 5–6, 8–9, 13–14, 16–21 |
| 16.1.2(B) | 1 | 5–6, 8–9, 13–14, 16–21 |
| 16.1.3 | 1 | 6, 9, 11–13, 22–23, 25 |
| 16.1.3(A) | 1 | 5–6, 8–11, 16–22 |
| 16.1.3(B) | 1 | 11–13, 25 |
| 16.1.4 | 1 | 5–6, 8–16 |
| 16.1.4(A) | 1 | 4–6, 8–13 |
| 16.1.4(B) | 1 | 4, 15–16 |
| 16.2 | 4 | 100 |
| 16.2.1 | 4 | 103–106, 111 |
| 16.2.1(A) | 4 | 103–106, 111 |
| 16.2.1(B) | 4 | 103–106, 111 |
| 16.2.2 | 4 | 106–109, 120–123 |

| NFPA 1006, *Standard for Technical Rescue Personnel Professional Qualifications*, 2017 Edition | Corresponding Chapter(s) | Corresponding Page(s) |
|---|---|---|
| 16.2.2(A) | 4 | 106–109, 120–123 |
| 16.2.2(B) | 4 | 106–109, 120–123 |
| 16.2.3 | 4 | 103–106, 109–111 |
| 16.2.3(A) | 4 | 103–106, 109–111, 122–123 |
| 16.2.3(B) | 4 | 106–111 |
| 16.2.4 | 4 | 102–111 |
| 16.2.4(A) | 4 | 102–111 |
| 16.2.4(B) | 4 | 102–111 |
| 16.2.5 | 4 | 110 |
| 16.2.5(A) | 4 | 106–108, 110–111 |
| 16.2.5(B) | 4 | 108–111 |
| 16.2.6 | 4 | 111, 113–114 |
| 16.2.6(A) | 4 | 106–111, 113–114 |
| 16.2.6(B) | 4 | 106–111, 113–114 |
| 16.2.7 | 4 | 115–118 |
| 16.2.7(A) | 4 | 106–111, 115–118 |
| 16.2.7(B) | 4 | 106–111, 115–118 |
| 16.2.8 | 2, 4 | 37–49, 53–61, 68–75, 120 |
| 16.2.8(A) | 2, 4 | 53–54, 120 |
| 16.2.8(B) | 2, 4 | 53–54, 120 |
| 16.2.9 | 4, 16 | 120, 409–410, 412–418 |
| 16.2.9(A) | 4, 16 | 120, 407–410, 412–418 |
| 16.2.9(B) | 4, 16 | 120, 409–410, 412–418 |
| 16.2.10 | 2, 4 | 58–60, 120 |
| 16.2.10(A) | 2 | 35–41, 43–49, 52–53, 58–60 |

| NFPA 1006, *Standard for Technical Rescue Personnel Professional Qualifications,* 2017 Edition | Corresponding Chapter(s) | Corresponding Page(s) |
| --- | --- | --- |
| 16.2.10(B) | 2 | 58–60 |
| 16.2.11 | 4 | 121–123 |
| 16.2.11(A) | 4 | 122 |
| 16.2.11(B) | 4 | 121–123 |
| 16.2.12 | 4 | 124–128 |
| 16.2.12(A) | 4 | 124–128 |
| 16.2.12(B) | 4 | 124–125 |
| 16.2.13 | 4 | 111 |
| 16.2.13(A) | 4 | 111 |
| 16.2.13(B) | 4 | 111 |
| 16.2.14 | 4 | 129–132 |
| 16.2.14(A) | 4 | 129–132 |
| 16.2.14(B) | 4 | 129–132 |
| 16.3 | 5 | 135 |
| 16.3.1 | 5 | 135–137 |
| 16.3.1(A) | 5 | 136, 139–143, 145–153 |
| 16.3.1(B) | 5 | 135–137 |
| 16.3.2 | 5 | 138–143 |
| 16.3.2(A) | 5 | 138–143, 145–153 |
| 16.3.2(B) | 5 | 138–143, 145–153 |
| 16.3.3 | 5 | 139, 141–143 |
| 16.3.3(A) | 5 | 141–143 |
| 16.3.3(B) | 5 | 141–143 |
| 16.3.4 | 5 | 153–154 |
| 16.3.4(A) | 5 | 153–154 |

| NFPA 1006, *Standard for Technical Rescue Personnel Professional Qualifications*, 2017 Edition | Corresponding Chapter(s) | Corresponding Page(s) |
| --- | --- | --- |
| 16.3.4(B) | 5 | 153–154 |
| 17.1 | 1 | 4 |
| 17.1.1 | 1 | 6, 8–14 |
| 17.1.1(A) | 1 | 10–14, 21–26 |
| 17.1.1(B) | 1 | 6, 8–11, 13–14 |
| 17.1.2 | 1 | 6, 8–9, 13–14, 16–21 |
| 17.1.2(A) | 1 | 6, 8–9, 13–14, 16–21 |
| 17.1.2(B) | 1 | 6, 8–9, 13–14, 16–21 |
| 17.1.3 | 1 | 6, 9, 11–13, 22–23, 25 |
| 17.1.3(A) | 1 | 5–6, 8–11, 16–22 |
| 17.1.3(B) | 1 | 11–13, 25 |
| 17.1.4 | 1 | 5–6, 8–16 |
| 17.1.4(A) | 1 | 4–6, 8–13 |
| 17.1.4(B) | 1 | 4, 15–16 |
| 17.2 | 6 | 159 |
| 17.2.1 | 6 | 163–165, 175, 177–194 |
| 17.2.1(A) | 6 | 175, 177–194 |
| 17.2.1(B) | 6 | 175, 177–194 |
| 17.2.2 | 6 | 160–161, 163–165, 173–174 |
| 17.2.2(A) | 6 | 163, 165, 181, 191 |
| 17.2.2(B) | 6 | 173–174 |
| 17.2.3 | 6 | 161–163, 166–170 |
| 17.2.3(A) | 6 | 161–163, 166–170 |
| 17.2.3(B) | 6 | 161–163 |
| 17.2.4 | 6 | 163–171, 177–189 |

| NFPA 1006, *Standard for Technical Rescue Personnel Professional Qualifications*, 2017 Edition | Corresponding Chapter(s) | Corresponding Page(s) |
|---|---|---|
| 17.2.4(A) | 6 | 161–170, 177–189 |
| 17.2.4(B) | 6 | 163–171, 177–189 |
| 17.2.5 | 6 | 194 |
| 17.2.5(A) | 6 | 194 |
| 17.2.5(B) | 6 | 194 |
| 17.3 | 7 | 198 |
| 17.3.1 | 7 | 199–203, 205–211 |
| 17.3.1(A) | 7 | 199–201, 203 |
| 17.3.1(B) | 7 | 199–203, 205–211 |
| 17.3.2 | 7 | 199–200, 203, 205 |
| 17.3.2(A) | 7 | 199–200, 203, 205, 207 |
| 17.3.2(B) | 7 | 199–200, 203, 205 |
| 17.3.3 | 7 | 199–200, 203, 211–212 |
| 17.3.3(A) | 7 | 211–212 |
| 17.3.3(B) | 7 | 211–212 |
| 18.1 | 1 | 4 |
| 18.1.1 | 1 | 5–6, 8–14 |
| 18.1.1(A) | 1 | 10–14, 21–26 |
| 18.1.1(B) | 1 | 5–6, 8–11, 13–14 |
| 18.1.2 | 1 | 6, 8–9, 13–14, 16–21 |
| 18.1.2(A) | 1 | 6, 8–9, 13–14, 16–21 |
| 18.1.2(B) | 1 | 6, 8–9, 13–14, 16–21 |
| 18.1.3 | 1 | 6, 9, 11–13, 22–23, 25 |
| 18.1.3(A) | 1 | 5–6, 8–11, 16–22 |
| 18.1.3(B) | 1 | 11–13, 25 |

| NFPA 1006, *Standard for Technical Rescue Personnel Professional Qualifications,* 2017 Edition | Corresponding Chapter(s) | Corresponding Page(s) |
| --- | --- | --- |
| 18.1.4 | 1 | 5–6, 8–16 |
| 18.1.4(A) | 1 | 4–13 |
| 18.1.4(B) | 1 | 4, 15–16 |
| 18.2 | 14 | 330 |
| 18.2.1 | 14 | 333–336, 339 |
| 18.2.1(A) | 14 | 333–336, 339 |
| 18.2.1(B) | 14 | 332–335 |
| 18.2.2 | 14 | 341–343 |
| 18.2.2(A) | 14 | 341–343 |
| 18.2.2(B) | 14 | 341–343 |
| 18.2.3 | 14 | 362–363 |
| 18.2.3(A) | 14 | 362–363 |
| 18.2.3(B) | 14 | 362–363 |
| 18.2.4 | 14 | 343–359, 361–367 |
| 18.2.4(A) | 14 | 352–355 |
| 18.2.4(B) | 14 | 339–340, 355–359, 361–362 |
| 18.2.5 | 14 | 332 |
| 18.2.5(A) | 14 | 332, 366 |
| 18.2.5(B) | 14 | 332, 335–336, 339–340 |
| 18.2.6 | 13–14 | 313–316, 318–326, 343–348 |
| 18.2.6(A) | 13–14 | 314–315, 322–326, 343–352 |
| 18.2.6(B) | 13–14 | 313–316, 318–326, 343–352 |
| 18.2.7 | 14 | 358–359 |
| 18.2.7(A) | 14 | 358 |
| 18.2.7(B) | 14 | 358 |

| NFPA 1006, *Standard for Technical Rescue Personnel Professional Qualifications,* 2017 Edition | Corresponding Chapter(s) | Corresponding Page(s) |
| --- | --- | --- |
| 18.2.8 | 14 | 367–368 |
| 18.2.8(A) | 14 | 367–368 |
| 18.2.8(B) | 14 | 367–368 |
| 18.3 | 15 | 373 |
| 18.3.1 | 15 | 387, 389–392 |
| 18.3.1(A) | 15 | 375, 387, 389–393 |
| 18.3.1(B) | 15 | 382–383, 387, 389–393 |
| 18.3.2 | 15 | 378, 389–391 |
| 18.3.2(A) | 15 | 376–378, 381–383, 394–395 |
| 18.3.2(B) | 15 | 378, 382–383, 389–391 |
| 18.3.3 | 15 | 374, 378 |
| 18.3.3(A) | 15 | 376–378, 381–383, 394–395 |
| 18.3.3(B) | 15 | 378, 382–383, 389–391 |
| 18.3.4 | 15 | 376–378 |
| 18.3.4(A) | 15 | 376–378 |
| 18.3.4(B) | 15 | 376–378 |
| 18.3.5 | 15 | 402–403 |
| 18.3.5(A) | 15 | 402–403 |
| 18.3.5(B) | 15 | 402–403 |
| 18.3.6 | 15 | 378, 381–383, 390–391 |
| 18.3.6(A) | 15 | 376–378, 381–383, 394–395 |
| 18.3.6(B) | 15 | 376–378, 381–383, 394–395 |
| 18.3.7 | 15 | 376–387 |
| 18.3.7(A) | 15 | 376–387 |
| 18.3.7(B) | 15 | 376–387 |

| NFPA 1006, *Standard for Technical Rescue Personnel Professional Qualifications,* 2017 Edition | Corresponding Chapter(s) | Corresponding Page(s) |
| --- | --- | --- |
| 18.3.8 | 15 | 384–387 |
| 18.3.8(A) | 15 | 384–387, 390–391 |
| 18.3.8(B) | 15 | 384–387 |
| 18.3.9 | 15 | 379–387, 389–391 |
| 18.3.9(A) | 15 | 379–387, 389–391 |
| 18.3.9(B) | 15 | 379–387, 389–391 |
| 18.3.10 | 15 | 389, 392–393 |
| 18.3.10(A) | 15 | 392–393 |
| 18.3.10(B) | 15 | 392–393 |
| 19.1 | 1 | 4 |
| 19.1.1 | 1 | 5–6, 8–14 |
| 19.1.1(A) | 1 | 10–14, 21–26 |
| 19.1.1(B) | 1 | 5–6, 8–11, 13–14 |
| 19.1.2 | 1 | 6, 8–9, 13–14, 16–21 |
| 19.1.2(A) | 1 | 6, 8–9, 13–14, 16–21 |
| 19.1.2(B) | 1 | 6, 8–9, 13–14, 16–21 |
| 19.1.3 | 1 | 6, 9, 11–13, 22–23, 25 |
| 19.1.3(A) | 1 | 5–6, 8–11, 16–22 |
| 19.1.3(B) | 1 | 11–13, 25 |
| 19.1.4 | 1 | 5–6, 8–16 |
| 19.1.4(A) | 1 | 4–6, 8–13 |
| 19.1.4(B) | 1 | 4, 15–16 |
| 19.2 | 9 | 236 |
| 19.2.1 | 9 | 244–246, 250–252 |
| 19.2.1(A) | 9 | 240, 242, 250–252 |

| NFPA 1006, *Standard for Technical Rescue Personnel Professional Qualifications*, 2017 Edition | Corresponding Chapter(s) | Corresponding Page(s) |
| --- | --- | --- |
| 19.2.1(B) | 9 | 251–252 |
| 19.2.2 | 9 | 236–239, 244–246 |
| 19.2.2(A) | 9 | 236–239, 242 |
| 19.2.2(B) | 9 | 242, 252–253 |
| 19.2.3 | 9 | 244–246, 248–252 |
| 19.2.3(A) | 9 | 239–240, 244–246, 248–253 |
| 19.2.3(B) | 9 | 236, 244–246, 248–250 |
| 19.2.4 | 9 | 253 |
| 19.2.4(A) | 9 | 253 |
| 19.2.4(B) | 9 | 253 |
| 19.3 | 10 | 257 |
| 19.3.1 | 10 | 257–259, 261–262, 266 |
| 19.3.1(A) | 10 | 257–258, 266, 269 |
| 19.3.1(B) | 10 | 257–258, 261–262 |
| 19.3.2 | 10 | 257–259, 261–267 |
| 19.3.2(A) | 10 | 257–258, 264 |
| 19.3.2(B) | 10 | 257–259, 261–267 |
| 20.1 | 1 | 4 |
| 20.1.1 | 1 | 5–6, 8–14 |
| 20.1.1(A) | 1 | 10–14, 21–26 |
| 20.1.1(B) | 1 | 5–6, 8–11, 13–14 |
| 20.1.2 | 1 | 6, 8–9, 13–14, 16–21 |
| 20.1.2(A) | 1 | 6, 8–9, 13–14, 16–21 |
| 20.1.2(B) | 1 | 6, 8–9, 13–14, 16–21 |
| 20.1.3 | 1 | 6, 9, 11–13, 22–23, 25 |

| NFPA 1006, *Standard for Technical Rescue Personnel Professional Qualifications*, 2017 Edition | Corresponding Chapter(s) | Corresponding Page(s) |
| --- | --- | --- |
| 20.1.3(A) | 1 | 5–6, 8–11, 16–22 |
| 20.1.3(B) | 1 | 11–13, 25 |
| 20.1.4 | 1 | 5–6, 8–16 |
| 20.1.4(A) | 1 | 4–6, 8–13 |
| 20.1.4(B) | 1 | 4, 15–16 |
| 20.2 | 11 | 273 |
| 20.2.1 | 11 | 274–275 |
| 20.2.1(A) | 11 | 274–275 |
| 20.2.1(B) | 11 | 274–275 |
| 20.2.2 | 11 | 275 |
| 20.2.2(A) | 11 | 273–281 |
| 20.2.2(B) | 11 | 274–281 |
| 20.2.3 | 11 | 285–286 |
| 20.2.3(A) | 11 | 274–281, 285–286 |
| 20.2.3(B) | 11 | 274–281, 285–286 |
| 20.2.4 | 11 | 281–285 |
| 20.2.4(A) | 11 | 275–285 |
| 20.2.4(B) | 11 | 274–275, 281–285 |
| 20.2.5 | 11 | 288 |
| 20.2.5(A) | 11 | 288 |
| 20.2.5(B) | 11 | 288 |
| 20.3 | 12 | 295 |
| 20.3.1 | 12 | 295–296 |
| 20.3.1(A) | 12 | 297–301, 303–307 |
| 20.3.1(B) | 12 | 295–301, 303–307 |

| NFPA 1006, *Standard for Technical Rescue Personnel Professional Qualifications*, 2017 Edition | Corresponding Chapter(s) | Corresponding Page(s) |
| --- | --- | --- |
| 20.3.2 | 12 | 296–301, 303–308 |
| 20.3.2(A) | 12 | 296–301, 303–308 |
| 20.3.2(B) | 12 | 295–301, 303–308 |
| 20.3.3 | 12 | 295–297, 300–301, 303–307 |
| 20.3.3(A) | 12 | 296–301, 303–307 |
| 20.3.3(B) | 12 | 295–301, 303–307 |
| 21.1 | 1 | 4 |
| 21.1.1 | 1 | 5–6, 8–11, 21–26 |
| 21.1.1(A) | 1 | 10–11, 22–23 |
| 21.1.1(B) | 1 | 10–11, 22–23, 25 |
| 21.1.2 | 1 | 6, 8–9, 13–14, 16–21 |
| 21.1.2(A) | 1 | 6, 8–11, 13–21 |
| 21.1.2(B) | 1 | 6, 8–11, 13–16 |
| 21.1.3 | 1 | 10–13, 24–25 |
| 21.1.3(A) | 1 | 4–5, 10–13, 16, 24–25 |
| 21.1.3(B) | 1 | 10–13 |
| 21.1.4 | 1 | 6, 8–10, 13–21, 24–25 |
| 21.1.4(A) | 1 | 14–16 |
| 21.1.4(B) | 1 | 14–16 |
| 21.1.5 | 1, 16 | 23, 407–410, 412–417 |
| 21.1.5(A) | 1, 16 | 23, 408–410, 412–413 |
| 21.1.5(B) | 1, 16 | 23, 408–410, 412–413 |
| 21.1.6 | 1 | 9–10, 23 |
| 21.1.6(A) | 1 | 23 |
| 21.1.6(B) | 1 | 23 |

| NFPA 1006, *Standard for Technical Rescue Personnel Professional Qualifications,* 2017 Edition | Corresponding Chapter(s) | Corresponding Page(s) |
|---|---|---|
| 21.1.7 | 1 | 16–21 |
| 21.1.7(A) | 1 | 16–22 |
| 21.1.7(B) | 1 | 16–21 |
| 21.1.8 | 1 | 22 |
| 21.1.8(A) | 1 | 22 |
| 21.1.8(B) | 1 | 22 |
| 21.2 | 2 | 33–34 |
| 21.2.1 | 2 | 38–46 |
| 21.2.1(A) | 2 | 38–46, 73 |
| 21.2.1(B) | 2 | 38–53 |
| 21.2.2 | 2 | 33–34, 37–61 |
| 21.2.2(A) | 2 | 37–38 |
| 21.2.2(B) | 2 | 37–38 |
| 21.2.3 | 2 | 51–52, 71 |
| 21.2.3(A) | 2 | 51–52, 71 |
| 21.2.3(B) | 2 | 51–52, 71 |
| 21.2.4 | 2 | 54, 58–59, 63 |
| 21.2.4(A) | 2 | 63 |
| 21.2.4(B) | 2 | 63 |
| 21.2.5 | 2 | 68–69 |
| 21.2.5(A) | 2 | 68–69 |
| 21.2.5(B) | 2 | 68–69 |
| 21.2.6 | 2 | 72–73 |
| 21.2.6(A) | 2 | 72–73 |
| 21.2.6(B) | 2 | 72–73 |

| NFPA 1006, *Standard for Technical Rescue Personnel Professional Qualifications*, 2017 Edition | Corresponding Chapter(s) | Corresponding Page(s) |
|---|---|---|
| 21.2.7 | 2 | 35–37 |
| 21.2.7(A) | 2 | 35–37 |
| 21.2.7(B) | 2 | 35–37 |
| 21.2.8 | 2 | 33, 65, 70–71 |
| 21.2.8(A) | 2 | 70–71, 73 |
| 21.2.8(B) | 2 | 70–71 |
| 21.2.9 | 2 | 65–67, 73 |
| 21.2.9(A) | 2 | 57, 65–67, 73 |
| 21.2.9(B) | 2 | 65–67 |
| 21.2.10 | 2 | 63–65, 73 |
| 21.2.10(A) | 2 | 63–65, 73 |
| 21.2.10(B) | 2 | 65–67, 73 |
| 21.2.11 | 2 | 72–73 |
| 21.2.11(A) | 2 | 72–73 |
| 21.2.11(B) | 2 | 72–73 |
| 21.2.12 | 2 | 74–75 |
| 21.2.12(A) | 2 | 74–75 |
| 21.2.12(B) | 2 | 74–75 |
| 21.2.13 | 2 | 73–75 |
| 21.2.13(A) | 2 | 47–48, 73–75 |
| 21.2.13(B) | 2 | 73–75 |
| 21.2.14 | 2 | 34–37, 56–60 |
| 21.2.14(A) | 2 | 58–60 |
| 21.2.14(B) | 2 | 52, 58–60 |
| 21.2.15 | 2 | 57, 63–65, 73 |

| NFPA 1006, *Standard for Technical Rescue Personnel Professional Qualifications*, 2017 Edition | Corresponding Chapter(s) | Corresponding Page(s) |
| --- | --- | --- |
| 21.2.15(A) | 2 | 57, 63–65, 73 |
| 21.2.15(B) | 2 | 63–65 |
| 21.2.16 | 2 | 53–55 |
| 21.2.16(A) | 2 | 53–55 |
| 21.2.16(B) | 2 | 53–55 |
| 21.2.17 | 2 | 55–56 |
| 21.2.17(A) | 2 | 55–56 |
| 21.2.17(B) | 2 | 55–56 |
| 21.2.18 | 2 | 75–76 |
| 21.2.18(A) | 2 | 75–76 |
| 21.2.18(B) | 2 | 75–76 |
| 21.3 | 3 | 82 |
| 21.3.1 | 3 | 82 |
| 21.3.1(A) | 3 | 82 |
| 21.3.1(B) | 3 | 82 |
| 21.3.2 | 3 | 82–87, 89–98 |
| 21.3.2(A) | 3 | 82–85, 91 |
| 21.3.2(B) | 3 | 84–87, 89–91 |
| 21.3.3 | 3 | 85–86 |
| 21.3.3(A) | 3 | 85–86 |
| 21.3.3(B) | 3 | 85–86 |
| 21.3.4 | 3 | 84, 90, 95 |
| 21.3.4(A) | 3 | 84 |
| 21.3.4(B) | 3 | 84, 91–96 |
| 21.3.5 | 3 | 82–84, 91–96 |

| NFPA 1006, *Standard for Technical Rescue Personnel Professional Qualifications*, 2017 Edition | Corresponding Chapter(s) | Corresponding Page(s) |
| --- | --- | --- |
| 21.3.5(A) | 3 | 82–84, 91–96 |
| 21.3.5(B) | 3 | 82–84, 91–96 |
| 21.3.6 | 3 | 86–87 |
| 21.3.6(A) | 3 | 86–87 |
| 21.3.6(B) | 3 | 86–87 |
| 21.3.7 | 3 | 96–97 |
| 21.3.7(A) | 3 | 96–97 |
| 21.3.7(B) | 3 | 87, 96–97 |
| 21.3.8 | 3 | 96–97 |
| 21.3.8(A) | 3 | 96–97 |
| 21.3.8(B) | 3 | 96–97 |
| 21.3.9 | 3 | 97 |
| 21.3.9(A) | 3 | 97 |
| 21.3.9(B) | 3 | 97 |
| 21.3.10 | 3 | 90–96 |
| 21.3.10(A) | 3 | 90–96 |
| 21.3.10(B) | 3 | 90–91 |
| 21.3.11 | 3 | 85–87, 89 |
| 21.3.11(A) | 3 | 85–87, 89 |
| 21.3.11(B) | 3 | 85–87, 89 |
| 21.3.12 | 3 | 98 |
| 21.3.12(A) | 3 | 98 |
| 21.3.12(B) | 3 | 98 |
| 22.1 | 1 | 4 |
| 22.1.1 | 1 | 6, 8–14 |

| NFPA 1006, *Standard for Technical Rescue Personnel Professional Qualifications*, 2017 Edition | Corresponding Chapter(s) | Corresponding Page(s) |
| --- | --- | --- |
| 22.1.1(A) | 1 | 10–14, 21–26 |
| 22.1.1(B) | 1 | 5–6, 8–11, 13–14 |
| 22.1.2 | 1 | 6, 8–9, 13–14, 16–21 |
| 22.1.2(A) | 1 | 6, 8–9, 13–14, 16–21 |
| 22.1.2(B) | 1 | 6, 8–9, 13–14, 16–21 |
| 22.1.3 | 1 | 6, 9, 11–13, 22–23, 25 |
| 22.1.3(A) | 1 | 5–6, 8–11, 16–22 |
| 22.1.3(B) | 1 | 11–13, 25 |
| 22.1.4 | 1 | 5–6, 8–16 |
| 22.1.4(A) | 1 | 4–6, 8–13 |
| 22.1.4(B) | 1 | 4, 15–16 |
| 22.2 | 8 | 217 |
| 22.2.1 | 8 | 217 |
| 22.2.1(A) | 8 | 217 |
| 22.2.1(B) | 8 | 217 |
| 22.2.2 | 8 | 218–221, 223–224 |
| 22.2.2(A) | 8 | 217–219, 224 |
| 22.2.2(B) | 8 | 217–219, 223–224 |
| 22.2.3 | 8 | 219–220, 228 |
| 22.2.3(A) | 8 | 219–225, 227 |
| 22.2.3(B) | 8 | 219–224 |
| 22.2.4 | 8 | 227–228 |
| 22.2.4(A) | 8 | 217–221, 227–228 |
| 22.2.4(B) | 8 | 227–228 |
| 22.2.5 | 8 | 227 |

| NFPA 1006, *Standard for Technical Rescue Personnel Professional Qualifications,* 2017 Edition | Corresponding Chapter(s) | Corresponding Page(s) |
|---|---|---|
| 22.2.5(A) | 8 | 227 |
| 22.2.5(B) | 8 | 227 |
| 22.2.6 | 8 | 221–223 |
| 22.2.6(A) | 8 | 221–223 |
| 22.2.6(B) | 8 | 221–223 |
| 22.2.7 | 8 | 224–225 |
| 22.2.7(A) | 8 | 224–225 |
| 22.2.7(B) | 8 | 244–225 |
| 22.2.8 | 8 | 223–224 |
| 22.2.8(A) | 8 | 217–218, 220–221, 223–224 |
| 22.2.8(B) | 8 | 217–218, 220–221, 223–224 |
| 22.2.9 | 8 | 220–221, 227–228 |
| 22.2.9(A) | 8 | 227–228 |
| 22.2.9(B) | 8 | 227–228 |
| 22.2.10 | 8 | 232–233 |
| 22.2.10(A) | 8 | 232–233 |
| 22.2.10(B) | 8 | 219–220, 222–223, 232–233 |
| 22.2.11 | 8 | 219–220, 228 |
| 22.2.11(A) | 8 | 227–228 |
| 22.2.11(B) | 8 | 227–228 |
| 22.3 | 8 | 217 |
| 22.3.1 | 8 | 229–232 |
| 22.3.1(A) | 8, 15 | 218–220, 223–225, 227, 397–398 |
| 22.3.1(B) | 8 | 218–224 |

| NFPA 1670, *Standard on Operations and Training for Technical Search and Rescue Incidents*, 2017 Edition | Corresponding Chapter(s) | Corresponding Page(s) |
| --- | --- | --- |
| 16.1 | 1 | 5 |
| 16.2.1 | 1 | 4–5 |
| 16.2.2 | 1 | 5 |
| 16.2.3 | 1 | 5–6, 8–14 |
| 16.2.3(1) | 1 | 5–6 |
| 16.2.3(2) | 1 | 5–6 |
| 16.2.3(3) | 1 | 10–11 |
| 16.2.3(4) | 1 | 11–13 |
| 16.2.3(5) | 1 | 6, 8–9, 13–14 |
| 16.2.3(6) | 1 | 6 |
| 16.2.3(7) | 1 | 9–10 |
| 16.3.1 | 4 | 100 |
| 16.3.2 | 4 | 101 |
| 16.3.3 | 4 | 117 |
| 16.3.4 | 2, 4 | 33–34, 119 |
| 16.3.5 | 4 | 103, 110 |
| 16.3.5(1) | 4 | 110 |
| 16.3.5(2) | 4 | 103 |
| 16.3.6 | 4 | 106, 120, 129 |
| 16.3.6(1) | 4 | 106, 120 |
| 16.3.6(2) | 4 | 106, 129 |
| 16.3.6(3) | 4 | 106 |
| 16.3.7 | 4, 10, 15 | 103–106, 108–111, 113–124, 127–129, 268, 396–400 |
| 16.3.7(1) | 4 | 108–110 |

| NFPA 1670, *Standard on Operations and Training for Technical Search and Rescue Incidents*, 2017 Edition | Corresponding Chapter(s) | Corresponding Page(s) |
|---|---|---|
| 16.3.7(2) | 4 | 106–107, 109, 124, 127–129 |
| 16.3.7(3) | 4 | 109–110 |
| 16.3.7(4) | 4 | 110 |
| 16.3.7(5) | 4 | 110 |
| 16.3.7(6) | 4 | 108 |
| 16.3.7(7) | 4 | 110, 114–119 |
| 16.3.7(8) | 4 | 114–119 |
| 16.3.7(9) | 4 | 103 |
| 16.3.7(10) | 4 | 120–123 |
| 16.3.7(11) | 4 | 103, 109, 119 |
| 16.3.7(12) | 4 | 127 |
| 16.3.7(13) | 4 | 123 |
| 16.3.7(14) | 4 | 119 |
| 16.3.7(15) | 4 | 103–106 |
| 16.3.7(16) | 4 | 111, 113–119 |
| 16.3.7(17) | 4 | 109–110 |
| 16.3.7(18) | 10, 15 | 268, 396–400 |
| 16.3.7(19) | 4 | 128 |
| 16.4.1 | 5 | 135–136 |
| 16.4.2 | 5 | 135–136 |
| 16.4.3 | 5, 15 | 135–138, 154, 396–400 |
| 16.4.3(1) | 5 | 135 |
| 16.4.3(2) | 5 | 154 |
| 16.4.3(3) | 5 | 154 |
| 16.4.3(4) | 5 | 136–138 |

| NFPA 1670, *Standard on Operations and Training for Technical Search and Rescue Incidents*, 2017 Edition | Corresponding Chapter(s) | Corresponding Page(s) |
| --- | --- | --- |
| 16.4.3(5) | 15 | 396–400 |
| 16.4.3(6) | 5 | 154 |
| 17.1 | 1 | 5 |
| 17.2.1 | 1 | 4–5 |
| 17.2.2 | 1 | 5 |
| 17.2.3 | 1 | 5–6, 8–14 |
| 17.2.3(1) | 1 | 5–6 |
| 17.2.3(2) | 1 | 5–6 |
| 17.2.3(3) | 1 | 10–11 |
| 17.2.3(4) | 1 | 11–13 |
| 17.2.3(5) | 1 | 6, 8–9, 13–14 |
| 17.2.3(6) | 1 | 6 |
| 17.2.3(7) | 1 | 9–10 |
| 17.3.1 | 6 | 159 |
| 17.3.2 | 6 | 159 |
| 17.3.3 | 6 | 159 |
| 17.3.4 | 6 | 163–165 |
| 17.3.4(1) | 6 | 163–164 |
| 17.3.4(2) | 6 | 164–165 |
| 17.3.4(3) | 6 | 164 |
| 17.3.4(4) | 6 | 165 |
| 17.3.4(5) | 6 | 165 |
| 17.4.1 | 7 | 198 |
| 17.4.2 | 7 | 198 |
| 17.4.3 | 7 | 198 |

| NFPA 1670, *Standard on Operations and Training for Technical Search and Rescue Incidents*, 2017 Edition | Corresponding Chapter(s) | Corresponding Page(s) |
|---|---|---|
| 17.4.4 | 7 | 207, 210–211 |
| 17.4.5 | 6–7 | 175, 177–194, 207–208 |
| 17.4.5(1) | 6 | 175, 177–194 |
| 17.4.5(2) | 6 | 185–186 |
| 17.4.5(3) | 6–7 | 186, 207–208 |
| 17.4.5(4) | 6 | 186, 189–193 |
| 17.4.6 | 6 | 182 |
| 17.4.7 | 6–7 | 182–183, 186, 189–194, 211–212 |
| 18.1 | 1 | 5 |
| 18.2.1 | 1 | 4–5 |
| 18.2.2 | 1 | 5 |
| 18.2.3 | 1 | 5–6, 8–14 |
| 18.2.3(1) | 1 | 5–6 |
| 18.2.3(2) | 1 | 5–6 |
| 18.2.3(3) | 1 | 10–11 |
| 18.2.3(4) | 1 | 11–13 |
| 18.2.3(5) | 1 | 6, 8–9, 13–14 |
| 18.2.3(6) | 1 | 6 |
| 18.2.3(7) | 1 | 9–10 |
| 18.3.1 | 14 | 330 |
| 18.3.2 | 14 | 330 |
| 18.3.3 | 14 | 330 |
| 18.3.4 | 13–14 | 312–315, 323–326, 343–348, 358 |
| 18.3.4(1) | 13–14 | 323–324, 343–348 |

| NFPA 1670, *Standard on Operations and Training for Technical Search and Rescue Incidents*, 2017 Edition | Corresponding Chapter(s) | Corresponding Page(s) |
|---|---|---|
| 18.3.4(2) | 13–14 | 313–315, 346 |
| 18.3.4(3) | 13–14 | 312, 358 |
| 18.3.4(4) | 13–14 | 325–326, 358 |
| 18.3.5 | 14 | 333, 335, 340–364 |
| 18.3.5(1) | 14 | 340–351, 355–363 |
| 18.3.5(2) | 14 | 343–349 |
| 18.3.5(3) | 14 | 333 |
| 18.3.5(4) | 14 | 350–357 |
| 18.3.5(5) | 14 | 342, 364 |
| 18.3.5(6) | 14 | 348–349 |
| 18.3.5(7) | 14 | 335, 346, 364 |
| 18.3.5(8) | 14 | 358, 363–364 |
| 18.3.5(9) | 14 | 358–359, 363–364 |
| 18.3.5(10) | 14 | 358, 363 |
| 18.3.5(11) | 14 | 355–358 |
| 18.3.5(12) | 14 | 349–352 |
| 18.3.5(13) | 14 | 340 |
| 18.4.1 | 15 | 373 |
| 18.4.2 | 15 | 373 |
| 18.4.3 | 15 | 373 |
| 18.4.4 | 15 | 373–374 |
| 18.4.5 | 15 | 376–377 |
| 18.4.6 | 15 | 375–376 |
| 18.4.6(1) | 15 | 376 |
| 18.4.6(2) | 15 | 376 |

| NFPA 1670, *Standard on Operations and Training for Technical Search and Rescue Incidents*, 2017 Edition | Corresponding Chapter(s) | Corresponding Page(s) |
| --- | --- | --- |
| 18.4.6(3) | 15 | 375–376 |
| 18.4.6(4) | 15 | 375–376 |
| 18.4.7 | 14–15 | 355, 358, 380, 387, 391, 403 |
| 18.4.7(1) | 15 | 387, 391 |
| 18.4.7(2) | 15 | 387 |
| 18.4.7(3) | 15 | 387 |
| 18.4.7(4) | 15 | 403 |
| 18.4.7(5) | 15 | 380 |
| 18.4.7(6) | 14 | 358 |
| 18.4.7(7) | 14 | 355 |
| 18.4.8 | 15 | 375, 378–380, 382–383, 387, 389–391, 399–400, 403 |
| 18.4.8(1) | 15 | 387, 389–390 |
| 18.4.8(2) | 15 | 387, 389 |
| 18.4.8(3) | 15 | 389 |
| 18.4.8(4) | 15 | 389 |
| 18.4.8(5) | 15 | 380, 389 |
| 18.4.8(6) | 15 | 387, 389–390 |
| 18.4.8(7) | 15 | 389 |
| 18.4.8(8) | 15 | 387, 403 |
| 18.4.8(9) | 15 | 380 |
| 18.4.8(10) | 15 | 382–383 |
| 18.4.8(11) | 15 | 390–391 |
| 18.4.8(12) | 15 | 378 |
| 18.4.8(13) | 15 | 379 |

| NFPA 1670, *Standard on Operations and Training for Technical Search and Rescue Incidents*, 2017 Edition | Corresponding Chapter(s) | Corresponding Page(s) |
| --- | --- | --- |
| 18.4.8(14) | 15 | 387, 403 |
| 18.4.8(15) | 15 | 399–400 |
| 18.4.8(16) | 15 | 375–376, 379 |
| 18.4.8(17) | 15 | 383 |
| 18.4.8(18) | 15 | 389 |
| 18.4.9 | 15 | 387 |
| 18.4.10 | 2 | 33–34 |
| 18.4.11 | 2 | 33–34 |
| 18.4.12 | 15 | 376–379 |
| 19.1 | 1 | 5 |
| 19.2.1 | 1 | 4–5 |
| 19.2.2 | 1 | 5 |
| 19.2.3 | 1 | 5–6, 8–14 |
| 19.2.3(1) | 1 | 5–6 |
| 19.2.3(2) | 1 | 5–6 |
| 19.2.3(3) | 1 | 10–11 |
| 19.2.3(4) | 1 | 11–13 |
| 19.2.3(5) | 1 | 6, 8–9, 13–14 |
| 19.2.3(6) | 1 | 6 |
| 19.2.3(7) | 1 | 9–10 |
| 19.3.1 | 9 | 236 |
| 19.3.2 | 9 | 236 |
| 19.3.3 | 9 | 236 |
| 19.3.4 | 9 | 244–246 |
| 19.3.4(1) | 9 | 244 |

| NFPA 1670, *Standard on Operations and Training for Technical Search and Rescue Incidents*, 2017 Edition | Corresponding Chapter(s) | Corresponding Page(s) |
|---|---|---|
| 19.3.4(2) | 9 | 244–245 |
| 19.3.4(3) | 9 | 244 |
| 19.3.4(4) | 9 | 245–246 |
| 19.3.5 | 9 | 236–240 |
| 19.4.1 | 10 | 257 |
| 19.4.2 | 10 | 257 |
| 19.4.3 | 10 | 257 |
| 19.4.4 | 10 | 259, 261–269 |
| 19.4.5 | 10 | 259, 261, 264, 266 |
| 19.4.6 | 10 | 261 |
| 20.2 | 1 | 5 |
| 20.2.1 | 1 | 5 |
| 20.2.2 | 1 | 5 |
| 20.2.3 | 1 | 5–6, 8–15, 24–25 |
| 20.2.3(1) | 1 | 5–6, 14–15 |
| 20.2.3(2) | 1 | 5–6, 8–10 |
| 20.2.3(3) | 1 | 10–11 |
| 20.2.3(4) | 1 | 11–13 |
| 20.2.3(5) | 1 | 13–14, 24–25 |
| 20.2.3(6) | 1 | 6 |
| 20.2.3(7) | 1 | 9–10 |
| 20.3.1 | 11 | 273 |
| 20.3.2 | 11 | 273 |
| 20.3.3 | 11 | 273 |
| 20.3.4 | 11 | 274 |

| NFPA 1670, *Standard on Operations and Training for Technical Search and Rescue Incidents*, 2017 Edition | Corresponding Chapter(s) | Corresponding Page(s) |
|---|---|---|
| 20.3.5 | 11 | 275–280 |
| 20.4.1 | 12 | 295 |
| 20.4.2 | 12 | 295 |
| 20.4.3 | 12 | 295 |
| 20.4.4 | 12 | 296–301, 303–308 |
| 20.4.5 | 12 | 305–307 |
| 20.4.6 | 12 | 305–307 |
| 21.1.1 | 1 | 5 |
| 21.1.2 | 1 | 21–22 |
| 21.1.3 | 1 | 21–22 |
| 21.1.4 | 1 | 21 |
| 21.2.1 | 1 | 21–22 |
| 21.2.2 | 1 | 5 |
| 21.2.3 | 1 | 5–6, 8–15 |
| 21.2.3(1) | 1 | 5–6, 10–11 |
| 21.2.3(2) | 1 | 5–6, 8–10 |
| 21.2.3(3) | 1 | 10–11, 14–15 |
| 21.2.3(4) | 1 | 11–13 |
| 21.2.3(5) | 1 | 13–14 |
| 21.2.3(6) | 1 | 6 |
| 21.2.3(7) | 1 | 9–10 |
| 21.3.1 | 2 | 33–34 |
| 21.3.2 | 2 | 33–34 |
| 21.3.3 | 2 | 33–34 |
| 21.3.4 | 2 | 35, 53 |

| NFPA 1670, *Standard on Operations and Training for Technical Search and Rescue Incidents*, 2017 Edition | Corresponding Chapter(s) | Corresponding Page(s) |
| --- | --- | --- |
| 21.3.4(1) | 2 | 35 |
| 21.3.4(2) | 2 | 53 |
| 21.3.4(3) | 2 | 35 |
| 21.3.5 | 2 | 33–35, 38–41, 43–46, 53–61, 63–65, 68, 70–75 |
| 21.3.5(1) | 2 | 38–41, 43–46 |
| 21.3.5(2) | 2 | 65, 68 |
| 21.3.5(3) | 2 | 35, 53–56, 74 |
| 21.3.5(4) | 2 | 33–34 |
| 21.3.5(5) | 2 | 33–34, 65, 68 |
| 21.3.5(6) | 2 | 53–55 |
| 21.3.5(7) | 2 | 57, 61, 63 |
| 21.3.5(8) | 2 | 73–75 |
| 21.3.5(9) | 2 | 58–60 |
| 21.3.5(10) | 2 | 41, 64–65 |
| 21.3.5(11) | 2 | 74–75 |
| 21.3.5(12) | 2 | 70, 71, 73 |
| 21.3.5(13) | 2 | 64–65, 68 |
| 21.3.5(14) | 2 | 70–71 |
| 21.3.5(15) | 2 | 53–55, 72, 74–75 |
| 21.3.5(16) | 2 | 55–56 |
| 21.4.1 | 3 | 82, 84 |
| 21.4.2 | 3 | 82, 84 |
| 21.4.3 | 2–3 | 53–54, 74–75, 83, 85–87, 89–91, 95–97 |
| 21.4.3(1) | 3 | 90–91 |

| NFPA 1670, *Standard on Operations and Training for Technical Search and Rescue Incidents*, 2017 Edition | Corresponding Chapter(s) | Corresponding Page(s) |
| --- | --- | --- |
| 21.4.3(2) | 3 | 96–97 |
| 21.4.3(3) | 3 | 95 |
| 21.4.3(4) | 3 | 85–86 |
| 21.4.3(5) | 2–3 | 53–54, 74–75, 83, 87 |
| 21.4.3(5) | 3 | 85–87 |
| 21.4.3(7) | 3 | 95–97 |
| 21.4.3(8) | 3 | 85–87, 89 |
| 22.1 | 1 | 5 |
| 22.1.1 | 1 | 14–15 |
| 22.1.2 | 1 | 14–15 |
| 22.2.1 | 1 | 4–5 |
| 22.2.2 | 1, 8 | 5–6, 11–13, 16, 217–219, 224–225, 227, 231–232 |
| 22.2.2(1) | 1 | 5–6 |
| 22.2.2(2) | 1 | 11–13 |
| 22.2.2(3) | 1 | 11 |
| 22.2.2(4) | 1 | 6 |
| 22.2.2(5) | 8 | 217–218 |
| 22.2.2(6) | 8 | 217–218 |
| 22.2.2(7) | 8 | 218–219 |
| 22.2.2(8) | 1 | 16 |
| 22.2.2(9) | 8 | 217–218, 224–225, 227 |
| 22.2.2(10) | 8 | 231–232 |
| 22.3.1 | 8 | 217 |
| 22.3.2 | 8 | 217 |

| NFPA 1670, *Standard on Operations and Training for Technical Search and Rescue Incidents*, 2017 Edition | Corresponding Chapter(s) | Corresponding Page(s) |
| --- | --- | --- |
| 22.3.3 | 8 | 217 |
| 22.3.4 | 8 | 219–220 |
| 22.3.4(1) | 8 | 219 |
| 22.3.4(2) | 8 | 220 |
| 22.3.4(3) | 8 | 220 |
| 22.3.4(4) | 8 | 220 |
| 22.3.4(5) | 8 | 220 |
| 22.3.5 | 8 | 216–217 |
| 22.3.6 | 8 | 217 |
| 22.3.7 | 8 | 217–223, 227–228 |
| 22.3.7(1) | 8 | 217–219 |
| 22.3.7(2) | 8 | 220 |
| 22.3.7(3) | 8 | 227–228 |
| 22.3.7(4) | 8 | 227–228 |
| 22.3.7(5) | 8 | 221–222 |
| 22.3.7(6) | 8 | 222–223 |
| 22.4.1 | 8 | 228 |
| 22.4.2 | 8 | 228 |
| 22.4.3 | 8 | 228 |
| 22.4.4 | 8 | 228–232 |
| 22.4.4(1) | 8 | 228–232 |
| 22.4.4(2) | 8 | 229–232 |
| 22.4.4(3) | 8 | 229 |
| 22.4.4(4) | 8 | 229 |
| 22.4.5 | 8 | 216, 227–229 |

| NFPA 1670, *Standard on Operations and Training for Technical Search and Rescue Incidents*, 2017 Edition | Corresponding Chapter(s) | Corresponding Page(s) |
| --- | --- | --- |
| 22.4.5(1) | 8 | 216, 228–229 |
| 22.4.5(2) | 8 | 216, 228–229 |
| 22.4.5(3) | 8 | 227 |
| 22.4.6 | 8 | 227 |
| 22.4.7 | 8 | 227 |

# Glossary

**90 percent diver, or safety diver** The backup's backup, dressed, but usually staged with the mask off.

**500-year flood** A flood event that has a 1 in 500 chance of occurrence.

**Abeam** A point directly perpendicular to the port or starboard side of a boat.

**Accountability** A system or process to track resources at an incident scene. (NFPA 1561)

**Active drowning stage** The drowning stage in which the victim is still fighting to keep the head above water by doing anything necessary.

**Active search measures** This phase of search measures includes those that are formalized and coordinated with other agencies. (NFPA 1006)

**Aerated water** Water mixed with a great deal of air, resulting in bubbles or foam. This water does not provide very good thrust from the propeller.

**Aerial extrication evolution** The use of an aerial ladder or platform to provide a lift point to recover victims; it uses a rope system in conjunction with the ladder itself.

**Airboats** Rigid-hull watercraft driven by a large fan blade powered by a car engine. There is no propeller system to foul, and these boats can cross water, ice, swampland, and ground.

**Ambient breathing valves** Small valves on a full-face mask used to breathe outside air so a suited up backup diver does not exhaust his or her gas supply.

**Anchor** A device designed to engage the bottom of a waterway and, through its resistance to drag, maintain a vessel within a given radius. (NFPA 1925)

**Anchor scope** The amount of line deployed relative to the depth of the water while using an anchor. The stronger the pull of the boat, the more scope is used.

**Approach zones** With the nose of the aircraft at the 12 o'clock position, the sides of the helicopter.

**Arterial gas embolism (AGE)** A bubble in the bloodstream lodging in an artery, creating a blockage of blood flow to the surrounding tissues or organs.

**Authority having jurisdiction (AHJ)** An organization, office, or individual responsible for enforcing the requirements of a code or standard, or for approving equipment, materials, an installation, or a procedure. (NFPA 1006)

**Awareness level** This level represents the minimum capability of individuals who provide response to technical search and rescue incidents. (NFPA 1006)

**Backboard** A hard, flat piece of wood or plastic used to provide a firm platform to secure patients suspected of having cervical spine injuries.

**Banana boat** An inflatable rescue craft specific to rescue work. It is very stable and resembles a large banana, with rescuers sitting inside the craft. Also called a *snout rig*.

**Barotrauma** Pressure-related injury. See *decompression illness*.

**Basic safety plan** The layout of the present and potential hazards found around an incident, the strategy to mitigate them, and assignment of duties to rescuers involved.

**Beam** The breadth (i.e., width) of a ship at its widest point. (NFPA 1405)

**Belaying** The act of guiding and controlling rescuers or victims in the water via a tether rope.

**Biological contaminants** Water pollution from living or dead organic tissue or matter, including microorganisms and toxins derived from natural substances.

**Blackwater** Slang term used by many public safety divers to describe no-visibility water.

**Body harness** Webbing or strap harness worn by the diver and attached to a tether line.

**Booties** See *wet shoes*.

**Bow** The front end of a boat or vessel. (NFPA 1005)

**Boyle's law** Law of physics that shows an airspace will decrease in volume as pressure increases and increase when pressure is reduced.

**Breaking waves** Waves that curl up to the point that they collapse back onto the water with force.

**Bubblers** Pumps that blow air or circulate water around structures or boats to keep ice from forming.

**Buoyancy control device (BCD)** Jacket or vest that contains an inflatable bladder for the purposes of controlling buoyancy.

**Burping an immersion suit** Squeezing the air from the lower to the higher areas of the suit, releasing trapped air from the neck area. This reduces the chance of a suit inversion in the water.

**Butyl rubber drysuits** Drysuits constructed of butyl rubber and glued together. Offer excellent all-around chemical protection, but must be worn with insulation for warmth. Usually durable, easy to clean, and field repairable.

**Canoe** A small watercraft made of metal or fiberglass; it has a very small beam compared to its length, resulting in poor stability. Canoes are usually human powered, but can accept very small motors if equipped properly.

**Cargo area** The section behind the control seats, carrying crew, cargo, or patients.

**Chemical contamination** Water pollution from human-made substances found in the water column or the bottom materials such as mud or silt.

**Cinch straps** A quick, self-tightening sling to provide a good grip on the victim.

**Clear ice** Strong, transparent ice, with no rotting or melting. Strongest of all ice, it offers good working conditions if the thickness of the ice is matched to the function being performed on it.

**Clearing a mask** Expelling water inside a dive mask while underwater in order to return it to a dry state.

**Clearing the snorkel** Blowing forcefully through a snorkel full of water with exhaled breath to expel water, so no water is inhaled on the next breath.

**Color sonar** Electronic unit using a transmitted beam aimed at the bottom surface to form pictures of the bottom density; similar to fish-finders found in many boats.

**Complex rope systems** Belay or hauling rope systems that use hardware or multiple lines to provide larger mechanical advantage systems or safety systems.

**Conduction melting** Melting of ice around objects that carry heat into the water. A prime example is metal stairs being warmed by sunlight and melting the ice immediately around the metal; it may weaken surrounding ice, even in below-freezing temperatures.

**Contamination** The process by which protective clothing or equipment has been exposed to hazardous materials or biological agents. (NFPA 1852)

**Contingency cylinders** Tanks only used for an out-of-air diver trapped underwater. Delivered by the backup diver, they are used to buy time while other problems are worked on.

**Control seats** The front seats of the aircraft, where the pilots sit.

**Crampons** Small spikes that can be attached to the sole of a boot or fin to provide traction to walking rescuers on ice.

**Crest** The top, or highest point, of a wave, just before the wave breaks.

**Crew resource management (CRM)** A communication system that encourages all crew members to verbalize their concerns, especially when dealing with safety, regardless of rank or authority.

**C-spine treatment** Cervical spine injury management, which relies on the general practice of keeping the head, neck, and body immobilized to prevent further injury. The head is kept in a neutral, centered position.

**Cummerbund** An elastic strap that holds the bottom of the BCD securely around the waist area.

**Current diving** Diving in water that is moving at more than 1 knot.

**Decompression illness** A syndrome due to evolved gas in the tissues resulting from a reduction in ambient pressure. (NFPA 99)

**Decompression stops** Stopping an ascent from depth for a prescribed period of time to allow excess nitrogen to leave the diver's body. Decompression stops must be followed faithfully, or injury can occur.

**Defensive swimming position** A position used by swimmers in fast current; the body is positioned to protect from injury in the water with the back down, head in the rear, feet forward, and knees bent to absorb shock.

**Defensive swimming** Self-preservation by using blocks, thrusts, and swimming maneuvers in the water to ensure rescuer safety.

**Differential pressure** The force exerted on a drain intake when water is draining to lower elevation.

**DIN regulators** A type of first-stage regulator in which the sealing O-ring is completely covered in the cylinder when attached. It is common in highly technical dives.

**Dive** Type of rescue where victims are submerged in a liquid, normally water.

**Dive computers** Devices that track a diver's depth, maximum depth, time of dive, and time allowed to dive. Can be mounted on the SCUBA rig console or worn on the body.

**Dive mask** A device consisting of a silicone skirt and glass lenses that attaches around the head, keeping water from the wearer's eyes and nose.

**Divemaster** A professional level of dive rank, capable of supervising and leading safe dive operations and assisting instructors.

**Diver checklist** A sheet the surface tender can use to check off diver preparations, thus making sure needed items are noted as ready to dive.

**Diver encapsulation** Personal protective equipment that offers a high degree of facial, hand, and body protection from thermal, biohazard, and chemical contamination by enveloping the user in nonpermeable suits, gloves, and masks.

**Dive supervisor** The member of a dive team who has the authority and expertise to manage and direct all aspects of the diver operation and has been trained to meet all nondiving job performance requirements of technician-level dive rescue. (NFPA 1006)

**Dive tables** Tools used to calculate a diver's nitrogen loading based on depth, length of exposure to a hyperbaric environment, and intervals between exposures of an actual or a planned dive. (NFPA 1006)

**Diving drysuit** A drysuit adapted for SCUBA use by installation of an inflator valve and an exhaust valve.

**Downstream (down current)** The direction in which water is traveling.

**Downstream V** Current feature that forms a "V" shape, with the "V" pointing downstream; often formed when water flows between two solid objects, normally rocks.

**Downwash** Water returning to the open area, but along the path the wave came to shore.

**Draft** (1) The vertical distance between the water surface and the lowest point of a vessel. (2) The depth of water a vessel needs in order to float. (NFPA 1005)

**Drag hooks** A set of small, clustered sharpened hooks used to snag a body on the bottom of a water area.

**Drain plug** A small rubber or threaded plug that installs into a port or hole to drain a boat of unwanted water.

**Drift ice** A large piece of ice not directly attached to a shore. Also called *floe ice.*

**Droop line** A rope stretched across a span, held at two points, and maintained just above the water. The rope can be dropped to the water level just as the victim touches it, and one end released to let the victim swing to one shore.

**Drownproofing** Letting the body float limply on the water, with the head lying in the water to save strength; breathing is done by raising the head.

**Dry drowning** A death by suffocation, as the lungs are sealed off forcibly by a laryngospasm.

**Dry gloves** Gloves worn by divers to keep the hands dry and free of water exposure.

**Dry hoods** An attached hood worn with a drysuit, offering similar protection as the suit.

**Drysuit** Clothing designed to operate on or around wet conditions, keeping the wearer dry. It may include rubber seals in the neck and wrist area and may require thermal insulation.

**Drysuit fins** Standard SCUBA fins, but with a larger foot pocket to allow larger boots to be used. Very effective with an immersion suit when swimming in open water.

**Drysuit insulation** Thermal garments the diver may use under the drysuit to offer warmth.

**Dynamic belay** Holding a rope steady, but moving with the current to absorb the load more gently.

**Eddies** River features that form when water curls around a solid object and creates a pool of calmer water flowing upstream; often found just downstream of rocks and bridge piers, but can also form behind trees, automobiles, and shore features.

**Eddy turn** Driving a watercraft bow first into an eddy, using the current differential to quickly turn the boat 180 degrees and then be held against the object forming the eddy.

**Encapsulation gear** Dive PPE consisting of a drysuit, dry gloves, dry hood, and full-face mask.

**Equipment specialist** Position that spearheads the status, field repair, and staging of all gear used in an operation.

**Family liaison** Position attached to the surviving family for channeling information between the family and the responders; usually filled by one person so the family does not have too many people to deal with.

**Ferry angle** Lateral movement across moving water, caused by angling an object away from being perpendicular to the current. The resulting force, which is applied at an angle, forces the object to move river right or river left.

**Ferrying** Use of the force of a current to drive a boat laterally from side to side, by changing the angle of the boat relative to the current.

**Ferry lines** Tight ropes stretched diagonally across the current, which let swimmers or boats use the current force to move laterally across the water; also called *tension diagonals.*

**Field neurological exam** Rapid assessment of a diver's physical condition as it relates to pressure-related injuries; usually done on exit of the water or shortly thereafter.

**First-stage regulator** Apparatus that attaches to the tank valve and delivers air at 125 to 150 psi to the other lines on the SCUBA rig.

**Fixed line haul** Using the helicopter to provide a lift for a line attached to the aircraft, with no hoisting capability.

**Flash flood** A very fast-rising flood, often localized from heavy precipitation.

**Flip lines** Shorter ropes or straps used to right a capsized boat from in the water.

**Float coats** Coats or jackets that offer surface warmth and flotation in the same garment.

**Flood** Type of rescue where victims are in areas not normally inundated with water.

**Fluke anchors** Traditional-looking anchors consisting of two spears on either side of a metal bar. The spears will dig into the bottom mud or snag an object, anchoring a boat.

**Foot entrapment** The inability to free a foot or leg wedged under crevices and rocks in current, which may potentially be worsened by the force of the current.

**Frazil ice** Ice that is in the formation stages and looks very ragged. Weak ice, but continued cold temperatures change it to clear ice.

**Free diving** Submerging and swimming down into the water column, without any breathing air other than what is in the lungs.

**Free flowing** A regulator condition that allows breathing gas to escape the cylinder quickly and uncontrolled; it is a common occurrence in cold water.

**Fuel bulb** A small, squeezable bulb in the fuel line used to prime a motor. The bulb is squeezed to provide priming fuel for start-up of the motor; it contains a one-way valve to prevent backflow of the line, making restarting easier.

**Full-face diving masks** A regulator that delivers air to a diver while covering the face and attaching to the head with strapping, similar to a fire-fighting SCBA mask.

**Fuselage** The main body of an aircraft. (NFPA 403)

**Fuselage doors** The entrance to the cargo area.

**Golden triangle** The area on a diver consisting of the neck to both hips. Cutting tools and the octopus should be carried in this area.

**Go/no-go decision** The initial decision to deploy or hold personnel, after considering all aspects of the impending mission.

**Grappling hook** A metal object with curved ends that is used to snag objects; it may be used as an anchor or victim retrieval tool in water rescue.

**Gross decontamination** The phase of the decontamination process during which the amount of surface contaminants is significantly reduced. (NFPA 472)

**Gunwale** The upper edge of a side of a vessel or boat designed to prevent items from being washed overboard. (NFPA 1405)

**Head-up running leap** An entry into surface water in which the rescuer kicks the legs together and sweeps with the arms in an effort to keep the head above water and maintain sight of the victim.

**Helical flow** Water that tends to circle into the middle of the flow, due to the friction of the flowing water against the bottom surfaces of the stream or river.

**Helmsman** A rescuer who works at the operations level and steers the boat under the guidance of another individual, usually the person in charge of the vessel.

**High surf** Surf higher than its normal condition.

**Hoisting** Using a winch onboard the aircraft to pick up the rescuer and/or victim, placing them into the fuselage.

**Hose inflator systems** Small kits comprising a fire hose cap, plug, and valve to allow inflation of a section of hose. This hose is then snaked or pushed out to a victim on the ice.

**Hot exit** Diver coming off the truck ready to dive.

**Hovercraft** Watercraft carried on a cushion of air blown from a motor-driven fan and propelled by a fan in the stern; similar to an airboat, but without friction on the hull.

**Hull** The main structural frame or body of a vessel below the weather deck. (NFPA 1005)

**Hurricane anchoring** See *three-point anchoring*.

**Hydraulic** Water that is forced down in the water column vertically, then recirculated in a rolling motion back into the vertical fall. A hydraulic is very dangerous, as it constantly recirculates objects, including humans.

**Hypothermia** Condition in which one's body temperature is lower than 95 degrees Fahrenheit (35°C).

**Ice awls** Small, handheld tools used to dig into the ice to provide a grip that helps a rescuer pull himself or herself out of the water and onto the ice surface. Also called *picks*.

**Ice diving** Any SCUBA operation under an ice canopy that blocks a normal emergency ascent.

**Ice-rescue-pontoon–style sleds** Small, ice-specific craft pushed onto the ice by rescuers. It spreads the weight effectively and lets rescuers operate with flotation, which is built into the pontoon.

**Ice rescue suits** Coverall-type suits used by rescuers or workers around ice that provide warmth, flotation, and the ability to stay dry.

**Ice screw** A small device twisted into the ice surface to be used an as anchor. Very quick to deploy and offers good strength in fairly thin ice.

**Ice** Type of rescue where victims are in frozen surfaces, even if surrounded by surface water.

**Impact zone** The area where plunging waves crash down onto the water or land surface.

**Inboard motor** A motor in which both the motor and the propeller are mounted and plumbed through the hull.

**Inboard/outboard motor** A hybrid motor in which the engine is mounted in the boat in front of the transom and the propeller is located behind the transom, similar to an outboard motor.

**Incident action plan (IAP)** A verbal or written plan containing incident objectives reflecting the overall strategy and specific control actions where appropriate for managing an incident or planned event. (NFPA 1026)

**Incident command system (ICS)** A standardized on-scene emergency management construct specifically designed to provide for the adoption of an integrated organizational structure that reflects the complexity and demands of single or multiple incidents, without being hindered by jurisdictional boundaries. ICS is a combination of facilities, equipment, personnel, procedures, and communications operating within a common organizational structure, designed to aid in the management of resources during incidents. It is used for all kinds of emergencies and is applicable to small as well as large and complex incidents. ICS is used by various jurisdictions and functional agencies, both public and private, to organize field-level incident management operations. (NFPA 1006)

**Incident safety officer (ISO)** A member of the command staff responsible for monitoring and assessing safety hazards or unsafe situations and for developing measures for ensuring personnel safety. (NFPA 1521)

**Incident size-up** The ongoing observation and evaluation of factors that are used to develop strategic goals and tactical objectives. (NFPA 1006)

**Incident termination** The act of safely returning crews and equipment to a safe area, documenting any damage or injury to crew or property, analyzing the mitigation effort, and rendering a scene safe before vacating it.

**Inflatable boats** Any boat that achieves and maintains its intended shape and buoyancy through the medium of inflation. (NFPA 1925)

**Inflatable rafts** Larger watercraft commonly used for whitewater rafting, which rely on air for flotation.

**Inshore holes** Depressions dug out by wave or current action in shallow water.

**Jet drive** A propulsion unit that generates thrust in reaction to a water stream. (NFPA 1925)

**Job performance requirements (JPRs)** A written statement that describes a specific job task, lists the items necessary to complete the task, and defines measurable or observable outcomes and evaluation areas for the specific task. (NFPA 1000)

**Johnboats** Field name for smaller (usually less than 16 feet) metal boats with a square bow shape.

**Jungle penetrator** A steel, hook-like device lowered through tree canopies, then unfolded to provide a seat for the rescuer and victim.

**Kayaks** Smaller canoe-type watercraft, which usually hold one person. The operator can sit in or on the kayak; most are human powered.

**Keel** The principal structural member of a ship, running fore and aft on the centerline, extending from bow to stern, forming the backbone of the vessel to which the frames are attached. (NFPA 1405)

**Kill switch** A safety measure in which a small clip is attached to the motor operator, with a leash connecting the clip to a switch on the motor that will stop the motor from running if the operator is thrown from the operating position.

**Laminar flow** Water flowing faster at the surface than the bottom, due to the same friction.

**Landing skids (or landing gear)** Steel feet the helicopter sits on when on the ground.

**Landing zone (LZ)** An area utilized for landing helicopters.

**Laryngospasm** The involuntary reflex of the body forcibly clamping the airway shut in an effort to deny water entry.

**Length** The total length of the boat from the bow to the transom.

**Live bait rescue** A tethered swimmer, controlled from shore, who enters the water to effect the rescue of a victim.

**Longshore currents** Water flowing along a shore, pushed by the waves incoming behind it. May feed rip currents.

**Lower unit** The section of an outboard motor enclosing the gears, drive shaft, and propeller.

**Low-head dam** A dam that has no effective means of regulating water flow. Water frequently flows over it and creates a hydraulic, which recirculates objects or people for long periods of time.

**Low surf** Surf at its normal condition.

**Lull** The period of time between set waves, creating an opportunity for safer and more effective movement by rescuers.

**Main rotor** The top blades on the helicopter that provide lift and some forward thrust.

**Manifold blocks** Units that permit a diver to control which tank the gas is supplied from; it may be mask mounted or hung on the body.

**Marker buoys** Small, can-shaped floating objects used to mark locations of objects underwater or to provide visual reference to an area needing to be marked; also called *pelican buoys*.

**Mission analysis** A type of postincident analysis that includes all responders, regardless of training level or agency affiliation.

**Most probable point (MPP)** In case of no witnesses, the location where the victim was likely to have been; determined by clues and logic.

**Movable control point (MCP)** A rope system that allows rescuers on shore to control all directions of a watercraft from one or both shores.

**Mushroom anchors** Boat anchors resembling a mushroom shape, which are mostly used in smaller watercraft.

**Navigation lights** Lighting that must be displayed while a watercraft is on the water during nighttime operations or anchoring.

**Negatively buoyant** The condition in which any object, including a person, sinks in the water column.

**Neoprene drysuits** Drysuits constructed of crushed neoprene; they can be damaged by chemicals and are hard to clean after exposures, but they do offer more inherent warmth than other kinds of drysuits.

**Nitrox** A special blend of breathing gas with higher than normal oxygen content, used to change dive decompression requirements.

**Normalization of deviance** The acceptance of risk based on the premise that previous similar circumstances did not result in any problems, normally in the form of injury. Rescuers often know risk is present, but choose to continue an evolution based on the fact that nothing happened in the past.

**No-wake speed** A boat speed slow enough as to not make large waves behind the craft; generally, an idle speed.

**No-wake zones** Areas where speeds faster than an idle, or causing waves behind a boat, are prohibited.

**Oarlocks** A pin-and-hole system used to hold oars at the side of a boat; it gives the user leverage to operate the oars.

**Oars** Long, paddle-like planks, which can also be used as a lever to provide propulsion; typically used two at a time, with oarlocks.

**Octopus** Nickname for the alternate air source carried by divers as a backup source or emergency source for other divers.

**One-skid landings** Placing one skid at a distance so that rescuers or victims can get onto the skid or into the fuselage, but not fully landing the aircraft.

**Open water** Any large, natural water environment.

**Operations level** This level represents the capability of individuals to respond to technical search and rescue incidents and to identify hazards, use equipment, and apply limited techniques specified in this standard to support and participate in technical search and rescue incidents. (NFPA 1006)

**Outboard motor** A motor for which the engine, gears, and prop are all mounted on the transom.

**Paddles** Stick-like instruments used to propel watercraft by hand; they are held by the user, with a wide blade used in the water.

**Parbuckling** A technique for moving a load utilizing a simple 2:1 mechanical advantage system in which the load is placed inside a bight formed in a length of rope, webbing, tarpaulin, blanket, netting, and so forth that creates the mechanical advantage, rather than being attached to the outside of the bight with ancillary rope rescue hardware. (NFPA 1006)

**Passive drowning stage** The stage of drowning in which the victim is no longer fighting to stay above the water.

**Passive search measures** Search efforts that do not require active searching by the rescuers. (NFPA 1006)

**Peel-out** Use of current forces to quickly turn and enter a current; usually done when leaving an eddy.

**Personal flotation device (PFD)** A device manufactured in accordance with U.S. Coast Guard specifications that provides supplemental flotation for persons in the water. (NFPA 1006)

**Personal protective equipment (PPE)** The equipment provided to shield or isolate a person from the chemical, physical, or thermal hazards that can be encountered at a specific rescue incident. (NFPA 1006)

**Personal throw bags** Smaller, belt-style throw bags meant for rescuer use, and short tethers used to belay shore crews. They are often deployed for rescuer assistance.

**Personal watercraft (PWC)** A vessel less than 13 ft (4 m) in length that uses an internal combustion engine powering a water jet pump as its primary source of propulsion and is designed to be operated by a person or persons sitting, standing, or kneeling on rather than within the confines of the hull. (NFPA 302)

**Personnel accountability report (PAR)** Periodic reports verifying the status of responders assigned to an incident or planned event. (NFPA 1026)

**Plane** A boat moving at a faster speed, such that it operates "on top" of the water; marked by making waves behind the boat during operation.

**Plunging waves** Waves that flow onto a steep bottom or object, resulting in faster and harder breaks.

**Point last seen (PLS)** The last geographical location where the victim was physically witnessed to be; can be on the water or shore.

**Poling a boat** Using pike poles or paddles to push a boat across the hard ice surface.

**Pontoon boats** Larger watercraft with a deck area floated on two or more sealed cylinders underneath. They do not handle current well, but provide great dive or staging platforms.

**Pony bottles** Smaller SCUBA bottles used as a backup air supply, usually carried with the diver. Also called *ponies*.

**Porpoising** A faster way to enter the surf zone, by jumping in headfirst, using shallow dives over shallow waves, instead of running.

**Port** The left side of the boat, looking forward to the bow.

**Positively buoyant** The condition in which any object, including a person, floats in the water column.

**Postincident analysis** A critique that can be done after any water rescue; it is usually performed within a few days after the event ends.

**Primary diver** The diver who is first to dive, or the diver actually in the process of searching at that time.

**Prop guard** An attachment to the lower unit that encircles the propeller, encasing it somewhat to reduce prop strikes.

**Prop** The propeller of a motor.

**Public information officer (PIO)** A member of the Command Staff responsible for interfacing with the public and media or with other agencies with incident-related information requirements. (NFPA 1026)

**Purging a regulator** Expelling water inside a dive regulator in order to use the breathing gas supply underwater.

**Push method** A way to get a tired or injured diver to shore. The injured diver is made positively buoyant by ditching weight and inflating the BCD, and the pushing diver pushes the injured diver by the feet. This provides good swim ability for the pushing diver.

**Rapid intervention crew (RIC)** A minimum of two fully equipped personnel on site, in a ready state, for immediate rescue of disoriented, injured, lost, or trapped rescue personnel. (NFPA 1006)

**Reach tools** Any device for water rescue that can be extended to a person in the water so that he or she can grasp it and be pulled to safety without physically contacting the rescuer. (NFPA 1006)

**Recompression chamber** Small room or tube used to control pressures on the body; used for treating decompression illness.

**Remotely operated vehicles (ROVs)** Small, unmanned units used to survey areas underwater, controlled by an operator on the surface. Has advantages over divers in that depth, air, and decompression concerns are not issues to ROVs.

**Rescue basket** A stainless steel floating cage in which the victim sits while being lifted to the helicopter.

**Rescue board** A long, floating rescue device, designed to be paddled by hand, resembling a sport surfboard.

**Rescue buoy** A hard plastic float, resembling a football, that is equipped with molded handles and a tether for the rescuer.

**Rescue flotation device (RFD)** A tethered floating tube or buoy used by rescuers for self- and victim rescue.

**Rescue sling** An assembly used to haul ice rescue victims up and across the ice, with some commercial models offering built-in flotation.

**Rescue tube** A needle-shaped tube made of foam rubber, which is towed by the rescuer and used for victim flotation; it may also be used as an accessory float due to its shape and ability to bend.

**Rigid-hull boats** Watercraft made mostly of a solid, rigid material, usually fiberglass or metal, that achieve buoyancy from water displacement.

**Rigid-hull inflatables (RHIs)** Solid-shaped hull mated with a flexible multicompartment buoyancy tube(s) at the gunwale. (NFPA 1925)

**Ring buoys** Circle-shaped flotation devices that may be thrown to, and provide buoyancy for, victims in the water.

**Rip currents** The water flowing back out to open water past a surf line, after being pushed onshore. Very strong, and can easily overcome swimmers. Also called *rips*.

**Rip tides** The water flowing back out to open water past a surf line, after being pushed onshore; also called *rip currents*.

**River flood** A flood event based in larger basins that collect water from tributaries.

**River ice** Ice formed on moving waters in creeks, streams, or rivers. It can be very hazardous to rescuers in that the water levels under the ice may vary, weakening it, and may subject rescuers to moving water under the ice surface.

**River left** When looking downstream, the land mass on the left.

**River right** When looking downstream, the land mass on the right.

**River shoes** See *wet shoes*.

**Rogue wave** A higher wave than others found in set waves.

**Rope tug signals** A set of predetermined pulls on a line between a diver and tender signaling different communications.

**Rotational capsizing force** Current force that acts to overturn a watercraft that is perpendicular to the current when the watercraft is not moving with the current.

**Rotational turning force** Current force that is applied to the bow of a watercraft when entering current from another direction.

**Rotten ice** The typical ice that rescuers have to deal with on missions. It can be several inches thick, yet still weak enough to fail without warning. It usually has been subjected to melting, sunlight, warmer temperatures, or other environmental factors to destabilize it.

**Sail area** The area of the ship that is above the waterline and that is subject to the effects of wind, particularly a crosswind on the broad side of a ship. (NFPA 1405)

**Scene sketch** A rough drawing of an area involved in a rescue, detailing landmarks, area probability, rescuer progress, and any safety hazards.

**Scene sketching** Making a written record of the scene, conditions, witness information, and any other pertinent activity.

**Screamer slings** A diaper-like bag the victim sits in to be lifted.

**Scuppers** An opening in the side of a vessel through which rain, sea, or fire-fighting water is discharged. (NFPA 1405)

**Search markings** A separate and distinct marking system used to identify information related to the location of a victim(s). (NFPA 1670)

**Search patterns** Different methods of ensuring the diver covers bottom areas thoroughly during a search.

**Secondary, or backup diver** The diver standing by at the water's edge to assist the primary, or to take over the search pattern when the primary is relieved.

**Second-stage regulators** The mouthpiece from which a diver breathes. They deliver air at a usable pressure and volume as needed.

**Self-rescue** Escaping or exiting a hazardous area under one's own power. (NFPA 1006)

**Set waves** Waves showing different traits than normal. Often formed from merging energies from storms, earthquakes, and other factors.

**Shallow water blackout** Passing out from reduced partial oxygen pressures in the brain while free diving.

**Shore box** A case containing the essentials of dive safety, usually containing oxygen, throw bags, signaling or recall device, and a flashlight, at the least.

**Side scan sonar** Electronic unit similar to color sonar, but the beam is directed out to the side of the boat, giving lateral pictures from the boat position.

**Simple 2:1 advantage** A basic rope system using a movable pulley on the load.

**Simple rope systems** Belay or hauling rope systems that include minimal hardware and have no more than a 2:1 mechanical advantage.

**Sit-in kayaks** Small, one-person, paddle-powered watercraft that the user sits in, with the shell of the craft covering his or her legs and lap.

**Sit-on-top kayaks** Paddle-powered watercraft that the user sits on, instead of in. They can be large enough to hold three boaters, all on top of the boat.

**Sked** A litter-like hard plastic sheet used to package victims and provide immobilization, carrying points, and sliding and hoisting ability.

**Skull cap** Tight-fitting, beanie-style thermal protection worn under a water rescue helmet.

**Snorkel** A plastic or rubber breathing tube through which the wearer draws breath from above the head.

**Snow ice** Ice with a layer of snow on top.

**Sonar units** Electronic devices used to examine the bottom contours, depth, and density, depending on the unit chosen; useful for searching high-hazard areas, or for covering larger areas more quickly than divers can.

**Spatial awareness** A deliberate and focused mindfulness of the entire scope and picture of the environment where a rescuer is working in or around during an incident.

**Spilling waves** Waves that flow onto a gradual slope bottom, resulting in a slow, controlled break.

**Spotters** Personnel positioned to scout for or search for missing victims, watercraft, and other issues.

**Squeeze** Pressure felt in or on the body from an increase in the surrounding water pressure. Unless equalized, pressure differentials can cause serious injury.

**Standard operating procedures (SOPs)** A written organizational directive that establishes or prescribes specific operational or administrative methods to be followed routinely for the performance of designated operations or actions. (NFPA 1670)

**Standing waves** Actual waves that form when water is forced up vertically in the flow and then falls back upstream. Unlike in a hydraulic, current passes underneath the waves.

**Starboard** The right-hand side of a ship as one faces forward. (NFPA 1405)

**Static line** A rope system stretched across a span, usually perpendicular to the current. It is used primarily in water rescue for movable control point operations involving watercraft.

**Sterile cockpit** A reduction of noise or conversation to only mission- or operation-crucial communication during times of stress or critical evolutions.

**Stern** The rear section of a boat or vessel.

**Stopper knot** A knot that keeps a rope from slipping or moving entirely out of the object controlling it; most commonly created by tying a knot in the end of the line, making it impossible to pass through a pulley.

**Storm surge** Water pushed ahead of a storm area. The energy from these waves does not dissipate readily.

**Strainer** Any object that lets water through it, but does not let larger objects pass.

**Streamlining** Ensuring each diver does not present entanglement issues, such as tucking in or securing dangling objects, and taping or securing straps.

**Submersible pressure gauge (SPG)** A unit attached directly to the first-stage regulator, showing how much gas remains in a tank.

**Surf** Type of rescue where victims are in water areas with waves or wave action.

**Surface air consumption (SAC) rates** The rates at which divers use their gas supply at the surface of the water. This rate can be used to calculate approximate times a known gas supply should last underwater at different depths.

**Surface supplied air (SSA) system** A SCUBA system delivering air to the diver via a hose controlled from the surface.

**Surface** Type of rescue where victims are in areas of open water with no current, ice, or waves, and water moving slower than 1 knot.

**Surf line** The line at the offshore edge of the surf zone. Swells are outside this line; waves are inside it.

**Surf zone** The area in which waves are actively building or breaking.

**Surging waves** Waves that do not break, but pile into obstructions in the water, such as cliffs, seawalls, etc.

**Survivability profile** Determination, based on a thorough risk-benefit analysis and other incident factors, that addresses the potential for a victim to survive or die with or without rescue intervention.

**Swells** Waves that rise up, but do not crest or break; found offshore in deeper water.

**Swiftwater helmets** Rescue helmets designed for water operations, which usually include a smaller visor to reduce flexion at the neck.

**Swiftwater** Water moving at a speed of greater than 1 knot (1.15 mph [1.85 km/h]). (NFPA1006)

**Swim boards** Small, flat flotation aids used for recreational swimming, flotation, or swiftwater rescue swimming.

**Swim fins** Rubber or plastic blade-like propulsion, worn and kicked by the feet.

**Swim test** An in-water exam to test needed skills, comfort level, and stamina to start a water rescue training class.

**Tagline** Any rope tied or fastened to an object.

**Tail rotors** The smaller blades at the tail.

**Technician level** This level represents the capability of individuals to respond to technical search and rescue incidents and to identify hazards, use equipment, and apply advanced techniques specified in this standard necessary to coordinate, perform, and supervise technical search and rescue incidents. (NFPA 1006)

**Tender** A member of the dive team responsible for assisting divers with assembly and donning of equipment, communicating with divers, tracking the diver's status and location, and managing subsurface search operations, and trained to meet all the job performance requirements of operations-level diver rescue. (NFPA 1006)

**Tension diagonal system** A rope that is tightly pulled across flowing water, but at an angle to the current (not perpendicular). Objects can pass laterally with the water flow from shore to shore.

**Tethered line rescues** Rescue evolutions in which the swimmer is physically attached to a rope or line and controlled from shore or a boat.

**Tether line** Rope used to physically connect the rescuer to the shore, with control of the rope by the shore crew. Used to pull the ice rescuer back to shore when signaled to, and to help remove the victim and rescuer from the ice.

**Tether switch** See *kill switch*.

**Three-point anchoring** Using three anchors off a watercraft to provide a solid, nondrifting platform in the water for operations.

**Throw bag** A water rescue system that includes 50 ft to 75 ft (15.24 m to 22.86 m) of water rescue rope, an appropriately sized bag, and a closed-cell foam float. (NFPA 1006)

**Throw tools** Anything that can be thrown or launched to a victim to provide a means of flotation.

**Tiller steer** Smaller outboard motors (less than 50 HP) that have throttle and steering controls on a short handle attached to the motor itself.

**Tilt and trim unit** A small hydraulic unit used to change the angle of the boat motor to provide better performance, shallow water clearance, and maintenance angles.

**Topsides** The sides of a boat between the water surface and the deck.

**Transom** The rear vertical surface of a boat.

**Triage tags** Tags used in the classification of casualties according to the nature and severity of their injuries. (NFPA 1006)

**Triangulation** Using different points of view to produce a narrowed down point last seen. Intersecting lines of sight from these views may provide the PLS.

**Tri-glide** A quick-release buckle found on most dedicated swiftwater flotation vests, which allows the wearer to release himself or herself from a tensioned rope with one quick and easy movement.

**Trilaminate drysuits (tri-lams)** Drysuits constructed of nylon layers bonded with cement that offer protection against some chemicals, but require insulation to provide warmth.

**Trimming** Adjusting the angle of the engine or outdrive of a watercraft to maximize the applied thrust from the propulsion system.

**Trough** The bottom of a wave.

**Turbulent flow** Water flowing in irregular fluctuations, or mixing.

**Two-boat tether** A method of approaching hazards, usually low-head dams, with two boats attached to each other. The downstream boat guards the upstream boat's safety, as it has the ability to peel out and pull the upstream boat from the boil.

**Two-line tagline** Two ropes fastened to two opposite sides of a flotation object to move the object into the current, controlled laterally and upstream or downstream by the crews on each shore.

**Two-line tethering** An evolution with one piece of flotation equipment being tied between two lines. The lines are then used on opposite banks, with one bank pulling the flotation out to the victim.

**Type** The level of a team's capability before an incident callout, which takes into account the team's training, equipment, and response times. The authority having jurisdiction can benefit by determining which team will best suit the response needs based on these credentials.

**Upline** A line divers deploy to hold onto while ascending.

**Upstream (up current)** The direction from which flowing water is coming.

**Upstream V** A current feature that forms a "V" shape, with the "V" pointing upstream; it often forms where water flows around a solid object, normally a rock.

**Upstream V formation** A rescue technique in which multiple rescuers walk through the current, with the lead walker forming the point of a "V," pointing upstream. The rest of the rescuers fall in behind the lead rescuer to create a wedge shape that makes walking easier for everyone.

**Upwash** The water rushing into and onto a shore area from wave action.

**Vise grip hold** Using the victim's arms as an impromptu cervical spine brace by raising both arms up above the head and holding them in place.

**Wake** Waves formed at the stern of a watercraft, caused by the boat moving through the water at a speed higher than idle.

**Watercraft** Manned vessels that are propelled across the surface of a body of water by means of oars, paddles, water jets, propellers, towlines, or air cushions and are used to transport personnel and equipment while keeping their occupants out of the water. (NFPA 1006)

**Waterline** The area where the surface of the water meets the gunwale.

**Water muffs** A small suction cup tool that supplies water to the lower unit of an outboard motor.

**Water rescue mannequin** A training prop with the flotation characteristics adjusted to provide more realistic training. It will sink if contact is lost by the rescuer.

**Water-type recovery jaws** A long set of hydraulically operated hooks, which are spread out as they are forced to the bottom and closed as they are brought up; they are used to hook and retrieve victims in deeper water, usually 8–20 feet.

**Wave period** The amount of time it takes for two crests to pass a set point.

**Wave train** Waves from the same source of energy, showing similar speeds and traits.

**Weight and balance** The figurement of the helicopter's total lift capabilities, and the weighting of all loads as not to upset the attitude of the aircraft in flight.

**Weight system** Releasable system that holds lead weights the diver uses to counteract positive flotation.

**Wet drownings** Asphyxiation in which the lungs of the victim fill with fluids.

**Wet gloves** A hand covering offering thermal protection; usually made of neoprene, with varying thicknesses.

**Wet shoes** Neoprene shoes that provide traction and thermal protection to the wearer; also called *booties* or *river shoes*.

**Wetsuit** A body-hugging suit that provides thermal protection to the wearer; it ranges in thickness depending on thermal protection desired.

**Yoke-and-screw system** Device that attaches a first-stage regulator to the SCUBA tank.

# Index

Note: Page numbers followed by *f* and *t* denote figures and tables respectively.